CSR, Sustainability, Ethics & Governance

In recent years the discussion concerning the relation between business and society has made immense strides. This has in turn led to a broad academic and practical discussion on innovative management concepts, such as Corporate Social Responsibility, Corporate Governance and Sustainability Management. This series offers a comprehensive overview of the latest theoretical and empirical research and provides sound concepts for sustainable business strategies. In order to do so, it combines the insights of leading researchers and thinkers in the fields of management theory and the social sciences and from all over the world, thus contributing to the interdisciplinary and intercultural discussion on the role of business in society. The underlying intention of this series is to help solve the world's most challenging problems by developing new management concepts that create value for business and society alike. In order to support those managers, researchers and students who are pursuing sustainable business approaches for our common future, the series offers them access to cutting-edge management approaches. CSR, Sustainability, Ethics and Governance is accepted by the Norwegian Register for Scientific Journals, Series and Publishers, maintained and operated by the Norwegian Social Science Data Services (NSD).

Franzisca Weder
Editor

Strategic Sustainability Communication

Principles, Perspectives, and Potential

Springer

Editor
Franzisca Weder
Department of Business Communication
WU Vienna University of Business and Economics
Vienna, Austria

ISSN 2196-7075 ISSN 2196-7083 (electronic)
CSR, Sustainability, Ethics & Governance
ISBN 978-3-031-89485-5 ISBN 978-3-031-89486-2 (eBook)
https://doi.org/10.1007/978-3-031-89486-2

This Springer imprint is published by the registered company Springer Nature Switzerland AG
The registered company address is: Gewerbestrasse 11, 6330 Cham, Switzerland

Preface

Look at climate change as the old story of separation, human superiority over nature, individual consumerism, and ecological degradation. And then look at sustainability as the new story, as a mutual narrative of the future, defined by inter-being, cooperation, and harmony. Doesn't it seem only natural to explore the central role of strategic communication in transforming the old to the new story and thus in communicatively cultivating sustainability as a new norm?

Until now, strategic communication in practice and academic disciplines like public relations, corporate and organizational communication, or even sustainability management has been approached predominantly with an instrumental understanding of communication and focuses mostly on organizations and their goals. Quite naturally, strategic communication is described as the purposeful use of communication by an organization to fulfill its mission (Hallahan et al., 2007, p. 3; Holtzhausen & Zerfass, 2014). From this functionalist perspective, the aim of strategic communication is further defined as understanding "how a certain set of audience attitudes, behaviors, or perceptions will support those objectives" is what makes communication strategic (Paul, 2011, p. 5).

Following the functionalist paradigm, *strategic sustainability communication* can be described as communication *of* sustainability, directed toward a certain audience with information and knowledge transfer as the central communication processes (Weder & Erikson, 2023). This includes all corporate reporting or marketing efforts using (and sometimes abusing) sustainability as a key term and concept in their communication, sometimes only as a language token representing a notion of responsibility toward the society, sometimes as an economic narrative expressing the will to take part in the "Great transformation" (Polanyi, 2002).

Next to these sustainability communication practices that put organizations and thus the sender of communication in the center of all considerations and strategic planning, there are approaches focusing on communication as more than one-directional information process, in fact as a meaning-making and therefore constitutive process of organizations and stakeholder relationships. This can also be described as communication *about* sustainability, which consequently includes all communication processes that potentially change or reinforce the relationship between people or between an organization and individuals (Stiff & Mongeau, 2016; Weder & Erikson, 2023). The goal of communication about sustainability is to participate in stakeholder and public discourses and to (co)create common narratives and a related consensus (Burkart, 2004; Habermas, 2021; Markell, 1997). Herein grounds the idea from above, that a climate change narrative of human superiority over nature as the origination of the ecological crisis can potentially be changed into the new, ecocultural narrative of sustainability. Then, strategic sustainability communication includes the management of sustainability-related issues (Coombs & Holladay, 2018; Coombs et al., 2019; Jaques, 2007), the active participation in related (media) discourses, and the co-creation of forums and spaces to talk about sustainability, which is institutionalized in industry networks, CSR or SDG hubs (SDG, 2024) or think tanks (Hautamäki & Oksanen, 2015).

However, there is a third perspective on strategic sustainability communication that goes beyond treating sustainability as an issue or issue field and beyond promoting sustainability as the counter-narrative of the climate crisis narrative. Communication *for* sustainability (Weder & Erikson, 2023) comprises all communication processes with the potential to stimulate, impact, and facilitate societal transformation processes. From this perspective, strategic sustainability communication includes all constitutive and organizing flows of communication within organizations (McPhee et al., 2014; Putnam & Nicotera, 2009; Schoeneborn et al., 2019) and also formative communication at the boundaries of organizations that renegotiates and re-frames existing narratives (Christensen & Cornelissen, 2013; Schoeneborn et al., 2020). With this understanding, sustainability can't be one alternative narrative or counter-narrative to the climate crisis narrative; instead, sustainability sneaks in as a normative principle or moral compass guiding communication, guiding the negotiation of transformative behavior within organizations and beyond (Weder et al., 2019), without necessarily knowing in advance where this process (should, must) lead(s) to. It thus means that strategic communication is rather open for new narratives and includes related negotiation processes and imaginative processes of co-creation of new narratives of the future.

Strategic communication as a co-creational process, as a process that has the potential to cultivate sustainable practices in organizations and in public discourses, requires professional communication management, a participatory understanding of stakeholder relations, and innovative analytical and planning concepts as well as specific competencies and skillsets. Consequently, strategic sustainability communication does not only link to the threefold communication of, about, and for sustainability (Golob et al., 2023; Newig et al., 2013; Weder & Erikson, 2023) but also goes beyond and includes negotiations about the quality of communication and therefore how sustainability as a norm can guide strategic communication (Röttger & Rettler, in this volume; Weder, 2023).

This volume entitled Strategic Sustainability Communication discusses the evolution of the sustainability story in corporate, political, and environmental discourses as well as paradigms and theoretical concepts used to better understand communication about, of, and for sustainability. It also offers concepts exploring and theorizing the cultivation of sustainability through communication and the character and potential impact of "sustainable communication." Much more, it frees space to think in new directions: to focus more on re-negotiating the goals of strategic communication, especially within organizations; to not forget the constitutive role of communication in re-defining the "common good," in re-focusing on the issues that are actually important for internal and external stakeholders; and to be open for re-calibration of organizational structures, processes, and goals in a crisis narrative.

With the manifold of contributions from the core of the disciplines of strategic and organizational communication as well as PR research, complemented by interdisciplinary and intercultural perspectives, this book is a world-leading reader for academics, scholars, students as well as practitioners in environmental management, communication management, constructive and solution journalism as well as sustainability management and campaigning. Effective communication about, of, and particularly for sustainability is critical for social change and transformation; therefore, the book not only elaborates various perspectives on strategic sustainability communication but also offers practical examples, case studies, and related exercises for making sustainability and related issues accessible and comprehensible, for co-creating social change and taking authorship in new emerging professions like corporate sustainability officer (CSO), ESG consultant, CSR commissioner, or corporate diversity manager.

The foundations for innovative and critical approaches to strategic sustainability communication going beyond the functional approaches are laid in the first introductory chapter, which is a springboard for the conceptual approaches and theoretical discussions presented in the first part of this volume. In the second and third part, the authors explore methodological, technological as well as pedagogical innovations, complemented by case studies and interviews around existing strategies and tactics to create sustainable solutions (communication planning, campaigning, messaging, etc.).

Vienna, Austria Franzisca Weder

References

Burkart, R. (2004). Consensus-oriented public relations (COPR): A concept for planning and evaluation of public relations. In B. Van Ruler, & D. Vercic (Eds.), *Public relations and communication management in Europe* (pp. 459–466).

Christensen, L. T., & Cornelissen, J. (2013). Bridging corporate and organizational communication: Review, development and a look to the future. In *Organisationskommunikation und Public Relations: Forschungsparadigmen und neue Perspektiven* (pp. 43–72).

Coombs, W. T., & Holladay, S. J. (2018). Social issue qua wicked problems: The role of strategic communication in social issues management. *Journal of Communication Management, 22*(1), 79–95.

Coombs, W. T., Holladay, S. J., & Tachkova, E. (2019). Crisis communication, risk communication, and issues management. In *Public relations theory: Application and understanding* (pp. 31–48).

Golob, U., Podnar, K., & Zabkar, V. (2023). Sustainability communication. *International Journal of Advertising, 42*(1), 42–51.

Habermas, J. (2021, August). Überlegungen und Hypothesen zu einem erneuten Strukturwandel der politischen Öffentlichkeit. In *Ein neuer Strukturwandel der Öffentlichkeit?* (pp. 470–500). Nomos Verlagsgesellschaft mbH & Co. KG.

Hallahan, K., Holtzhausen, D., Van Ruler, B., Verčič, D., & Sriramesh, K. (2007). Defining strategic communication. *International Journal of Strategic Communication, 1*(1), 3–35.

Hautamäki, A., & Oksanen, K. (2015). *Sustainable innovation: Competitive advantage for knowledge hubs* (pp. 87–102). Orchestrating Regional Innovation Ecosystems. Espoo Innovation Garden, Aalto University in cooperation with Laurea University of Applied Sciences and Built Environment Innovations RYM Ltd.

Holtzhausen, D. R., & Zerfass, A. (2014). *Routledge handbook of strategic communication.* Routledge.

Jaques, T. (2007). Issue management and crisis management: An integrated, non-linear, relational construct. *Public Relations Review, 33*(2), 147–157.

Markell, P. (1997). Contesting consensus: Rereading Habermas on the public sphere. *Constellations: An International Journal of Critical & Democratic Theory, 3*(3).

McPhee, R. D., Poole, M. S., & Iverson, J. (2014). Structuration theory. In *The SAGE handbook of organizational communication: Advances in theory, research, and methods* (Vol. 3, pp. 75–100).

Newig, J., Schulz, D., Fischer, D., Hetze, K., Laws, N., Lüdecke, G., & Rieckmann, M. (2013). Communication regarding sustainability: Conceptual perspectives and exploration of societal subsystems. *Sustainability, 5*(7), 2976–2990.

Paul, C. (2011). *Strategic communication: Origins, concepts, and current debates*. Praeger.

Polanyi, K. (2002). The great transformation. *Readings in Economic Sociology*, 38–62.

Putnam, L. L., & Nicotera, A. M. (2009). *Building theories of organization: The constitutive role of communication*. Routledge.

Schoeneborn, D., Kuhn, T. R., & Kärreman, D. (2019). The communicative constitution of organization, organizing, and organizationality. *Organization Studies, 40*(4), 475–496.

Schoeneborn, D., Morsing, M., & Crane, A. (2020). Formative perspectives on the relation between CSR communication and CSR practices: Pathways for walking, talking, and t(w)alking. *Business & Society, 59*(1), 5–33.

SDG. (2024). *The 17 goals*. https://sdgs.un.org/goals

Stiff, J. B., & Mongeau, P. A. (2016). *Persuasive communication*. Guilford Publications.

Weder, F., Lemke, S., & Tungarat, A. (2019). (Re)storying sustainability: The use of story cubes in narrative inquiries to understand individual perceptions of sustainability. *Sustainability, 11*(19), 5264.

Weder, F., & Erikson, M. (2023). *CSR communication and cultures of sustainability*. Pressbook, UQ.

Weder, F. (2023). Cultivation of sustainability in a discourse of change: Perspectives on communication for sustainability as new "norm" and principle of action in socio-ecological transformation processes. *Journal of Language and Politics, 22*(5), 577–600.

Acknowledgments

This volume is based on three wonderful conference-related expert workshops, one on local perspectives on sustainability (ICA2023, post-conference, Toronto, Canada), and two on Public Relations and sustainability and the question of how (much) sustainability challenges existing PR theory, methodologies, and pedagogy and vice versa (PROK conference 2022, Bamberg, Germany; PROK conference 2023, Darmstadt, Germany). The meetings can be seen as a starting point for an international discourse on how to deal with sustainability as a normative horizon for strategic communication in the future, also guiding our behavior as scholars and teachers.

Next to the contributions from a variety of disciplinary backgrounds, the high quality of the chapters could not have been achieved without the tremendous support of the reviewers and the great coordination of the editorial team from Springer. Thanks heaps!

Finally, I want to express my gratitude for all the inspirational and critical discussions and debates I had with my amazing colleagues across the globe, who continuously stepped on my toes, challenged me in my thinking about sustainability, sustainable communication, the notion of care and kinship, performative and formative communication and provocation! Thank you!

Contents

Introduction

The In(de)finition of Sustainability: Strategic Communication for Sustainability as Conversational Contestation

Franzisca Weder

Abstract Sustainability communication is an increasingly institutionalized field of research and has developed into a vivid and impactful part of corporate communications. Strategic sustainability communication aims at informing relevant audiences (communication of sustainability), creating a common understanding and sense of a "good life" (communication about sustainability), and seeks to coordinate, create, and maintain relationships that enable transformation (communication for sustainability). However, despite these efforts, sustainability has not (yet) been cultivated as a universal value in our society. Still, sustainability is used (and abused) as a language token and has been established as a synonym for technological innovation and solution-based approaches—simply: sustainability is the happy face that individuals but mostly corporates put on covering the ugly face of the climate crisis.

This is part of sustainability reporting and sustainability storytelling, mostly strategically planned and implemented to support a company in fulfilling their mission in regard to what is called "sustainable development" or the "great transformation." This introduction takes a critical look at strategic sustainability communication and challenges that sustainable development and the related 17 political-economic goals (SDGs) have been simply accepted and adopted as goals particularly for corporate sustainability communication. Based on the critique, the chapter conceptualizes strategic sustainability communication as *sustainable* itself if the genuine purpose of strategic sustainability communication is to spark conversations *about* sustainability and, thus, stimulate and facilitate the re-negotiation and conversational contestation of sustainability. The following contributions of the volume are briefly introduced.

F. Weder (✉)
Department of Business Communication, WU Vienna University of Business and Economics, Vienna, Austria
e-mail: Franzisca.weder@wu.ac.at

F. Weder (ed.), *Strategic Sustainability Communication*, CSR, Sustainability, Ethics & Governance, https://doi.org/10.1007/978-3-031-89486-2_1

1 Introduction to the Introduction

Sustainability is a principle of action that guides individual and collective action. Communication plays a central role in this. Media discourses about sustainable and, above all, nonsustainable entrepreneurial action have sharpened our sense of value in relation to a world in a climate crisis. The (over)moralization of the concept of sustainability works as a moral compass for collective as well as individual (consumer) behavior.

Derived from the verb to hold, to stay, to last, and act over a longer period of time, the core of today's use of the term *sustainability is the idea that no more may be consumed, as it can regrow or regenerate*. This makes it clear that the term refers primarily to the human–nature relationship or the relationship of humankind to natural resources—and to how we deal with them. Since the term was first used in 1713 and applied in forestry (Grober, 2014), sustainability as a noun has become synonymous with behavior that protects nature, the environment, and resources, and aims to counteract climate change. Sustainability, therefore, stands for harmony between humanity and nature, for protectionism and conservatism, and has become increasingly established over the last 100 years. It has been institutionalized in the 1983 definition of sustainability ("Brundtland Report," 1987) and the 17 Sustainable Development Goals, which were developed and manifested by the United Nations throughout the COPs (Conferences of the Parties) in Rio de Janeiro (1992) and Rome (2012) (SDGs: 2015; UN, 2024).

At the heart of the idea of sustainable development today is a bundle of goals of ecological, social, and economic sustainability (triple bottom line), further developed in environmental sciences and economic discourses into circular process models or concepts that are oriented toward the—again mainly resource related—"planetary boundaries" (Doughnut Model, Raworth, 2017). The models also speak of specific strategies of sufficiency (reduction of production and consumption), efficiency (more abundant use of energy and materials), and consistency (recycling, nature-friendly material cycles, waste avoidance) that need to be applied to reach the 17 SDGs. The SDGs guide organizational actions and legitimizations, Therefore, they need to be communicated to the corporate stakeholders, which is why broadcasting an organization's commitment to sustainability has become fundamental—and is increasingly professionalized as sustainability, CSR, or, more recently, ESG communication (Weder, 2024).

Simultaneously, sustainability communication is an emerging area of research and has developed into a breeding ground for new theoretical thinking—not only but mainly in all areas of professional communication (Godemann & Michelsen, 2011; Robertson, 2018; Weder et al., 2021; Golob et al., 2023). However, pretty much none of the authors so far re-thinks the genuine goal orientation of strategic and in particular promotional sustainability communication and challenges the predefined goals for sustainable development as the ultimate goal taken on by organizations in the form of the SDGs. The introduction to this volume on strategic sustainability communication wants to do exactly that, wants to challenge the currently very narrow goal and purpose orientation of strategic communication. It advocates for a new purpose of professional communication in a world in crisis, where sustainability has been defined and consequently manifested as one narrative of human–nature harmony within a business framework, too quickly. Re-thinking our disciplinary ontologies is demanded by an increasing number of scholars from communication studies and, in particular, colleagues from environmental science and climate change communication (Brüggemann et al., 2023; Kannengießer, 2021). What they describe as transformative communication, transformative PR, or transformative journalism (see also Pleil et al., 2025; Brüggemann et al., 2022, or Valentini, 2021) is here further developed into a new understanding of strategic communication guided by sustainability as a universal value. The introduction proposes that the new purpose of strategic sustainability communication is to *not* perpetuate the definition of sustainability and further narrow down the sustainability story to one master narrative, but instead to foster sense-making, re-open sustainability for debate, and to re-negotiate its meaning and potential to become a core value in organizations and beyond, in the society. Consequently, it can be said, if strategic sustainability communication itself is guided by sustainability as an ethical principle, then *strategic sustainability communication must spark conversations and bring regeneration as the principle of communicative action back into organizational and societal discourses and debates, and keep sustainability definitions and narratives continuously challenged and questioned.*

In the following chapter, sustainability communication as a field of research is introduced with a specific focus on how (much) PR and professional communication have adopted the Sustainable Development Goals perpetuated in academic research. Based on the reflections on the status quo, the conceptual idea of in(de)finition is introduced, not only but mainly to challenge the "goal fixation" in strategic communication mentioned above.

2 Serving the SDGs: Strategic Communication and Its "Goal Fixation"

While Public Relations, marketing, advertisement, and internal communication are mostly approached separately—conceptually rather than in practice actually—there is a new term that integrates and unifies previously separated thinking: strategic communication. The concept of strategic communication seems to end the "decade of definitions" (Frandsen & Johansen, 2022, p. 15), driven predominantly by PR scholarship and now manifested in recently published volumes and journals (e.g., Falkheimer & Heide, 2022; Botan, 2017; Heath et al., 2018; Zerfaß et al., 2018). However, instead of a separation along the disciplines or the fine line between internal and external communication, there is still a massive "wall" between organization and communication-focused approaches, following Hoffjann's metaphor (Hoffjann, 2024). His first diagnosis: the majority of today's approaches focus on organizations and—which is of interest for the considerations in this introduction—on the possible and necessary means to achieve a central organizational *goal*. This means that strategic communication supports either the achievement of all organizational goals or certain strategically relevant goals—for example, communicating a recently applied sustainability strategy, a new ESG management program, or simply the publication of the annual sustainability report. To fulfill an organization's mission, that's what strategic communication does, according to Hallahan et al. (2007, p. 3), or in other words to "reach set goals" (Holtzhausen & Zerfaß, 2014, p. 284), which is a very functionalist, transmission-oriented understanding of communication (Heide et al., 2020).

From this perspective or "side of the wall," the Sustainability Development Goals (SDG, 2025) were a gift! Introduced in 2012 in Rome (UN, 2024) and further institutionalized in 2015, the SDGs present 17 issues or areas of action—and thus 17 issue-specific ways to frame corporate and organizational behavior. Accompanied by long lists of dimensions and criteria for the evaluation of sustainability-related behavior, today manifested in guidelines, frameworks, and policies, the SDGs have also stimulated, fostered, and established a functional approach to strategic communication, mainly focused on and thus limited to sustainability reporting with the strategic role of adding sustainability-related goals to the organizational goals.

Accordingly, sustainability communication research—happening at the intersections between management and communication disciplines, influenced by sustainability studies, environmental and science communication with an apparent overlap with the field of CSR communication (Diehl et al., 2017; Golob et al., 2013, 2023; Weder, 2023; Weder et al., 2021; Schaltegger et al., 2016, 2024; Schoeneborn et al., 2022)—has predominantly focused on the communication *of* sustainability, mainly reporting (Hahn & Kühnen, 2013; Baumüller & Sopp, 2021; Weder & Erikson, 2023; Benameur et al., 2024). Delving deeper into the literature, it becomes obvious that in most of the papers and books the main question to be answered is how communication can be planned, set up, managed, implemented, and evaluated to reach either a specific, strategic goal in relation to one of the Sustainable Development Goals and related issues, or societal or even "great transformation" as such (see, e.g., McLeod & Marshall, 2023; Servaes & Yusha'u, 2023).

However, looking at sustainability communication research done since 2015, there is a change from communication *of* sustainability as "classic strategic sustainability communication," focusing predominantly on reporting and external communication endeavors (Weder & Erikson, 2023), to communication *about* sustainability. This goes parallel to the shift from organizations taking responsibility and communicating their related activities to, more recently, organizations developing specific sustainability strategies and weaving sustainability as a principle of action deeply in their organizations structure. One example is the institutionalization of sustainability managers as new strategic communicator roles (Schaltegger et al., 2024; Weder et al., forthcoming). This shift has been also acknowledged by PR researchers, looking at sustainability, CSR and, more recently, ESG being part of an organization's strategic positioning, especially of larger corporations, and thus talking about a shift from sustainability communication to sustainable communication (Zerfaß et al.,

2018, 2021; Röttger & Rettler, 2025). However, even if we "tear down the walls" (Hoffjann, 2024) between a functionalist and more critical, constitutive perspective on communication, the goal of all communication endeavors still seems to be subordinated to the business or organizational goal, to the overall idea of sustainability as narrative of progress and economic success. But what if we crack the "goal fixation" of strategic sustainability communication?

3 As You Were! The In(de)finition of Sustainability

Goals drive strategic communication; they define the purpose, the desired results and outcomes; thus, they also limit the scope of action. The genuine goal fixation and the adaptation of the organizational communication goals to the SDGs—or vice versa—has created a challenge, which will be outlined in the next paragraphs, followed by a concept for a rather critical understanding of sustainable communication with *infinition of sustainability as the main purpose*.

3.1 Goal Fixation as Challenge

If sustainability and/or the SDGs are (part of) the organizational goal, sustainability needs to be defined, categorized, and made tangible. Nearly every chapter, article, and corporate report or website starts with a definition of sustainability with the most prevalent being that sustainable development is the "development that meets the needs of the present without compromising the ability of future generations to meet their own needs" (Brundtland, 1987; Barkemeyer et al., 2014). This doesn't leave much room for alternative interpretations and narratives beyond the resource-orientation and market logic of the business world, as claimed in, e.g., Weder et al. (2021) or Frank (2017). Since responsibility thinking started under the umbrella of business ethics as the—so far rather unsolved—question of what comes first, ethical or economic purposes, taking responsibility and sustainable development always includes *goal conflicts* (Dawes, 2020; Lahsen & Turnhout, 2021) and challenges through *path dependencies* (Turnheim & Sovacool, 2020). To avoid the conflicts, there is and has always been a strong tendency to define sustainability and put sustainable development in dimensions, and therefore to cling to the consensual three columns of sustainability (also called "triple bottom line," Elkington & Rowlands, 1999). The consequence of subordinating communication goals to the SDGs led to the point that *narrative openness* and deliberation—within organizations and with relevant stakeholders and audiences—has not really been part of sustainability communication so far.

To bring back deliberative thinking, firstly, leads to Habermas (2021), who sees communication as a process to change or reinforce the relationship between people groups or an organization and individuals (Stiff & Mongeau, 2016). Built on this understanding of strategic communication, strategic sustainability communication should encompass deliberative, rather than horizontal, but also purposive communication. Or in other words, instead of the main purpose to reach a (pre-)set goal, the focus on strategic communication could (and should perhaps) lay more on *negotiation processes about the meaning of sustainability as a normative principle that guides the allocation and taking of responsibility*, and about the issues that are relevant within a sustainable development framework and in the face of climate crisis related problems that urgently require solutions (Weder & Erikson, 2023). From a constructivist perspective, strategic communication includes all sensemaking and organizing processes of communication (Craig, 1999; Putnam & Nicotera, 2009; Coreen, 2018). Consequently, communication *about* sustainability needs to be rather diverse (ibid.) and therefore relate to different rationalities and relationalities and semantics and the inevitability of (sometimes contradictory) polyphony in and around organizations (Hoffjann, 2024).

3.2 Daring the Infinition of Sustainability

This introduction promised to dare something new, to stimulate critical thinking and initiate further theoretical and conceptual work on sustainability, sustainability communication, and sustainable communication.

Therefore, I focus on one aspect that hasn't been part of the considerations so far. Instead of asking *how* we reach the (sustainable development) goals, I propose to go one step back and re-define the goal of at least strategic *sustainability* communication as re-opening the meaning of sustainability, to *in(de)fine* sustainability. Indefinition, or short: infinition seems to be an odd word. From a language perspective, infinition presupposes the indefinability of its object (Epstein, 2015). At the core of this word is "finis," meaning a limit or boundary (latin). The term infinition has therefore two interesting aspects for radically new strategic communication thinking: firstly, it opens up innovative thinking about an object, narrative, principle, or concept because the term manifests the *inadequacy of definitions* in general, or at least puts them into question. It says that it is impossible to completely define an object, etc. Secondly, the concept of infinition assumes that there is a large variety of definitions of an object or new concept like sustainability. However, *it's a matter of keeping the conversations going to get closer to a definition, but also to be open for a re-definition.*

At this point, we refer to Weder's suggestion of sustainability communication as a dialectic process of stabilization and questioning, proposed in 2023. As one of the first attempts to offer a theory to explain and explore sustainability communication, Weder describes the oscillation process of the manifestation (institutionalization) of certain structural elements on the one hand, and processes that put these structures into question on the other hand. This points to the necessity to crack normative patterns again and again, to remove terminological or normative boundaries (Weder, 2023)—basically to *infinitize*.

True regenerative communication then does not mean a re-framing or re-definition in the sense that every organization needs to come up with their very own interpretation or definition of sustainability. However, the act of challenging existing and preset definitions and goals and the stimulation of negotiation processes is from my perspective the key of sustainable strategic communication because it's genuinely formative, following for example Christensen et al. (2021) and being rooted in the notion of procedural sustainability embedded in constructivist social theory (Robinson & Cole, 2015). This implies that sustainability communication is based on sustainability as a concept that cannot be defined within that particular system. However, sustainability can be used to define and guide other concepts derived from it and work as a *new normative horizon* (Waisbrod, 2019).

3.3 *A Call to Wrestle*

The proposition that I want to make in this introductory chapter is that embracing the multiplicity of communication processes that connect publics with an organization that is becoming increasingly acknowledged in PR thinking means to be open for a re-orientation and investigation of the guiding principle of strategic communication and related PR practices. The complex interplay of material practices, human decisions, technological advancements, communication strategies, and broader systematic dynamics implies furthermore that *there is no reason to limit strategic communication and related communication effort to a specific goal.* Thus, instead of searching for the most concrete and narrow definition, strategic sustainability communication should encompass all processes and structures that spark conversations and new conceptual thinking around our human relationship to the more-than-human world, which means to *infinitize sustainability through communication.*

I strongly advocate for looking and dealing with sustainability in a way that acknowledges the nondefiniteness—or rather unfinishedness of sustainability (Epstein, 2015). Then, **strategic sustainability communication means conversational contestation of sustainability.** This means that sustainability communication is not (only) intentional communication toward what already exists but rather communication that intentionally searches for what doesn't exist yet (infinition). To re-conceptualize sustainability communication as contestational challenges mainstream work in sustainability management, CSR communication, and strategic communication, as discussed. But critical scholarship needs variety in standpoints as alternative ways of thinking, especially if we deal with uncertainty and ambiguity, and the interaction of individuals and collectives with lived experiences. This means not only ontological but also epistemological work that follows the need to decentering ideas and standpoints and stepping sideways and decenters our role as researchers. It includes a call to action—which wonderfully has been picked up by the authors of the chapters in this volume, as the following overview of the contributions shows.

4 Outline of the Book

This volume, entitled **Strategic Sustainability Communication,** neither seeks to demarcate a research area nor define a field, nor give Public Relations and/or strategic communication research a new purpose by claiming the sustainable development goals as the overarching purpose of strategic communication in a transforming society. Then, it would limit itself by emphasizing a functional perspective on strategic communication and rather information-based approaches to predominantly external corporate communication or "only" communication *of* sustainability from an organization toward their stakeholder or the perception of sustainability being an issue (field) or field of discourse that organizations can talk *about*.

Instead, the volume wants to open the door to think more creatively about communication being the key in socio-cultural transformation processes; it also wants to stimulate critical thinking about the need for re-negotiating sustainability within organizations, the formative potential of organizational communication in cultivating sustainability as new norm, as guiding principle of organizational activities as well as strategic communication, and, more broadly, part of promotional culture.

The contributions to this volume are important stepping stones on the way to re-framing sustainable development from a communication perspective. They are not exclusively critical, and not every chapter challenges sustainability or the SDGs as guiding frameworks. However, all contributions dare something new, are courageous in terms of challenging their own theoretical concepts and approaches (if it's from linguistics, communication, psychology, or media studies and PR research) and thus, the authors are somehow boundary spanners, which always was the main idea when the book was conceptualized.

(I) The first part of the book unites four conceptual contributions, complementing a formative perspective on communication in and beyond organizations and thus rather constructivist approaches in strategic communication with a more genuine Public Relations perspective. *L.T. Christensen and E. Christensen* stimulate critical thinking about what sustainability communication is, what it can be, and what it should be with their chapter entitled **When Talk Is Not Cheap. The Performative Potentials of Strategic Sustainability Communication**. Based on their previous work on Corporate Social Responsibility Communication as "aspirational talk," they further develop their thinking and performative perspective on communication further and challenge common-sense understandings of sustainability communication as "cheap" and lacking consequentialism. With a case study, the authors show the importance of strategic sustainable communication in advancing sustainability agendas and shaping responsible corporate behavior. The second chapter in this theoretical and conceptual part offers a deep reflection on Public Relations as a specific form of strategic communication and its crucial role in the transformation toward sustainability. *S. Kannengießer, L. Rettler, U. Röttger, and H. Stehle* discuss two different modes of PR for sustainability, accompanied by a third dimension, the dimension of sustainability of PR itself. The chapter

Public Relations and Sustainability: Points of Friction in a Complex Relationship creates a fruitful ground for the following chapter in which *T. Pleil, P.S. Helferich, and T. Otsa* develop their **Typology of Strategic Sustainability Communication: From Reporting to Transformation**. The typology offered also works with the differentiation between communication of sustainability (reporting), sustainability PR, and marketing, and a third dimension that captures their concept of strategic communication for sustainability: transformative strategic communication (TSC). After these deep dives, the strong conceptual part of the book is completed by a chapter written by *J. Anderson, S. Kazmi, N. Yu, L. Atkinson, P. Jamar, and J. Tao* on Environmental Justice and their discussion of (in)accessibility sustainable goods. Based on insights from an empirical study, the authors put up the tensions and ambiguity that are created by presenting sustainable products as solution for environmental problems. Thus, the chapter **Sustainable and Just? Access to Sustainable Goods and Environmental Justice** supports the idea of re-negotiating sustainability on a public discourse level that was presented in the first part of the introduction.

(II) The second part of the book at hand takes an organizational perspective on sustainability communication. It brings together empirical work as well as case studies, offering various perspectives on communication of, about and for sustainability in different organizational contexts (internal to CEO and external communication and stakeholder relations) and in various issue fields (from energy to music). This part of the volume starts with the chapter **Striving for Legitimacy, Strategic Energy Project Communication and Its Characteristics**, written by *J. Lührmann, B. Uth, and H. Stehle*. The authors focus on energy projects and the role of sustainability as a framework that builds legitimacy for these projects—with an outlook for actual communication planning and practices.

Creating legitimacy and gaining the license to operate through strategic sustainability communication is not only important in relation to external stakeholders but also internal stakeholders. *Y. Botha and L. O'Mahony* offer **Considerations for Sustainable Employee Engagement as Enablers for Communication About Sustainability**. Related to the overarching idea of re-opening sustainability for discussion in stakeholder relationships, the authors show the potential to acculturate internal and external contexts through strategic sustainability communication and particularly through building collaborative networks toward societal, environmental, and economic well-being.

One of the challenges of communication of sustainability or risk, looking at the imbalance between external and internal sustainability communication, is intransparency and a lack of trust, which is the starting point for an in-depth look at greenwashing and **The Role of Greenwashing Suspicion in Advertising Effectiveness**, written by *A.-E. Waldeier, S. Weihnacht, and C. Morasch*. This chapter builds the groundwork for *J. Florén*'s contribution, which puts the Business Case Frame and thus the dominant perspective on sustainability, especially among practitioners and researchers into question. In his chapter entitled **Strategic Communication Beyond the Business Case: A Study of Investors Framing Sustainability Tensions,** the author builds on a qualitative study exploring the paradoxical tensions to then focus on formative approaches and new insights and ideas about how sustainability-related meaning is continuously negotiated to go beyond the market logic also driving sustainability communication so far.

A. Schwinges and M. Mpadanes introduce **Corporate Environmental Efficacy Beliefs and with that A Stakeholder-Centric Perspective on Strategic Sustainability Communication**. This chapter adds to the previous chapters develop a concept of purposeful communication based on a set of beliefs of the capability and effectiveness of corporate environmental action. Complementary to their focus on corporate environmental communication, the next chapter showcases an issue of social responsibility: *R. Palmieri and E. Balabanova*'s linguistic-rhetorical analysis of modern slavery statements of UK firms supports the authors in **Debating Sustainability or Avoiding Responsibility**. Their study of micro-level—and apparently rather minimalistic—**Framing and Argumentative Strategies in Corporate Modern Slavery Statements in the UK.**

All the chapters in this section point to the relevance of not only strategic planning but also strong (issue) leadership when it comes to organizational sustainability communication. *A. Duarte and P. Dias* show that **Sustainability** works **as CEO Activism Trigger** and present **the Portuguese Case**. In-depth interviews show that the sustainability framework apparently guides CEOs not only in their own behavior but also in their organizational communication, renegotiation of corporate values, and collaborative value creation.

Lastly, we open up the discussion about internal and external communication, environmental and social engagement, and the role of organizational leadership beyond the corporate context and include learnings from universities as a different type of organization. The phenomena observed in the corporate world point toward a similar logic in less business-driven contexts. *M. Rochyadi-Reetz and J. Wolling* conclude this section of the book with their chapter on **Strategic Communication of and for Sustainability in Higher Education Institutions. Conceptional Considerations and Results of a Case Study.** The authors show that corporates could potentially learn from "whole institution approaches" applied—for example—in universities across the globe, analyzed with a specific case study in this chapter. An important conclusion from this chapter is that strategic sustainability communication should be flexible, opportunity-driven, and rather interpersonal than public—which leads from the more established concepts of communication of sustainability toward new methods, strategies, and concepts of communication for sustainability. These learnings are also supported by the work presented in the third part of the book, bringing together local perspectives on strategic sustainability communication.

(III) The first chapter in this third part of the book starts with a reflection on the tensions between a global and local perspective on sustainability: **From**

Global to Local: Communication of, for and about Sustainability in the Local are(n)as. *B. Dernbach* points to the fact that each region, country, and especially locality must define its own challenges and understand the transformation process as a holistic cultural change. The authors case from Germany (Nuernberg) is complemented by a second case from Italy (Venice), presented by *M. Porzionato* under the title of **The Strategic Making of Sustainable Venice.** Supported by a constitutive understanding of strategic communication, the chapter examines how the municipality of Venice's strategic communication constitutes what a sustainable Venice "means" in the first place—another wonderful example of the significance of formative communication approaches and the key role of sustainability-related negotiations and, thus, sense and meaning making guided by sustainability.

F. Höhs also follows the idea that place matters—as well as meaning making of sustainability through sustainability-related negotiation processes. In her chapter on the **Country of Origin Effect as Part of Strategic Sustainability Communication**, *F. Höhs* introduces a marketing perspective and offers an example for creating local, especially cultural and social—proximity as part of communication about sustainability—and about what sustainability actually is. Communicatively created local proximity is also key in the next cases, the first presented by *K. Lee, N. Watt, J. Burdon,* and *F. Weder*. The authors share experiences from local and place-based energy communication, especially on **Community Identity and Evaluation of Oppositional Framing of Renewable Energy Projects**, presented with **Learnings for Strategic Sustainability Communication** from Australia. This is followed by a chapter written by *C. Itdhiamornkulchai,* and *P. Sthapitanonda*, describing their **Case Study of Stakeholder's Co-Creation of Dove's Award Winning #LetHerGrow Campaign in Thailand**, and *M. K. Hassan,* complementing this third part of the book with a perspective on **Food Security and Farmer's Political Participation in Rural Communication** from **Bangladesh.**

(IV) The fourth part of the book embarks on the cases and focuses on various forms of media and technology that is used to communicate about and for sustainability. In the chapters the concepts and foci from previous chapters shine through, predominantly a critical constitutive and thus (per)formative perspective on communication as well as a strategic lens emphasizing the need to renegotiate sustainability to make it more tangible.

M. Rocchi presents a chapter entitled **Embedding Sustainability in Film and Television Industries: Framing Practices and Narratives for Strategic Sustainability Communication**, where the author underscores the role of the film and television industries within strategic sustainability communication, especially when it comes to making sustainability more tangible through storytelling. Tying in with the chapter on CEO activism and leadership, *F. Tigre Moura and B. Huber* present a study on **Musicians as Sustainability Communication Agents: Insights from a Structured Survey and Content Analysis.** The authors particularly discuss the key role of artists and other leaders next to corporate leadership figures in promoting sustainability.

M. Reumont, C. Déméné, and G. Raiche-Savoie envision a **Canadian Durability Index for Electronic Devise Through Communication Design: The Case of the Smartphone**. And from there, *M. Pope, E. Hauke, M. Páczai, B. Pinar, and E. Stoclet* shift the perspective toward new technologies with their chapter on web applications supporting what M. Rochyadi-Reetz and J. Wolling described as a whole institution approach. In their chapter on **Boundary Crossing Approaches Through Education, Sustainability and Communications**, the authors identify communication strategies enabling sustainable thinking and practice within a Higher Education institution, recognizing the contingent nature of knowledge about sustainability. Their chapter also points to the key role of visual communication, which is further discussed in the final chapter on **Visual Storytelling for Sustainability Communication**, in which *A. Birt* offers Practice-Based Design Techniques.

(V) The final part of the book marks an outlook and actually opens the doors to future research based on the core idea of strategic sustainability communication as processes that open sustainability up for re-negotiation. *U. Röttger and L. Rettler* make the start with their conceptual paper on **Sustainability as a Guiding Principle of Strategic Communication. A Heuristic for Future-Proof Strategic Communication**. The authors mainly discuss how communication itself can be sustainable and promote sustainable transformation. With their comprehensive heuristic, they offer a systematization of various relationships of strategic communication to sustainability issues, which offers, combined with Pleil et al.'s typology (part I), the groundwork for future strategic sustainability communication research.

Hand in hand with their considerations, *S. Walke* and *H. Stehle* discuss sustainability communication that is strategically set up and implemented to delegitimate corporations in their chapter entitled **How Civil Society Actors Employ Strategic Communication to Delegitimate Corporations in Public Debates on Sustainability**.

This chapter also points to the ethical character of sustainability communication itself and thus the need for further research on sustainability as guiding principle of communicative action—from corporates and collective actors to individuals. Here, *S. McLaughlin* offers **An Ethical Framework for Constructive Sustainability Debate in Political Forums,** based on **Aristotle's principles of "deliberative rhetoric" for analysis of rhetorical argumentation**. The author poses this analytical framework as a model for further research in sustainability communication and beyond. Finally, *C. Troy* takes an actual look into the future with the chapter on **Communicating Sustainable Futures**. She creates a link to theoretical concepts of other chapters in this book to motivate future studies delving deeper into constructive responses to depictions of desirable futures. She also claims—similar to other authors—that sustainability itself provides moral guidance for the operationalization of sustainable futures messaging in strategic communication contexts.

Enjoy reading!

Any thoughts? Get in touch: Franzisca.weder@wu.ac.at

References

Barkemeyer, R., Holt, D., Preuss, L., & Tsang, S. (2014). What happened to the 'development' in sustainable development? Business guidelines two decades after Brundtland. *Sustainable Development, 22*(1), 15–32.

Baumüller, J., & Sopp, K. (2021). Double materiality and the shift from non-financial to European sustainability reporting: Review, outlook and implications. *Journal of Applied Accounting Research, 23*(1), 8–28.

Benameur, K. B., Mostafa, M. M., Hassanein, A., Shariff, M. Z., & Al-Shattarat, W. (2024). Sustainability reporting scholarly research: A bibliometric review and a future research agenda. *Management Review Quarterly, 74*(2), 823–866.

Botan, C. H. (2017). *Strategic communication theory and practice: The cocreational model*. Wiley.

Brüggemann, M., Carvalho, A., Brevini, B., & Downey, J. (2023). Still watching from the sidelines? The case for transformative environmental communication scholarship. *International Journal of Communication, 17*, 5039–5052.

Brüggemann, M., Frech, J., & Schäfer, T. (2022). Transformative journalisms: How the global ecological crisis is transforming journalism. In *The Routledge handbook of environment and communication* (pp. 221–236). Routledge.

Brundtland, G. H. (1987). What is sustainable development. *Our Common Future, 8*(9).

Christensen, L. T., Morsing, M., & Thyssen, O. (2021). Talk–action dynamics: Modalities of aspirational talk. *Organization Studies, 42*(3), 407–427.

Cooren, F. (2018). Materializing communication: Making the case for a relational ontology. *Journal of Communication, 68*(2), 278–288. https://doi.org/10.1093/joc/jqx014

Craig, R. T. (1999). Communication theory as a field. *Communication Theory, 9*(2), 119–161. https://doi.org/10.1111/j.1468-2885.1999.tb00355.x

Dawes, J. H. (2020). Are the Sustainable Development Goals self-consistent and mutually achievable? *Sustainable Development, 28*(1), 101–117.

Diehl, S., Karmasin, M., Mueller, B., Terlutter, R., & Weder, F. (Eds.). (2017). *Handbook of integrated CSR communication*. Springer International Publishing.

Elkington, J., & Rowlands, I. H. (1999). Cannibals with forks: The triple bottom line of 21st century business. *Alternatives Journal, 25*(4), 42.

Epstein, M. (2015). Paradoxical speech acts: Transformatives and counterformatives. *Parallax, 21*(2), 134–142.

Falkheimer, J., & Heide, M. (2022). Introduction: The emergent field of strategic communication. In *Research handbook on strategic communication* (pp. 1–12). Edward Elgar.

Frandsen, F., & Johansen, W. (2022). Strategic communication: A discipline in the making? In *Research handbook on strategic communication* (pp. 14–32). Edward Elgar.

Frank, A. K. (2017). What is the story with sustainability? A narrative analysis of diverse and contested understandings. *The Journal of Environmental Studies and Sciences, 7*, 310–323. https://doi.org/10.1007/s13412-016-0388-3

Godemann, J., & Michelsen, G. (2011). *Sustainability communication–an introduction* (pp. 3–11). Springer Netherlands.

Golob, U., Podnar, K., Elving, W. J., Ellerup Nielsen, A., Thomsen, C., & Schultz, F. (2013). CSR communication: Quo vadis? *Corporate Communications: An International Journal, 18*(2), 176–192.

Golob, U., Podnar, K., & Zabkar, V. (2023). Sustainability communication. *International Journal of Advertising, 42*(1), 42–51.

Grober, U. (2014). The discovery of sustainability: The genealogy of a term. In *Theories of sustainable development* (pp. 6–15). Routledge.

Habermas, J. (2021, August). Überlegungen und Hypothesen zu einem erneuten Strukturwandel der politischen Öffentlichkeit. In *Ein neuer Strukturwandel der Öffentlichkeit?* (pp. 470–500). Nomos Verlagsgesellschaft mbH & Co. KG.

Hahn, R., & Kühnen, M. (2013). Determinants of sustainability reporting: A review of results, trends, theory, and opportunities in an expanding field of research. *Journal of Cleaner Production, 59*, 5–21.

Hallahan, K., Holtzhausen, D., Van Ruler, B., Verčič, D., & Sriramesh, K. (2007). Defining strategic communication. *International Journal of Strategic Communication, 1*(1), 3–35.

Heath, R. L., Johansen, W., Hallahan, K., Steyn, B., Falkheimer, J., & Raupp, J. J. C. (2018). Strategic communication. In R. L. Heath & W. Johansen (Eds.), *The international encyclopedia of strategic communication* (pp. 1–24). Wiley. https://doi.org/10.1002/9781119010722.iesc0172

Heide, M., von Platen, S., Simonsson, C., & Falkheimer, J. (2020). Expanding the scope of strategic communication: Towards a holistic understanding of organizational complexity. In *Future directions of strategic communication* (pp. 124–140). Routledge.

Hoffjann, O. (2024). "Tear down this wall". A theory of strategic communication that integrates both communication and organizational theory. *Journal of Communication Management.*

Holtzhausen, D., & Zerfaß, A. (2014). Strategic communication: Opportunities and challenges of the research area. In D. Holtzhausen & A. Zerfaß (Eds.), *The Routledge handbook of strategic communication* (pp. 3–17). Routledge.

Kannengießer, S. (2021). Sozial-ökologische Transformationsforschung: Desiderata der Kommunikations-und Medienwissenschaft. In N. S. Borchers (Ed.), *Transformation der Medien Medien der Transformation. Verhandlungen des Netzwerks Kritische Kommunikationswissenschaft*.

Lahsen, M., & Turnhout, E. (2021). How norms, needs, and power in science obstruct transformations towards sustainability. *Environmental Research Letters, 16*(2), 025008.

McLeod, S., & Marshall, J. (2023). Communication for all and the Sustainable Development Goals. *International Journal of Speech-Language Pathology, 25*(1), 1–8.

Pleil, T., Helferich, P. S., & Otsa, T. (2025). A typology of strategic sustainability communication: From reporting to transformation. In F. Weder (Ed.), *Strategic sustainability communication*. Springer.

Putnam, L. L., & Nicotera, A. M. (Eds.). (2009). *Building theories of organization: The constitutive role of communication*. Routledge.

Raworth, K. (2017). *Doughnut economics: Seven ways to think like a 21st-century economist*. Chelsea Green Publishing.

Robertson, M. (2018). *Communicating sustainability*. Routledge.

Robinson, J., & Cole, R. J. (2015). *Building Research & Information, 43*(82), 133–143.

Röttger, U., & Rettler, L. (2025). Sustainability as a guiding principle of strategic communication. A heuristic for future-proof strategic communication. In F. Weder (Ed.), *Strategic sustainability communication*. Springer.

Schaltegger, S., Hansen, E. G., & Lüdeke-Freund, F. (2016). Business models for sustainability: Origins, present research, and future avenues. *Organization & Environment, 29*(1), 3–10.

Schaltegger, S., Girschik, V., Trittin-Ulbrich, H., Weissbrod, I., & Daudigeos, T. (2024). Corporate change agents for sustainability: Transforming organizations from the inside out. *Business Ethics, the Environment & Responsibility, 33*(2), 145–156.

Schoeneborn, D., Glozer, S., & Trittin-Ulbrich, H. (2022). Constitutive views on CSR communication: The communicative constitution of responsible organization, organizing, and organizationality. In *The Routledge handbook of corporate social responsibility communication* (pp. 73–84). Routledge.

SDG. (2025). Sustainable Development Goals. https://sdgs.un.org/goals, accessed on 1.7.2025.

Servaes, J., & Yusha'u, M. J. (Eds.). (2023). *SDG18 communication for all, Vol. 2: Regional Perspectives and Special Cases*. Springer Nature.

Stiff, J. B., & Mongeau, P. A. (2016). Persuasive communication. *Guilford Publications*.

Turnheim, B., & Sovacool, B. K. (2020). Forever stuck in old ways? Pluralising incumbencies in sustainability transitions. *Environmental Innovation and Societal Transitions, 35*, 180–184.

UN. (2024). *Transforming our world: The 2030 Agenda for Sustainable Development*. 21252030 Agenda for Sustainable Development web.pdf (un.org)

Valentini, C. (Ed.). (2021). *Public relations* (Vol. 27). Walter de Gruyter GmbH & Co KG.

Waisbord, S. (2019). *Communication: A post-discipline*. Wiley.

Weder, F. (2023). Cultivation of sustainability in a discourse of change: Perspectives on communication for sustainability as new "norm" and principle of action in socio-ecological transformation processes. *Journal of Language and Politics, 22*(5), 577–600.

Weder, F. (2024). Sustainability as guiding principle of communicative action the transformative and transformational potential of corporate sustainability communication as niche construction, a case from the energy sector. *Corporate Communications: An International Journal, 30*(2), 241–259.

Weder, F., & Erikson, M. (2023). *CSR communication and cultures of sustainability*. Pressbook.

Weder, F., Krainer, L., & Karmasin, M. (2021). *Sustainability communication reader*. Springer Fachmedien Wiesbaden.

Zerfaß, A., Stieglitz, S., Clausen, S., Ziegele, D., & Berger, K. (2021). *Communications Trend Radar 2021. Denialism, virtual corporate communications, sustainable communications, digital nudging & voice interaction* (No. 10). Communication Insights.

Zerfaß, A., Vercic, D., Nothhaft, H., & Werder, K. P. (2018). Strategic communication: Defining the field and its contribution to research and practice. *International Journal of Strategic Communication, 12*(4), 487–505. https://doi.org/10.1080/1553118X.2018.1493485

Franzisca Weder, Chair for International Organizational Communication at the Vienna University of Economics and Business, Department of Business Communication, Institute for Strategic Organizational Communication, is a leading scholar in the field of Sustainability Communication. She is an experienced qualitative researcher with over 20 years of experience in exploring, studying, and teaching communication of, about, and for sustainability and transformative communication. With her background in journalism and strategic communication, she also evaluates mediated sustainability discourses, internal Corporate Social Responsibility management and communication, and sustainability-related policies and their impact on business, NGOs, and political institutions.

Conceptualizing Strategic Sustainability Communication

When Talk Is Not Cheap: The Performative Potentials of Strategic Sustainability Communication

Lars Thøger Christensen and Emma Christensen

Abstract Corporate sustainability communication is often met with stakeholder skepticism, focusing on what is perceived as insincerity and a lack of alignment with current sustainability practices. While acknowledging the significance of such skepticism, this chapter challenges common-sense understandings of sustainability communication as inherently cheap or inconsequential. Amidst urgent calls for substantial sustainability action, we explore the potential of strategic sustainability communication to catalyze further sustainability practices. Drawing on a performative perspective on communication, we focus on the conditions under which such communication can unfold its performative potential. We take our point of departure in a recent sustainability campaign by Apple Inc. in which the audience is invited to participate in a simulated meeting between the company's sustainability staff and Mother Nature. Analyzing this campaign, we identify and discuss a set of processual boundary conditions for such communication to bring about more sustainable practices. Our performative communication perspective, in other words, offers an alternative to the conventional a priori denigration of sustainability claims. Ultimately, our analysis underscores the importance of strategic sustainable communication in advancing sustainability agendas and shaping responsible corporate behavior. At the same time, we do not suggest that such communication automatically or in any easy way unfolds into more sustainable corporate practices. Several factors, such as aspirational inflation, corporate self-absorption, stakeholder cynicism, and jamming, can disrupt the performative process and, eventually, prevent society from attaining its sustainability goals.

L. T. Christensen (✉)
Department of Management, Society & Communication, The Copenhagen Business School, Frederiksberg, Denmark
e-mail: ltc.msc@cbs.dk

E. Christensen
Department of Communication and Arts, Roskilde University, Roskilde, Denmark
e-mail: emch@ruc.dk

F. Weder (ed.), *Strategic Sustainability Communication*, CSR, Sustainability, Ethics & Governance, https://doi.org/10.1007/978-3-031-89486-2_2

1 Introduction

For several good reasons, corporate talk is often considered cheap, insincere, or outright manipulative. This is especially the case when such talk strategically employs venerated ideals such as responsibility and sustainability (e.g., Ansari et al., 2011; Banerjee, 2008; Fleming & Jones, 2013; Jahdi & Acikdilli, 2009). Strategic sustainability communication may refer to all types of communication—retrospective and prospective—involved in deliberately advancing the sustainability missions of organizations (cf. Hallahan et al., 2007). As such, strategic sustainability communication foregrounds the communicative intentions of senders rather than their impact on receivers (cf. Christensen & Christensen, 2022b). While strategic communication in the context of sustainability is often past-oriented and compliance-driven, we focus in this chapter on its prospective dimension because this is where its change potential is most likely to play out. Thus, we consider corporate sustainability talk strategic whenever it systematically conveys corporate ideals, ambitions, and plans in efforts to build a reputation for sustainability. Such talk rarely reflects existing sustainability practices. Yet, it may still have an impact on sustainability practices. Given the gravity of climate change and the urgency for real sustainability action, the need to understand the role of sustainability talk in shaping sustainable practices is more acute than ever before. While it may seem logical to suggest that corporations simply start *acting* instead of just talking, research has in various ways indicated that talk about sustainability might be able to stimulate more sustainable practices (Christensen et al., 2013; Haack et al., 2012; Koep, 2017; Penttila, 2019; Winkler et al., 2020). The optimal conditions for such processes, however, remain to be unpacked.

Hoping to avoid or at least reduce greenwashing and other types of hypocrisy in the sustainability arena, it might be tempting to outlaw corporate sustainability claims that cannot be fully documented by practice. The European Union's recent "Green Claims Directive" that bans deceptive advertising and other unfair communication practices points in that direction. Without denying that such initiatives might eliminate certain types of cheap sustainability talk, this chapter explores an alternative possibility. Drawing on a performative perspective on communication (e.g., Ashcraft et al., 2009; Cooren et al., 2011; Schoeneborn et al., 2019), we discuss the likelihood that sustainability talk might be less than cheap. Specifically, we examine the conditions under which such talk has the potential to contribute to more sustainable corporate practices. To that purpose, we take our point of departure in a recent strategic sustainability communication campaign from Apple Inc. Following a brief description of the campaign, we present and discuss a set of processual boundary conditions that might be considered essential for strategic sustainability communication to be performative in terms of stimulating sustainable corporate practices.

2 Constructing the Critical Audience

In a mockumentary video called "2030 Status|Mother Nature" Apple simulates a status meeting between its sustainability staff and Mother Nature, played by Octavia Spencer. In the meeting, in which CEO Tim Cook participates, Mother Nature asks tough questions ostensibly formulated to seriously challenge Apple's existing sustainability goals and accomplishments. Her opening statement sets the scene: "This is my third corporate responsibility gig today, so who wants to disappoint me first?". Everyone present, except Mother Nature, is very nervous, eager to convince her that the company's efforts and aspirations in the sustainability arena are sincere and significant. A dry home plant, for example, is moved out of sight, and a leather jacket on a staff member is acknowledged as an imperfection in a company that plans to phase out leather. Mother Nature retains a skeptical attitude throughout, although she lightens up a bit when presented with what the company calls "carbon-free products." "I wanna see you do more of this," she says, to which Tim Cook replies, "You will!". Toward the end of the video, a staring contest between Mother Nature and the CEO follows his repeated proclamation that the company *will* live up to its 2030 ambitions. While leaving and announcing that she will be back next year, Mother Nature says, "Don't disappoint your mother!"

The video is obviously designed to convince a broader audience, including NGOs, consumers, and the company's own employees, that Apple is on the right track toward more sustainable products and practices. What is especially interesting in the video from a strategic communication perspective is the construction of a critical other, embodied by Mother Nature, whose questioning is meant to preempt the inquiries of the company's most critical NGOs. "Let's cut to the chase," she says up front, calling for "materials" and "status" and emulating the sentiments of many critics, even those who may refrain from articulating their suspicions so clearly. Reminding the company of its promises and calling out sustainability clichés such as "carbon neutrality" and the planting of trees ("Everybody's planting trees"), Mother Nature not only represents intelligent but absent critics. She simultaneously allows the company to feature its most remarkable achievements, such as the planting of forests (as opposed to trees) in several places in South America and Africa. Her critical and almost arrogant attitude, in other words, provides a perfect opportunity for the company to emphasize its claimed qualities and ambitions that allegedly set it apart from its competitors. Its ambition to "permanently remove carbon from the atmosphere," for example, is staged toward the end of the video as a promise that significantly trumps all other sustainability ambitions.

Thus, it is not difficult to deconstruct its overall strategic intentions. At the same time, it reflects a number of more generic corporate insights, hopes, and concerns when it comes to the strategic communication of sustainability. Firstly, corporations like Apple know that their audiences are sophisticated and not easily manipulated. Secondly, they are aware that no matter what they communicate about sustainability, they are likely to be criticized and perhaps called out as hypocrites. Thirdly, they realize that critics will focus especially on gaps between corporate sustainability talk and corporate action. Despite the risks of announcing sustainability ambitions under these conditions, they recognize that public mediation and exposure may be necessary, not only to build a particular brand identity but to apply the necessary pressure on themselves. In the remainder of this chapter, we unfold these dimensions in some detail.

3 When Sustainability Talk Is Performative

In organization and management theory, discussions of communicative performativity usually take their point of departure in Austin's (1962) Speech Act Theory and its notion that things can be accomplished with words (Gond et al., 2016). According to Austin and his student Searle (1969), talk can *do* something provided the circumstances or "felicity conditions" such as participants, procedures, preparations, and intentions are right. While similar conditions may also apply to corporate sustainability communication, it is important to acknowledge that the examples of speech acts discussed by Austin and Searle describe speech situations where roles and expectations are well defined and where reactions are often scripted in some detail. Outside such conventionalized situations—for example, when corporations talk about their hopes, visions, or ideals—talk's performative potential is usually unstable and uncertain (Butler, 2010; McKinlay, 2011).

The temporality of performativity differs, too. While much strategic sustainability communication, such as reporting, is past-oriented and describes practices already accomplished, some communication about sustainability is future-oriented without precise conventions for fulfillment. Such communication belongs to a particular genre that has been labeled aspirational talk, that is, "organizational self-descriptions to which current practices cannot yet live up" (Christensen et al., 2021, p. 412; see also Christensen et al., 2013). Whereas the speech acts of Austin and Searle mostly perform in the very moment they are uttered, aspirational talk extends into an unknown future that can seriously affect its final performativity. Although strategic sustainability talk, following a speech act perspective, is a type of action in and of itself, the talk may not accomplish anything beyond the moment in which the words are uttered (Christensen et al., 2021). The challenge, thus, is to specify the optimal conditions for such talk to stimulate further and more material types of corporate action. By specifying those conditions and their possible interactions, we

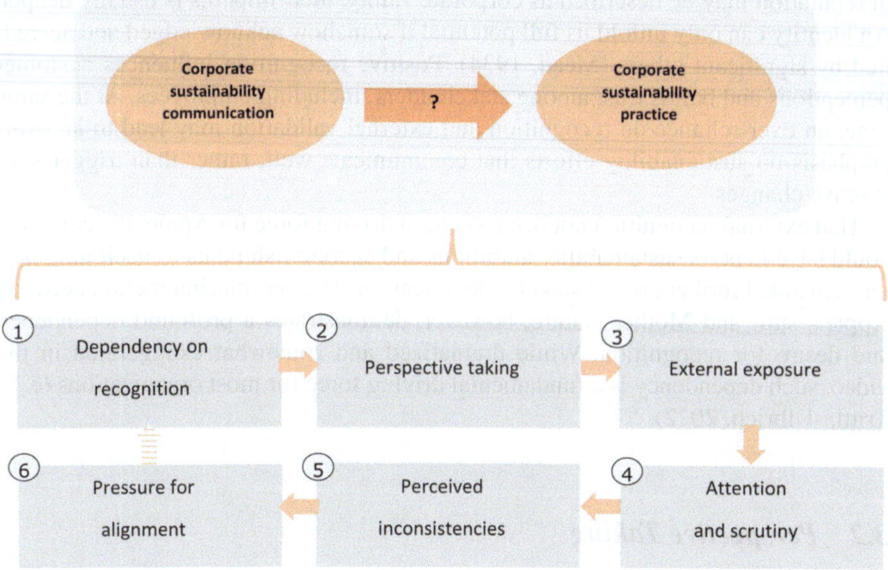

Fig. 1 Boundary conditions for the performativity of sustainability communication

not only contribute to the growing literature on aspirational talk but also and simultaneously demonstrate its relevance for the study of strategic sustainability communication.

Specifically, we argue that the following boundary conditions are essential for contemporary sustainability communication to stimulate better and more sustainable corporate practices: (1) Dependency on external recognition; (2) Perspective taking; (3) External exposure; (4) Attention and scrutiny; (5) Perceived inconsistencies; and (6) Pressure for alignment. Without assuming that this list is exhaustive of all possible conditions for communicative performativity in the sustainability arena, we will use the Apple video described above to illustrate how these conditions and their mutual interplay can help us understand how major corporations use strategic communication to cultivate sustainable practices (Fig. 1).

3.1 Dependency on External Recognition

While some communicators seem unaffected by what the surrounding world thinks of them (see e.g., Hatch & Schultz, 2002), most social actors, including corporations, depend on social recognition and approval (Athens, 1994; Goffman, 1959; Honneth, 1995; see also Suchman, 1995). If such recognition is absent, corporations are likely to take deliberate steps to be officially acknowledged by select audiences, for example, by improving their positions in corporate reputation rankings or obtaining positive media coverage. While the desire for recognition and the interest in reputation may be described as corporate vanity, their impetus is usually deeper. An identity can only unfold its full potential if somehow acknowledged and accredited by significant others (Mead, 1934). Positive recognition influences customer perceptions and builds trust among stakeholders, including employees. At the same time, an over-reliance on recognition and external validation may lead to an overemphasis on sustainability efforts that communicate well, rather than trigger substantive changes.

Had external recognition not been a central driving force for Apple, the company could have kept its sustainability ambitions and accomplishments to itself and, perhaps, avoided further critical stakeholder attention. The nervous interaction between Apple's staff and Mother Nature, however, demonstrates a profound dependency and desire for recognition. While dramatized and somewhat exaggerated in the video, such dependency is a fundamental driving force for most organizations (e.g., Trittin-Ulbrich, 2022).

3.2 Perspective Taking

One thing is to desire and pursue social recognition, another is to be able to critically look at oneself, imagining how others might evaluate one's behavior. The social psychologist Charles Horton Cooley (1902/1983) coined the notion of "the looking-glass self" to describe how social actors develop their self-images through the gazes of others. Social actors, Cooley pointed out, look at others looking at them, trying to figure out what these others see. Even if nobody is actually looking or making assessments of one's behavior, the capacity to imagine that such looking is taking place is a vital social skill (Athens, 1994).

For corporations, the ability to assume the perspective of NGOs and other critical stakeholders can help them predict and, perhaps, preempt critical assessments of their practices. In Apple's video, perspective taking seems to be the primary message. Using Mother Nature as an apotheosis of the critical stakeholder, the video claims that the company has taken precautionary steps to imagine what significant others might think of the corporation. In most organizations, such reflections take place in backstage settings, such as marketing or corporate communication departments, where official messages are devised and prepared. In such settings, external audiences are usually absent, although images of such audiences might figure prominently in discussions and deliberations (Christensen & Christensen, 2022a). The Apple video, however, claims to take the audience backstage. In its effort to demonstrate that the company takes the concerns of external audiences very seriously, the video simulates external access into the very "engine room" where the company's sustainability communication is devised and tested.

Without dependency on external recognition and the ability to assume the perspective of critical others, corporate communicators can claim virtually anything they might want without feeling committed to their own words. In such cases, the communication is unlikely to have any substantial effect beyond the moment in which it is uttered. Dependency on external recognition and perspective taking are therefore indispensable conditions for sustainability communication to perform.

3.3 *External Exposure*

Corporate sustainability aspirations are likely to be discussed at great length in internal settings before they are made public. And while such discussions can be very creative and invigorating, the reality status of such aspirations changes dramatically when publicized. Suddenly, they are for real for management and other organizational members who may be held to the words. Even if the words are ignored or ridiculed by some members, their authority has increased by their official announcement (cf. Christensen, 1997). External exposure, thus, is an essential step toward corporate commitment.

Communicating publicly about sustainability, of course, is not without risk. In today's communication environment, sustainability claims—especially from major corporations—are likely to be fiercely challenged by stakeholders, met with recurrent hypocrisy charges, and denigrated with reference to shortcomings of actual sustainability practices (Christensen et al., 2020). In this environment, corporations may be expected or advised to keep quiet unless they already practice what they preach. Some organizations take this advice one step further by engaging in "greenhushing" (Font et al., 2017) or other types of strategic silence (Carlos & Lewis, 2018)—even when they have made significant sustainability advances. Such self-restraint might seem reasonable when critical stakeholders are looking for and exposing gaps and inconsistencies between sustainability claims and more substantial sustainability actions. Yet, the call for self-restraint or strategic silence ignores the role of external exposure in facilitating and supporting real sustainability transformations. Again, this potential hinges on corporate concern with external recognition and the ability for perspective taking discussed above. Knowing or at least expecting that their sustainability aspirations will be exposed to critical audiences is likely to raise their significance to the corporate senders (Junge, 2006; Koep, 2017; see also Taylor & Cooren, 1997). The anxiety among the staff in the Apple video—even if somehow overstated—illustrates how such awareness can be a driving force for a collectivity when planning how to publicly present itself and its accomplishments.

3.4 *Attention and Scrutiny*

From a strategic communication perspective, the significance of aiming for external exposure is to generate attention and scrutiny. Interest and involvement in corporate messages, however, cannot be taken for granted. In a world saturated with corporate claims, signs, and other messages, the giving and receiving of attention is constrained by what has been labeled an "attention economy" (e.g., Franck, 2019). Since attention is a scarce resource, only matters of significant collective concern are likely to attract more than passing consideration. Thus, many corporations publicize ideals, plans, and news without attracting much attention. Yet, when major corporations speak publicly about sustainability, attention is likely to be widespread, substantial, and critical.

Being aware of such risk, Apple will be inclined to study its own communication meticulously in that particular arena. Described as autocommunication (Lotman, 1990), such engagement in one's own communication is foundational to any identity, individual or collective. At the same time, autocommunication may help communicators explore new ideas and possibilities (see also Christensen, 1997). Corporations speak to themselves whenever they are contemplating and preparing identity statements or claims, for example, about responsibility and sustainability. In the process of trying out various formulations, corporations are systematically referring to their own images of what might happen if they claim this or that (Christensen & Christensen, 2022a). As Apple's video illustrates, such preparatory talk is both precautionary and autocommunicational, especially in this particular case where the critical audience is constructed by the corporation itself. While the autocommunicational effects are likely to increase if the claims simultaneously attract actual attention and scrutiny from outside audiences, the ability of communicators to imagine how their claims might be received by such audiences may be enough to trigger an organization's deep engagement with its own messages. While constructing and interacting with their own image of a critical "other" in the shape of Mother Nature, the video from Apple is essentially illustrating such autocommunicative potential.

3.5 Perceived Inconsistencies

In the hope of circumventing greenwashing accusations, hypocrisy charges, or other types of negative stakeholder attention, organizations may seek to proactively eliminate gaps or inconsistencies between sustainability talk and sustainability action (Carlos & Lewis, 2018; Font et al., 2017). Yet, gaps or differences may be necessary to stimulate better practices. As a number of studies have indicated, perceived gaps can create attention and trigger the stakeholder skepticism necessary to force organizations to practice what they preach (e.g., Girschik, 2018; Koep, 2017; Penttilä, 2019). If no discrepancies between talk and action are noted by relevant stakeholders, corporations might not need to improve their practices. Christensen et al. (2013) therefore suggest that corporations need to "nurse" such gaps: "On the one hand, it implies keeping the gap open, so that stakeholders are constantly motivated to devote time and energy to ideals that have not yet been realized. If the gap closes in, it must be re-opened with new aspirations. On the other hand, it means nursing the gap with visible and effective achievements aimed at closing the gap" (p. 385).

The Apple video describes how the fear of being labeled hypocritical can act as a collective impetus for the organization to try harder. At the same time, the video strategically uses inconsistencies in its pursuit of authenticity and credibility. The presented flaws, such as the dry home plant and the misplaced leather jacket, are included to give the impression that the company is aware that its practices are less than perfect. And while this confession may serve to obscure more substantial imperfections, it simultaneously helps generate sympathy for the company whose members are ordinary people like everybody else. The promise from Tim Cook to Mother Nature toward the end of the video, emphasizing that she *will* see more "carbon-free products" from the company, shows further awareness of how perceived gaps can be used strategically by corporations to apply pressure on themselves.

3.6 Pressure for Alignment

One thing is to observe and acknowledge gaps or inconsistencies between corporate ideals and current reality. Another is to use such gaps to demand real changes. When gaps are multiple and commonplace, they may pass unnoticed or engender cynicism and apathy among stakeholders (e.g., Christensen et al., 2021). When that is the case, pressure for alignment is likely to be absent or negligible. After all, organizations face many different pressures from their surroundings and will, in most cases, need to negotiate compromises or develop partial solutions and responses to them (e.g., Brunsson, 2003; Oliver, 1992). Whether it produces full alignment or not, pressure from stakeholders, including NGOs, employees, regulators, and media, is arguably the *sine qua non* for performativity in the sustainability arena (e.g., Haack et al., 2012; Livesey & Graham, 2007; Winkler et al., 2020).

Outside pressure is likely to be directed especially toward major, well-known corporations. Because of its substantial size and influence, Apple clearly knows that its sustainability aspirations will engender significant demands and expectations, both among NGOs, employees, regulators, and the media. With its video, the company seeks to integrate and preempt such demands and expectations by proactively imposing them on itself. What the video illustrates above all is, in other words, the energy and efforts that major corporations put into the preparation of the "show," that is, the planning of the important meeting where their aspirations meet the critical public (see Christensen & Christensen, 2022a).

4 Conclusion and Final Reflections

Recognizing that some conditions are conducive to the performativity of sustainability communication does not suggest that corporate insight into the nature of these conditions and their interactions automatically or inevitably generate more sustainable practices. Apple's strategic use of these dimensions in its video, in other words, does not necessarily serve the sustainability agenda. Our discussion does not imply such automatism. What we hope to convey is that each of these conditions is indispensable for corporations to take their own strategic sustainability communication seriously. At the same time, they are mutually interrelated and co-dependent. The absence of any of these dimensions will likely interrupt the productive relationship between sustainability talk and sustainability action. When, for example, the "nursing" of gaps between sustainability aspirations and sustainability actions (Christensen et al., 2013, p.385) degenerates into overbidding and aspirational inflation, the performative process is aborted. The same will be true when perspective taking becomes self-absorbed and fails to capture how outside audiences will interpret corporate claims. Or when stakeholder attention and scrutiny are replaced by ignorance and cynicism.

Adding to these risks, which are all realistic outcomes in today's communication environment, critical audiences may learn to deconstruct corporate messages in ways that undermine their performativity. Corporations like Apple may, for example, hope to appear more authentic and sincere by portraying its staff sweating nervously and making a real effort when confronted with an intelligent and critical audience. Yet, that very same audience can easily see through such efforts and is unlikely to be *significantly* influenced by the message. Even though the video reflects insight into the dynamics between corporate exposure and stakeholder pressure, the company risks ignoring that most critics are enlightened readers of corporate claims and will be inclined to consider the video, in spite of its sophistication and explicit promises, as yet another example of impression management.

Moreover, strategic sustainability communication may be jammed by critics, not only in efforts to expose inconsistencies between talk and action, which—as we have argued—*could* serve the performativity process, but sometimes also to ridicule corporate aspirations. No matter how strategically corporations approach the sustainability agenda, they do not own their "own" messages. Once the messages are publicized, they can be manipulated in various ways. This has also befallen Apple's Mother Nature video. Free the People—an independent production company that creates documentary films, videos, and podcasts focused on pro-liberty values—has produced a video spoof that taunts Apple's Mother Nature message. In the video called "Apple fights climate change with child labor," two satirical characters posing as Apple employees are inserted into the Mother Nature video. Mocking Apple's sustainability claims, the video suggests that Apple has managed to reduce its emissions in America by moving its production to China, India, and Vietnam. The spoof also highlights Apple's (in)direct use of child labor in cobalt mining. One of the inserted employee remarks, "it's much cheaper to mine it over there [Congo] because of all modern slavery and child labor stuff." The other employee agrees, noting that "Apple's child labor has a very small carbon footprint, and that their actual footprints are really small too, because they're children," to which Mother Nature replies, "I want to see more of this."

How such critical stakeholder receptions, including jamming efforts, affect further corporate efforts in the sustainability arena remains to be studied. Will a lack of external recognition in the long run discourage further sustainability initiatives and aspirations, as Trittin-Ulbrich (2022) seems to suggest? Or will negative reactions only affect how corporations *communicate* about their initiatives and aspirations. While some stakeholders, including the initiators behind the EU's "Green Claims Directive," may hope for the latter, there is a risk that restrictions on bold sustainability claims might delay or, perhaps, even terminate important projects in that arena, partly because organizations use public claims to impose stricter demand on themselves.

This line of argumentation begs the obvious question of whether corporate talk is never cheap. The immediate answer to that question is "yes, of course it is!" But first of all, and primarily, for organizations that are able to operate under the radar. Large, well-known corporations are likely to attract, as we have argued, intense stakeholder scrutiny and critique. And while such critique, even when most sharp and precise, may fail to eradicate the gaps and inconsistencies that triggered stakeholder critique to begin with, it plays a significant role in holding corporations to the fire, to remind them of their ambitions and to press them toward better practices.

Finally, the Mother Nature video calls for further reflection on corporate aspirational talk as a particular genre of strategic sustainability communication. Does it make a difference to the performativity of such communication that its boundary conditions are anticipated by the communicator and proactively integrated into corporate messages? Will the performative potential dwindle by such proactivity? Are the boundary conditions, in other words, more effective in stimulating communicative performativity if *un*anticipated by the strategic communicator? These questions obviously call for further theoretical and empirical work and cannot be fully addressed in this chapter. It is possible to argue, however, that increased organizational precaution when formulating sustainability claims and ambitions can be performative too. Where most writings on aspirational talk assume that organizations learn to take their own words seriously when the aspirations are publicized, that is, exposed to external publics, Christensen and Christensen (2022a) argue that "preparatory talk" that take place in backstage settings where sustainability messages are contemplated and devised can have significant effects on performance, provided of course that communicators are able to imagine and adapt to possible stakeholder expectations.

References

Ansari, S. M., Gray, B., & Wijen, F. (2011). Fiddling while the ice melts or a more active role for organizational scholars in the climate change debate? *Strategic Organization, 9*(1), 70–76.

Ashcraft, K. L., Kuhn, T. R., & Cooren, F. (2009). Constitutional amendments: "Materializing" organizational communication. *The Academy of Management Annals, 3*(1), 1–64.

Athens, L. (1994). The self as soliloquy. *The Sociological Quarterly, 35*(3), 521–532.

Austin, J. L. (1962). *How to do things with words*. Oxford University Press.

Banerjee, S. B. (2008). Corporate social responsibility. The good, the bad and the ugly. *Critical Sociology, 34*(1), 51–79.

Brunsson, N. (2003). Organized hypocrisy. In B. Czarniawska & G. Sevón G. (Eds.), *The Northern lights – organization theory in Scandinavia* (pp. 201–222). Copenhagen Business School Press.

Butler, J. (2010). Performative agency. *Journal of Cultural Economy, 3*(2), 147–161.

Carlos, W. C., & Lewis, B. W. (2018). Strategic silence: Withholding certification status as a hypocrisy avoidance tactic. *Administrative Science Quarterly, 63*(1), 130–169.

Christensen, L. T. (1997). Marketing as auto-communication. *Consumption, Markets and Culture, 1*(3), 197–227.

Christensen, L. T., & Christensen, E. (2022a). Preparing the show: Organizational ventriloquism as autocommunication. *Organization Theory, 3*, 1–21.

Christensen, L. T., & Christensen, E. (2022b). The saying and the doing. When communication is strategic. In J. Falkheimer & M. Heide (Eds.), *Research handbook of strategic communication* (pp. 33–45). Edward Elgar.

Christensen, L. T., Morsing, M., & Thyssen, O. (2013). CSR as aspirational talk. *Organization, 20*(3), 372–393.

Christensen, L. T., Morsing, M., & Thyssen, O. (2020). Timely hypocrisy? Hypocrisy temporalities in CSR communication. *Journal of Business Research, 114*, 327–335.

Christensen, L. T., Morsing, M., & Thyssen, O. (2021). Talk-action dynamics: Modalities of aspirational talk. *Organization Studies, 42*(3), 407–427.

Cooley, C. H. (1902/1983). *Human nature and the social order*. Transaction Books.

Cooren, F., Kuhn, T., Cornelissen, J. P., & Clark, T. (2011). Communication, organizing and organization: An overview and introduction to the special issue. *Organization Studies, 32*(9), 1149–1170.

Fleming, P., & Jones, M. T. (2013). *The end of corporate social responsibility*. Sage.

Franck, G. (2019). The economy of attention. *Journal of Sociology, 55*(1), 8–19.

Font, X., Elgammal, I., & Lamond, I. (2017). Greenhushing: The deliberate under communicating of sustainability practices by tourism businesses. *Journal of Sustainable Tourism, 25*(7), 1007–1023.

Girschik, V. (2018). Shared responsibility for societal problems: The role of internal activists in reframing corporate responsibility. *Business & Society, 59*(1), 34–66.

Goffman, E. (1959). *The presentation of self in everyday life*. Penguin Books.

Gond, J.-P., Cabantous, L., Harding, N., & Learmonth, M. (2016). What do we mean by performativity in organizational and management theory? The uses and abuses of performativity. *International Journal of Management Reviews, 18*(4), 440–463.

Haack, P., Schoeneborn, D., & Wickert, C. (2012). Talking the talk, moral entrapment, creeping commitment? Exploring narrative dynamics in corporate responsibility standardization. *Organization Studies, 33*(5–6), 815–845.

Hallahan, K., Holtzhausen, D., van Ruler, B., Vercic, D., & Sriramesh, K. (2007). Defining strategic communication. *International Journal of Strategic Communication, 1*(1), 3–35.

Hatch, M. J., & Schultz, M. (2002). The dynamics of organizational identity. *Human Relations, 55*(8), 989–1018.

Honneth, A. (1995). *The struggle for recognition. The moral grammar of social conflicts*. The MIT Press.

Jahdi, K., & Acikdilli, G. (2009). Marketing communications and corporate social responsibility (CSR): Marriage of convenience or shotgun wedding? *Journal of Business Ethics, 88*, 103–113.

Junge, K. (2006). The promise of performance and the problem of order. In J. Alexander, B. Giesen, & J. Mast (Eds.), *Social performance: Symbolic action, cultural pragmatics, and ritual* (pp. 283–314). Cambridge University Press.

Koep, L. (2017). Investigating industry expert discourses on aspirational CSR communication. *Corporate Communication: An International Journal, 22*(2), 220–238.

Livesey, S. M., & Graham, J. (2007). Greening of corporations? Eco-talk and the emerging social imagery of sustainable development. In S. May, G. Cheney, & J. Roper (Eds.), *The debate over corporate social responsibility* (pp. 336–350). Oxford University Press.

Lotman YM (1990). *Universe of the mind. A semiotic theory of culture*. London: Tauris.

McKinlay, A. (2011). Performativity: From J. L. Austin to Judith Butler. In P. Armstrong & G. Lightfoot (Eds.), *'The leading journal in the field': Destabilizing authority in the social sciences of management* (pp. 119–142). Mayfly.

Mead, G. H. (1934). *Mind, self, and society*. University of Chicago Press.

Oliver, C. (1992). Strategic responses to institutional processes. *Academy of Management Review, 16*(1), 145–179.

Penttilä, V. (2019). Aspirational talk in strategy texts: A longitudinal case study of strategic episodes in corporate social responsibility communication. *Business & Society, 59*(1), 67–97.

Schoeneborn, D., Kuhn, T. R., & Kärreman, D. (2019). The communicative constitution of organization, organizing and organizationality. *Organization Studies, 40*(4), 475–496.

Searle, J. R. (1969). *Speech acts: An essay in the philosophy of language*. Cambridge: Cambridge University Press.

Suchman, M. C. (1995). Managing legitimacy: Strategic and institutional approaches. *Academy of Management Review, 20*(3), 571–610.

Taylor, J. R., & Cooren, F. (1997). What makes communication 'organizational'? How the many voices of a collectivity become the one voice of an organization. *Journal of Pragmatics, 27*(4), 409–438.

Trittin-Ulbrich, H. (2022). From the substantive to the ceremonial: Exploring interrelations between recognition and aspirational CSR talk. *Business & Society, 62*(5), 917–949.

Winkler, P., Etter, M., & Castelló, I. (2020). Vicious and virtuous circles of aspirational talk: From self-persuasive to agonistic CSR rhetoric. *Business & Society, 59*(1), 98–128.

Further Reading

Butler, J. (1999). Performativity's social magic. In R. Shusterman (Ed.), *Bourdieu: A critical reader* (pp. 113–128). Blackwell.

Finnemore, M., & Jurkovich, M. (2020). The politics of aspiration. *International Studies Quarterly, 64*(4), 759–769.

Lars Thøger Christensen (Ph.D.), Professor of Communication and Organization at Copenhagen Business School (Denmark). In addition to corporate social responsibility and sustainability, his research focuses on identity, autocommunication, transparency, hypocrisy, talk-action dynamics, and voicing. He approaches these issues from a performative perspective on communication focusing on how organizations talk themselves into new realities. In addition to six books and contributions to many edited volumes, his research appears in *Organization Studies, Human Relations, Communication Theory, Organization, Management Communication Quarterly,* and elsewhere.

Emma Christensen (Ph.D.), Associate Professor at Roskilde University (Denmark). Her overall research interests are communicative constitution, performativity, temporality, and imaginaries. From sociological and communicative perspectives, she currently studies the expectations and uses of artificial intelligence, and the materialization and organizing of local food systems. In addition to contributions to edited volumes, her research appears in *Organization Theory, Management Communication Quarterly, Corporate Communication: An International Journal, International Journal of Strategic Communication*, and elsewhere.

Public Relations and Sustainability: Points of Friction in a Complex Relationship

Sigrid Kannengießer, Lennart Rettler, Ulrike Röttger, and Helena Stehle

Abstract We model public relations as a specific form of strategic communication that plays a crucial role in the transformation toward sustainability that encompasses social, ecological, and economic aspects in equal measure. Such transformation is urgently needed because of the multiple crises that societies are facing worldwide. Drawing on existing literature, we discuss three different modes of the relationship between sustainability and public relations, namely (1) public relations *about* sustainability, (2) public relations *for* sustainability, and (3) the sustainability *of* public relations. Analyzing the interrelationship of these different modes, we reveal points of friction between them that must be acknowledged if the crucial role and responsibility of public relations in the transformation are to be impactful.

1 Introduction

In the context of the multiple crises that societies are facing worldwide (Demirović et al., 2011), especially the climate crisis, there is a shared responsibility of all people, including organizations of all kinds, to become an active part of the transformation toward sustainability. This transformation is defined by the Brundtland Report as "socio-ecological-economic" justice for present and future generations (United Nations (UN), 1987). Transformation means more than merely societal change; it is a process with a systematic and substantial character shaped intentionally by different actors (Kollmorgen et al., 2014). In this intentionality lies a specific

Sigrid Kannengießer, Lennart Rettler, Ulrike Röttger and Helena Stehle contributed equally with all other contributors.

S. Kannengießer (✉) · L. Rettler · U. Röttger · H. Stehle
Department of Communication, University of Münster, Münster, Germany
e-mail: sigrid.kannengiesser@uni-muenster.de; lennart.rettler@uni-muenster.de; ulrike.roettger@uni-muenster.de; helena.stehle@uni-muenster.de

F. Weder (ed.), *Strategic Sustainability Communication*, CSR, Sustainability, Ethics & Governance, https://doi.org/10.1007/978-3-031-89486-2_3

responsibility of and potential for organizations and their public relations to contribute to a transformation toward sustainability.

Public relations is a communication function that addresses the issue of organizational legitimacy (e.g., Crumley et al., 2022). Following a post-positivist paradigm in this chapter that emphasizes organizational efficiency and effectiveness and the contribution of public relations to organizational success among others (Putnam, 1982; Trujillo & Toth, 1987; Sandhu, 2012; Stehle, 2015), public relations can be described as a specific form of strategic communication by organizations, primarily aiming at establishing, shaping, and maintaining communication relationships with socio-political stakeholders in order to support the achievement of objectives (e.g., Hallahan et al., 2007; Ki & Hon, 2009). Following this definition, we understand public relations as an organizational function which crucially contributes to the overall success of an organization. Public relations that refer specifically to sustainability can be embedded in the broader concept of strategic sustainability communication. Strategic sustainability communication, in turn, is defined as all intentionally used forms of communication with which organizations participate in conversations and opinion-forming processes. These conversations and processes are critical to achieving the organization's objectives (e.g., Nothhaft & Zerfaß, 2020) and deal with any aspect of sustainability, whether in relation to the ecological, economic, or social dimensions (e.g., Stehle, 2022).

In the following, rather than focusing on specific organizations and their public relations, we argue that organizations in general have a responsibility to become an active part of the transformation toward sustainability, and that organizations can use public relations to address this responsibility (see also Rademacher et al., 2024, for a normative model that emphasizes the responsibility of public relations with regard to sustainability).

Organizations have long been understood as being a part of their environment. Since the 1990s, research on organizations (especially corporations) and their communication has been concerned with responsibility and sustainability, in particular in the form of the management concept of corporate social responsibility (CSR) (McWilliams & Siegel, 2011; Carroll, 2021). For many years, corporations have been making increasing efforts to communicate about their responsibility and sustainability activities, due to social and political pressure. However, there are still many examples where organizational actions and communication diverge or even contradict each other, for example, when corporations claim to contribute to sustainability, but develop or produce nonsustainable goods.

In its "boundary spanning" role between an organization and its environment (Verhoeven, 2018), public relations can be seen as bearing a special responsibility and having a unique potential to support—or hinder—the transformation toward sustainability: Public relations can sensitize people to or convince people of the relevance of sustainability.

In the relationship between sustainability and public relations, different modes occur. In this chapter, we discuss these different modes, namely (1) public relations *about* sustainability, (2) public relations *for* sustainability, and (3) the sustainability *of* public relations (Weder, 2022; Rettler & Röttger, 2023; Weder & Erikson, 2023).

While public relations *about* sustainability refers to those communicative practices that report about the sustainability of an organization, public relations *for* sustainability relates to those practices which advocate for sustainability. Finally, the sustainability *of* public relations touches on the question of whether the public relations processes themselves are sustainable. While research so far has started to differentiate between these modes (Weder, 2022; Rettler & Röttger, 2023; Weder & Erikson, 2023), public relations *about* sustainability has traditionally been the focus of the research field dealing with strategic sustainability communication, analyzing and conceptualizing the organization's practices to report on its sustainability efforts. In particular, the sustainability *of* public relations is only now starting to gain the attention of the research field, while public relations *for* sustainability remains under-researched. Furthermore, what has not been acknowledged is the interplay between the different modes. When looking at this interplay, however, several points of friction emerge. We hypothesize that these points of friction result from the diverse objectives and practices of various actors inside and outside an organization, which is itself embedded in a socio-political context.

This chapter reveals and discusses these points of friction, which need to be acknowledged when analyzing and practically developing public relations *about* and *for* sustainability, and the sustainability *of* public relations. Firstly, based on existing literature in the field of strategic communication, we argue that public relations can play a crucial role in the "socio-ecological-economic" transformation toward the much-needed sustainability. Secondly, we show how public relations relates to sustainability by discussing three different modes that characterize the relationship between the two. Thirdly, we reveal the points of friction between these different modes. These points of friction need to be acknowledged if the potential of public relations is to be realized.

This analysis aims to develop a framework that, first, identifies and models the three modes that characterize the relationship between sustainability and public relations and, second, critically assesses and systematically discusses the points of friction between these modes.

As a team of authors from different research fields of communication studies—strategic communication and media sociology—, but with a common focus on sustainability, we analyze the relationship between public relations and sustainability from different perspectives. We bring together literature stemming from public relations research as well as research on media practices and digital infrastructures that focus on sustainability.

2 Public Relations and Sustainability: Three Relationship Modes

Both public relations and sustainability are multidimensional, complex phenomena, and the relationship between the two is correspondingly diverse. Existing models of this relationship are scarce (Weder, 2023). In this chapter, we follow Rettler and Röttger (2023) and propose three modes of the relationship.

We define modes as different practices in which media is used as a facilitator in communication processes. A mode of action defines the conditions under which something can be approached. We thus propose three different ways in which the approach toward sustainability by public relations may be described, these being:

- Public relations *about* sustainability
- Public relations *for* sustainability
- The sustainability *of* public relations.

While the modes "public relations *about* sustainability" and "public relations *for* sustainability" use media to report and advertise in support of sustainability, the mode "the sustainability *of* public relations" focuses on the sustainability of public relations itself. Kannengießer (2020) differentiates these modes through the concepts of acting *with* and acting *on* media, respectively. While the first two modes use media as facilitators and refer to practices of acting with media, the latter puts the sustainability of media used for public relations in the center of acting and is therefore a practice that can be characterized as "acting on media." Using this distinction helps to clarify the different modes that occur in the interplay of public relations and sustainability and the potential that public relations can have in support of sustainability.

In the following, the three modes are described and discussed. It becomes clear that they arise from different ways of looking at sustainability from a public relations perspective, which have their origins, for example, in different dominant paradigms and understandings of public relations in relation to sustainability (see Fig. 1).

2.1 *Public Relations About Sustainability*

With regard to the relationship between public relations and sustainability, public relations *about* sustainability has so far been comprehensively established in practice and has also been intensively researched (Ihlen et al., 2011; Golob et al., 2013, 2023; Diehl et al., 2017).

Public relations *about* sustainability is studied under various labels and concepts, the most common being CSR communication and sustainability communication. This mode is the most prominent one in the research field dealing with strategic communication.

	Public relations about sustainability	**Public relations for sustainability**	**The sustainability of public relations**
Dominant paradigm	(Post-)Positivist	Interpretive	No dominant paradigm
Dominant understanding	Public relations as a means of presenting organizational sustainability to its stakeholders	Public relations as a means of problematizing and negotiating sustainability issues between an organization and its stakeholders	Public relations as resource-consuming processes and products

Fig. 1 The three modes and their dominant paradigms and objects of investigation (own illustration)

There is a disagreement in the research literature, particularly with regard to the scope of CSR and the relationship between CSR and sustainability. However, there is a consensus that the two concepts are closely related. For example, sustainability is sometimes understood as the guiding principle of CSR (Hentze & Thies, 2014) or sustainable engagement is described as a component of CSR (Panapanaan, 2003). Since this chapter focuses on the relationship between public relations and sustainability, we use the term sustainability, but also take into account relevant literature from the field of CSR communication.

Public relations about organizational sustainability commitment and activities has been a focus of previous research, as shown by Golob et al. (2023). In their meta-narrative systematic review, Golob et al. (2023, p. 47) identify four key forms of public relations about sustainability: public relations as "transmission of information about sustainability," "disclosure of information" in the context of sustainability reports, "persuasive/commercial communication" in the context of green advertising, and "misleading communication" (greenwashing).

Previous research has primarily addressed the ecological dimension of sustainability, while socio-economic aspects have rarely been analyzed (Golob et al., 2023). Research on public relations *about* sustainability focuses on content and messages, media and channels, and techniques, as well as on the intended and achieved effects of communication.

Overall, this mode is dominated by an understanding of public relations that characterizes public relations as a means of successfully presenting an organization's commitment to sustainability to its stakeholders (Verk et al., 2021). While the positivist paradigm used to dominate this mode, today it is the post-positivist paradigm that focuses, among other things, on the contribution of public relations to organizational goals, effectiveness, and efficiency (Putnam, 1982; Trujillo & Toth, 1987; Sandhu, 2012; Stehle, 2015). Thus, this mode focuses on the meso level of analysis of organizations and is concerned with the achievement of organizational objectives.

2.2 *Public Relations for Sustainability*

The "socio-ecological-economic" transformation toward sustainability is linked to a fundamental social change that requires a comprehensive rethinking in all areas of society. In this context, the role of public relations has recently come under critical scrutiny. The transformative role of public relations *for* sustainable development is increasingly coming into focus (Weder, 2022).

Public relations *for* sustainability aims to actively contribute to change toward a more sustainable society using information and activation of individuals who are addressed by public relations. Accordingly, public relations *for* sustainability is based on sustainability as a guiding principle or moral compass (Prexl, 2010; Weder et al., 2019b; Weder, 2023). In this context, sustainability serves as a mode of evaluation with which public relations processes and means are assessed in terms of the extent to which they contribute to sustainable development or make a contribution to societal communication on the topic of sustainable development (Prexl, 2010; Weder, 2022).

As sustainability issues have always been associated with negotiations and conflicts of interest, problematization is a key mode of public relations *for* sustainability (Weder et al., 2019a; Weder, 2021b, 2022): "Problematization is a process of permanent contestation, the stimulation of dissent in lively stakeholder relationships and an ongoing review and legitimization of (hegemonic) arguments" (Weder, 2022, p. 342). With regard to the socially necessary negotiations on sustainability, public relations has the task of creating communicative spaces that allow for dissent and agonism (Weder, 2021a), as in this way existing patterns of interpretation are questioned and problematized. This means that public relations ultimately "assumes the role of controversy and authorship in the creation of a narrative of sustainability" (Weder, 2021a, p. 194).

Overall, this mode is dominated by an interpretive paradigm that characterizes public relations as a means of problematizing and negotiating sustainability issues between an organization and its stakeholders. It is often illustrated in public relations of socio-political organizations such as environmental groups or citizens' initiatives. This mode focuses more on the role of public relations for society and its transformation toward sustainability (macro level of analysis) but includes influences on individuals such as citizens (micro level of analysis).

2.3 *Sustainability of Public Relations*

Finally, in addition to public relations *about* and *for* sustainability, the sustainability *of* public relations has to be considered separately. This mode is concerned with the extent to which the production environment or processes and the products of public relations are sustainable. While the first two modes focus on the issues and intended effects of public relations from a sustainability perspective, the third mode, the sustainability *of* PR, refers to the characteristics and features of the public relations process and products themselves.

The sustainability *of* public relations is conceived and designed to contribute to long-term inter- and intragenerational justice in ecological, social, and economic terms. First approaches in research (Prexl, 2010) and practice (Dillmann, 2023) are addressing this idea. Prexl (2010), for example, points out that sustainability should be considered in every step of the public relations process, and emphasizes that the sustainability of communication media and measures is also important, as they act as a point of contact with both external and internal stakeholders.

Despite these initial approaches, there is still a lack of broad research and discussion. In this context, some authors even speak of "a lack of transfer of the sustainability principle in the field of communication" (Bittner-Fesseler & Weicht, 2019, p. 38). For example, the socio-ecological effects of the production, use, and disposal of digital media technologies and communication practices on the sustainable development of a society need to be given greater consideration, focusing on the question of whether they are, or should be, sustainable in themselves. These effects comprise processes such as the destructive extraction of minerals needed for digital technologies, which are used in PR, the carbon dioxide emission caused by online communication, or the problems of e-waste (Kannengießer, 2020).

Such "socio-ecological-economic" effects and sustainable media practices (Kannengießer, 2022) are the subject of an interdisciplinary field of research (for an overview, see Kannengießer & McCurdy, 2021), but so far only a secondary aspect in public relations research. However, this subject can and should be transferred to public relations research, as it is important to consider how public relations is produced and what resources are used. A shift in awareness that public relations itself requires all kinds of scarce resources, such as time (Görland & Kannengießer, 2021) or attention (Goldhaber, 1997), is of great importance but often overlooked. Public relations that strives for sustainability should take such considerations into account.

Overall, there is so far no dominant paradigm in this mode. Rather, there are post-positivist, interpretive, and postmodern-critical perspectives (for an overview of the paradigms see, e.g., Putnam, 1982; Trujillo & Toth, 1987; Sandhu, 2012; Stehle, 2015), each of which brings its own facets to the discussion. But there is a focus in terms of the level of analysis. The focus in this mode is so far on the individuals in organizations who engage in public relations actions (micro level of analysis), that means, it is directed inward, to those responsible for public relations and their actions and responsibilities, and to the professional field of public relations. This mode views public relations as resource-consuming processes and products.

3 Points of Friction in the Relationship Between Public Relations and Sustainability

After having distinguished and discussed the different modes—public relations *about* sustainability, public relations *for* sustainability, and the sustainability *of* public relations—the relationship between public relations and sustainability is further analyzed and reflected upon in the following. Particular attention is paid to potential points of friction between the three modes.

As described above, the three modes of the relationship between public relations and sustainability arise from different ways of looking at sustainability from a public relations perspective, which have their origins, for example, in different dominant paradigms and understandings of their objects of investigation (see Fig. 1).

If we argue that post-positivistically understood public relations, which is still dominant in strategic sustainability communication research, must also contribute to sustainable transformation, all three modes will need to be taken into account. It is not sufficient for public relations to communicate only about sustainability, but also advocate for sustainability and reflect on its own sustainability. Thus, in this chapter, we seek to analyze all three modes from a single paradigmatic standpoint, post-positivist public relations. However, in juxtaposing the three modes on one another, points of friction arise between them. Such points of friction can be identified at all cross-sections between all three modes and will be systematically discussed in the following for each cross-section.

In order to analyze the points of friction, we examine the following guiding questions:

- How can the points of friction be described (e.g., in terms of potential causes or stakeholder arenas)?
- Where can modern, post-positivist public relations begin to address the points of friction?

3.1 *Points of Friction Between Public Relations About and for Sustainability*

Points of friction between public relations *about* and *for* sustainability arise primarily when organizations communicate about sustainability, but there are differences between an organization's objectives and the aim of societal transformation toward sustainability. These differences may be real or only suspected or perceived by stakeholders from within an organization (e.g., employees) or from the external environment (e.g., public authorities, customers, citizens, nonprofit organizations). Sustainability efforts might not be seen as a commitment to a cause, based on "public serving motives" (Kim & Lee, 2012, p. 170; see also Viererbl & Koch, 2022), but as an act of self-interest, based on "firm-serving motives" (Kim & Lee, 2012, p. 170;

see also Viererbl & Koch, 2022). Stakeholders can question whether the organization actually pursues and implements sustainability in the sense of a societal transformation. This is especially true for companies whose business model, products, or services are—in the eyes of their stakeholders—not (yet) sustainable or even stand in the way of a sustainable transformation. Real, suspected, or perceived differences can, for example, lead to boycott practices by consumers (Baringhorst, 2010).

Where differences exist, points of friction can arise from public relations communicating about sustainability anyway, rather than coaching the organization to be more sustainable. Where differences between an organization's objectives and the aim of societal transformation are suspected or perceived by stakeholders, points of friction can arise from public relations, that, for example, have not communicated the organization's sustainability efforts convincingly, credibly, or transparently enough, or have not linked them sufficiently to the societal aim of sustainability. In addition, suspected and perceived differences may be due to different understandings and visions of sustainability between the organization and its internal or external stakeholders, or to different ideas about strategies and approaches to pursuing sustainability. Points of friction can result from the fact that public relations has not fulfilled its function as a boundary spanner between the organization and its stakeholders and has not mediated sufficiently between them.

When differences exist from a stakeholder perspective, an organization's public relations about its own sustainability efforts is mostly seen as a form of self-promotion and criticized as not contributing to sustainability. The accusation of "greenwashing" (Seele & Gatti, 2017) assumes that organizations communicate about their own sustainability efforts but ultimately continue to pursue their own—not necessarily sustainable—objectives, rather than aligning themselves with the objectives of sustainability as societal transformation. In the eyes of their stakeholders, organizations then communicate *about* sustainability rather than *for* sustainability.

Points of friction between public relations *about* and *for* sustainability can also occur when organizations that communicate for sustainability do not adequately communicate about their own actions concerning sustainability.

To further illustrate the points of friction between public relations *about* and *for* sustainability, we describe an example and focus on an organization whose core business is seen as unsustainable: Imagine a public relations department of an oil company. This department decides to put their communication about sustainability second, trying to advocate for sustainable development, thus concentrating on public relations *for* sustainability. However, they might face critique for trying to be an advocate of sustainable development, while appearing to conceal their own activities. Such critique can be seen in the case of British Petroleum (BP) and the "invention" of the *carbon footprint*. In 2004, BP launched the "carbon footprint calculator" as an interactive online tool, helping individuals to calculate how much pollution they produce daily, by their means of transportation, their shopping of groceries, or travel habits (Kaufman, 2020). Individuals may use the calculator to monitor their daily behavior and to further adjust behavior to more sustainable practices. However, the calculator and the accompanying campaign by BP have not been seen as a commendable corporate initiative to foster sustainable behavior, but "a "successful, deceptive" PR campaign" (Kaufman, 2020, n. p.). Kaufman (2020) cites media and communications scholar Doyle (2011), who discusses the BP campaign on the carbon footprint calculator in a book chapter. Doyle (2011) states that the advertisement of the carbon footprint by BP "assigns responsibility for climate impact to the individual, while BP registers its own concerns by appearing already to be doing something about it" (p. 210). The critique of Kaufman (2020) and Doyle (2011) points to the fact that by introducing a calculator, which focuses on individual pollution, BP shifts sustainability discourses away from the huge amount of pollution that big companies (especially in the oil industry) are responsible for. The responsibility to work toward a more sustainable development thus appears to primary be a problem of individual behavior (Doyle, 2011). Whereas on the surface BP's launch of the carbon footprint calculator appears to be a measure of public relations *for* sustainability, its environment, and stakeholders questioned the intentions and motives of the company and may label it as a—potentially intentional—way to forego communicating *about* their own sustainability.

For post-positivist public relations, these points of friction between public relations *about* and *for* sustainability raise the question of how organizations can communicate their objectives and activities in a way that is compatible with societal objectives—to external and internal stakeholders. They also raise the question of how public relations, in its function as a boundary spanner, can better contribute to ensuring that organizations move in the direction of sustainable transformation and that organizational objectives are adapted in such a way that they correspond to social sustainability objectives.

3.2 Points of Friction Between Public Relations About Sustainability and the Sustainability of Public Relations

Points of friction between public relations *about* sustainability and the sustainability *of* public relations can arise when there are differences between public relations efforts to communicate about an organization's sustainability activities and its own sustainability. This might be the case, for example, when the unit or individuals responsible for public relations communicate about their organization's sustainability, but do not reflect on the sustainability of their own actions. In this case, an organization continues to communicate about sustainability, but the public relations itself is not sustainable because, for example, it uses technologies that require resources that are extracted or produced under unethical conditions. In other cases, units or individuals responsible for public relations may think about the sustainability of their own actions, but are not able to implement them, or the organization's management, for example, is still focused on communicating about sustainability. Similar discrepancies can also arise between a public relations unit and other units within an organization or within a public relations unit. Points of friction can occur when the people running an organization's public relations disagree on how important the sustainability of their own work is to them.

While these points of friction are primarily an internal issue that can rarely be perceived by external stakeholders, sometimes stakeholders can indeed notice that an organization communicates about sustainability, but its own public relations is not particularly sustainable—leading to points of friction between public relations *about* sustainability and the sustainability *of* public relations. Furthermore, points of friction can occur when an organization may engage in very sustainable public relations but does not address this in its public relations about sustainability.

To further illustrate the points of friction between public relations *about* sustainability and the sustainability *of* public relations, we use the following example: In Germany, the *Institute for Ecological Economy Research* and the *future e. V.* (an association for strengthening the sustainability of medium-sized companies) evaluated and ranked annual sustainability reports of major and medium-sized German companies.[1] In the ranking of 2015, the authors indicate that companies heavily switch from print versions of sustainability reports to digital outlets to save on resources (Hoffmann et al., 2016). This can be seen as a measure to secure the sustainability *of* public relations (reducing resource consumption) in doing public relations *about* sustainability (publishing a sustainability report). In the 2021 ranking, the authors grant the sustainability report of the *Pure Taste Group GmbH & Co KG* (operating under the name *Lebensbaum*) the second place regarding medium-sized companies (Hoffmann et al., 2022). This company decided to produce and promote a print version of their sustainability report that is accompanied by a small online version. The company is part of the *true cost initiative*, a network that tries to estimate the "real" costs of food and agricultural products. In a similar vein, *Lebensbaum* estimated the costs of their communicational output. In their 2021 sustainability report, they state that "digital is not better," comparing costs of print with power consumption and the IT hardware needed to produce digital outlets (Hoffmann et al., 2022). This example highlights an interesting point of friction between the public relations *about* sustainability and the sustainability *of* public relations: Stakeholders judge sustainability according to their embedded standards. Most will judge a digital alternative to be more sustainable than an analog output. Therefore, printing the sustainability report on resource-intensive paper can lead to a loss of credibility as stakeholders might perceive this as unsustainable behavior.

For post-positivist public relations, the points of friction between public relations *about* sustainability and the sustainability *of* public relations raise the question of how the units or individuals responsible for public relations can reflect on and change their own objectives, processes, activities, and products in a way that is compatible with sustainability objectives. They also raise the question of how these units or individuals can gain support for these changes within their organization and coordinate with similarly minded units and individuals. This is also true for their communication with external stakeholders. Here, if public relations stick to

[1] From 1994 to 2021, they published 11 rankings.

seemingly unsustainable measures, they at least—as *Lebensbaum* in the example—need to "justify" or explain their decision in order to prevent loss of credibility. Furthermore, public relations needs to estimate the "real costs" of its processes and outputs—despite the difficulties to do so.

3.3 Points of Friction Between Public Relations for Sustainability and the Sustainability of Public Relations

Points of friction can not only be found when considering public relations *about* sustainability in conjunction with the other two modes. They can also be found when public relations *for* sustainability is considered in conjunction with the sustainability *of* public relations. Points of friction between these two modes can arise when there are (perceived) differences between an organization's efforts for sustainability and its own public relations' sustainability. While these points of friction may also initially be perceived within an organization, external stakeholders may become aware of them, too. This is the case, for example, when an organization communicates for sustainability, but the public relations work itself is in an obvious way not sustainable, for example, due to the extensive use of digital instruments consuming a lot of energy.

Similar to Sect. 3.1, claims of *hypocrisy*—saying A but doing B—also mark a possible point of friction between the two modes (for discussions on hypocrisy in communication, see Christensen et al., 2013, 2020, 2021). Communication is generally susceptible to hypocrisy claims, as people experience a dichotomy between talk and action, and communication is usually perceived as being of less value than action (Christensen et al., 2021). With regard to sustainability, stakeholders may judge organizations for "not practicing what they preach" when an organization tries to advocate for sustainable development, but at the same time relies on unsustainable means of public relations. Furthermore, as all forms of communication are resource intensive (or at least use some kind of resources), but are often seen as something "worse" than action (or less needed/effective), communication has a general vulnerability to hypocrisy claims. An organization may get confronted with claims of hypocrisy for doing public relations *for* sustainability (talk), while lacking sustainability *of* public relations (action). Stakeholders compare the content of communication (talk) with the means by which it is communicated (action).

To further illustrate the points of friction between public relations *for* sustainability and the sustainability *of* public relations, we use the following example: Especially in the early years of *Fridays For Future (FFF)*, the social movement—at least in the German public sphere—was often confronted with accusations of hypocrisy (e.g., Böhme, 2019). Not only were FFF activists confronted with "climate hypocrisy" in the form that, for example, "some media commentators have denigrated participants of the Fridays for Future student movement for using fossil-fuel powered transportation" (Beck, 2023, p. 4). Similar "outcalls" might also accompany public relations. Since FFF relies heavily on posters and flyers to draw attention to its upcoming protests, hypocrisy claims could address a talk-action-difference in that FFF advocates practices for sustainable development (its protests) but does so through unsustainable means of public relations (flyers, posters, etc.). Actors such as FFF could thus be criticized for lacking the sustainability *of* their public relations while trying to do public relations *for* sustainability.

For post-positivist public relations, the points of friction between public relations *for* sustainability and the sustainability *of* public relations raise the question of how these modes can be systematically taken into account and integrated into public relations experts' own thoughts and work—despite the underlying divergent paradigms and objectives of public relations. Furthermore, as seen in this example of FFF, even marginalities from an organization's perspective (such as a printed public relations product) can lead stakeholders to perceive and proclaim "gotcha" (Beck, 2023, p. 4), when they feel they are witnessing hypocritical behavior. Therefore, frictions are likely to arise between public relations *for* sustainability and the sustainability *of* public relations.

4 Discussion and Conclusion

Based on a systematization that distinguishes between public relations *about* and *for* sustainability and the sustainability *of* public relations, we have identified and discussed examples of points of friction that can arise within and between these three different modes. To illustrate these points of friction, we used examples from public relations practices. Our analysis showed that points of friction can occur in the interplay of different objectives and practices of organizational public relations, especially when the objectives or practices of different members of an organization, the organization itself, or the societal context diverge.

From a post-positivist point of view, the potential of public relations as a boundary spanner to feed social sustainability goals and expectations into an organization as decision-relevant information and thus make a significant contribution to the sustainable transformation of organizations becomes clear. Furthermore, in order to avoid perceived discrepancies and contradictions or doubts about credibility, it is also crucial that public relations professionals take sustainability issues into account and integrate them into their own thoughts and work processes.

As our argument is a theoretical one, using practical examples for illustration, further empirical research is needed to better understand these points of friction in order to further develop the relationship between sustainability and public relations, especially from a post-positivist perspective, integrating all three modes of said relationship. Empirical research can also uncover the actual practices and objectives of different actors within an organization and of an organization itself. Capturing points of friction also requires empirical data that allow for understanding the different modes of an organization and potentially conflicting interrelations.

Moreover, further theoretical and empirical work is needed to identify potential effects of points of friction between the two modes on the respective third mode. In addition, the question arises as to how the three modes arc interrelated and influence each other. After all, the modes are analytical in nature, and there are interfaces and interactions between all three of them.

The conceptual work in this chapter is intended not only to complement existing research but also to sensitize public relations researchers and practitioners to address and overcome the contradictions discussed in their research and work practices, with the objective of contributing to a sustainable society. At the beginning of this chapter, we emphasized the special responsibility that public relations can have in the context of the "socio-ecological-economic" transformation toward sustainability. Taking into account the contradictions we have explored in this chapter can help to further discuss the responsibility of strategic communication in general and public relations in particular, to expose pitfalls and deceptions, and to highlight the potential and possibilities of public relations to act for sustainability.

References

Baringhorst, S. (2010). Anti-Corporate Campaigning – neue mediale Gelegenheitsstrukturen unternehmenskritischen Protests [Anti-corporate campaigning—new structures of opportunity for anti-corporate protests in the media]. In S. Baringhorst et al. (Eds.), *Unternehmenskritische Kampagnen: Politischer Protest im Zeichen digitaler Kommunikation* (pp. 9–31). VS. https://doi.org/10.1007/978-3-531-92417-5_1

Beck, V. (2023). Climate hypocrisy and environmental integrity. *Journal of Social Philosophy*, 1–20. https://doi.org/10.1111/josp.12522. online first.

Bittner-Fesseler, A., & Weicht, J. F. (2019). Mehr als Nachhaltigkeitskommunikation: Auf dem Weg zu einer nachhaltigkeitsintegrierten Unternehmenskommunikation [More than just sustainability communication: Getting started with sustainability-integrated corporate communications]. In S. R. H. Fernhochschule (Ed.), *Nachhaltigkeit im interdisziplinären Kontext: Herausforderung für Wirtschaft und Gesellschaft* (pp. 33–50). Springer. https://doi.org/10.1007/978-3-658-24288-6_3

Böhme, H. (2019). Germany's climate activism is so hypocritical. *DW*. https://www.dw.com/en/opinion-germanys-climate-activism-is-full-of-hypocrisy/a-48292725.

Carroll, A. B. (2021). Corporate social responsibility: Perspectives on the CSR construct's development and future. *Business & Society, 60*(6), 1258–1278. https://doi.org/10.1177/00076503211001765

Christensen, L. T., Morsing, M., & Thyssen, O. (2013). CSR as aspirational talk. *Organization, 20*(3), 372–393. https://doi.org/10.1177/1350508413478310

Christensen, L. T., Morsing, M., & Thyssen, O. (2020). Timely hypocrisy? Hypocrisy temporalities in CSR communication. *Journal of Business Research, 114*, 327–335. https://doi.org/10.1016/j.jbusres.2019.07.020

Christensen, L. T., Morsing, M., & Thyssen, O. (2021). Talk–action dynamics: Modalities of aspirational talk. *Organization Studies, 42*(3), 407–427. https://doi.org/10.1177/0170840619896267

Crumley, E. T., et al. (2022). Media interviews as strategic external communication to maintain legitimacy for sustainability activities. *Corporate Communications: An International Journal, 27*(1), 148–166. https://doi.org/10.1108/CCIJ-05-2021-0053

Demirović, A., et al. (2011). *VielfachKrise: Im finanzmarktdominierten Kapitalismus [Multiple crises: In financial market-dominated capitalism]*. VSA.

Diehl, S., et al. (Eds.). (2017). *Handbook of integrated CSR communication*. Springer. https://doi.org/10.1007/978-3-319-44700-1

Dillmann, T. (2023). GPRA gründet Initiative für Nachhaltigkeit [GPRA launches sustainability initiative]. *PR-Journal*. Accessed August 15, 2023, from https://pr-journal.de/index.php?option=com_content&view=article&id=30489:gpra-gruendet-initiative-fuer-nachhaltigkeit&catid=395:organisationen

Doyle, J. (2011). Where has all the oil gone? BP branding and the discursive elimination of climate change risk. In N. Heffernan & D. A. Wragg (Eds.), *Culture, environment and ecopolitics* (pp. 200–225). Cambridge Scholars Publishers. https://research.brighton.ac.uk/en/publications/where-has-all-the-oil-gone-bp-branding-and-the-discursive-elimina

Goldhaber, M. H. (1997). The attention economy and the Net. *First Monday, 2*(4). https://doi.org/10.5210/fm.v2i4.519

Golob, U., et al. (2013). CSR communication: *Quo vadis* ? *Corporate Communications: An International Journal, 18*(2), 176–192. https://doi.org/10.1108/13563281311319472

Golob, U., Podnar, K., & Zabkar, V. (2023). Sustainability communication. *International Journal of Advertising, 42*(1), 42–51. https://doi.org/10.1080/02650487.2022.2144035

Görland, S. O., & Kannengießer, S. (2021). A matter of time? Sustainability and digital media use. *Digital Policy, Regulation and Governance, 23*(3), 248–261. https://doi.org/10.1108/DPRG-11-2020-0160

Hallahan, K., et al. (2007). Defining strategic communication. *International Journal of Strategic Communication, 1*(1), 3–35. https://doi.org/10.1080/15531180701285244

Hentze, J., & Thies, B. (2014). *Stakeholder-Management und Nachhaltigkeits-Reporting [Stakeholder management and sustainability reporting]*. Springer Gabler.

Hoffmann, E., et al. (2016). *Nachhaltigkeitsberichterstattung in Deutschland: Ergebnisse und Trends im Ranking der Nachhaltigkeitsberichte 2015 [Sustainability reporting in Germany: Results and trends in the sustainability reports ranking 2015]*. Institut für ökologische Wirtschaftsforschung, future e. V. https://www.ranking-nachhaltigkeitsberichte.de/fileadmin/ranking/user_upload/2015/Ranking_Nachhaltigkeitsberichte_2015_Ergebnisbericht.pdf

Hoffmann, E., Schöpflin, P., & Lautermann, C. (2022). *CSR-Reporting in Deutschland: Ergebnisse im Ranking und Trends in der Berichterstattung von Großunternehmen und KMU [Csr reporting in Germany: Ranking results and trends in the reporting of large companies and SMEs]*. Institut für ökologische Wirtschaftsforschung, future e. V. https://www.ranking-nachhaltigkeitsberichte.de/fileadmin/ranking/user_upload/2021/Ranking_Nachhaltigkeitsberichte_2021_Ergebnisbericht_1.pdf

Ihlen, Ø., Bartlett, J. L., & May, S. (Eds.). (2011). *The handbook of communication and corporate social responsibility*. Wiley-Blackwell. https://doi.org/10.1002/9781118083246

Kannengießer, S. (2020). Nachhaltigkeit und das „gute Leben": Zur Verantwortung der Kommunikations- und Medienwissenschaft in digitalen Gesellschaften [Sustainability and the "good life": The responsibility of communication and media studies in digital societies]. *Publizistik, 65*(1), 7–20. https://doi.org/10.1007/s11616-019-00536-9.

Kannengießer, S. (2022). *Digitale Medien und Nachhaltigkeit: Medienpraktiken für ein gutes Leben [Digital media and sustainability: Media practices for a good life]*. Springer. https://doi.org/10.1007/978-3-658-36167-9

Kannengießer, S., & McCurdy, P. (2021). Mediatization and the absence of the environment. *Communication Theory, 31*(4), 911–931. https://doi.org/10.1093/ct/qtaa009

Kaufman, M. (2020). The carbon footprint sham: A "successful, deceptive" PR campaign. *Mashable*. Accessed June 18, 2024, from https://mashable.com/feature/carbon-footprint-pr-campaign-sham

Ki, E.-J., & Hon, L. (2009). Causal linkages between relationship cultivation strategies and relationship quality outcomes. *International Journal of Strategic Communication, 3*(4), 242–263. https://doi.org/10.1080/15531180903218630

Kim, S., & Lee, Y.-J. (2012). The complex attribution process of CSR motives. *Public Relations Review, 38*(1), 168–170. https://doi.org/10.1016/j.pubrev.2011.09.024

Kollmorgen, R., Merkel, W., & Bucher, H.-J. (2014). Transformation und Transformationsforschung: Zur Einführung [Transformation and transformation research: An introduction]. In R. Kollmorgen, W. Merkel, & H.-J. Bucher (Eds.), *Handbuch Transformationsforschung* (pp. 11–27). Springer VS. https://doi.org/10.1007/978-3-658-05348-2_1

McWilliams, A., & Siegel, D. S. (2011). Creating and capturing value: Strategic corporate social responsibility, resource-based theory, and sustainable competitive advantage. *Journal of Management, 37*(5), 1480–1495. https://doi.org/10.1177/0149206310385696

Nothhaft, H., & Zerfaß, A. (2020). Strategische Kommunikation: Begriffliche und konzeptionelle Grundlagen aus internationaler Perspektive [Strategic communication: Terminology and conceptual foundations from an international perspective]. In A. Zerfaß, M. Piwinger, & U. Röttger (Eds.), *Handbuch Unternehmenskommunikation: Strategie – Management – Wertschöpfung* (pp. 1–27). Springer. https://doi.org/10.1007/978-3-658-03894-6_5-1

Panapanaan, V. M. (2003). Roadmapping corporate social responsibility in Finnish companies. *Journal of Business Ethics, 44*(2), 133–148. https://doi.org/10.1023/A:1023391530903

Prexl, A. (2010). *Nachhaltigkeit kommunizieren – nachhaltig kommunizieren: Analyse des Potenzials der Public Relations für eine nachhaltige Unternehmens- und Gesellschaftsentwicklung [Communicating sustainability—sustainable communication: Analyzing the potential of public relations for sustainable corporate and societal development]*. VS. https://doi.org/10.1007/978-3-531-92471-7.

Putnam, L. L. (1982). Paradigms for organizational communication research: An overview and synthesis. *Western Journal of Speech Communication, 46*(2), 192–206. https://doi.org/10.1080/10570318209374077

Rademacher, L., Helferich, P., & Pleil, T. (2024). Die Verantwortung der strategischen Kommunikation in der sozial-ökologischen Transformation: Ein normatives Modell [Strategic communication and its responsibility in the social-ecological transformation: A normative model]. *Annual conference of the network 'Media ethics'*, Munich, February 22.

Rettler, L., & Röttger, U. (2023). Is this responsible communication? Characterizing sustainable strategic communication for professionals and researchers. In *Annual Conference, European Public Relations Education and Research Association (EUPRERA)*, Prague.

Sandhu, S. (2012). *Public Relations und Legitimität: Der Beitrag des organisationalen Neo-Institutionalismus für die PR-Forschung [Public relations and legitimacy: The contribution of organizational neo-institutionalism to PR research]*. Springer VS. https://doi.org/10.1007/978-3-531-19304-5

Seele, P., & Gatti, L. (2017). Greenwashing revisited: In search of a typology and accusation-based definition incorporating legitimacy strategies. *Business Strategy and the Environment, 26*(2), 239–252. https://doi.org/10.1002/bse.1912

Stehle, H. (2015). *Unternehmenskommunikation in Geschäftsbeziehungen: Business-to-Business-Kommunikation als Teil der funktionalen PR-Forschung [Corporate communication in business relationships: Business-to-business communication as part of functional PR research]*. Springer VS. https://doi.org/10.1007/978-3-658-07294-0

Stehle, H. (2022). Citizens' perceptions of dialogue: A Q study on the communication of energy projects. In *Annual Conference, European Public Relations Education and Research Association (EUPRERA)*, Vienna.

Trujillo, N., & Toth, E. L. (1987). Organizational perspectives for public relations research and practice. *Management Communication Quarterly, 1*(2), 199–231. https://doi.org/10.1177/0893318987001002004

United Nations (UN). (1987). *Report of the World Commission on Environment and Development: Our common future*. Annex to Document A/42/427—Development and International Co-operation: Environment. United Nations (UN). http://www.un-documents.net/wced-ocf.htm

Verhoeven, P. (2018). On Latour: Actor-networks, modes of existence and public relations. In Ø. Ihlen & M. Fredriksson (Eds.), *Public relations and social theory: Key figures, concepts and developments* (2nd ed., pp. 99–116). Routledge. https://doi.org/10.4324/9781315271231-6

Verk, N., Golob, U., & Podnar, K. (2021). A dynamic review of the emergence of corporate social responsibility communication. *Journal of Business Ethics, 168*(3), 491–515. https://doi.org/10.1007/s10551-019-04232-6

Viererbl, B., & Koch, T. (2022). The paradoxical effects of communicating CSR activities: Why CSR communication has both positive and negative effects on the perception of a company's social responsibility. *Public Relations Review, 48*(1), 102134. https://doi.org/10.1016/j.pubrev.2021.102134

Weder, F. (2021a). (P)Rebels needed! Transformative potential of problematization for social change. In D. Pompper (Ed.), *Public relations for social responsibility: Affirming DEI commitment with action* (pp. 191–206). Emerald. https://doi.org/10.1108/978-1-80043-167-620211014

Weder, F. (2021b). Sustainability as master frame of the future? Potency and limits of sustainability as normative framework in corporate, political and NGO communication. In F. Weder, L. Krainer, & M. Karmasin (Eds.), *The sustainability communication reader: A reflective compendium* (pp. 103–119). Springer. https://doi.org/10.1007/978-3-658-31883-3_7

Weder, F. (2022). Strategic problematization of sustainability reframing dissent in strategic communication for transformation. *Public Relations Inquiry, 11*(3), 337–360. https://doi.org/10.1177/2046147X211026857

Weder, F. (2023). Communication of, about and for CSR and sustainability. In F. Weder & M. B. Erikson (Eds.), *CSR communication and cultures of sustainability* (pp. 61–102). University of Queensland. https://doi.org/10.14264/428d81b

Weder, F., Einwiller, S., & Eberwein, T. (2019a). Heading for new shores: Impact orientation of CSR communication and the need for communicative responsibility. *Corporate Communications: An International Journal, 24*(2), 198–211. https://doi.org/10.1108/CCIJ-02-2019-0020

Weder, F., & Erikson, M. B. (2023). *CSR communication and cultures of sustainability*. UQ Pressbook. https://doi.org/10.14264/428d81b

Weder, F., Lemke, S., & Tungarat, A. (2019b). (Re)storying sustainability: The use of story cubes in narrative inquiries to understand individual perceptions of sustainability. *Sustainability, 11*(19), 5264. https://doi.org/10.3390/su11195264

Further Reading

Kannengießer, S. (2019). 'Engaging with and reflecting on the materiality of digital media technologies: Repair and fair production. *New Media and Society, 22*(1), 123–139. https://doi.org/10.1177/1461444819858081

Weder, F., Krainer, L., & Karmasin, M. (Eds.). (2021). *The sustainability communication reader: A reflective compendium*. Springer.

Sigrid Kannengießer is a professor of communication science with a focus on media sociology at the University of Münster. In her research, she focuses on sustainability, digital infrastructures and artificial intelligence, critical data practices, social movements, and gender media studies. https://www.uni-muenster.de/Kowi/en/personen/sigrid-kannengiesser.shtml

Lennart Rettler is a PhD candidate at the University of Münster. His research takes a relational approach to strategic communication and public relations. He discusses concepts such as community, reciprocity, cooperation, and social capital. His interests further lie in social media-based strategic communication. https://www.uni-muenster.de/Kowi/personen/lennart-rettler.shtml

Ulrike Röttger is a professor of communication science with a focus on strategic communication at the University of Münster. In her research, she focuses on strategic communication and specifically on the areas of sustainability and strategic communication, university communication, communication consulting, and (strategic) communication at the local level. https://www.uni-muenster.de/Kowi/en/personen/ulrike-roettger.shtml

Helena Stehle is a professor of communication science with a focus on strategic communication at the University of Münster. Her research interests include strategic and organizational communication, especially sustainability and energy communication, and digital journalism. https://www.uni-muenster.de/Kowi/en/personen/helena-stehle.shtml

A Typology of Strategic Sustainability Communication: From Reporting to Transformation

Thomas Pleil, Pia Sue Helferich, and Teresa Otsa

Abstract In an era of ecological and social crises, the question of the influence of strategic sustainability communication has become increasingly significant. Current discourse in communication science explores how organizations must align their sustainability communication to achieve transformative effects—not only for organizations but also for society and therefore the common good. This chapter presents a typology that systematically records different patterns of strategic sustainability communication between the poles of "reporting" and "transformation." The findings presented in this chapter are based on a literature review and two expert workshops. Adapted to the interactive process of the grounded theory method, the protocols of the workshops were clustered, abstracted, and consolidated into three types of strategic sustainability communication. (Transparency note: For this chapter, the AI tool DeepL was used to translate some passages and FlotAI was used for linguistic improvements in individual passages.)

1 Introduction

The sustainability transformation ranks among the most pressing societal challenges of our time. Addressing this complex process of change requires not only technological innovations and policy decisions but also communication efforts. Strategic communication can make a valuable contribution in this context because it is a societal understanding process (Weder et al., 2019) and has the task of critically questioning the relationship between people and the environment and bringing it into the social debate (Newig et al., 2013; Voci & Karmasin, 2023). Strategic communication can raise awareness about sustainability, foster dialogue between

T. Pleil (✉) · P. S. Helferich · T. Otsa
Research Center Digital Communication & Media Innovation, Darmstadt University of Applied Sciences, Darmstadt, Germany
e-mail: thomas.pleil@h-da.de; pia-sue.helferich@h-da.de; teresa.otsa@h-da.de

F. Weder (ed.), *Strategic Sustainability Communication*, CSR, Sustainability, Ethics & Governance, https://doi.org/10.1007/978-3-031-89486-2_4

relevant actors, and drive behavioral changes in society. There is no doubt that the way we communicate about sustainability is important.

However, the scientific discussions on strategic communication and sustainability show that the organizational perspective dominates. At the same time, however, it is clear from climate science (e.g. IPCC, 2023) and the international political agreements of the world climate conferences that far-reaching changes to all social subsystems are necessary. Weder (2021), therefore, conceptualizes sustainability as a master frame and an absolute obligation. So far, according to Weder's analysis of examples from communication practice, this understanding corresponds to a normative principle with limited potential for mobilization (Weder, 2021, p. 114). Against this background, we have developed a typology of strategic sustainability communication based on a literature analysis. Two types follow the organizational logic, while the third type, transformative strategic communication (TSC), establishes the reference to the societal level that is often missing in the specialist discussion. This new type of sustainability communication is normatively construed and has rarely existed in practice to date. TSC is conceptualized as an approach to support the sustainable transformation of societies with the means of strategic communication.

In the following, we first create an understanding of strategic communication and then look at the discussion on sustainability communication. We then consolidate the perspectives and present a typology of strategic sustainability communication. The three types of sustainability reporting, sustainability PR and marketing, and transformative strategic communication can, on the one hand, offer perspectives for the scientific analysis of practical sustainability communication. On the other hand, the typology can help in practice to design strategic sustainability communication.

2 Strategic Communication

While for many years the discussion among communication scholars focused on the term "public relations" to describe purposeful public communication that addresses various stakeholders, the concept of strategic communication has recently become more important (Hallahan et al., 2007; Hoffjann, 2024). As Hoffjann (2024) and van Ruler (2018) put it, the idea of the newer concept is to regard the most diverse communication disciplines and processes together. Van Ruler (2018) takes a deeper look at the term "communication" from a theoretical perspective and states that there are three different definitions of what communication is at present in the literature:

1. Communication as a one-way process from sender to receiver,
2. Communication as a two-way process between actors,
3. Communication as an omnidirectional diachronic process of meaning development itself.

However, Hoffjann (2024) shows in a detailed analysis that organization-centric definitions dominate this field by far. Thus, there is mainly an interest in the contribution of strategic communication to the achievement of (strategic) organizational goals. In many discussions on strategic communication, however, it remains vague what exactly constitutes the strategic part or what such communication aims to achieve. Holtzhausen and Zerfass describe a wide range of possible goals of strategic communication: "Goals can be as diverse as winning market share, winning a political campaign, building a positive reputation, or effecting social change" (Holtzhausen & Zerfass, 2013). This understanding can be linked to the concept of a "social integration model of public relations," as proposed by Macnamara (2016). This is associated with a shift that moves strategic communication "from being a function of management serving the singular reality and universalized truths of an organization to being informed by pluralism and serving as a function of society and for society as well as organizations" (Macnamara, 2016).

Another shortcoming in the discussion of strategic communication, as Hoffjann (2024) points out, is that many definitions related to communication fail to consider the organizational context. While Hoffjann combines communication and organizational theory perspectives of strategic communication, the typology presented in this chapter, and in particular the type of TSC, is intended to show also perspectives of strategic communication that go beyond the organizational level and are therefore oriented toward the common good.

Particularly in the German-speaking debate on the self-image of communication science, which is shaping us, there is increasing discussion of the contribution the discipline can make to solving societal problems (Pleil et al., 2021). Against this backdrop and in view of the enormous challenges posed by the climate crisis, it is necessary to discuss how strategic communication efforts should be characterized to contribute to sustainable development—not only of corporations but of society at all. This is not a matter of course: strategic communication practices in many cases have supported big business and elite groups (Logan, 2023) and have also contributed to harmful developments such as overconsumption, excessive resource consumption, and social problems caused by globalization, and continue to do so today. On the other hand, strategic communication practices by NGOs, civil society movements, and public institutions have always focused on social developments. However, the society-oriented perspectives of strategic communication are usually neglected in the debate. If the research field of strategic communication follows less the functionalist-pragmatic paradigm and more the constructivist/critical paradigm (Weder, 2021), we see a better basis for science to answer questions that are oriented toward the common good. The reason is that the functionalist-pragmatic paradigm is often driven by the interest of single organizations. A critical perspective opens the view for an orientation of the consequences, e.g., for the environment that a behavior has. This then creates a basis for practical implementation.

3 Sustainability Communication

Sustainability communication has been the subject of academic discussion for several years (e.g., Michelsen, 2005). Communication studies, as a discipline that originally deals with public communication and media, was initially hesitant to explore the subject area and until recently has hardly dealt with social transformation (Kannengießer, 2021).

As reviewed by Golob et al. (2023), the term sustainability communication brings together a variety of different approaches that deal with environmental or ecological issues in some way (see Fig. 1). In this sense, sustainability communication can take place, for example, in the context of education for sustainable development, through journalism or through strategic communication of different actors. They all have several things in common: the typically normative motivation, the reference to the public, and the challenges of the subject matter, which on closer

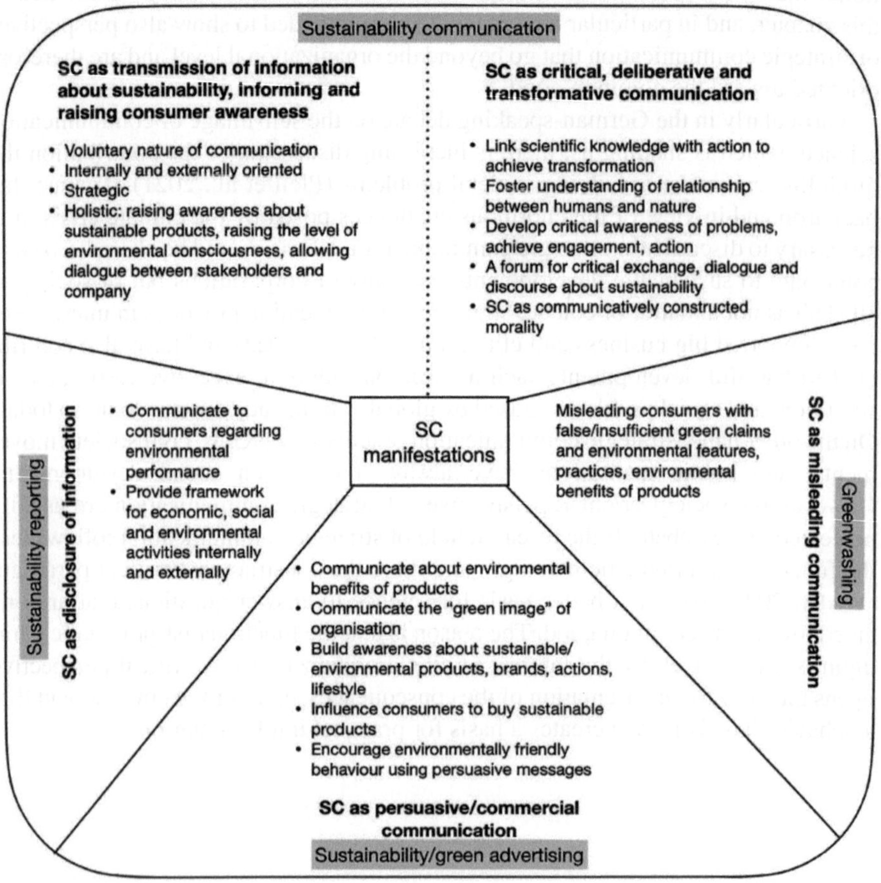

Fig. 1 Manifestations of sustainability communication (Golob et al., 2023, p. 47)

observation turns out to be complex, abstract, invisible, controversial, and uncomfortable (Bilandzic & Kalch, 2021). This affects different actors with different levels of knowledge, values, and goals (Karmasin et al., 2021).

In order to get a broader picture, we will now take a look at important approaches to sustainability communication in general. The discussion very often refers to three types of sustainability communication presented by Newig et al. (2013). They outline sustainability communication based on environmental, risk, and science communication (Newig et al., 2013). They differentiate:

(a) Communication about sustainability: Within Civil Society, this type describes discourses on sustainable development in media, assemblies, or social networks.
(b) Communication of sustainability: This type uses a sender/recipient model, typical are awareness and information campaigns, but also "mass media marketing."
(c) Communication for sustainability: Here, the focus is on normative aspects of sustainable development, which may be achieved through political pressure but also participation and cooperation in Sustainable Development governance.

This typology is very rich and can also be applied to the various social subsystems like education, science, politics, economy, and media. The dissertation by Prexl (2010) is the first to develop a theory of sustainability communication from a communication science perspective and with a background in PR theory. However, the author herself limits that she chooses a corporate approach. In other words, she sees sustainability communication (only) as an aspect of corporate communication. As is paradigmatic in PR theory, she describes three perspectives of sustainability communication (Prexl, 2010):

(a) The organizational theory perspective: important goals here are securing trust, the license to operate, and ultimately the contribution of communication to supporting sustainability management.
(b) The marketing theory perspective: Here, the focus is on achieving marketing goals, i.e., essentially the sale of sustainable products.
(c) The social theory perspective: This is based primarily on ethics and looks at the contribution sustainability communication makes to social developments, for example, in terms of purchasing decisions, knowledge, or awareness. However, the author qualifies that she sees such results as a "pleasant side effect" (Prexl, 2010) of corporate communication, which in the best case improves the image of a company.

In terms of strategy and content, two fundamentally different approaches can also be identified in practice (Groffman et al., 2010; Godemann, 2021):

(a) The information deficit approach: This approach, which originates primarily from science communication, assumes a lack of science literacy in society and supposes that there is often a lack of understanding of interrelationships when it comes to sustainability and that these deficits should be remedied by communicating information and people then adapt their behavior on the basis of better information.

(b) Public engagement: This approach is based on the importance of central values, trust, and identity and links environmental problems to these. The decisive factor in this approach is that people should be empowered to participate in the development of solutions through dialog and mutual exchange.

However, it has long been shown that the deficit model is not effective (e.g., Schrader, 2020; Groffman et al., 2010) and only encourages a few people to change their behavior. One reason: the large amount of information alone can lead to excessive pressure and therefore helplessness and passivity. Another reason at the time of publication of Groffman et al. (2010): Although an increasing proportion of people have felt well informed about sustainability and climate change over the years, the effects of the climate crisis such as droughts, forest fires, record temperatures, or migration were little felt, especially by the people who have the largest carbon footprint. This has now changed in many regions of the world and, despite all the tragedy, communication can now draw more on people's experiences and promote the discussion of values and visions for the future. However, there is relatively little discussion in the literature on sustainability communication when it comes to the question of how decisions, once negotiated, should be disseminated and how they should be kept in the public eye, even if loud minorities mobilize against what has already been achieved, for example, with the help of social media mechanisms. It is also challenging that climate change is often the subject of conspiracy myths, especially in right-wing populist thinking (Meijers et al., 2022). Sustainability communication must also be discussed against the backdrop of such developments and debates, and it should therefore promote trust (Meijers et al., 2022) in (climate and sustainability) science and clarify the benefits for all citizens.

Overall, it is clear that the majority of the literature on sustainability communication is either aimed at specific individual actions, for example, when it comes to consumer decisions—or, as described in the section on strategic communication, it focuses on the perspective of an isolated organization and its goals. In our view, however, communication in larger transformative contexts—beyond the boundaries of organizations and aimed at social change—must be considered more intensively. This could be a building block toward an understanding of strategic communication in wicked problems.

4 A Typology of Strategic Sustainability Communication

In our typology presented in the following, we build on the work presented earlier. However, unlike Newig et al. (2013), we do not take an overall approach to sustainability communication, nor, like Prexl (2010), to corporate communication. Instead, we choose the perspective of strategic communication as our starting point and do not view strategic sustainability communication exclusively as an organizational function. In other words, we do not remain solely at the meso level, but also strive for a societal perspective. This happens when we ask about the contribution that strategic communication can make to transformation, not only in organizations, but also in society. This explicitly includes fields such as health communication, advocacy, activism, communication for social change, and others. Instead of focusing on the management perspective that is widespread in the field (Munshi & Kurian, 2016), we believe it is necessary to integrate other perspectives (Taylor & Kent, 2023) in line with the critical perspective, such as those that take a closer look at the relationship level in communication and also use psychological findings, for example (Fig. 2).

Overall, our typology outlines and structures widespread sustainability communication discussions in the literature. It was found that the company perspective

Fig. 2 Typology of strategic sustainability communication (own illustration)

dominates the literature, as already described above. Within this, we see a strong thread of discussion in the tradition of reporting, which we see in close connection with corporate social responsibility. Another strand of the discussion stems from the discussions on public relations and marketing, such as the subfield of green marketing. We are aware that these fields are described very broadly in the second category and that the discussions do not always refer to each other. However, in the context of integrated communication, we believe that a joint approach for PR and marketing is appropriate. By contrast, a third discussion is only just beginning: it goes beyond the narrow organizational level and asks how strategic communication can have a transformative effect, including on society.

The typology was developed using an agile approach reminiscent of grounded theory: the first step was to work on a theoretical basis and to develop an initial draft from a literature review. This was then presented, discussed, and further developed in two workshops with practitioners and scholars on strategic communication. In both workshops, which took place online, the draft typology was presented on a digital whiteboard. The board was first used to collect comprehension questions from the participants, after which the participants commented on and discussed the substance of the proposals. All participants were initially able to note their comments individually on the board. During the discussion, the results were then documented by the research team for all to see. The results were then verified again in a final discussion. The participants of the second workshop received the modified version of the typology from the first workshop. The first workshop with five practitioners lasted approx. 1.5 h, the second workshop with eight scholars and two practitioners from the field of strategic communication lasted 2.5 h. The typology presented in this chapter integrates the suggestions from these workshops.

The typology presented here has three types, which describe strategic sustainability communication: (1) sustainability reporting, (2) sustainability PR and marketing, and (3) transformative strategic communication (TSC). From the perspective of professional strategic communication, which is strongly characterized by a corporate perspective, the development of these three types can be seen historically, following Prexl (2010): According to this, reporting on sustainability has the oldest roots. With the development of sustainable products and corporate strategies, the need arose to develop sustainability PR and marketing, also to differentiate oneself from the competition. At the same time, civil society groups, NGOs and institutions have traditionally also made use of strategic communication—albeit with little mutual attention from the respective specialist communities. Now, an understanding of sustainable transformation is gradually emerging—both within organizations and in society as a whole. This is where the type of transformative strategic sustainability communication (TSC) comes in, which views communication as systematically embedded in multidisciplinary activities in order to support transformative processes. On the other hand, the three types also express an opening of perspectives: According to these, reporting is very much about the formalized self-presentation of an organization and a relatively narrow focus on selected stakeholders. Sustainability PR and marketing, on the other hand, relate more strongly to markets and competition and have the task of focusing on a broader spectrum of stakeholders and their requirements. Transformative strategic communication, on the other hand, focuses on the societal perspective: the aim is to make a contribution to social change—whereby different actors such as companies, authorities, NGOs, or networks can be considered for this type of strategic communication. In the following, the three types of strategic sustainability communication are presented in more detail.

4.1 Sustainability Reporting

Reporting has a long tradition within strategic communication. Starting with general annual reports, which have to comply with the conventions of the financial world, different types of reporting have developed over the years. Initially, the focus was placed more on employees and, where applicable, business partners, and social reports were published (Prexl, 2010). They can be interpreted as a response to critical discussions about the role of companies in society. A little later, this discussion did not only refer to the employees in companies—i.e., an internal perspective—but the social responsibility of companies was discussed more broadly. This gave rise to the field of corporate responsibility and the associated reporting in the form of CSR reports. Environmental scandals in the 1970s and 1980s initially required intensive crisis communication, especially from large corporations; environmental reporting emerged against this background at the end of the 1980s (Prexl, 2010). Formalized external audits and reporting standards were then intended to ensure public trust and traceability of the reports and create comparability. In the area of environmental reporting, the Global Reporting Initiative (GRI) standard dominates today (de Villiers et al., 2022). Voluntary guidelines have been continuously developed since the GRI was founded in 1997, and reporting in relation to the Sustainable Development Goals (SDG) has now also been formally integrated (de Villiers et al., 2022). For some countries and regions, such as the European Union, binding sustainability reporting standards have now been adopted or are under discussion. Against this background, it is understandable that the legitimacy theory has been the dominant perspective of reporting from the outset (Hooghiemstra, 2000).

If we consider sustainability reporting as a type of strategic sustainability communication, we define it as formalized reporting on the development of sustainability in an organization. Typically, this is a company, possibly also a municipality. In terms of content, reporting includes aspects such as specific targets relating to the carbon footprint. According to Herbohn et al. (2022), this includes details on the strategies to achieve this, e.g., specific investment plans, planned product developments, divestments of invested financial resources, and financial scenarios—including risks for the reporting organization due to climate change, but also changing markets and political framework conditions (Herbohn et al., 2022).

This is strongly linked to impression management, understood as self-presentation on a specific topic (Hooghiemstra, 2000; Herbohn et al., 2022; Talbot & Boiral, 2018). The objectives are quite broad and always include reputation (Hooghiemstra, 2000). In terms of investor relations (Piwinger & Hoffmann, 2022), a company should also be presented as a worthwhile investment or its share price should be secured. Particularly with regard to activists, NGOs and supervisory authorities, securing the legitimacy of the sender and presenting it as a good citizen plays a special role—it is therefore a matter of making its position in society visible from the perspective of the respective organization and securing the license to operate (Sandhu, 2022).

The internal objectives of sustainability reporting are less frequently addressed. However, we increasingly see it as an instrument for communicating change, as the goals and milestones made transparent in the reports also serve as a roadmap for employees regarding the future direction of the company in terms of sustainability. Sustainability reporting and its translation into internal communication strategies can therefore also contribute to the much-coveted cultural change in companies (Weder & Erikson, 2023). Overall, however, the communication mode should essentially be understood as information or one-way communication (see also van Ruler, 2018).

The strong links between this type of reporting and CSR on the one hand and controlling on the other result in a self-image that is also explained by these disciplines. Clearly comparable key figures that include both targets and interim results along the way are crucial. This reduces risks that can occur in CSR reporting, such as rainbow-washing, disinformation, or greenwashing (Cappizo & Luisi, 2024). Overall, the objectives of this sustainability reporting are strongly related to the company's own organization: it is about demonstrating social responsibility, securing trust, and thus promoting reputation, but at the same time the requirements of investors and supervisory authorities also play a special role, so that sustainability reporting is ultimately an indication of a company's future viability.

4.2 *Sustainability PR and Marketing*

We identify sustainability PR and sustainability marketing as the second type of strategic sustainability communication. The latter is often discussed under the term "green marketing." Sustainability is communicated as an important topic for an organization and is intended to support its development. The associated objectives are to support the sustainability goals of an organization, provided that the communication, in contrast to reporting, actively involves stakeholders. We argue in the sense of integrated communication. Zerfass et al. (2022) base the integrated perspective on increasingly diversified media and communication channels, whereby topic management and storytelling are determined in practice by corporate communication. The development of credible messages is particularly sensitive in the area of sustainability in order to reduce the risks of greenwashing. In contrast to conventional communication strategies, the inclusion of psychological insights is of particular importance here (Ross, 2024; Schrader, 2020).

To distinguish this type of communication from sustainability reporting, it is useful to compare it with digital communication: While sustainability reporting tends to focus on pull communication (Pleil, 2015), in which stakeholders become active themselves according to their interests and inform themselves about an organization's sustainable developments, we see sustainability PR and marketing mainly as push communication (Pleil, 2015), which prepares messages in line with stakeholder needs—whereby timing, framing, intensity, channels, etc. are determined as part of the communication or marketing strategy.

Green marketing aims to ensure the economic success of a company by prioritizing sustainability (Alzi'bi & Kontor, 2023). The focus of green marketing is on addressing customers in order to establish sustainable consumption (Alzi'bi & Kontor, 2023) and develop a sustainable corporate image (Mukonza & Swarts, 2020). From a corporate perspective, branding plays a fundamental role after the development of a sustainability strategy (Santos et al., 2023). The decisive factor here is how to create a sustainable, credible image, differentiate oneself from the competition, and ensure purchase preference through messages tailored to the target group (Santos et al., 2023). Green marketing strategies are also used in the business-to-business environment, where, for example, price, product quality, and corporate image are key success factors (Gelderman et al., 2021)—whereby quality here refers to aspects of sustainability such as energy use and recyclability (Gelderman et al., 2021). The green corporate image is all the more important the more environmentally damaging the activities of an industry tend to be (Gelderman et al., 2021).

Addressing employees in the context of sustainability communication has also become increasingly important in recent years—the aims here are, on the one hand, the support of sustainability goals by employees and, on the other hand, the retention and recruitment of employees who value sustainability (Alzi'bi & Kontor, 2023). Strottner and Huck-Sandhu (2021) see sustainability in the company as a central challenge of change communication: the decisive factor here is to reach the hearts and minds of employees, i.e., to address awareness, attitudes, behavior, and knowledge (Strottner & Huck-Sandhu, 2021). Other important aspects in the expert discussion on internal sustainability communication are the development of organizational culture, organizational learning, health, safety, but also the role of trade unions—the latter also with regard to participation and empowerment (Mahdy et al., 2023). Studies also show a positive correlation between emotional attachment of employees, trust, transformative leadership and innovative behavior, and commitment of employees to sustainability (Kara et al., 2023).

With regard to external stakeholders, aspects of sustainability communication are discussed and examined in relation to all conceivable fields, for example, in relation to industries or strategic approaches or areas of work such as brand communities (Lin et al., 2024), neuromarketing (Martinez-Levy et al., 2023), and many more.

4.3 Transformative Strategic Communication: Supporting Pathways to Sustainable Societies

With the type of Transformative Strategic Sustainability Communication (TSC), we describe an approach to strategic communication that goes beyond the management approach and the organizational function. We understand transformative strategic communication (TSC) as strategic communication that accompanies and supports transformations. Transformation is understood as a far-reaching, long-term change process that is geared toward an overarching goal such as sustainability and encompasses all relevant aspects such as culture, work, learning, production, and consumption. TSC can be limited to the transformation of an organization, but is usually aimed at larger systems and social development.

This scalability of TSC results from network theory: according to this, we understand an organization as a relatively small network and society as the totality of all networks of social actors and see the communicative power that arises as a result (Castells, 2007). Accordingly, there is no single approach to TSC for social change; instead, it takes place in different networks and constellations. By considering organizations as parts of large networks, the connection between the meso and macro levels can be established and TSC has the task of supporting the development of relationships (Ihlen, 2023) and, together with stakeholders, developing common understandings, visions, and narratives, and synchronizing them in the different parts of the networks.

In contrast to the primarily organization-related mindset of sustainability PR and marketing, the focus here is on the social effect. In our understanding, TSC approaches have a normative character: Sustainability is therefore a moral compass (Weder et al., 2019; Godemann, 2021) or as described above it is a master frame (Weder, 2021) to serve the common good. This requires not only an ethical repositioning but also a changed strategic understanding of communication. Adi and Stoeckle (2023) describe the background: "Public relations is unprepared to face increasingly complex issues such as pandemics, global warming, and polarization amid globalization." The VUCA world and wicked problems in particular lead to radical uncertainty for strategic communication. Traditional organizational structures are proving unsuitable, especially in the context of super wicked problems, for which solutions must be developed step by step (Wexler & Oberlander, 2020). Based on this, Adi and Stoeckle (2023) call for "a new paradigmatic and normative approach" to guide "scholarly research and practice toward achieving greater legitimacy and credibility." Up to now, they argue, PR has been working on tame problems with tame means such as goal-oriented approaches, thinking in terms of KPIs and best practices—all in the spirit of protecting the interests of the respective organization (Adi & Stoeckle, 2023).

In order to describe TSC in more detail, it is helpful to first take a closer look at the concepts of transformation and transdisciplinarity. Transformation is usually discussed in the context of major global challenges (Hennen et al., 2023). The concept of transformation encompasses social processes, institutions or systems, political structures and power, economic production, but also cultural patterns such as consumption, attitudes, and even the question of the good life—in short, it is about social innovation (Hennen et al., 2023).

Transformation is often viewed at an organizational level and described as a kind of far-reaching and continuous change process. This is discussed particularly intensively with regard to digital transformation (e.g., Pleil & Helferich, 2020). Such a transformation affects all aspects of the organization—from the purpose, the organizational culture, and the brand to the actual core tasks. For companies, this includes aspects such as the business model and all associated processes from the supply chain to sales. For other types of organizations, similarly far-reaching changes to the core tasks would have to be considered. At the same time, the term transformation is used for social developments, typically in relation to digitalization and/or sustainability. According to Malik (2018), the networking of previously closed or separate systems in organizations, industries, and societies and the associated growth in complexity are typical of transformations.

While digitalization is seen here as the starting point from which aspects of sustainability are also considered, others, such as the German Advisory Council on Global Change (WBGU, 2011) or Schneidewind (2018), relate the concept of the great transformation directly to socio-ecological changes. This transformation is seen as a super wicked problem (Wexler & Oberlander, 2020) and thus as a challenge that can only be treated successfully if politics, business, and civil society work together (Wagner, 2022, p. 44). Transdisciplinary research is increasingly recognized as a fundamental approach for addressing the highly complex problems and fostering comprehensive social transformations. This approach not only involves multiple scientific disciplines but also engages stakeholders from various sectors in co-developing solutions aimed at social change (Klein et al., 2001). Anderson and McLachlan (2016) further emphasize the goal of such research to drive societal shifts.

The project-based mindset prevalent in transformation science facilitates the delineation of problems and fosters the formulation of solutions. Murphy (2023) illustrates how this encourages "collaborative problem solving shaped by multiple stakeholders." This collaboration, enriched by new partnerships and communications strategies, may include public consultation, citizen deliberation, and participation. Projects characterized by their lighthouse effect have the potential to spur public debate and necessitate strategic communication efforts to achieve social change, argue Bohn and Braun (2021). The importance of strategic communication also extends to significant events, such as natural or nuclear disasters, suggesting its critical role in shaping public discourse and response.

The concept of social tipping points, developed by Otto et al. (2020), further illuminates this discussion. Described as the moment when a minor trend becomes a major societal practice, these tipping points highlight the pivotal role of belief, behavior, or technology adoption within a community. Otto et al. (2020) assert the importance of strategic communication elements in reaching these tipping points, including the influential role of opinion leaders. Their studies indicate that social tipping points are often achieved when 17–20% of a population adopts a new behavior, underscoring the transformative power of targeted communication strategies. Despite the acknowledged significance of strategic communication in the context of sustainable transformation, the methodologies for its effective implementation often remain undefined, as Pleil et al. (2021) note. Perspectives such as engagement theory (e.g., Taylor & Kent, 2023) or critical theory provide a good basis for bringing this social perspective to life.

TSC, how we conceptualize it can firstly play a supporting role within specific transformative projects and help to develop solutions. Here, we follow Weder's (2022) understanding of transformative PR: In this way, facts, but also norms, cultural patterns, and opposing interests become objects of reflection—on the basis of which new values and new behavior can emerge (Weder, 2022). However, we assume that this does not automatically result in overall social dynamics. Secondly, TSC may therefore be necessary to contribute to dissemination and to help make findings, solutions, and decisions from projects (e.g., a transformative project for the development of an urban neighborhood) accessible to all stakeholders in the sense of broader transferability. Even if the transformative process was started openly and inclusively, it is likely that only a (small) proportion of stakeholders will actually have been involved, and all others will subsequently need to be picked up in terms of communication. Moreover, in similar situations, it is beneficial to leverage existing knowledge and prior problem identifications. This approach can streamline the process of developing new solutions. TSC can therefore have an educational effect here, create awareness, and support the next loops of problematization elsewhere.

Thirdly, we see that TSC in particular can support transformative social dynamics in the normative framework described, for example by developing alternative narratives and frames. Furthermore, in times of highly dynamic public spheres (Weder & Weaver, 2023), which are also influenced by populism and disinformation (Pleil, 2018), it makes sense to provide reminder loops for the consensus once it has been reached. At this point, the deliberative process would be supplemented by traditional approaches of strategic communication—as are common in healthcare communication, for example. This means that TSC can also use traditional communication tools depending on the situation. We see TSC as a flexible toolbox of different approaches and instruments. These include, for example, approaches from psychology or interdisciplinary fields such as health communication or communication for social change (C4SC). In C4SC is a consensus that different strategic approaches to communication make sense depending on the situation. We consider the five strategic levels of communication developed in C4SC to be particularly helpful (Hamidi & Mielke Möglich, 2021; Servaes & Malikhao, 2007):

1. Behavioral change communication with the goal of behavior changes in individuals and communities
2. Mass communication as an approach to address citizens through the media
3. Advocacy communication, which aims to influence framework conditions such as regulations
4. Participatory communication, which supports social design processes by citizens

5. Communication for structural and sustainable social change, which focuses on changes in society and social and political institutions.

All of these approaches must be viewed independently of the interests of an individual organization and serve the common good. Under these conditions, TSC does not only have to be deliberative, but can also use strategies that Adi and Stoeckle (2023) refer to as "responsible persuasion" depending on the situation. However, the focus on the common good is central to this (Adi & Stoeckle, 2023).

In the realm of sustainability communication, the significance of participatory communication approaches is increasingly emphasized, as highlighted by Wade et al. (2024). These approaches stand in stark contrast to traditional unilateral communication strategies, which depend on one-way or top-down information dissemination and therefore run the risk of favoring privileged experts or interest groups. In contrast, critical PR theory underscores the importance of a communication model that facilitates the exchange of ideas and even allows for the problematization of certain issues to ultimately support change (Weder, 2022). Here, questions of social justice play a pivotal role.

The concept of "sustainable citizenship," coined by Munshi and Kurian (2016), explicitly addresses issues of power dynamics and marginalization. This represents a departure from conventional practices: instead of prioritizing stakeholders based on their power, this approach advocates for the inclusion of those who lack power. This contrasts with the management approach to strategic communication, which traditionally is part of the "dominant coalition" and aligns itself with the most powerful stakeholders, as described by Grunig and Hunt (1984).

Unlike traditional approaches in this field, we recognize that the complex nature of wicked problems makes TSC challenging to plan. This complexity necessitates the adoption of agile methodologies, often requiring a trial-and-error process. Consequently, setting SMART goals[1] for transformations is not always realistic. Instead, we find it more practical to aim for target ranges or envision desirable futures. Our perspective aligns with Adi and Stoeckle (2023), who advocate practitioners for incorporating metamodern concepts, embracing a form of relativism in strategies and special reflection on values and responsibilities, e.g., in terms of loyalties. If we argue that TSC goes beyond organizational interests, it can be asked who can run TSC at all. We are not excluding any group of actors: Companies can engage in TSC, as can NGOs, cities and public institutions, churches and, of course, networks of the aforementioned actors. To illustrate this with three examples: This can be done, for example, through cooperation between companies and other actors such as civil society groups, NGOs, municipalities, or through network-like campaigns in which different groups of actors participate. Companies that form new coalitions in the spirit of the circular economy could also use TSC. Strategic communication is thus supplemented by decentralized and flexible approaches (Adi & Stoeckle, 2023) and is also given an enabling role in this context (Pleil, 2015).

[1] SMART formula: Goals should always be formulated Specific, Measurable, Achievable, Relevant, and Time-Bound (Zerfass & Volk, 2022).

4.4 TSC in Practice

Below, we highlight examples from Germany and Switzerland, illustrating the application of Transformative Strategic Communication (TSC) as the versatile toolkit previously described:

- *Patagonia*, the outdoor gear manufacturer, not only financially backs environmental groups but also provides them with training. This support aims to amplify their (and consequently, societal) objectives. Moreover, Patagonia adopts a comprehensive narrative in its strategic communication by spotlighting less recognized viewpoints, including those of indigenous communities, thereby bringing them to a wider audience (patagonia.com; Pleil et al., 2023).
- *Vaude*, another outdoor gear producer, has been dedicated to establishing fair supply chains and sustainable production practices for years. The company shares its accumulated knowledge within the industry, contributing to broader sectoral improvements (vaude.com).
- The consultancy *quäntchen + glück* excels in connecting local environmental groups. It employs facilitation techniques to foster collaborations among previously disparate entities, enhancing their collective impact on environmental goals (Pleil et al., 2023).
- *Energie Schweiz* and the Swiss Federal Office of Energy have implemented strategies including local community building and the deployment of change agents. These initiatives were particularly designed to encourage energy-efficient behaviors among apartment dwellers, demonstrating the power of targeted, grassroots approaches in driving behavioral change (Schmid et al., 2020).

These instances underscore communication strategies that transcend the traditional limits of strategic communication practices, targeting societal benefits. Specifically, the Swiss example showcases the pivotal role of change agents—drawn from the campaign's target audience—in ensuring the initiative's success (Schmid et al., 2020).

5 Summary and Future Perspectives

As shown, strategic sustainability communication practices can take very different forms (see also Chan-Olmsted et al., 2023)—depending on the actors involved and their objectives. We assume that all three types of strategic sustainability communication practices presented here make a contribution to sustainable development—albeit with different starting points and scopes. With Transformative Strategic Communication (TSC), we are attempting to describe a type of strategic communication practice that goes beyond its use as an organizational function and which can contribute to democratically legitimized change in a society.

The typology presented here is, of course, ideal-typical. Further research is required in the next step, for example, to examine the state of the art of the respective types in more detail. In practice, there is likely to be a fluid boundary between the described types and between sustainability PR and TSC in particular. The stronger the social reference and the intention to support social change, the more clearly we would assign concrete communication strategies to TSC.

More recent transformative approaches, such as that of the "safe system for sustainable development" (Lah, 2024), should also be considered intensively with regard to being a TSC practice. The core of this approach, conceived by the Wuppertal Institute, is not to focus primarily on individual behavior to achieve sustainable development in the transformation process. Instead, the safe system approach concentrates on developing sustainability as a default and not as a choice, e.g., in a system such as regional mobility (Lah, 2024). At the same time, Maeseele (2022) clearly views sustainability as a social conflict, i.e., something political, in which conflict, power, exclusion, and division always play a role. For this reason, communication practitioners need to be aware of power and network structures.

Like other authors, we see a significant potential for strategic communication efforts in supporting multi-stakeholder collaboration in transformative contexts, as requested, e.g., in a report by the Club of Rome (Weizsäcker & Wijkman, 2019). This increases the chance of advancing the great transformation as a successful innovation process. The perspective of critical theory plays a special role here and can help to shape such processes and the associated communication in an inclusive manner and detached from the usual power constellations.

On the other hand, actors who are committed to a path to sustainable societies also cannot underestimate social dynamics. We have addressed norms and values, individual decisions, and political decision-making. Fisher (2024) derives very clear findings from her research on climate movements and activism: she states that politicians worldwide do not do enough to stop the climate crisis—the results of global climate conferences such as COP28 in 2023 and "the continued expansion of fossil fuels" are particularly important evidences for her. She has also an ambivalent view of the economy and distinguishes between companies that continue to profit massively from climate-damaging business, and companies that are on the side of sustainable transformation. Fisher also sees climate shocks (e.g., droughts, floods) as strong triggers for social conflicts. In her view, the best way out of these situations is strong climate activism and social movements that are not afraid of conflict in order to build up sufficient pressure on political decisions. These aspects can be linked very well with Macnamara's approach of strategic communication with a "humanistic, subjectivist, social constructionist perspective of how knowledge, meaning and understanding are created" (Macnamara, 2016, p. 341). As discussed above, it can also be helpful to systematically discuss the social tipping elements presented by Otto et al. (2020) in terms of the supporting contribution that TSC efforts can make to achieving them. The toolbox of transformative strategic communication can provide suitable support in individual cases—and at the same time, it must be constantly scrutinized and expanded through interdisciplinary perspectives.

References

Adi, A., & Stoeckle, T. (2023). Public relations as responsible persuasion – activism and social change. In D. Pompper, K. R. Place, & K. C. Weaver (Eds.), *The Routledge companion to public relations* (pp. 302–314). Routledge Handbooks Online. https://doi.org/10.4324/9781003131700-5

Alzu'bi, R. M., & Kontor, E. (2023). The Impact of green marketing strategies on employees green performance: An HRM perspective. *Economica, 14*(3–4), 83–92. https://doi.org/10.47282/economica/2023/14/3-4/13573

Anderson, C. R., & McLachlan, S. M. (2016). Transformative research as knowledge mobilization: Transmedia, bridges, and layers. *Action Research, 14*(3), 295–317. https://doi.org/10.1177/1476750315616684

Bilandzic, H., & Kalch, A. (2021). Fictional narratives for environmental sustainability communication. In F. Weder, L. Krainer, & M. Karmasin (Eds.), *The sustainability communication reader: A reflective compendium* (pp. 123–142. Springer Fachmedien). https://doi.org/10.1007/978-3-658-31883-3_8

Bohn, S., & Braun, T. (2021). Field-configuring projects: How projects shape the public reflection of electric mobility in Germany. *International Journal of Project Management, 39*(6), 605–619.

Capizzo, L., & Luisi, M. (2024). Responsibility, sustainability, or environmental justice? Strategic communication and evolving expectations for stewardship and citizenship. *Public Relations Inquiry, 0*, 2046147X241232811. https://doi.org/10.1177/2046147X241232811

Castells, M. (2007). Communication, power and counter-power in the network society. *International Journal of Communication, 1*, 238–266.

Chan-Olmsted, S., Wolter, L.-C., & Adam, E. D. (2023). Rethinking sustainability communication: Meeting new audience needs with a comprehensive framework. In G. Schuster & L.-C. Wolter (Eds.), *Nachhaltiges Markenmanagement: Innovative Unternehmenspraxis: Insights, Strategien und Impulse* (pp. 53–70). Springer Fachmedien. https://doi.org/10.1007/978-3-658-42569-2_4

de Villiers, C., La Torre, M., & Molinari, M. (2022). The Global Reporting Initiative's (GRI) past, present and future: Critical reflections and a research agenda on sustainability reporting (standard-setting). *Pacific Accounting Review, 34*(5), 728–747. https://doi.org/10.1108/PAR-02-2022-0034

Fisher, D. (2024). A realistic path forward for climate action. *Next Big Idea Club*. Retrieved March 28, 2024, from https://nextbigideaclub.com/magazine/realistic-path-forward-climate-action-bookbite/47964/

Gelderman, C. J., Schijns, J., Lambrechts, W., & Vijgen, S. (2021). Green marketing as an environmental practice: The impact on green satisfaction and green loyalty in a business-to-business context. *Business Strategy and the Environment, 30*(4), 2061–2076. https://doi.org/10.1002/bse.2732

Godemann, J. (2021). Communicating sustainability. Some thoughts and recommendations for enhancing sustainability communication. In F. Weder, L. Krainer, & M. Karmasin (Eds.), *The Sustainability communication reader: A reflective compendium* (pp. 15–29). Springer Fachmedien. https://doi.org/10.1007/978-3-658-31883-3_2

Golob, U., Podnar, K., & Zabkar, V. (2023). Sustainability communication. *International Journal of Advertising, 42*(1), 42–51. https://doi.org/10.1080/02650487.2022.2144035

Groffman, P. M., Stylinski, C., Nisbet, M. C., Duarte, C. M., Jordan, R., Burgin, A., Previtali, M. A., & Coloso, J. (2010). Restarting the conversation: Challenges at the interface between ecology and society. *Frontiers in Ecology and the Environment, 8*(6), 284–291. https://doi.org/10.1890/090160

Grunig, J. E., & Hunt, T. (1984). *Managing public relations*. Holt, Rinehart and Winston.

Hallahan, K., Holtzhausen, D., van Ruler, B., Verčič, D., & Sriramesh, K. (2007). Defining strategic communication. *International Journal of Strategic Communication, 1*(1), 3–35. https://doi.org/10.1080/15531180701285244

Hamidi, K., & Mielke Möglich, A. (2021). Kommunikation für Sozialen Wandel: Ein aktueller Blick in das internationale Forschungs- und Praxisfeld mit Schwerpunkt auf Deutschland. *Publizistik, 66*(3–4), 565–588. https://doi.org/10.1007/s11616-021-00679-8

Hennen, L., Kopfmüller, J., Maia, M., Nierling, L., & Scherz, C. (2023). Ways towards transformation—conceptual approaches and challenges. *Sustainability, 15*(20), 14867. https://doi.org/10.3390/su152014867

Herbohn, K., Clarkson, P., & Wallis, M. (2022). The state of climate change-related risk disclosures and the way forward. In C. A. Adams (Ed.), *Handbook of accounting and sustainability*. Edward Elgar. https://doi.org/10.2139/ssrn.3958452

Hoffjann, O. (2024). Tear down this wall. A theory of strategic communication that integrates both communication and organizational theory. *Journal of Communication Management*. ahead-of-print. https://doi.org/10.1108/JCOM-09-2023-0103

Holtzhausen, D. R., & Zerfass, A. (2013). Strategic communication – pillars and perspectives of an alternative paradigm. In A. Zerfaß, L. Rademacher, & S. Wehmeier (Eds.), *Organisationskommunikation und Public Relations* (pp. 73–94). Springer Fachmedien. https://doi.org/10.1007/978-3-531-18961-1_4

Hooghiemstra, R. (2000). Corporate communication and impression management – new perspectives why companies engage in corporate social reporting. *Journal of Business Ethics, 27*(1/2), 55–68. https://doi.org/10.1023/A:1006400707757

Ihlen, Ø. (2023). A sociological lens on public relations. In D. Pompper, K. R. Place, & K. C. Weaver (Eds.), *The Routledge companion to public relations* (pp. 125–135). Routledge Handbooks Online. https://doi.org/10.4324/9781003131700-5

IPCC. (2023). Climate Change 2023: Synthesis Report. In Core Writing Team, H. Lee, & J. Romero (Eds.) *Contribution of Working Groups I, II and III to the Sixth Assessment Report of the Intergovernmental Panel on Climate Change* (pp. 35–115). IPCC. https://doi.org/10.59327/IPCC/AR6-9789291691647.

Kannengießer, S. (2021). Sozial-ökologische Transformationsforschung: Desiderata der Kommunikations- und Medienwissenschaft. In N. S. Borchers, S. Güney, U. Krüger, & K. Schamberger (Eds.) *Transformation der Medien – Medien der Transformation* (pp. 245–258). https://doi.org/10.53291/JFPN1141

Kara, E., Akbaba, M., Yakut, E., Çetinel, M. H., & Pasli, M. M. (2023). The mediating effect of green human resources management on the relationship between organizational sustainability and innovative behavior: An application in Turkey. *Sustainability, 15*(3), 2068. https://doi.org/10.3390/su15032068

Karmasin, M., Voci, D., Weder, F., & Krainer, L. (2021). Future perspectives: Sustainability communication as scientific and societal challenge. In F. Weder, L. Krainer, & M. Karmasin (Eds.), *The sustainability communication reader: A reflective compendium* (pp. 585–591). Springer Fachmedien. https://doi.org/10.1007/978-3-658-31883-3_31

Klein, J. T., Häberli, R., Scholz, R. W., Grossenbacher-Mansuy, W., Bill, A., & Welti, M. (Eds.). (2001). *Transdisciplinarity: Joint problem solving among science, technology, and society*. https://doi.org/10.1007/978-3-0348-8419-8

Lah, O. (2024). Safe system for sustainable development. *Sustainable Earth Reviews, 7*(1), 9. https://doi.org/10.1186/s42055-024-00072-z

Lin, J., Long, C., & Liu, B. (2024). The impact of shared value proposition on consumer engagement through a sense of brand community. *International Journal of Consumer Studies, 48*(2), 13026. https://doi.org/10.1111/ijcs.13026

Logan, N. (2023). Public relations and the racial reckoning. In D. Pompper, K. R. Place, & K. C. Weaver (Eds.), *The Routledge companion to public relations* (pp. 46–60). Routledge Handbooks Online. https://doi.org/10.4324/9781003131700-5

Macnamara, J. (2016). Socially integrating PR and operationalizing an alternative approach. In J. L'Etang, D. McKie, N. Snow, & J. Xifra (Eds.), *The Routledge handbook of critical public relations* (pp. 335–348). Routledge. https://doi.org/10.4324/9781315852492

Maeseele, P. (2022). Beyond the post-political zeitgeist 2.0. In A. Hansen & R. Cox (Eds.), *The Routledge handbook of environment and communication* (pp. 441–454). Routledge. https://doi.org/10.4324/9781003119234

Mahdy, F., Alqahtani, M., & Binzafrah, F. (2023). Imperatives, benefits, and initiatives of Green Human Resource Management (GHRM): A systematic literature review. *Sustainability, 15*(6), 4866. https://doi.org/10.3390/su15064866

Malik, F. (2018). Digital society? The great transformation. In C. Linnhoff-Popien, R. Schneider, & M. Zaddach (Eds.), *Digital marketplaces unleashed* (pp. 7–12). Springer. https://doi.org/10.1007/978-3-662-49275-8_2

Martinez-Levy, A. C., Sasso, D., Vozzi, A., Menicocci, S., Trettel, A., Babiloni, F., Cartocci, G., & Cherubino, P. (2023). How to promote and communicate an effective green and sustainable communication: A neuromarketing study. In M. Brandstrup, L.-P. Dana, D. Ryding, G. Vignali, & M. Caratù (Eds.), *The garment economy: Understanding history, developing business models, and leveraging digital technologies* (pp. 191–208). Springer International. https://doi.org/10.1007/978-3-031-33302-6_11

Meijers, M. J., van Drunen, Y., & Jacobs, K. (2022). It's a hoax! The mediating factors of populist climate policy opposition. *West European Politics, 0*, 1–24. https://doi.org/10.1080/01402382.2022.2145447

Michelsen, G. (Ed.). (2005). *Handbuch Nachhaltigkeitskommunikation: Grundlagen und Praxis*. Oekom-Verl., Ges. für Ökologische Kommunikation.

Mukonza, C., & Swarts, I. (2020). The influence of green marketing strategies on business performance and corporate image in the retail sector. *Business Strategy and the Environment, 29*(3), 838–845. https://doi.org/10.1002/bse.2401

Munshi, D., & Kurian, P. (2016). Public relations and sustainable citizenship: Towards a goal of representing the unrepresented. In J. L'Etang, D. McKie, N. Snow, & J. Xifra (Eds.), *The Routledge handbook of critical public relations* (pp. 405–4149). Routledge. https://doi.org/10.4324/9781315852492

Murphy, P. (2023). The place of the environment in the field of communication for development and social change. In A. Hansen & R. Cox (Eds.), *The Routledge handbook of environment and communication* (pp. 82–98). Routledge. https://doi.org/10.4324/9781003119234

Newig, J., Schulz, D., Fischer, D., Hetze, K., Laws, N., Lüdecke, G., & Rieckmann, M. (2013). Communication regarding sustainability: Conceptual perspectives and exploration of societal subsystems. *Sustainability, 5*(7), 2976–2990. https://doi.org/10.3390/su5072976

Otto, I. M., Donges, J. F., Cremades, R., Bhowmik, A., Hewitt, R. J., Lucht, W., Rockström, J., Allerberger, F., McCaffrey, M., Doe, S. S. P., Lenferna, A., Morán, N., van Vuuren, D. P., & Schellnhuber, H. J. (2020). Social tipping dynamics for stabilizing Earth's climate by 2050. *Proceedings of the National Academy of Sciences, 117*(5), 2354–2365. https://doi.org/10.1073/pnas.190057711

Piwinger, M., & Hoffmann, C. P. (2022). Kommunikation mit Kapitalgebern: Grundlagen von Finanzkommunikation und Investor Relations. In A. Zerfaß, M. Piwinger, & U. Röttger (Eds.), *Handbuch Unternehmenskommunikation: Strategie – Management – Wertschöpfung* (pp. 799–816). Springer Fachmedien. https://doi.org/10.1007/978-3-658-22933-7_35

Pleil, T. (2015). Online-PR. In R. Fröhlich, P. Szyszka, & G. Bentele (Eds.), *Handbuch der Public Relations: Wissenschaftliche Grundlagen und berufliches Handeln. Mit Lexikon* (pp. 1017–1038). Springer Fachmedien. https://doi.org/10.1007/978-3-531-18917-8_62

Pleil, T. (2018). Zum Umgang mit Populismus aus PR-Sicht. Ein Essay mit Tipps für Praktiker_innen. *Communicatio Socialis (ComSoc), 51*(2), 169–181. https://doi.org/10.5771/0010-3497-2018-2-169

Pleil, T., & Helferich, P. S. (2020). Unternehmenskommunikation in der digitalen Transformation. In A. Zerfaß, M. Piwinger, & U. Röttger (Eds.), *Handbuch Unternehmenskommunikation* (pp. 1–19). Springer Fachmedien. https://doi.org/10.1007/978-3-658-03894-6_49-1

Pleil, T., Helferich, P. S., & Otsa, T. (2023, June 30). *Strategic communication as a transformative approach in the context of sustainable development*. BledCom 2023 - 30th International Public Relations Research Symposium, Bled.

Pleil, T., Müller, S., & Matusiak, L. (2021). Public Relations in der transformativen Wissenschaft: Der Beitrag Strategischer Kommunikation zum Gelingen eines transformativen Forschungsparadigmas. In N. S. Borchers, S. Güney, U. Krüger, & Kerem Schamberger (Eds.) *Transformation der Medien – Medien der Transformation* (pp. 259–282). https://doi.org/10.53291/JFPN1141.

Prexl, A. (2010). *Nachhaltigkeit kommunizieren – nachhaltig kommunizieren*. VS Verlag für Sozialwissenschaften https://doi.org/10.1007/978-3-531-92471-7.

Ross, E. (2024). Communicating climate. How to transmit your climate message and avoid greenwashing. In *Communicating climate*. Emerald Group. https://doi.org/10.1108/978-1-83753-640-520241026.

Sandhu, S. (2022). Gesellschaftsorientierte Unternehmenskommunikation: Stakeholderorientierung und Legitimation als Ziel der Public Relations. In A. Zerfaß, M. Piwinger, & U. Röttger (Eds.), *Handbuch Unternehmenskommunikation: Strategie – Management – Wertschöpfung* (pp. 859–880). Springer Fachmedien. https://doi.org/10.1007/978-3-658-22933-7_38

Santos, L. D. V., Holanda, F. S. R., Oliveira, C. V. D. S., De Menezes, P. V. B., Villwock, A. P. S., & Pedrotti, A. (2023). Consumer perception of brand equity based on environmental sustainability – development and validation of a scale. *OBSERVATÓRIO DE LA ECONOMÍA LATINOAMERICANA, 21*(6), 4778–4819. https://doi.org/10.55905/oelv21n6-087

Schmid, C., Rosenberger Staub, N., Klopfenstein Frei, N., Widmer, H., Zollinger, T., & Koch, C. (2020). *Zusammenarbeit und Kommunikation für das energieoptimierte Bewirtschaften und Nutzen von Wohngebäuden*. https://doi.org/10.21256/zhaw-21212.

Schneidewind, U. (2018). *Die große Transformation: eine Einführung in die Kunst gesellschaftlichen Wandels (Originalausgabe)*. Fischer Taschenbuch.

Schrader, C. (2020). Klimafakten Handbuch. Klimafakten Handbuch. https://klimakommunikation.klimafakten.de/

Servaes, J., & Malikhao, P. (2007). Communication and sustainable development. In J. Servaes & S. Liu (Eds.), *Moving targets: Mapping the paths between communication, technology, and social change in communities* (pp. 11–44). Southbound. http://lib.ugent.be/catalog/rug01:002828218

Strottner, L., & Huck-Sandhu, S. (2021). Mit Herz und Verstand: Rolle der internen Kommunikation für die Etablierung neuer Nachhaltigkeitsstrategien. *Sustainability Management Forum | NachhaltigkeitsManagementForum, 29*(3), 197–216. https://doi.org/10.1007/s00550-021-00523-7

Talbot, D., & Boiral, O. (2018). GHG reporting and impression management: An assessment of sustainability reports from the energy sector. *Journal of Business Ethics, 147*(2), 367–383. https://doi.org/10.1007/s10551-015-2979-4

Taylor, M., & Kent, M. L. (2023). Paradigm shifts in public relations theory. In D. Pompper, K. R. Place, & K. C. Weaver (Eds.), *The Routledge companion to public relations* (pp. 103–114). Routledge. https://doi.org/10.4324/9781003131700-5

van Ruler, B. (2018). Communication theory: An underrated pillar on which strategic communication rests. *International Journal of Strategic Communication, 12*(4), 367–381. https://doi.org/10.1080/1553118X.2018.1452240

Voci, D., & Karmasin, M. (2023). Sustainability communication: How to communicate an inconvenient truth in the era of scientific mistrust. *Journal of Communication Management, 28*(1). https://doi.org/10.1108/JCOM-05-2022-0060

Wade, B., Bharadwaj, B., Kambo, A., Jensen, M., Witt, K., Weder, F., Phelan, A. (Anya), & Ashworth, P. (2024). Stakeholder engagement: The role of facilitators and gender in 'opening up' conversations and enabling participation. *Australasian Journal of Environmental Management, 0*(0), 1–33. https://doi.org/10.1080/14486563.2023.2298195

Wagner, R. (2022). Überholt von der Realität: Formt die Nachhaltigkeitskommunikation PR der Zukunft? In S. Pranz, H. Heidbrink, F. Stadel, & R. Wagner (Eds.), *Journalismus und*

Unternehmenskommunikation: Zwischen Konvergenz und Konkurrenz (pp. 41–54). Springer Fachmedien. https://doi.org/10.1007/978-3-658-35471-8_3

WBGU. (Ed.). (2011). *Welt im Wandel: Gesellschaftsvertrag für eine Große Transformation [Hauptgutachten]* (2). Wissenschaftlicher Beirat der Bundesregierung Globale Umweltveränderungen (WBGU).

Weder, F. (2021). Sustainability as master frame of the future? Potency and limits of sustainability as normative framework in corporate, political and NGO communication. In F. Weder, L. Krainer, & M. Karmasin (Eds.), *The sustainability communication reader: A reflective compendium* (pp. 103–119). Springer Fachmedien. https://doi.org/10.1007/978-3-658-31883-3_7

Weder, F. (2022). Strategic problematization of sustainability reframing dissent in strategic communication for transformation. *Public Relations Inquiry, 11*(3), 337–360. https://doi.org/10.1177/2046147X211026857

Weder, F., & Erikson, M. (2023). *CSR communication and cultures of sustainability*. The University of Queensland. https://doi.org/10.14264/428d81b.

Weder, F., & Weaver, C. K. (2023). The complexities of contemporary communicator roles and the future of public relations. In D. Pompper, K. R. Place, & K. C. Weaver (Eds.), *The Routledge companion to public relations* (pp. 428–440). Routledge.

Weder, F., Lemke, S., & Tungarat, A. (2019). (Re)storying sustainability: The use of story cubes in narrative inquiries to understand individual perceptions of sustainability. *Sustainability, 11*(19), 5264. https://doi.org/10.3390/su11195264

Weizsäcker, E. U., & Wijkman, A. (2019). *Wir sind dran: was wir ändern müssen, wenn wir bleiben wollen: Club of Rome: der große Bericht : erstellt für das 50-jährige Bestehen des Club of Rome 2018 / Ernst Ulrich von Weizsäcker, Anders Wijkman u.a.* (Erste Auflage). Pantheon.

Wexler, M. N., & Oberlander, J. (2020). Public leadership and the wicked problem continuum. *International Journal of Public Leadership, 16*(4), 359–373. https://doi.org/10.1108/IJPL-04-2020-0025

Zerfaß, A., & Volk, S. C. (2022). Strategische Ausrichtung der Unternehmenskommunikation: Zieldefinition, Alignment mit Organisationszielen und Wertbeitrag. In A. Zerfaß, M. Piwinger, & U. Röttger (Eds.), *Handbuch Unternehmenskommunikation: Strategie – Management – Wertschöpfung* (pp. 469–492). Springer Fachmedien. https://doi.org/10.1007/978-3-658-22933-7_21

Zerfaß, A., Lautenbach, C., & Sass, J. E. (2022). Organisation der Kommunikationsfunktion: Grundmodelle und Benchmarking. In A. Zerfaß, M. Piwinger, & U. Röttger (Eds.), *Handbuch Unternehmenskommunikation: Strategie – Management – Wertschöpfung* (pp. 619–638). Springer Fachmedien. https://doi.org/10.1007/978-3-658-22933-7_28

Further Reading

Adi, A., & Stoeckle, T. (2023). Public relations as responsible persuasion - activism and social change. In D. Pompper, K. R. Place, & K. C. Weaver (Eds.), *The Routledge companion to public relations* (pp. 302–314). Routledge Handbooks Online. https://doi.org/10.4324/9781003131700-5

Murphy, P. (2023). The place of the environment in the field of communication for development and social change. In A. Hansen & R. Cox (Eds.), *The Routledge handbook of environment and communication* (pp. 82–98). Routledge.

Thomas Pleil is a Professor of Public Relations, particularly Online PR, at Darmstadt University of Applied Sciences and spokesperson for the Research Center Digital Communication and Media Innovation. He set up the online communication course (B.Sc.) and has carried out numerous research and knowledge transfer projects. His main topics are online communication, digital transformation, and sustainability communication.

Pia Sue Helferich is a professor of online communication, especially organizational communication, at Darmstadt University of Applied Sciences. She has worked on various projects and companies with the main focus on online communication, public relations, and adult education. Her work also focuses on lifelong learning, digital transformation and communication in SMEs.

Teresa Otsa is a research associate at Darmstadt University of Applied Sciences and Ph.D. student at Munster Technological University Cork. Her research focuses on sustainability and environmental communication.

Sustainable and Just? Access to Sustainable Goods and Environmental Justice

Josh Anderson, Saima Kazmi, Na Yu, Lucy Atkinson, Patrick Jamar, and Jingyue Tao

Abstract Promoted as a solution to environmental challenges, sustainable consumer goods can be inaccessible to many consumers because of their high price points compared to similar products. In this work, we implement a concept analysis that explores the lack of access to these products as an environmental justice issue. As a final step of this analysis, we present the results of a focus group-style interview study as empirical referents to our proposed concept. Throughout these sessions, we heard from participants who were concerned about the environment and aware of products that were promoted as solutions but found many of these to be inaccessible. We also heard these participants describe their situation in terms of commercial entities, placing on them the burden of answering environmental challenges and unjustly profiting from doing so. Overall, we argue that there is value in considering inaccessible sustainable goods as an environmental justice issue.

J. Anderson (✉)
School of Journalism, University of Arizona, Tucson, AZ, USA
e-mail: jtlanderson@arizona.edu

S. Kazmi
School of Journalism and Communication, University of Oregon, Eugene, OR, USA
e-mail: skazmi@uoregon.edu

N. Yu
Department of Communication, School of Humanities and Social Sciences, College of Charleston, Charleston, SC, USA
e-mail: nyu02@wm.edu

L. Atkinson
School of Advertising and Public Relations, The University of Texas at Austin, Austin, TX, USA
e-mail: lucyatkinson@austin.utexas.edu

P. Jamar
College of Agriculture & Life Sciences, Texas A&M University, College Station, TX, USA
e-mail: patrick.jamar@tamu.edu

J. Tao
Digital Media, University of Arkansas, Fayetteville, AR, United States
e-mail: jingyue.tao@utexas.edu

F. Weder (ed.), *Strategic Sustainability Communication*, CSR, Sustainability, Ethics & Governance, https://doi.org/10.1007/978-3-031-89486-2_5

1 Intro

Collectively, humans are consuming natural resources at a greater rate than they can be replenished. Moreover, human activities produce by-products that can have negative lasting changes on the climate and natural environments. One measure that partly captures the amount of resources needed for human life is "biocapacity," or the resources that a given area has to support life (e.g., Yue et al., 2013). This is useful because it can be used to describe the environmental consequences of human lifestyles, not just human life overall, since not all lifestyles are equal. For instance, advocacy groups have used biocapacity to argue that if all people in the world lived like the average United States citizen, they would need 5.1 times the amount of natural resources available on planet Earth (Footprint Network, 2024). Similar results can be found across most wealthy nations (Footprint Network, 2024).

Thinking in terms of lifestyle support is an important way to understand sustainability issues because this gets at the heart of the challenge of sustainable living: consumption. Consumer goods are key contributors to the overextension of Earth's natural resources and space due to resources used and byproducts emitted during production, packaging, transportation, and disposal after usage (Jaller & Pahwa, 2020). Many environmental justice scholars, who study how negative environmental consequences are distributed in society, have pointed out that while people in wealthy societies are the most likely to contribute to environmental hazards, the people who are most likely to face problems related to negative environmental issues live in comparatively less wealthy societies (Malin & Ryder, 2018; McDermott, 2022). A main issue in addressing environmental damage due to consumption is that the practice of consumption is inevitable; while people may make consumer choices that are more or less damaging to the natural environment, they must still engage in some degree of resource consumption to meet their basic needs. Consumers have agency to choose consumer goods based on their environmental impacts, but their choices are structured according to the material and social contexts of these products. For example, consumers may pay a premium for sustainably labeled consumer goods, and brands may use strategic communication to justify higher prices.

Sustainable strategic communication often entails communication strategies designed to promote products that are comparatively better for the environment than similar products (Atkinson, 2012) or to highlight the environmental policies of brands (Sroufe, 2017). However, there are several concerns with this development, many of which focus on price. One major concern is that consumer goods marketed as sustainable tend to cost more than similar products, and brands that emphasize good environmental practices tend to sell their products at higher price points than their competitors (Griskevicius et al., 2010; Pieters et al., 2022). In essence, commercial entities can be incentivized to pursue environmentally friendly practices because they can then charge more for their products (Pieters et al., 2022). The tension of consumers paying a premium to lessen the environmental burden of the products they consume can come with an unintended negative consequence: it may alienate low-income consumers who cannot afford the premium. However, this chapter argues that to address the global issue of sustainable consumption, strategic communication should support sustainable goods and services which are accessible to low- and medium-income consumers rather than promoting them as aspirational premium goods. Because if not addressed, sustainable consumption otherwise may be an issue of environmental justice. Throughout this chapter, we explore whether promoting relatively sustainable products and services as premium commercial goods is a form of environmental injustice.

2 Methodology

This chapter explores whether the inability of low-income consumers to pay a premium for sustainable goods, which are supported by sustainable strategic communication, may be considered an issue of environmental justice. To address this question, this chapter is structured as a concept analysis, as defined by Walker and Avant (2018). A concept analysis is a multi-step, critical, and integrated approach to research questions that focus on the existence and extent of a proposed concept (Walker & Avant, 2018). In Walker and Avant's (2018) description of concept analyses, there are eight steps that researchers may take to explore their concept of interest: (1) the researchers select a concept, (2) they describe the purpose of the analysis, (3) they analyze applications of the concept, and (4) they describe essential characteristics of the concept. After these steps, the researchers may further understand the concept by (5) proposing and describing a model case in which a concept is used, (6) describing cases that may be considered contrary to the proposed concept or edge cases for the concept, (7) analyze antecedents necessary for the concept and potential outcomes of the concept's existence, and (8) analyze empirical data relevant to the concept.

Primarily used in the field of nursing, concept analyses have been used to explore concepts such as LGBTQAI+ centered healthcare (Laporte & De Santis, 2023). Concept analyses can also be useful for a detailed critical description of concepts across social science disciplines. We engaged in an abridged concept analysis that does not engage with a model case, contrary cases, or edge cases when applying our concept of interest. Previous works have used concept analyses abridged as such and have still achieved thick descriptions of their concepts of interest (DeBastiani & De Santis, 2018; Fortunatti et al., 2021; Laporte & De Santis, 2023).

Accordingly, this chapter is structured sequentially from the steps listed above, excluding steps five and six. Steps one through seven will take the form of critical engagement with the concept of income disparities in access to sustainable consumer goods as an issue of environmental justice. In step eight, we draw upon empirical referents for this concept by presenting data from focus group-style interview sessions with low-income consumers that centered on sustainable strategic communication and consumer goods that are promoted using sustainable strategic communication. These focus group sessions were carried out in September 2023 on-site at a low-income housing facility in Austin (Texas, United States of America). The researchers obtained informed consent from the participants (who are referred to using pseudonyms in this chapter), as well as permission to conduct the study from the organization administering the housing facility and approval from their respective IRBs (institutional review boards). The interviews were analyzed using a constructivist approach to generate an understanding of sustainable strategic communication as it exists in the minds of the participants (see Mills et al., 2006). Data were analyzed through collective discussion by the authors following the sessions. The results presented here represent the content of the focus group conversations that were relevant to the concept that we are analyzing.

2.1 Concept Selection

The concept that we will be discussing throughout this chapter is income-based disparity in access to communicated sustainable consumer goods as an issue of environmental justice. In effect, we will be exploring an expansion of the concept of environmental justice to encompass this issue. To begin, we briefly review the present work on environmental justice. The Environmental Protection Administration (EPA) Office of Environmental Justice (2021) broadly defines *environmental justice* as

> The fair treatment of all people, regardless of race, national origin, or income, with respect to the development, implementation, and enforcement of environmental laws, regulations, and policies. Fair treatment means that no population, due to policy or economic disempowerment, is forced to bear a disproportionate share of the negative human health or environmental consequences resulting from industrial, municipal, and commercial operations or execution of federal, state, local or tribal programs and policies.

As the United States (USA) federal government's definition reflects, environmental justice research has historically centered on questions of whether economically disadvantaged communities are disproportionately exposed to environmental hazards, such as toxic chemical waste, climate-based food insecurity, and negative impacts of climate change (Kallhoff, 2021; Motta, 2022). The environmental justice movement can be traced back to the 1982 protests in Warren County, North Carolina (USA), following the dumping of 31,000 gallons of chemical waste (McGurty, 1997). Two sites were considered for dumping, but Warren County was selected, as opposed to a more affluent and white Chatham County (McGurty, 1997). The decision drew significant criticism about race, class, and lack of inclusive decision-making because Warren County at the time was populated by 60% Black residents and 25% of families living below the poverty line (McGurty, 1997).

Following the initial activism, environmental justice research developed into an interdisciplinary field, with researchers using many different perspectives to understand how pollution and other environmental issues disproportionately affect lower social classes and marginalized racial groups. For example, a recent forecast of coastal flooding due to climate change identified 100-year floodplains in the United

States as areas with higher than a 1% chance of annual flooding (Wobus et al., 2017). Demographic research of these areas indicated that coastal flooding is likely to disproportionately impact poor and non-white residents compared to adjacent areas (McDermott, 2022; Wobus et al., 2017). These data suggest that climate-related flooding risks are unequally distributed, highlighting the correlation between environmental risks and existing patterns of social and demographic disparities (Mohai et al., 2009). Moreover, climate change both mirrors and exacerbates social inequalities through various dimensions, such as who bears the brunt of its impacts, who is responsible for its cause, who is anticipated to respond, and who possesses the means to take action.

While the historical development of environmental justice applications has pertained to questions of environmental harm, it is important to separate these applications from the essential characteristics of the concept of environmental justice. The essential basis of most understandings of environmental justice is distributive justice, which in a Rawlsian sense analyzes the fairness of the distribution of rights, liberties, and opportunities in a society (Rawls, 1971). In the prior example, there was a violation of distributive justice because serious environmental harms were unequally distributed to Warren County as opposed to Chatham County, and other cases could be referred to in which Black communities disproportionately took on environmental harms (McGurty, 1997). This presents a distributive injustice because the people of Warren County were not given the same opportunity for healthful living as their neighbors in Chatham County. While environmental justice may also be conceptualized in reference to other theories of justice, such as procedural justice and recognition justice which examine the processes that determine environmental outcomes and the agency of identity groups with respect to environmental policy, most understandings of environmental justice hold questions of distribution of environmental outcomes at their core (Light & DeShalit, 2003). Keeping in mind these essential characteristics of environmental justice provides a guide for the application of our concept analysis.

To begin, we contend that the unfair distribution of an environmental outcome is enough to examine a question of environmental justice. This is because distribution is the essential question common to mainstream understandings of environmental justice, as opposed to the process or agency of environmental policy. An injustice defined along these lines, wherein a community had relatively poor environmental outcomes forced upon them without due process or with little agency in guiding the process, could make a more compelling case for environmental injustice. However, it is not necessary to answer how environmental outcomes are distributed nor the degree of agency of the communities involved to identify a situation as environmentally unjust. Following from this, we would also argue that public policy is not the only domain in which we may interrogate environmental justice. While most applications of environmental justice, especially those that suppose a procedural theory of justice, are focused on public environmental policy, nonstate actors may still engage in behaviors that create unfair distributions of environmental outcomes and are thus essentially environmental injustices. Lastly, we argue that environmental outcomes considered for questions of environmental justice should not be limited to environmental harms but should also include opportunities for, or access to, environmental benefits. This is because we also claim that the opportunity for individuals to make comparatively sustainable decisions is a social good, which may allow for those individuals to feel recognition for their environmentally conscious efforts. This point will be expanded on throughout this chapter.

3 Purpose of Analysis

The main purpose of this analysis is to discuss unequal access to sustainable consumer goods in terms of environmental justice. While the majority of discussions around climate change and other environmental challenges center on technological and economic tools used to reduce global emissions (Pink, 2018), sustainable consumer goods have been promoted by businesses and strategic communication professionals to address environmental challenges in the private sector (Atkinson, 2012). Much of sustainable strategic communication research is dedicated to commercial strategic communication applications, such as green advertising, which involves studying how to convince consumers to purchase goods marketed as sustainable (Sharma, 2021). While there are certainly many other applications of sustainable strategic communication that are not as obviously commercial, such as environmental public service campaigns (e.g., Maibach, 1993), our analysis centers on the sustainable strategic communication activities that are used to support commercial products and services.

Sustainable products, including garments, organic produce, and self-care items, are often priced higher than their counterparts, which may make them inaccessible to low-income consumers. For organic foods, producers may justify charging higher prices owing to the cost of production (Michalke et al., 2023). However, charging premium prices becomes problematic when companies use "sustainable" and "eco-friendly" branding as a marketing strategy to charge premium prices far beyond the increased production costs (Griskevicius et al., 2010). Furthermore, research has shown that consumers of sustainable goods respond better to moral motives or environmental saving, and these motivations tend to be more effective in environmental campaigns than emphasizing financial incentives (Bolderdijk et al., 2013). However, strategic communication practitioners and scholars overlook that many less-affluent consumers do not have the financial means to enact these altruistic behaviors. Consequently, low-income consumers with altruistic interests intending to purchase eco-friendly products may be exposed to marketing materials for sustainable products that are beyond their financial reach. Thus, while strategic communication presents a choice for consumers to purchase goods marketed as sustainable, it is not a choice that is reasonably available to low-income consumers. So, while commercial entities might consider this a matter of improper targeting, in the broad environmental justice movement, this lack of access is an important consideration for sustainability communication and may constitute an injustice.

4 Applications

The expansion of the concept of environmental justice to include income-based disparities in access to consumer goods can have applications for scholars of environmental justice and sustainable strategic communication alike. Recent scholarship on environmental justice can be conceptualized in two phases: (1) the first phase primarily documents environmental inequality through race and class and (2) the second phase addresses deeper considerations of theory, incorporating case studies that look at class, gender, historical context, and race as variables that apply across many environmental contexts (Malin & Ryder, 2018; Pellow, 2016). Drawing from understandings of intersectionality and critical environmental justice, this chapter builds on the second phase by orienting the concept toward an important environmental concern for commercial entities. For scholars of sustainable strategic communication, discussion in terms of environmental justice may provide a valuable normative framework to understand the implications of their work. The implications of these applications will be discussed further as the consequences of this concept.

5 Essential Characteristics

When considering which characteristics are essential for the issue of income-based access to sustainable goods and services to be an issue of environmental justice, it is important to think from the perspective of low-income consumers and interrogate the normative assumptions of the concept from this position. Thus, an essential characteristic is low-income consumers' desire to address environmental challenges with their product choices. Put another way, a person may not experience an environmental injustice about their inability to afford sustainable goods if the sustainability aspect of those goods is not appealing in the first place. In fact, the lack of low-income consumer interest in sustainable goods has been previously cited by strategic communicators of sustainable brands as to why these consumers are not considered a target audience (Finisterra do Paço et al., 2009). However, low-income consumers are not all alike and can vary according to their interest in sustainable product purchase intentions. Thus, we should consider this first essential characteristic as a qualifying condition that the consideration of income-based access to sustainable goods as an issue of environmental justice may only apply to low-income consumers who are interested in sustainable consumption.

Another essential characteristic of this concept is that low-income consumers perceive this lack of access. In other words, low-income consumers must actually regard sustainable goods and services as prohibitively expensive for disparities in access to be considered an environmental justice issue. This will likely be the case because sustainable goods tend to be priced higher than similar products (Griskevicius et al., 2010; Pieters et al., 2022). However, it is worth mentioning that this concept may only apply to consumers whose disposable income is below the price difference to have a substantial effect on their access. A final essential characteristic of this concept is that disparities in access to sustainable goods are understood by the participants to be undeserved. A necessary component of most understandings of justice is that unjust situations are essentially undue in their nature (e.g., Rawls, 1971). Accordingly, this has been a component in most environmental justice research, such as the field's focus on the undue disparities in exposure of communities to environmental hazards.

6 Antecedents and Consequences

The antecedents of this concept are structuring forces that precede environmental and income disparities in societies like the United States. In these societies, privileged communities are more insulated from environmental harm, but long legacies of racist policy and practice have resulted in racialized economic disparities (Motta, 2022). In the context of our concept, economically privileged (and largely white) communities disproportionately contribute to environmental harm due to their consumption habits. Consequently, less economically privileged (and largely non-white) communities may be disproportionately affected by the consequences of that environmental harm (Motta, 2022). Antecedent also to this concept is the trend of mass-marketed consumer goods being more accessible to economically privileged communities over others (Stern, 2008), which creates a disparity in access to environmental solutions.

In terms of the consequences of considering disparities in access to sustainable goods as an issue of environmental justice, there are potential benefits and risks. A central potential benefit of this concept, as described, is that doing so would allow the field of sustainable strategic communication to discuss normative implications of consumers being excluded from the audience of most sustainable goods and services. While price differentials are a purchasing barrier to sustainable goods, they are often considered to be primarily messaging challenges for sustainable strategic communicators who work to promote them despite their comparatively high prices (Sharma, 2021). However, this conception of price differences does not engage with the normative question of how sustainable strategic communicators should consider their impact on people who are financially unable to purchase the products they are promoting.

While this is the main benefit of this concept, we should also acknowledge that this conceptualization may risk trivializing the wider concept of environmental justice. The field of environmental justice has done much to illuminate the serious environmental issues that historically marginalized communities are disproportionately exposed to (McGurty, 1997). The thesis of this chapter should not be taken to mean that low-income consumers' inability to access sustainable goods is the same degree of injustice as these consumers being more likely to face the negative health effects of toxic chemical exposure than others. Instead, our argument is that sustainable good price disparities may affect low-income consumers via similar social structures that have driven disparities in environmental hazard exposure rates. That being said, we believe this is an important concern to scholars of sustainable strategic communication and professionals who promote sustainable products.

7 Empirical Referents

For this chapter, a primary guide to making sense of our interview data was seeing how they may or may not fit the essential characteristics of income-based lack of access to sustainable products as an environmental justice issue. One essential characteristic of the aforementioned concept was that low-income consumers actually desired sustainable goods. Another essential characteristic was that these consumers perceived sustainable goods as inaccessible due to comparatively higher prices than other goods. A final essential characteristic was that low-income consumers must have recognized that this disparity was happening to them because of an undue negative situation. From these essential characteristics, we may begin to see if disparities in access to these products and strategic communication used to promote them are an environmental justice issue.

To begin, there was wide interest in acting in environmentally sustainable ways among participants across the four sessions. Participants often spoke about sustainable issues in ways that related to health, with "poisons" being frequently discussed as a negative environmental consequence. Additionally, organic foods were the most commonly described consumer goods that are promoted using sustainable strategic communication. Not considering pesticides used in nonorganic foods, the sources of pollution functioning as human contaminants that were most prominent in participants' minds were waste products from disposable consumer goods, such as plastic containers. To many participants, consumer goods were often sustainable because they could be used multiple times (thus reducing consumption and waste). These products were preferred by the participants who mentioned them, but they also typically noted that consumer goods that were sustainable in this way could also be prohibitively expensive.

These themes, and the participants' affective responses to their understood reality, were more or less captured in this exchange in the third session:

Sycamore:
"There's a lot of stuff that's like one-and-done. So anything that you can reuse again is sustainable. Like my shoes—sustainable!"
Rowan:
"You get what you pay for is what it is. You get what you pay for."

This conversation then continued on the point of sustainability in terms of long-lasting qualities:

Oak
"I'm just talking about a cast iron skillet. Now that's a bad son of a gun."

Cedar

"Now cast iron. That'll last 'till you go six feet under."

Oak

"You get something like that and it lasts! You know you take care of it and you can use it for years. Coffee machines are not as sustainable as they used to be! You remember using like the coffee pots and it would last you like 10 years?"

However, participants returned to the issue of price:

Sycamore

"But (sustainable goods) are also more expensive."

Oak

"Okay, but I don't understand *why* they're more expensive." (emphasis participant's)

Furthermore, while some participants gave different reasons why some high-quality, sustainable goods are prohibitively expensive, most participants appeared to have negative experiences with this shared reality. As Oak put it to conclude his thoughts:

"(Sustainable goods producers can) kiss my tushie"

While this was one exchange, similar ideas could be taken from exchanges in each of the four sessions, and it appeared to the researchers that there was a rough consensus among participants that the expensive, sustainable goods being marketed to them left them in an unfavorable situation. That said, we also explored themes related to environmental justice that emerged beyond our identified essential characteristics.

One theme that emerged from each session was a lack of trust in companies that make sustainable goods and the actors that promote them. At least one participant in each session was doubtful that the premium prices asked for sustainable goods were fully due to higher production costs, believing instead that these goods cost more simply because some consumers would be willing to pay the difference. This belief was often intertwined with the belief that goods marketed as sustainable are not as good for the environment as they were described in related strategic communication. Some participants noted distrust in companies producing sustainable goods as a constraint of working within a capitalist system, as in the following exchange in the second session:

Cypress

"My understanding is that the USA is a capitalist nation and the goal is profit so if the goal is profit—then its profit. It doesn't matter what they say to you. To me they're gonna try to enhance me, to encourage me to purchase because they want profit. So I don't believe what they say."

Maple

"You kinda have to pick and choose what you wanna believe in or what you choose to believe in to what you just need to get by."

Participants commonly mentioned a lack of trust, and several noted feelings of low ability to parse fact from fiction, as Maple noted above. When we followed up in this exchange, asking participants how commercial entities can be more responsible actors, one participant, Alder, had something to say. Alder related how he was pursuing legal action against a tobacco brand for allegedly adding chemicals to their products to make them more addictive, despite their natural and relatively environmentally-friendly branding:

> Call them on their BS (lies). I am calling American Spirit on their BS, I am joining in a class action lawsuit against them and I hope they stick it to them. In the meantime I'm gonna try to quit smoking, but it's irritating to do that after the fact (of using them and getting addicted). But now hearing about the enhanced addictiveness of their products? Oh I am furious that they're blatantly lying. It probably isn't better (for the environment), it's more addictive, and it's dangerous ... and I wonder how many other products do that where they're just blatantly lying to you.

Alder was not the only participant who expressed anger toward companies that produce goods that may be communicated as being deceptively better for the environment.

Other participants were skeptical that purchasing sustainably branded consumer goods would produce noticeable environmental benefits. A number of these participants viewed consumers who pay premiums for environmentally friendly goods as deceiving themselves by thinking their consumer choices are doing good for the environment. This came up in the first session when the group was discussing dietary meat substitutes as environmental product solutions. As Magnolia noted:

> It's the simple minded people who think if they eat this it's better for everyone. It's full of a bunch of crap is what it is ... when I said simple-minded people I mean they believe everything they hear and they think that they're doing themselves good because they're eating a veggie burger as opposed to a meat burger.

Participants like Magnolia were resistant to sustainable behavioral interventions, feeling they were being thrust upon them due to the high volume of strategic communication promoting them. Even though these participants may have felt a desire to act in environmentally friendly ways (as Magnolia noted earlier in the session), they did not feel that many of the products that are marketed as having a positive environmental impact were effective.

A final theme that we drew from these data, relating to disparities in access to sustainable goods as an issue of environmental justice, was the question of who should be responsible for producing positive environmental change. It should be emphasized again that there was a virtual consensus among participants that environmental change was needed, and some participants were concerned about personally experiencing the consequences of environmental damage. That said, in each session, at least one participant stated that they believed that commercial organizations or other powerful entities should be the ones to address environmental challenges. This theme first emerged in the first session, with Pine stating:

> "I think the kind of (sustainable consumer goods) they are coming up with should be *given* to the public." (emphasis participant's)

Upon hearing this, we asked directly if Pine meant that brands had an obligation to provide access to consumer goods, to which he confirmed:

> In a way. They owe it to us, to the public, to the community. It should be their responsibility.

This theme arose next in session two, where Maple tied dissatisfaction with consumers having to pay for environmental solutions to low trust in the companies that make environmentally sustainable goods:

> Well they lie about the product ... they dangle the products like 'you want (to be) healthy? Well you gotta pay for this' ... Well why don't we have products that are good for their environment that don't cost us an arm and a leg, so that we can maybe solve (environmental issues), hunger and homelessness at the same time?

As she explained, companies raising the price point of sustainable goods disproportionately higher than the increase in cost to make them effectively results in shifting the burden of environmental challenges onto consumers. Furthermore, in the final session, one participant (Spruce) described this burden shift when asked why he was skeptical about purchasing sustainable goods:

> It's because we're the ones who always gotta clean up the mess. Meanwhile they (executives for sustainable businesses) get to ride around and get luxuries and go where they wanna go and we live paycheck to paycheck and month to month.

This theme stood out to the research team because the burden of driving environmental changes is rarely discussed in sustainable strategic communication literature. It also stood out because it seemed to be close to the heart of unaffordable sustainable goods as an issue of environmental justice; participants described a situation in which powerful commercial entities are demanding money from them to address the harms they would otherwise cause.

It should be noted, however, that these empirical referents have their share of limitations. The most prominent limitation of these data, common among qualitative data, is that it is difficult to generalize to all low-income consumers in relatively wealthy countries. These data were collected from a single site in a major United States city, and the experiences of our participants may vary from other low-income consumers in other parts of the United States, let alone other nations. The value of these results for the concept analysis, however, is that they provide an illustrative case of how some low-income consumers may experience environmental injustice when presented with consumer goods sold at a premium due to their promoted environmental benefits. While concept analyses benefit from the inclusion of previously published empirical referents (Walker & Avant, 2018), there is no previous work to our knowledge that has examined this issue as an environmental justice issue. Thus, we argue that these data are appropriate empirical referents to support the case that the promotion of unaffordable sustainable goods to low-income consumers can be considered a concern of environmental justice.

8 Conclusion

Scholars of sustainable strategic communication often discuss sustainable products and sustainable practices by industry groups as a means of addressing environmental challenges. These products, however, are often priced higher than other similar products (Griskevicius et al., 2010; Pieters et al., 2022). Furthermore, a frequent goal of strategic communication research is to craft messaging that lets customers feel comfortable paying for this premium (Sharma, 2021). Throughout this chapter, we have discussed the issues of this reality for low-income consumers as a potential issue of environmental justice. These individuals observe strategic communications that proclaim the importance of addressing environmental issues by making the right purchases. However, consumers may not have the means to comfortably pay the premiums for environmentally-friendly consumer goods, which was supported by our empirical data.

The individuals we spoke with were largely concerned about environmental harm and particularly concerned with personally suffering the negative outcomes of environmental damage. Moreover, many of these consumers were interested in sustainable consumer goods not only because they were better for the environment, but also because they perceived them as having greater quality than other consumer goods they had access to. These individuals felt that these goods were inaccessible because of their high prices and regarded this disparity as unjust. Not only were these consumers concerned with their lack of access to these goods, but they found it unjust that the onus was placed on them, as consumers, to deal with environmental problems that they did not see as their fault. They commonly felt distrust in strategic communicators and commercial entities, seeing them as deceptive in the promotion of their sustainable products. They also thought these organizations were charging a premium that was not necessitated by an increase in production costs. Further, while facilitating the sale of premium products may be an important function of sustainable strategic communication for many commercial entities, communicating sustainable products as premium rather than accessible fuels the distrust of less economically advantaged consumers.

We made sense of these data as individuals sensing responsibilities had been put upon them by powerful entities that were profiting by offloading their sustainable obligations. Additionally, we saw a clear example of how low-income consumers can understand their lack of access to sustainable consumer choices as an environmental injustice. Commercial entities are an important stakeholder in addressing environmental challenges, and we do not mean to imply in this chapter that sustainable consumer choices are problematic. However, we would contend that commercial entities selling sustainable products, and scholars of sustainable strategic communication who support the promotion of these products, should question who can access these products and who is being asked to pay for addressing environmental damage. This is not to say that sustainable strategic communication is categorically unjust, and it is important to note that strategic communication may be employed in ways that do not promote consumption or even promote nonconsumption (e.g., Armstrong & Sekhon, 2022). The takeaway from this chapter should be that even well-intentioned commercial sustainable strategic communication may reinforce environmental injustice when used to promote products that are out of reach for many environmentally conscious consumers.

Acknowledgement We are grateful to the Waterhouse Family Institute for their financial support that made this research possible.

References

Armstrong Soule, C. A., & Sekhon, T. S. (2022). Signaling nothing: Motivating the masses with status signals that encourage anti-consumption. *Journal of Macromarketing, 42*(2), 308–325.

Atkinson, L. (2012). Buying in to social change: How private consumption choices engender concern for the collective. *The Annals of the American Academy of Political and Social Science, 644*(1), 191–206.

Bolderdijk, J. W., Steg, L., Geller, E. S., Lehman, P. K., & Postmes, T. (2013). Comparing the effectiveness of monetary versus moral motives in environmental campaigning. *Nature Climate Change, 3*(4), 413–416.

DeBastiani, S., & De Santis, J. P. (2018). Suicide lethality: A concept analysis. *Issues in Mental Health Nursing, 39*(2), 117–125.

Finisterra do Paço, A. M., Barata Raposo, M. L., & Filho, W. L. (2009). Identifying the green consumer: A segmentation study. *Journal of Targeting, Measurement and Analysis for Marketing, 17*, 17–25.

Footprint Network. (2024). How many Earths? How many countries? *Earth Overshoot Day*. https://overshoot.footprintnetwork.org/how-many-earths-or-countries-do-we-need/

Fortunatti, C. P., De Santis, J. P., & Munro, C. L. (2021). Family satisfaction in the adult intensive care unit: A concept analysis. *Advances in Nursing Science, 44*(4), 291–305.

Griskevicius, V., Tybur, J. M., & Van den Bergh, B. (2010). Going green to be seen: Status, reputation, and conspicuous conservation. *Journal of Personality and Social Psychology, 98*(3), 392.

Jaller, M., & Pahwa, A. (2020). Evaluating the environmental impacts of online shopping: A behavioral and transportation approach. *Transportation Research Part D: Transport and Environment, 80*, 102223.

Kallhoff, A. (2021). *Climate justice and collective action*. CU Boulder Catalog.

Laporte, R. L., & De Santis, J. P. (2023). Concept analysis: LGBT+ inclusive health care. *Advances in Nursing Science*. https://doi.org/10.1097/ANS.0000000000000467

Light, A., & De-Shalit, A. (Eds.). (2003). *Moral and political reasoning in environmental practice*. MIT Press.

Maibach, E. (1993). Social marketing for the environment: Using information campaigns to promote environmental awareness and behavior change. *Health Promotion International, 8*(3), 209–224.

Malin, S. A., & Ryder, S. S. (2018). Developing deeply intersectional environmental justice scholarship. *Environmental Sociology, 4*(1), 1–7.

McDermott, T. K. J. (2022). Global exposure to flood risk and poverty. *Nature Communications, 13*, 3529.

McGurty, E. M. (1997). From NIMBY to civil rights: The origins of the environmental justice movement. *Environmental History, 2*(3), 301–323.

Michalke, A., Köhler, S., Messmann, L., Thorenz, A., Tuma, A., & Gaugler, T. (2023). True cost accounting of organic and conventional food production. *Journal of Cleaner Production, 408*, 137134.

Mills, J., Bonner, A., & Francis, K. (2006). The development of constructivist grounded theory. *International Journal of Qualitative Methods, 5*(1), 25–35.

Mohai, P., Pellow, D., & Roberts, J. T. (2009). Environmental justice. *Annual Review of Environment and Resources, 34*, 405–430.

Motta, M. (2022). The pre-political origins and policy consequences of environmental justice concern. *Politics & The Life Sciences, 41*(2), 182–199.

Pellow, D. N. (2016). TOWARD A CRITICAL ENVIRONMENTAL JUSTICE STUDIES: Black lives matter as an environmental justice challenge. *Du Bois Review: Social Science Research on Race, 13*(2), 221–236.

Pieters, L., Novak, D. R., Pankratz, D., & Rogers, S. (2022). The cost of buying green. *Deloitte Insights*.

Pink, R. M. (2018). North America: Canada and the USA. In R. M. Pink (Ed.), *The climate change crisis: Solutions and adaption for a planet in peril* (pp. 15–49). Springer International.
Rawls, J. (1971). *A theory of justice.*
Sharma, A. P. (2021). Consumers' purchase behaviour and green marketing: A synthesis, review and agenda. *International Journal of Consumer Studies, 45*(6), 1217–1238.
Sroufe, R. (2017). Integration and organizational change towards sustainability. *Journal of Cleaner Production, 162*, 315–329.
Stern, N. (2008). The economics of climate change. *American Economic Review, 98*(2), 1–37.
US EPA, O. (2021, May 3). *Environmental justice in enforcement and compliance assurance [Other policies and guidance].* U.S. Environmental Protection Agency, Office of Environmental Justice. https://www.epa.gov/enforcement/environmental-justice-enforcement-and-compliance-assurance
Walker, L., & Avant, K. (2018). *Strategies for theory construction in nursing* (6th ed.). Pearson.
Wobus, C., Gutmann, E., Jones, R., Rissing, M., Mizukami, N., Lorie, M., Mahoney, H., Wood, A. W., Mills, D., & Martinich, J. (2017). Climate change impacts on flood risk and asset damages within mapped 100-year floodplains of the contiguous United States. *Natural Hazards and Earth System Sciences, 17*(12), 2199–2211.
Yue, D., Guo, J., & Hui, C. (2013). Scale dependency of biocapacity and the fallacy of unsustainable development. *Journal of Environmental Management, 126*, 13–19.

Further Reading

Bell, D. (2004). Environmental justice and Rawls' difference principle. *Environmental Ethics, 26*(3), 287–306.
Coolsaet, B., & Néron, P. Y. (2020). Recognition and environmental justice. In *Environmental justice* (pp. 52–63). Routledge.

Josh Anderson is an assistant professor in the School of Journalism in the College of Social & Behavioral Sciences at the University of Arizona.

Saima Kazmi is an assistant professor of advertising in the School of Journalism and Communication at the University of Oregon.

Na Yu is an assistant professor in the Department of Communication in the School of Humanities and Social Sciences, at the College of Charleston.

Lucy Atkinson is an associate professor in the Stan Richards School of Advertising & Public Relations at the University of Texas at Austin.

Patrick Jamar is a postdoctoral research associate in the Agricultural Leadership, Education, and Communications Department at Texas A&M University.

Jingyue Tao is an Assistant Professor in the Stan Richards School of Advertising & Public Relations at the University of Arkansas.

Organizational Sustainability Communication

Striving for Legitimacy: Strategic Energy Project Communication and Its Characteristics

Helena Stehle, Julia Lührmann, Bernadette Uth

Abstract Energy project communication is a crucial task and prerequisite for the success of energy projects, as it contributes to building legitimacy for these projects. In this chapter, we define energy project communication of corporations as a specific subfield of strategic sustainability communication. Based on a literature review, we discuss its characteristics, including communicators, stakeholders, objectives (specifically legitimacy), and implementation. We address the concept of legitimacy as a key objective of strategic energy project communication and discuss how strategic energy project communication attempts to achieve this key objective. We analyze how different types of legitimacy are addressed and which discursive legitimation strategies are of particular importance in this regard. By including literature on both strategic energy communication and legitimation communication, particularly from public relations research and related fields, our chapter brings together two strands of research that have rarely been considered together and provides theoretical implications and impetus for both strands. Our findings also have practical implications as they help communication practitioners to reflect on their communication about energy projects.

This chapter was developed as part of the research project "Energy projects in the Münsterland: Strategies of legitimacy communication and their effects," which is financially supported by the Universitätsgesellschaft Münster e. V. (https://www.uni-muenster.de/Kowi/en/forschen/projekte/energy-projects-in-the-muensterland.html). The authors are grateful for the funding of this study. They would also like to thank Johanna Heckert, M.A., for her support in researching and analyzing the literature.

H. Stehle(✉) · J. Lührmann
Department of Communication, University of Münster, Münster, Germany
e-mail: helena.stehle@uni-muenster.de; julia.luehrmann@uni-muenster.de

B. Uth
Department of Communication, University of Vienna, Vienna, Austria
e-mail: bernadette.uth@univie.ac.at

F. Weder (ed.), *Strategic Sustainability Communication*, CSR, Sustainability, Ethics & Governance, https://doi.org/10.1007/978-3-031-89486-2_6

1 Introduction

Energy projects are often accompanied by intense public debate, and corporations involved in and responsible for energy projects are faced with the need, or even the pressure, to engage in this debate (Schweizer et al., 2016; Latapí Agudelo et al., 2020; Stehle, 2022). These corporations have a high potential to contribute to sustainable development (Stjepcevic & Siksnelyte, 2017; Cozen et al., 2018) in terms of ensuring the generation and accessibility of affordable and clean energy worldwide (United Nations, 2023), and they are also held accountable for this (Mast & Stehle, 2016). At the same time, they are often under close scrutiny, for example, from activists (Fürst & Stehle, forthcoming, in this volume) or the local public, and must justify their decisions for or against an energy technology or project, or deal with accusations of greenwashing. To face such criticism, state their case, and ensure the attribution of legitimacy, they must communicate about their intentions, plans, and efforts. This is reflected, for example, in the extent to which energy suppliers address corporate social responsibility issues in their communication (Weder et al., 2019). For corporations involved in and responsible for energy projects, strategic energy project communication is a key field of activity.

Against this background, this chapter deals with strategic energy project communication by corporations involved in and responsible for energy projects. In this chapter, we model strategic energy project communication as a specific subfield of the strategic sustainability communication of corporations (Stehle, 2022). Such strategic sustainability communication is defined here as a corporation's goal-oriented, strategically planned communication about sustainability issues to support the organization's goals, especially its "license to operate" or legitimacy (Stehle, 2022). This definition has been adapted to strategic energy project communication concerning a corporation's own energy projects (Mast & Stehle, 2016; Stehle, 2022). One of the main objectives of such communication is to legitimate these specific—current and future—energy projects and the corporation itself (Mast et al., 2011). Against this background, this chapter addresses the research question of how to characterize corporate strategic energy project communication aimed at achieving legitimacy for an energy project and also for the corporation itself. Hereby, the concept of legitimacy is defined as the basic acceptance of an object (in this case, the energy project and corporation) by its environment (e.g., its stakeholders); it is the result of a process of perception, evaluation, and attribution on the part of that environment (Sandhu, 2012, 2014; Bartlett & Hurst, 2021). In the course of this chapter, the concept of legitimacy is introduced in more detail.

We address the research question based on a theoretical literature review (Paré et al., 2015), particularly from public relations research and related fields, and discuss how literature on strategic energy communication and on legitimation communication can be integrated.

The chapter is structured as follows: First, we characterize strategic energy project communication as a specific subfield of strategic sustainability communication. In particular, we discuss its different communicators, stakeholders, and contexts.

Second, we focus on legitimacy as an objective of strategic energy project communication. Third, we discuss how strategic energy project communication attempts to achieve legitimacy in its implementation.

2 Strategic Energy Project Communication and Its Characteristics

Energy projects are infrastructure projects for the generation of energy—especially renewable energy, such as wind power, biogas or solar plants—or for the transmission of energy, such as power transmission lines and grids (Stehle, 2022; Mast & Stehle, 2016). Energy project communication revolves around the object of its communication: energy projects. It "deals with and accompanies the planning, implementation, evaluation and/or reassessment of energy projects" (Stehle, 2022, p. 3). Due to the focus on *strategic* energy project communication, we specifically analyze a form of communication, which is planned, controlled, associated with an objective, and usually part of corporate communication (Mast et al., 2011; Röttger et al., 2013). This specific type of corporate communication is seen as a subfield of strategic sustainability communication as defined in the introduction.

Sustainability itself is defined as a development that "meets the needs of the present without compromising the ability of future generations to meet their own needs" (United Nations, 1987, n. p.) and is often said to consist of three interdependent dimensions: the environmental, social, and economic (Kuhlman & Farrington, 2010). Each of these three dimensions focuses on different aspects of sustainability: The environmental dimension refers to environmental quality and the protection of natural resources and ecosystems; the social dimension refers to aspects of society such as health and well-being; and the economic dimension focuses on the long-term creation and maintenance of essential goods and services (Thomsen, 2018; Geiger & Swim, 2021). At the core of the concept of sustainability is the objective of reconciling different goals from these three dimensions, which are often in conflict or at odds with each other (Thomsen, 2018).

When corporations talk about sustainability, their focus tends to be on their own sustainability efforts that are presented to stakeholders, particularly those outside the company (Brunton et al., 2017; Reilly & Larya, 2018; Golob et al., 2023). Public relations research, which predominantly explores this form of sustainability communication, especially in postpositivist studies, focuses on how organizations, especially corporations, strategically address and use sustainability efforts for their corporate purposes and goals (for an overview, see Stehle, 2022; for a discussion of the relationship between postpositivist public relations and sustainability, see Kannengießer et al., forthcoming, in this volume). Efforts in this area are also subsumed and discussed under the terms CSR or responsibility communication (Thummes, 2022; Weder & Erikson, 2023).

Strategic energy project communication can be seen as a part of the strategic sustainability communication of corporations if these energy projects are part of a corporation‘s sustainability efforts (e.g., in the case of renewable energy projects), or if the communication about a corporation’s energy projects that are not (fully) renewable emphasizes sustainability aspects (e.g., if the communication focuses on the ecological construction of power lines or on the social and economic importance of nuclear power plants for local employment). In the following, the first option is explored in particular, as renewable energy projects are one focus of attention in the climate debate.

Whereas strategic sustainability communication has been receiving attention in public relations research for quite some time, strategic energy project communication has received relatively little attention in public relations research (Mast et al., 2011; Mast & Stehle, 2016; Stehle, 2022). However, as a topic of interdisciplinary relevance, this specific communication is also addressed to varying extents in the growing field of environmental communication (Hay, 1979; Cox, 2013; Bräuer, 2017; Cozen et al., 2018; Elving, 2021) and in other disciplines, such as political science (Decker & Fleischer, 2012; Stadelmann-Steffen & Dermont, 2021) and business ethics (Ramirez, 2021). In the following, findings from public relations research as well as those from related fields of research and disciplines are included to characterize strategic energy project communication of corporations.

Strategic energy project communication is characterized by several specific features. In addition to the sensitive and specific nature of the projects themselves (Mast et al., 2011), these relate to communicators, stakeholders, objectives, especially legitimacy, and the implementation of strategic energy project communication.

Concerning the communicators of strategic energy project communication, research often focuses on corporations in the energy industry, for which energy project communication is an essential or even one of their primary tasks (Mast et al., 2011; Malynovska et al., 2022). Energy projects are mainly planned and implemented to support their objectives. These objectives, however, can vary depending on the communicator: While they are often primarily economic ones for private corporations, public companies, or corporations that are citizen-backed, for example, often have the local or regional social and political environment in mind (Stehle, 2022).

The corporations are always integrated into a complex network of other organizations and players who can potentially be involved in the planning and development of an energy project, or at least have a say in the public debate about an energy project. These include players such as government agencies, lobby associations, citizen interest groups, shareholders, and research institutions (Mast et al., 2011; Correljé et al., 2015; Mast & Stehle, 2016). Both from a research and a practice perspective, it is crucial for strategic energy project communication to identify the network of all possible players that are involved in decision-making or publicly comment on and communicate about a specific energy project. They are part of a corporation’s stakeholders and need to be considered when planning and implementing an energy project. As energy projects usually affect a wide range of stakeholders (Maqbool et al., 2020; Martinez & Komendantova, 2020), such players are not the only ones that have to be considered. Consumers and the civil society are usually the most important stakeholders of strategic energy project communication (Mast et al., 2011). They are the ones whose behavior can have far-reaching consequences for organizations that do not communicate in a sufficiently transparent and credible manner, and that have high expectations. With their interests in wider social issues and viewing themselves as part of a community and having a social responsibility, they want to be addressed as such, rather than just as customers or consumers (Mast et al., 2011).

For strategic energy project communication, understanding the points of view of all these stakeholders and addressing them appropriately are crucial for the successful completion of a project and can prevent undesirable situations (Maqbool et al., 2020). For energy projects at the local level to be successful, it can further be necessary to get stakeholders to actively participate (Kalkbrenner & Roosen, 2016).

Against this background, the importance of strategic energy project communication is strongly emphasized in research and practice, and it is important for energy companies to justify their actions in terms of sustainability goals (Santos & Silva Bastos, 2021) and make their objectives, decisions, and approaches to their energy projects transparent (Firestone et al., 2017). All steps associated with an energy project must be explained and their legitimacy ensured, as they are important for the stakeholders involved (Mast et al., 2011). One of the key objectives of strategic energy project communication is to legitimate the project and the corporation responsible for it.

3 Legitimacy as Key Objective of Strategic Energy Project Communication

Perceptions, evaluations, and attributions resulting in legitimacy—as defined in the introduction as the basic acceptance of an object by its environment (Sandhu, 2012, 2014; Bartlett & Hurst, 2021)—are collective and generalized, and judge the actions or statements of an organization or the organization itself (Sandhu, 2022). While legitimacy is often unquestioned, it comes into play when it is challenged, especially in situations of conflict or crisis (Sandhu, 2022). This illustrates the particular importance of legitimacy as a key objective of strategic energy project communication.

As described above, energy projects are often perceived as conflict-laden. Particularly, renewable energy projects are often met with public criticism and mistrust during their planning and implementation, which can be traced back to a public perception of insufficient information and communication on behalf of these projects (Hall & Jeanneret, 2015; Firestone et al., 2017). They face a lack of public acceptance or legitimacy, oftentimes resulting in protests and sometimes even the failure of the project (Kluge et al., 2021). Thus, the legitimacy of such projects and the organizations involved is seen as a prerequisite for achieving the goal of ensuring the generation and accessibility of affordable and clean energy globally (Nasiritousi & Faber, 2021). Compared to legitimacy, other traditional goals of strategic energy project communication that are linked to those of strategic sustainability communication and public relations, such as image or reputation, are therefore subordinate.

In strategic communication, legitimacy is seen not only as a basic acceptance but also as a "license to operate" (Sandhu, 2014). Such a license to operate is defined in a social sense, complementing an official license to operate based on laws. It describes an organization's privilege to be allowed to (continue to) operate on the basis of the acceptance (of the organization itself and its actions) ascribed by various stakeholders, such as the population in the project's neighborhood or the general public (Nielsen, 2013). Thus, legitimacy is also seen as an intangible success factor that is assumed to support an organization's ability to acquire resources and that determines and supports an organization's current and future scope of actions (Sandhu, 2014).

Looking to the future is both central and challenging in the energy sector: "[T]here is always a back and forth between justifying the 'old energy' production and legitimatizing 'new' energy forms together with new ways of energy sourcing" (Weder et al., 2019, p. 382). Therefore, legitimacy is seen as a prerequisite for organizations in general, but especially for organizations in the energy sector and for the planning and implementation of energy projects (Mast & Stehle, 2016).

Research on legitimacy as an objective of strategic communication categorizes and examines different types of legitimacy (Lock & Schulz-Knappe, 2019; Sandhu, 2022): pragmatic, moral, and cognitive legitimacy. Although all three types "involve a generalized perception or assumption that organizational activities are desirable, proper, or appropriate within some socially constructed system of norms, values, beliefs, and definition" (Suchman, 1995, p. 577), each type of legitimacy is characterized by a different focus.

Pragmatic legitimacy is attributed when organizations achieve their objectives and are perceived as predictable entities that meet the expectations placed on them (Suchman, 1995; Sandhu, 2014). Therefore, the goal orientation must be appropriate for the type of organization. For example, the attribution of pragmatic legitimacy to energy corporations is always based, at least in part, on the expectation that they are pursuing profit-oriented interests. This type of legitimacy can be summarized as being determined by the fulfillment of stakeholder expectations or whether "*the right requirements are being met*" (Sandhu, 2022, p. 875, emphasis in original, authors' translation).

Moral legitimacy is attributed when organizations fulfill the norms, values, and expectations of a particular culture (Sandhu, 2014, 2022). Thus, this legitimacy type is about "the right thing to do" (Suchman, 1995, p. 579) in terms of fulfilling the norms and values of the particular society in which an organization operates. For example, to be considered morally legitimate, energy corporations need to consider and respond to the moral demands of local politicians or environmental activists.

Cognitive legitimacy refers to “fundamental beliefs that cannot be questioned” (Sandhu, 2014, p. 1171, authors’ translation). Cognitive legitimacy is achieved when an organization’s right to exist is generally accepted by society, or when it appears logical and self-evident that the organization exists (Lock & Schulz-Knappe, 2019; Sandhu, 2022): “for things to be otherwise is *literally* unthinkable” (Suchman, 1995, p. 583, emphasis in original). Accordingly, for cognitive legitimacy, the organization has to be in line with cultural, non-value-based patterns of interpretation, or those of the organization and those of the stakeholders must be consistent with each other (Lock & Schulz-Knappe, 2019; Sandhu, 2022). For energy corporations, cognitive legitimacy means, for example, that they are seen as a central part of a society and that it is recognized, for instance, that energy production also earns (private) money. Cognitive legitimacy is a necessary precondition for achieving pragmatic legitimacy (Lock & Schulz-Knappe, 2019).

Since the three types of legitimacy occur simultaneously, are interdependent, and develop in a complex process of perception and evaluation (Lock & Schulz-Knappe, 2019), strategic energy project communication must take all three into account, not only when setting its objectives but also in implementation.

4 Implementing Strategic Energy Project Communication to Achieve Legitimacy

Not only is the implementation of renewable energy projects complex, but so is the strategic communication of said projects. Depending on the situation, it can be a matter of gaining legitimacy (e.g., for a new energy project), maintaining legitimacy (e.g., for an existing, well-known energy project), or repairing legitimacy, for example, when a corporation is in conflict with its stakeholders (Sandhu, 2014). To achieve this, the implementation of strategic communication is assigned an important role. In the literature, for example, campaigns (Schlichting, 2014), techniques, channels, and communicators are analyzed (e.g., Schlichting, 2014; Mast & Stehle, 2016; Hobbs, 2020). In addition, communicative legitimation strategies can be examined and analyzed.

Such strategies are starting points and approaches that corporations use to influence the attribution of legitimacy by their environment and stakeholders. Research differentiates between different approaches depending on the necessary objective: there are approaches for gaining, maintaining, or restoring legitimacy, each for pragmatic, moral, and cognitive legitimacy (Sandhu, 2014, 2022; Lock & Schulz-Knappe, 2019; Crumley et al., 2022). In order to gain pragmatic legitimacy, for example, an image campaign could be launched for an energy project that emphasizes how the project meets stakeholder expectations for climate protection. To maintain such legitimacy, an energy corporation could strive for a consistent and transparent process communication that uses many communication channels. To repair pragmatic legitimacy, strategic energy project communication can be used to downplay or reject stakeholder expectations. To gain moral legitimacy, for example, strategic communication can use techniques such as framing or storytelling to convince stakeholders that an energy project is aligned with societal norms and values. To maintain this type of legitimacy, an energy corporation can engage in discussions about the norms and values of its environment, for example through media interviews or participation in public events such as roundtables; and in times of crisis (e.g., following an accident at a biogas plant) a public apology can serve to restore moral legitimacy. To gain cognitive legitimacy, energy corporations can build on existing background beliefs about their activities and recall them in their strategic communication about an energy project. This also applies when such background beliefs change, as was, for example, the case in Germany with the political "Energiewende" (energy transition) after the Fukushima catastrophe. Introducing new background beliefs through strategic communication is considered to be much more complex and time-consuming. Instruments such as lobbying can be used for this purpose. In addition, the literature points out that organizations can maintain their cognitive legitimacy through communicating with a high degree of clarity and comprehensibility, and they must explain themselves and their actions when it comes to restoring this legitimacy (Sandhu, 2014, 2022).

In the literature on legitimation communication, particular attention is also paid to discursive legitimation strategies (Sandhu, 2014, 2022). These strategies characterize linguistic patterns that are used in strategic communication (e.g., in verbal statements or publications such as annual reports, press releases, or social media posts) to gain, maintain, or restore the attribution of legitimacy. These strategies are closely related to the three legitimacy types and are often classified as three ideal-typical main strategies (authorization, rationalization, and moralization) and narration as a combined strategy (Sandhu, 2012, 2014). The authorization strategy is characterized by the usage of habit (traditional or habitual practices), authority (personal or impersonal), or recommendation (incorporation of experts or role models) and relates to pragmatic legitimacy (Van Leeuwen, 2007; Sandhu, 2014). Applied to strategic energy project communication, such communication can, for example, point out that a certain activity has always been a good tradition, or allow technical experts to have their say as important voices. The rationalization strategy is related to pragmatic and cognitive legitimacy and is about linking an organization's objectives to its actions and results achieved (Van Leeuwen, 2007; Sandhu, 2014). This can be done by emphasizing objectives and results, or by using definitions and forecasts. For example, in the development and implementation of energy projects, an energy corporation can communicate its objectives and the results it wants to see in terms of climate protection. (Positive) projections of these desired organizational and societal outcomes can also be used to legitimate an energy project. The moralization strategy is associated with moral legitimacy and aims to achieve dominance in a dispute over whose interpretation is the correct one (Van Leeuwen, 2007; Sandhu, 2014). It uses linguistic patterns such as evaluation, generalization, and comparison. For an evaluation, adjectives with certain valuations are used and these are associated with objects or actions; generalization links actions to values or norms unrelated to the given situation, so that the focus is on general values and norms; and in the case of comparisons, metaphors or figures are applied for illustrative purposes (Van Leeuwen, 2007; Sandhu, 2014). To legitimate an energy project, corporations can, for example, evaluate their own actions (e.g., as the only possible way), generalize the importance of renewable energy, or compare themselves with an industry that has addressed and supposedly solved a major societal problem. The mixed strategy of narration can refer to any legitimacy type, depending on the situation, and can focus on different patterns such as personalization or dramatization (Sandhu, 2012, 2014). In their strategic communication about an energy project, corporations can apply this strategy, for example, when they tell a story about their CEO or another employee who is particularly interested in the project. Events, for example, from the planning history of the project, or messages can be dramatized to create an impact.

In their strategic communication, energy companies can draw on various strategies and techniques, as well as communication channels and instruments, which can be managed and optimized down to the linguistic level, to legitimate a specific energy project. Using communication as a component of symbolic action, they attempt to influence existing stakeholder expectations and perceptions regarding the specific project and renewable energy as a whole, or work toward the development of new expectations and perceptions. However, there are still some gaps and open questions in the research, such as the limits of this communication.

5 Discussion and Conclusion

This chapter focused on the question of how to characterize corporate strategic energy project communication aimed at achieving legitimacy for renewable energy projects. This question was addressed through a literature review and discussion of the literature on strategic energy communication and on legitimation communication, particularly from public relations research and related fields. The chapter thus brings together two strands of research that have rarely been considered together and provides impetus for both strands of research. This is of importance as there are only a few studies on strategic energy project communication in public relations research.

First, we highlighted the particularities and challenges of strategic communication about energy projects (as a subfield of strategic sustainability communication), discussing the corporations that apply such communication and the diversity of their stakeholders. Second, we showed that the definition of legitimacy as a key objective of strategic communication can be transferred to the context of energy projects, where it refers to the basic acceptance of an energy project and the corporation involved or responsible. We further analyzed corporate energy project communication by, for example, differentiating and typifying legitimacy into pragmatic, moral, and cognitive legitimacy. Third, we took a closer look at the implementation of strategic energy project communication (e.g., recurring linguistic patterns such as the discursive legitimation strategies authorization, rationalization, moralization, and narration).

These findings and the integration of both research on strategic energy project communication and research on legitimation communication can help communication practitioners to reflect on their communication about energy projects (e.g., their own linguistic patterns) and possibly to plan it better (e.g., in matching the linguistic patterns used with the desired types of legitimacy). Foremost, however, they advance our understanding of energy project communication as one significant research field of strategic sustainability communication. As research on strategic sustainability and energy communication, on the one hand, and research on legitimacy, on the other, are traditionally located in different strands and paradigms of public relations research, an integrated view can offer new insights. While, for example, the analysis of strategic sustainability and energy communication is still dominated by the postpositivist perspective of public relations research and is often expressed in a communicator-centered approach, the research on legitimacy is rooted in the interpretivist paradigm. The interpretivist paradigm emphasizes subjective and individual views, such as those of stakeholders. Our analysis offers insights for an integration of both research strands as an analysis from a postpositivist perspective should at least keep an eye on potential discrepancies between considering legitimacy as an organization's objective and as an attribute that is granted or not granted by stakeholders, and thus also on the limits of strategic energy project communication. The impact and especially the attribution of legitimacy depend on the expectations and perceptions of stakeholders. Here, we see starting points for future research, which could shed further light on the so far only scarcely analyzed stakeholder perspectives on energy project communication and point out contrasts or overlaps with corporate interests (for an overview on research on this, see Stehle, 2022).

Numerous questions also remain unanswered from a communicator-centered perspective. Future research could, for example, empirically analyze how strategic energy project communication is implemented in organizational practice in order to achieve legitimacy for an energy project. Content analyses could, for example, investigate which legitimation strategies and types of legitimacy are prevalent within such communication. This would add to the present literature-based approach to this question.

In addition to future empirical research, the theoretical and literature-based examination of the topic could also be expanded. This chapter focuses on legitimacy as a key objective of strategic energy communication. Other objectives and their relations to legitimacy, such as trust and reputation, could also be considered more closely within this context.

Other open questions that could be addressed by future theoretical and empirical research include focusing on specific renewable energy projects (e.g., solar versus wind power plants), comparing renewable energy projects with fossil ones (e.g., gas-fired power plants), and analyzing energy storage or conservation projects (e.g., dams) in contrast. As strategic energy project communication is not only carried out by energy companies, other types of organizations could also be explored (Mast et al., 2011; Malynovska et al., 2022).

Depending on the type of energy project, different actors and actor networks can emerge, the characteristics and consequences of which for strategic energy project communication still need to be explored. Furthermore, questions of where and why challenges can arise need to be examine in more detail (e.g., because other organizations engage in strategic delegitimation communication (Fürst & Stehle, forthcoming, in this volume) or because there are accusations of discrepancies between communication and the sustainability of the energy project). There has also been little research on the relationship between the legitimacy of an energy project and that of the responsible organization: What are the consequences if they refer to different dominant types of legitimacy or if the organization and the energy project are even opposed to each other in terms of attributed legitimacy? These and other questions are worthy of further investigation, as energy projects will continue to be central to sustainable societal development, and their planning and implementation will continue to be controversially discussed.

References

Bartlett, J., & Hurst, B. (2021). Public relations and legitimacy. In C. Valentini (Ed.), *Public relations* (pp. 107–122). De Gruyter. https://doi.org/10.1515/9783110554250-006

Bräuer, M. (2017). *Energiewende und Bürgerproteste: Eine Untersuchung der Kommunikation von Bürgerinitiativen im Themenfeld Netzausbau [Energy system transition and protests: A study of the communication of protestors against the extension of the German high voltage power grid]*. Universitätsverlag Ilmenau. https://nbn-resolving.org/urn:nbn:de:gbv:ilm1-2016000757

Brunton, M., Eweje, G., & Taskin, N. (2017). Communicating corporate social responsibility to internal stakeholders: Walking the walk or just talking the talk? *Business Strategy and the Environment, 26*(1), 31–48. https://doi.org/10.1002/bse.1889

Correljé, A., et al. (2015). Responsible innovation in energy projects: Values in the design of technologies, institutions and stakeholder interactions. In B.-J. Koops et al. (Eds.), *Responsible innovation 2* (pp. 183–200). Springer. https://doi.org/10.1007/978-3-319-17308-5_10

Cox, R. (2013). *Environmental communication and the public sphere*. Sage.

Cozen, B., et al. (2018). Energy communication: Theory and praxis towards a sustainable energy future. *Environmental Communication, 12*(3), 289–294. https://doi.org/10.1080/17524032.2017.1398176

Crumley, E. T., et al. (2022). Media interviews as strategic external communication to maintain legitimacy for sustainability activities. *Corporate Communications: An International Journal, 27*(1), 148–166. https://doi.org/10.1108/CCIJ-05-2021-0053

Decker, M., & Fleischer, T. (2012). Participation in "big style": First observations at the German citizens' dialogue on future technologies. *Poiesis & Praxis, 9*, 81–99. https://doi.org/10.1007/s10202-012-0119-0

Elving, W. J. L. (2021). Energy and sustainable communication. In F. Weder, L. Krainer, & M. Karmasin (Eds.), *The sustainability communication reader: A reflective compendium* (pp. 397–417). Springer. https://doi.org/10.1007/978-3-658-31883-3_22

Firestone, J., et al. (2017). Reconsidering barriers to wind power projects: Community engagement, developer transparency and place. *Journal of Environmental Policy & Planning, 20*(3), 370–386. https://doi.org/10.1080/1523908X.2017.1418656

Fürst, S., & Stehle, H. (forthcoming). How civil society actors employ strategic communication to delegitimate corporations in public debates on sustainability. In F. Weder (Ed.), *Strategic sustainability communication*. Springer.

Geiger, N., & Swim, J. K. (2021). Understanding lay individuals' mental models of sustainability. In F. Weder, L. Krainer, & M. Karmasin (Eds.), *The sustainability communication reader: A reflective compendium* (pp. 301–321). Springer. https://doi.org/10.1007/978-3-658-31883-3_17

Golob, U., Podnar, K., & Zabkar, V. (2023). Sustainability communication. *International Journal of Advertising, 42*(1), 42–51. https://doi.org/10.1080/02650487.2022.2144035

Hall, N. L., & Jeanneret, T. (2015). Social licence to operate: An opportunity to enhance CSR for deeper communication and engagement. *Corporate Communications: An International Journal, 20*(2), 213–227. https://doi.org/10.1108/CCIJ-01-2014-0005

Hay, K. G. (1979). Environmental communications and energy - A look around. *The Journal of Environmental Education, 11*(1), 30–32. https://doi.org/10.1080/00958964.1979.9941353

Hobbs, M. J. (2020). Conflict ecology: Examining the strategies and rationales of lobbyists in the mining and energy industries in Australia. *Public Relations Review, 46*(2), 101868. https://doi.org/10.1016/j.pubrev.2019.101868

Kalkbrenner, B. J., & Roosen, J. (2016). Citizens' willingness to participate in local renewable energy projects: The role of community and trust in Germany. *Energy Research & Social Science, 13*, 60–70. https://doi.org/10.1016/j.erss.2015.12.006

Kannengießer, S., et al. (forthcoming). Public relations and sustainability: Points of friction in a complex relationship. In F. Weder (Ed.), *Strategic sustainability communication*. Springer.

Kluge, J., Offermann-van Heek, J., & Ziefle, M. (2021). How will I know about local energy projects? A conjoint-analysis of communication preferences in Germany. *Energy Research & Social Science, 75*, 102016. https://doi.org/10.1016/j.erss.2021.102016

Kuhlman, T., & Farrington, J. (2010). What is sustainability? *Sustainability, 2*(11), 3436–3448. https://doi.org/10.3390/su2113436

Latapí Agudelo, M. A., Johannsdottir, L., & Davidsdottir, B. (2020). Drivers that motivate energy companies to be responsible: A systematic literature review of corporate social responsibility in the energy sector. *Journal of Cleaner Production, 247*, 119094. https://doi.org/10.1016/j.jclepro.2019.119094

Lock, I., & Schulz-Knappe, C. (2019). Credible corporate social responsibility (CSR) communication predicts legitimacy: Evidence from an experimental study. *Corporate Communications: An International Journal, 24*(1), 2–20. https://doi.org/10.1108/CCIJ-07-2018-0071

Malynovska, Y., et al. (2022). Enhancing the activity of amployees of the communication department of an energy sector company. *Energies, 15*(13), 4701. https://doi.org/10.3390/en15134701

Maqbool, R., Deng, X., & Rashid, Y. (2020). Stakeholders' satisfaction as a key determinant of critical success factors in renewable energy projects. *Energy, Sustainability and Society, 10*(1), 1–15. https://doi.org/10.1186/s13705-020-00259-0

Martinez, N., & Komendantova, N. (2020). The effectiveness of the social impact assessment (SIA) in energy transition management: Stakeholders' insights from renewable energy projects in Mexico. *Energy Policy, 145*, 111744. https://doi.org/10.1016/j.enpol.2020.111744

Mast, C., & Stehle, H. (2016). *Energieprojekte im öffentlichen Diskurs: Erwartungen und Themeninteressen der Bevölkerung [Energy projects in public discourse: Expectations and thematic interests of the general public]*. Springer. https://doi.org/10.1007/978-3-658-12711-4

Mast, C., Stehle, H., & Krüger, F. (2011). *Kommunikationsfeld Strom, Gas und Wasser – brisante Zukunftsthemen in der öffentlichen Diskussion [Electricity, gas, and water as a field of communication—Controversial issues in public discussion]*. LIT.

Nasiritousi, N., & Faber, H. (2021). Legitimacy under institutional complexity: Mapping stakeholder perceptions of legitimate institutions and their sources of legitimacy in global renewable energy governance. *Review of International Studies, 47*(3), 377–398. https://doi.org/10.1017/S0260210520000431

Nielsen, A. E. (2013). License to operate. In S. O. Idowu et al. (Eds.), *Encyclopedia of Corporate Social Responsibility* (pp. 1585–1591). Springer. https://doi.org/10.1007/978-3-642-28036-8_502

Paré, G., et al. (2015). Synthesizing information systems knowledge: A typology of literature reviews. *Information & Management, 52*(2), 183–199. https://doi.org/10.1016/j.im.2014.08.008

Ramirez, J. (2021). Contentious dynamics within the social turbulence of environmental (in)justice surrounding wind energy farms in Oaxaca, Mexico. *Journal of Business Ethics, 169*(3), 387–404. https://doi.org/10.1007/s10551-019-04297-3

Reilly, A. H., & Larya, N. (2018). External communication about sustainability: Corporate social responsibility reports and social media activity. *Environmental Communication, 12*(5), 621–637. https://doi.org/10.1080/17524032.2018.1424009

Röttger, U., Gehrau, V., & Preusse, J. (2013). Strategische Kommunikation: Umrisse und Perspektiven eines Forschungsfeldes [Strategic communication: Outlines and perspectives of the research field]. In U. Röttger, V. Gehrau, & J. Preusse (Eds.). (2013). *Strategische Kommunikation: Umrisse und Perspektiven eines Forschungsfeldes* (pp. 9–17). Springer. https://doi.org/10.1007/978-3-658-00409-5

Sandhu, S. (2012). *Public Relations und Legitimität: Der Beitrag des organisationalen Neo-Institutionalismus für die PR-Forschung [Public relations and legitimacy: The contribution of organizational neo-institutionalism to public relations research]*. Springer. https://doi.org/10.1007/978-3-531-19304-5

Sandhu, S. (2014). Public Relations und gesellschaftliche Kommunikation: Legitimation im Diskurs [Public Relations and societal communication: Legitimation in discourse]. In A. Zerfaß & M. Piwinger (Eds.), *Handbuch Unternehmenskommunikation: Strategie, Management, Wertschöpfung* (pp. 1161–1183). Springer. https://doi.org/10.1007/978-3-8349-4543-3_58

Sandhu, S. (2022). Gesellschaftsorientierte Unternehmenskommunikation: Stakeholderorientierung und Legitimation als Ziel der Public Relations [Societal corporate communication: Stakeholder orientation and legitimation as public relations objectives]. In A. Zerfaß, M. Piwinger, & U. Röttger (Eds.), *Handbuch Unternehmenskommunikation: Strategie – Management – Wertschöpfung* (3rd ed., pp. 859–880). Springer. https://doi.org/10.1007/978-3-658-03894-6_38

Santos, M. J., & Silva Bastos, C. (2021). The adoption of sustainable development goals by large Portuguese companies. *Social Responsibility Journal, 17*(8), 1079–1099. https://doi.org/10.1108/srj-07-2018-0184

Schlichting, I. (2014). Consumer campaigns in corporate public affairs management: The case of climate change and the German energy industry. *Journal of Communication Management, 18*(4), 402–421. https://doi.org/10.1108/jcom-02-2011-0020

Schweizer, P.-J., et al. (2016). Public participation for infrastructure planning in the context of the German "Energiewende". *Utilities Policy, 43*(Part B), 206–209. https://doi.org/10.1016/j.jup.2014.07.005

Stadelmann-Steffen, I., & Dermont, C. (2021). Acceptance through inclusion? Political and economic participation and the acceptance of local renewable energy projects in Switzerland. *Energy Research & Social Science, 71*, 101818. https://doi.org/10.1016/j.erss.2020.101818

Stehle, H. (2022). *Citizens' perceptions of dialogue: A Q study on the communication of energy projects.* Conference paper, Annual Conference, European Public Relations Education and Research Association (EUPRERA), Vienna.

Stjepcevic, J., & Siksnelyte, I. (2017). Corporate Social Responsibility in energy sector. *Transformations in Business & Economics, 16*(1), 21–33.

Suchman, M. C. (1995). Managing legitimacy: Strategic and institutional approaches. *The Academy of Management Review, 20*(3), 571–610. https://doi.org/10.2307/258788

Thomsen, C. (2018). Sustainability. In R. L. Heath & W. Johansen (Eds.), *The international encyclopedia of strategic communication* (pp. 1–9). Wiley. https://doi.org/10.1002/9781119010722.iesc0188

Thummes, K. (2022). CSR-Kommunikation von Unternehmen: Grundlagen der Verantwortungskommunikation [Corporate CSR communication: Responsibility communication and its basics]. In A. Zerfaß, M. Piwinger, & U. Röttger (Eds.), *Handbuch Unternehmenskommunikation: Strategie – Management – Wertschöpfung* (3rd ed., pp. 1023–1038). Springer. https://doi.org/10.1007/978-3-658-22933-7_44

United Nations. (1987). *Report of the world commission on environment and development: Our common future*. http://www.un-documents.net/wced-ocf.htm

United Nations. (2023). *The sustainable development goals report.* https://unstats.un.org/sdgs/report/2023/The-Sustainable-Development-Goals-Report-2023.pdf

Van Leeuwen, T. (2007). Legitimation in discourse and communication. *Discourse & Communication, 1*(1), 91–112. https://doi.org/10.1177/1750481307071986

Weder, F., & Erikson, M. (2023). Sustainability, CSR, ESG and more: Terminology clean up. In F. Weder & M. Erikson (Eds.), *CSR communication and cultures of sustainability* (pp. 12–22). The University of Queensland. https://doi.org/10.14264/428d81b

Weder, F., Koinig, I., & Voci, D. (2019). Antagonistic framing of sustainability by energy suppliers: Dissecting corporate CSR messages in a cross-cultural comparison. *Corporate Communications: An International Journal, 24*(2), 368–390. https://doi.org/10.1108/CCIJ-01-2018-0014

Further Reading

Elving, W. J. L. (2021). Energy and sustainable communication. In F. Weder, L. Krainer, & M. Karmasin (Eds.), *The sustainability communication reader: A reflective compendium* (pp. 397–417). Springer. https://doi.org/10.1007/978-3-658-31883-3

Weder, F., Koinig, I., & Voci, D. (2019). Antagonistic framing of sustainability by energy suppliers: Dissecting corporate CSR messages in a cross-cultural comparison. *Corporate Communications: An International Journal, 24*(2), 368–390. https://doi.org/10.1108/CCIJ-01-2018-0014

Julia Lührmann, M.A., is a PhD candidate at the University of Münster. In her research, she focuses on strategic and organizational communication, especially internal communication, and energy communication. https://www.uni-muenster.de/Kowi/en/personen/julia-luehrmann.shtml

Bernadette Uth is a postdoctoral researcher at the University of Vienna. In her research, she focuses on trust research, journalism studies, and energy communication. https://publizistik.univie.ac.at/en/department/staff/postdocs/uth-bernadette-sophie/

Helena Stehle is a professor of communication science with a focus on strategic communication at the University of Münster. Her research interests include strategic and organizational communication, and sustainability and energy communication. https://www.uni-muenster.de/Kowi/en/personen/helena-stehle.shtml

Considerations for Sustainable Employee Engagement as Enablers for Communication About Sustainability

Yolandi Botha and Lauren O'Mahony

Abstract Strategic sustainability communication has the potential to acculturate internal and external organisational contexts. Strategic sustainability communication could broaden societal values and social learning while building collaborative networks toward societal, environmental, and economic well-being. Arguably, internal and organisational communication is the starting point to convey and achieve an organisation's sustainability messages and goals. Thus, this chapter takes a two-pronged approach to strategic sustainability communication from an organisational perspective. Firstly, as we argue, a necessary precondition for communication for sustainability is several considerations for sustainable employee engagement. These considerations are grounded in an approach that allows for reflectivity and encourages radical leadership thinking towards a decentralised leadership approach. Such an approach may be facilitated by the strategic communication professional as the broker of collaboration. This, arguably, contributes to the establishment of an innovation culture for curiosity-driven knowledge within an organisation. Such a culture has the potential to build a transdisciplinary response to complex problems. Secondly, these considerations for sustainable employee engagement collectively serve as enablers for communication about sustainability, addressing the sustainability issues and goals of the organisation while encouraging a desire within employees to extend these sustainability practices outside the organisation.

1 Introduction and Contextualisation

In recent years, increasing numbers of organisations have sought to integrate sustainability practices and initiatives in their day-to-day operations. The introduction of the United Nations Sustainable Development Goals in 2012 (replacing the

Y. Botha (✉) · L. O'Mahony
School of Media and Communication, College of Law, Arts and Social Sciences, Murdoch University, Perth, WA, Australia
e-mail: Yolandi.Botha@murdoch.edu.au; L.OMahony@murdoch.edu.au

F. Weder (ed.), *Strategic Sustainability Communication*, CSR, Sustainability, Ethics & Governance, https://doi.org/10.1007/978-3-031-89486-2_7

Millennium Development Goals) aimed to "produce a set of universal goals [to] meet the urgent environmental, political and economic challenges facing our world" (UNDP, 2023a). While 193 countries have aimed to achieve the SDGs by 2030, as the UNDP argues, "achieving these goals require[s] collective action across governments, civil society, [the] private sector and dedicated individuals and communities and need to be matched with the necessary resources, innovation capacity and partnerships to drive implementation" (2023a). Thus, to achieve these ambitious targets, organisations, large and small, from the public, private and not-for-profit sectors, as well as communities and individuals must all engage with and aim for the SDG targets. As the UNDP (drawing on the OECD) states, "private sector operations constitute, [on] average, 60 percent of GDP, while generating 90 percent of jobs and 80 percent of capital inflows" (2023b). Talal Rafi writing for the World Economic Forum argues that sustainability should be considered as a "corporate strategy" that is "critical" for organisations that intend to "remain relevant and competitive in today's world" (Rafi, 2022). There are several drivers of sustainability in organisations, including pressure from investors and consumers and demands associated with regulations to attract talent and highly skilled workers and to increase productivity (Rafi, 2022). Moreover, the UNDP lists five drivers of financial value that can flow from businesses integrating the SDGs: "new revenue by creating new opportunities", increased "employer attractiveness", "increased supply chain resilience", increased investor interest and proactive measures to address likely regulatory requirements and risks (UNDP, 2023b). Alignment with the SDGs and sustainability measures, therefore, promises much for organisations, especially those in the private sector.

While much attention is directed towards externally facing sustainability initiatives by organisations, seemingly less attention is directed towards the process of embedding sustainability awareness, practices, and initiatives internally with organisational employees, let alone on the role of communication in facilitating that process. This chapter therefore discusses sustainable employee engagement (communication for sustainability) as a platform to communicate about sustainability. Our chapter is positioned from a strategic sustainability communication perspective. Sustainable employee engagement, as we explain below, should be approached with several key considerations in mind. These considerations can serve as enablers for communication about sustainability, in particular, an organisation's sustainability goals whereby employees exercise a sustainability mindset in their work context and, ideally, beyond. Firstly, we discuss strategic sustainability communication and explain the role of strategic communications professionals and, importantly, their role in facilitating employee collaboration within an organisation. Secondly, we explore what is meant by employee engagement, primarily in the context of strategic sustainability communication. We use this understanding of employee engagement to explain the key considerations we have identified as part of the process toward communication about sustainability including using a Communicative Constitution of Organisation (CCO) approach, reflective and radical leadership thinking, holacracy, developing a culture of innovation, curiosity-driven knowledge and the role of the strategic communication professional. As we argue, such considerations and an organisation's ability to demonstrate them in practice may lead to more effective communication about sustainability. An organisation's internal communication about sustainability has the potential to align with the organisation's outward-facing sustainability messaging. We unpack why such messaging internally may be important to employees, their broader organisation and ultimately beyond.

2 Strategic Sustainability Communication

The growing mediatisation of society, reinforced by the explosion of social media and digital technologies, supports increasing user interactivity and content creation (Zerfass & Link, 2022). In this milieu of co-creation and information sharing, strategic communication acts as "a transdisciplinary, holistic and inclusive field of knowledge" (Heide et al., 2018: 452) that is in continuous dialogue with happenings in society and organisations (Falkheimer & Heide, 2022). More specifically, strategic communication is "the practice of deliberate and purposive communication" to reach set goals (Holtzhausen et al., 2021: 4) while being "practice-based, relational and collaboratively constructed" (McGreavy et al., 2022: 2).

Strategic communication professionals often play a pivotal role in engaging in stakeholder conversations to foster transdisciplinary collaborations. Such engagement aims to meaningfully address complex environmental, social, and economic issues—the cornerstones of sustainability (McGreavy et al., 2022). Moreover, sustainability, especially managing stakeholder expectations related to sustainability issues, was listed as one of the top three issues for strategic communication in the 2023 European Communication Monitor (ECM, 2023). This, in essence, places a renewed emphasis on sustainability communication for organisations and communication professionals. From a broad perspective, sustainability communication is a global social process consisting of a "recursive sequence of contributions and debates" to improve the economy, the environment and society (Goleb et al., 2023: 42). Communication therefore helps to navigate the complex and ambiguous nature of sustainability towards the implementation of sustainability goals (Newig et al., 2013).

There are various perspectives or modes of sustainability communication. These perspectives are best differentiated as communication about, for and of sustainability (Newig et al., 2013; Fischer et al., 2016; Goleb et al., 2023; Saifudin, 2022; Weder et al., 2021; Weder, 2023). Communication *about* matters pertaining to sustainability is generally part of public discourse to construct meaning. Comparatively, communication *of* sustainability is instrumental or managerial and is focused on the one-way flow of communication to address a specific objective. It should be noted that communication *of* sustainability is a sender-orientated process aimed at persuading the recipient whereas communication *about* sustainability is discourse-orientated (Saifudin, 2022). Lastly, communication *for* sustainability is a call to action aimed at facilitating transformation towards the normative goals of sustainable development.

Sustainability communication is complex, diverse and multi-discursive in nature. Goleb et al., (2023: 43) argue that "[j]ust as sustainability is an inter- or transdisciplinary phenomenon, sustainability communication also lies at the intersection of different fields or disciplines". This chapter argues, however, that for organisations to truly be sustainable to contribute to improved sociological, economic and social life, such improvements need to be undertaken from a strategic communication perspective. However, how could strategic communication, as a transdisciplinary, holistic and inclusive field of knowledge, broaden societal values and increase social learning towards the implementation of complex and diverse sustainability goals? McGreavy et al. (2022) contend that this could be done by incorporating an interdisciplinary and rhetorical approach to strategic communication, one that emphasises the need to be mindful of the "connections between language, knowledge and power" in communication. McGreavy et al.'s (2022) approach resonates with the Communicative Constitution of Organisation (CCO) approach, contributing to situated and socially constructed meaning-making (Holtzhausen & Zerfass, 2015). It should be noted that integrating the principles of the CCO approach aims to promote a critical, meta-modern perspective to strategic communication and does not accept strategic communication as a functionalist, linear process (modernist perspective). Instead, strategic communication is viewed as a pluralistic approach that is contextually sensitive, drawing from both critical and functional perspectives.

To scaffold the kinds of organisational conditions that would support communication about sustainability, it is first important to define strategic sustainability communication. Strategic sustainability communication, from an organisational perspective, can be defined as *the means for organisations to create collaborative stakeholder networks of diverse expertise to facilitate interactivity, dissent, negotiation and debate to broaden societal values and social learning to identify, attain and maintain shared sustainability goals* for the continuous advancement of society. This definition implies that strategic sustainability communication:

- Integrates diverse areas of knowledge, skills and capabilities (is interdisciplinary) as well as different cultures and contexts (is transdisciplinary);
- Is bottom-up, promoting stakeholder inclusivity and engagement;
- Is multi-faceted by incorporating both emergent and planned strategies; and
- Facilitates and realises an organisation's commitment to contribute to economic, environmental and societal well-being while meeting its financial objectives.

This chapter now builds on this definition of strategic sustainability communication from an organisational and internal communication perspective to explore sustainable employee engagement as an enabler for communication about sustainability.

3 Strategic Sustainability Communication in Organisational and Internal Communication Contexts

Terminology regarding communication within and about an organisation is often convoluted and ambiguous. Internal communication is used interchangeably with concepts such as organisational communication, corporate communication, and employee communication (Tkalac Verčič et al., 2023). Organisational communication is defined as "the process of creating and negotiating collective, coordinated systems of meaning through symbolic practices oriented toward the achievement of organizational goals" (Mumby, 2018: 12). It is, however, important to note that organisational communication includes the organisation's processes and a consideration of the organisation's internal and external contexts (Tkalac Verčič et al., 2023). Organisational communication is therefore concerned with communication in terms of both internal and external stakeholders. Comparatively, corporate communication is communication initiated by an organisation's corporate communication department (Men, 2021). Internal communication is a strategic communication process that seeks to attain specified internal communication goals to contribute to the organisation's success; employee communication is a sub-category of internal communication and an example of peer or horizontal communication (Men, 2021). Effective internal communication "contributes to organizational commitment, employee identification with the organisation [and] employee ability, motivation, and commitment" as well as enhanced productivity (Tkalac Verĉiĉ & Men, 2023: 3). Moreover, when employees are well-informed, they may better serve as organisational ambassadors or advocates (Tkalac Verčič et al., 2023). Employees are carriers of the organisation's authentic voice (Men, 2021) and can shape external stakeholders' perceptions of that organisation. This chapter is concerned with both internal and organisational communication. Initially, the focus is on employee engagement facilitated by internal communication, building towards a broader sustainability perspective typified by a transdisciplinary response. A broader sustainability perspective necessitates an outside-in orientation, and consideration of the organisational context (internal and external to the organisation), structures, and processes as part of the organisational communication process.

Sustainability research in an organisational context is, however, often orientated towards external stakeholders despite the indispensable role that employees play in generating awareness of and conveying the organisation's sustainability messages (Brunton et al., 2017). Based on the view that internal communication is the starting point to achieve an organisation's sustainability goals (Barendson et al., 2021), this chapter proposes a twofold exploration of sustainability, encompassing both communication for and about sustainability. Firstly, considerations for sustainable employee engagement are proposed, which, in the context of this chapter, represent communication *for* sustainability (Newig et al., 2013; Saifudin, 2022). Secondly, considering a broader organisational communication perspective, these considerations could, conversely, serve as enablers for communication *about* sustainability (Saifudin, 2022).

3.1 *Employee Engagement: Communication* for *Sustainability*

3.1.1 Defining Employee Engagement

Employees in the context of this chapter include workers of all position levels, including top management, middle management, and subordinate (Arif et al., 2023). In line with the definition and discussion of internal communication in the previous section, employee engagement (also known as work/job engagement) (Auh et al., 2016; Chen et al., 2020) is affected (Ruck & Welch, 2012) or shaped (Tkalac Verĉiĉ, 2021) by internal communication. A prerequisite for employee engagement is that internal communication should facilitate the co-creation of meaning and relationship-building between employees and employee commitment to elicit advocative behaviours (Arif et al., 2023; Morehouse & Saffer, 2023; Botan, 2018). From this view, engagement manifests through interaction, exchange, and collaboration between the communicative parties (Dhanesh, 2017; Johnston & Taylor, 2018). Most importantly, engagement is regarded as "communicative action" between an organisation and its stakeholders with an "emphasis on listening, dialogue and openness" (Tkalac Verĉiĉ & Men, 2023: 3). In the context of this chapter, this implies that employee engagement allows employees to take part in organisational decision-making and problem-solving. Employees' input is valued which results in employee enablement and empowerment.

Since engagement is ambiguous and there is no clear definition of what it entails, it is often used synonymously (and somctimes inaccurately) with interaction, commitment, involvement, dialogue and participation. In this regard, Verčič and Vokić (2017) assert that engagement in public relations and related communication literature is used as an umbrella term to encapsulate the organisation's wide array of endeavours to involve stakeholders in its activities. To emphasise the unique nature of employee engagement, Saks (2006) defines employee engagement as a "distinct and unique construct that consists of cognitive, emotional, and behavioural components that are associated with individual role performance….and is distinguishable from several related constructs, most notably organizational commitment, organizational citizenship behaviour, and job involvement" (602). In line with Kahn's (1990) work on engagement, this definition underscores the notion that engagement involves physiological and physical presence when employees enact their roles in the organisation.

Sustainable employee engagement therefore embodies the process where employees are encouraged to apply their intrinsic problem-solving, leadership and organising skills which thereby elicits employee enablement, empowerment, and a sense of belonging (Villaluz & Hechanova, 2019).

3.1.2 Employee Engagement in the Context of Strategic Sustainability Communication

Employees themselves have become more diverse and globalised, leaving organisational leaders with the challenging task of developing unified messages for workforces with highly varied backgrounds (Men, 2021). Since employees tend to react emotionally to their work environment, they increasingly require expert guidance to "create, reinforce, stabilize, and challenge sustainable values and goals within the organisation" (Allen, 2016: 190). In this diverse and multi-cultural context, strategic communication professionals are essential in assisting employees to make better sense of daily organisational life. Instilling shared values, beliefs and an internal organisational identity helps to acculturate employees (Men, 2021) towards broader organisational objectives.

3.1.3 Considerations for Sustainable Employee Engagement

This section discusses various considerations for sustainable employee engagement which, in the context of this chapter, collectively comprise "communication *for* sustainability". Figure 1 depicts the proposed considerations for sustainable employee engagement, that may serve as enablers for communication about sustainability. These considerations, in essence, include key elements to instil sustainable employee engagement in an organisation, anchored in the CCO approach. To allow for a broader, sustainability perspective, these considerations include a reflective approach to strategic communication and radical leadership thinking; holacracy and

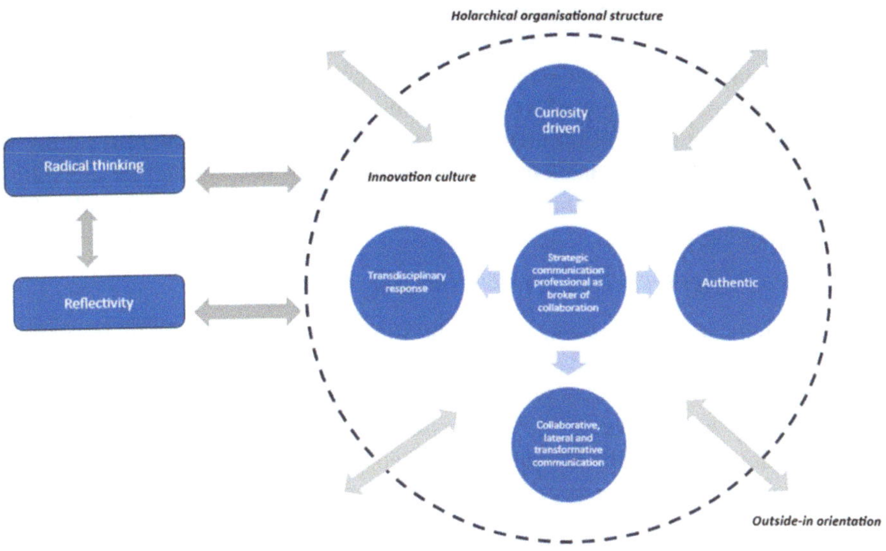

Fig. 1 Considerations for sustainable employee engagement

an innovation culture; curiosity-driven knowledge for a transdisciplinary response and the strategic communication professional acting as a broker of collaboration.

- *Communicative Constitution of the Organisation (CCO)*

The CCO approach emphasises the centrality of communication to an organisation's existence (Cooren & Martine, 2016; Cooren & Seidl, 2022). A CCO approach amalgamates diverse perspectives while affirming its focus on "how organizations as discursive-material configurations are reproduced and coproduced through ongoing interactions" (Brummans et al., 2014). A CCO explores how communication constitutes, conveys, and maintains the organisation (Basque et al., 2022), or is regarded as the foundation of organisational life (Mumby, 2018). In an earlier study, Cornelissen et al. (2015) concluded that the CCO emphasises that communication "is the means by which organisations are established, composed, designed and sustained" (p. 1150).

This chapter specifically draws from the influential Montreal School of Organisational Communication (Cooren et al., 2013), which underscores the heart of the CCO approach, the co-creation of meaning (Gregory & Halff, 2017). Scholars in this School argue that "communication is the means by which organizations are talked, written and acted into existence" (Vásquez et al., 2018: 420). An organisation is not a container within which employees can communicate but is established by people who create complex systems of meaning through communication (Mumby, 2018). Internal communication here refers to more than conversations between humans as well as to communication and meaning conveyed and derived from "feelings, concerns, principles, collectives, text, interests and artifacts…to literally and figuratively participate in communication events" (Vásquez et al., 2018: 421). This will, over time, allow an organisation to become "a collective actor capable of making itself known to its members and representing itself to others" (Basque et al., 2022: 27). This chapter argues that acculturating employees to establish a shared internal organisational identity is a communicative accomplishment. Internal communication should be used to inspire, motivate, and connect employees (Men, 2021). Such practices will ultimately enhance employee engagement resulting in employee empowerment enablement and collaboration. In a broader sustainability context, this will further require reflectivity in strategic communication and radical leadership thinking.

- *Reflectivity and Radical Leadership Thinking*

Reflectivity suggests viewing the organisation from the outside-in by integrating a stakeholder perspective and a societal-legitimacy orientation (Van Ruler & Verčič, 2005) which recognises the organisation as part of an "interdependent social structure" (Verwey & Benecke, 2021: 7). In essence, reflectivity means a consideration of the organisation's place in the larger social structure. From a strategic sustainability communication perspective, reflectivity implies an honest consideration of the impact that organisational behaviour might have in the public sphere related to societal, environmental, and economic issues. Van Ruler and Verčič (2005) maintain that communication professionals, in this capacity, should help "organizations by counselling the deliberations on legitimacy, by coaching its members in the development of their communication competencies, by conceptualizing communication plans, and by executing communication means, using informational, persuasive, relational, and discursive interventions" (Van Ruler & Verčič, 2005: 241).

In line with the perspective that current society is disruptive, pluralistic, and subject to continuous uncertainty and change, organisational leaders must surpass conventional and established ways of doing. They must adopt radical thinking approaches at times to promote agility and readiness for quick and innovative adaptation in response to change. Radical thinking infers that organisational leaders should undertake a critical introspection to establish their current status in terms of driving their organisation to noticeably contribute towards the overall well-being of society (Adams & Soudien, 2022: 2). Radical here does not refer to the extreme; rather, radical refers to searching and striving for an issue's core origin or source. Therefore, leaders may need a detailed critical analysis to determine what has contributed to certain issues towards identifying new ways of addressing these issues which may involve thinking out of the box and challenging the conventional.

- *Holacracy and an Innovation Culture*

In conjunction with the radical leadership thinking and reflectivity mentioned above, genuine sustainability may also require a reconsideration of the leadership approach. Often, an organisation's management systems and leadership style need to change, sometimes dramatically, but require "coordination" and the benefit of a "shared initiative that has the necessary content, credibility and sense of urgency" (Cardona & Rey, 2022: 148). Cardona and Rey point out that even the most efficient and innovative internal communications cannot bring together a workforce without authentic changes in these areas. Henriksson and Grunewold (2020) found in interviews with CEOs that organisations where sustainability was carefully built into the company purpose were more likely to have employees who understood and supported that vision while being "proud" and "committed" (44).

In the context of this chapter, a reconsideration of the leadership approach to identify innovative or unconventional ways of doing requires an integration of a decentralised leadership approach while accepting self-managing organisational structures such as holacracy. Holacracy is an approach where leadership is distributed among self-organising teams to enhance collaboration, productivity, and continuous change (Benecke & Phumo, 2021). Distributing authority and leadership aims to maximise flexibility and individualisation (Kumar & Mukherjee, 2018). Notably, to instil a decentralised leadership approach such as holacracy, a unified understanding and support from employees to achieve the organisation's goals towards the betterment of society should first be established. Only once such an understanding is established could employees, as societal members themselves, truly contribute and build forth toward sustainable employee engagement. Furthermore, to facilitate holacracy, an organisation should prioritise continuous employee education, mentorship, and the promotion of leadership skills (European Communication Monitor, 2023).

A holarchical organisational structure requires clearly articulated organisational values, high levels of trust, support, and confidence in employees' abilities which subsequently empowers and enables employees to devise innovative and creative outcomes (Lee et al., 2023; Villaluz & Henchanova, 2019; Snyder et al., 2017). From an organisational cultural perspective, innovation is contextual—the extent to which an organisation may be regarded as innovative depends on its culture. In other words, innovation is more likely to occur in organisational environments that encourage creativity, including the identification of novel ideas (Villaluz & Henchanova, 2019). Consequently, a supportive organisational environment that envelops organisational trust and empathic leadership is essential (European Communication Monitor, 2023). The process of sustainable employee engagement calls for an innovation culture, which "includes the intention to be innovative, the infrastructure to support innovation, operational level behaviours necessary to influence a market and value orientation and the environment to implement innovation" (Dobni, 2008: 542).

- *Curiosity-Driven Knowledge for a Transdisciplinary Response*

The above-mentioned considerations of sustainable employee engagement could arguably foster curiosity-driven knowledge among employees to co-create innovative transdisciplinary responses to complex issues and at times, "wicked problems" (Rittel & Webber, 1973a; Lee et al., 2023). A wicked problem (Churchman, 1967; Rittel & Webber, 1973b) is "ambiguous, contentious, has no definitive resolution, and is resistant to solutions". Moreover, as Coombs and Holladay note, "the greater the disagreement between stakeholders, the more wicked the problem" (2017: 83). Wicked problems pose both technical and social complexities, making them difficult to manage (Coombs & Holladay, 2017). This, according to Barendsen et al. (2021), encapsulates the goal of sustainability communication to "introduc[e] an understanding of the world, that is of the relationship between humans and their environment, into social discourse, developing a critical awareness of the problems about this relationship and then relating them to social values and norms" (Godeman & Michels, 2011: 6 in Barendsen et al., 2021).

- *Strategic Communication Professional as Broker of Collaboration*

A vital part of the proposed considerations for sustainable employee engagement is the strategic communication professional acting as the "broker" of collaboration (Overton-de Klerk, 2023: 3) to facilitate employee co-creation. Strategic communication professionals play an active role in promoting leadership approaches of participation and inclusion within the organisation to meet the demands and changes of contemporary society (European Communication Monitor, 2023). In this process, the strategic communication professional should avoid being solely focused on the reflexivity of organisational decision-making and accept communication as a social phenomenon. Communication should be collaborative, lateral, and transformative to "attempt at formulating an integrative, holistic process of knowledge production that moves beyond a disciplinary, multidisciplinary or interdisciplinary approach…it is the integration of knowledge paradigms that involve a multiplicity of stakeholders" (Overton-de Klerk, 2023: 16). Driving this process is authenticity, showcasing employees' true desire to make a continuous and meaningful contribution within the organisation towards broader societal, environmental, and economic improvement.

It should be noted that the above considerations for sustainable employee engagement are some of several elements that could be considered when approaching sustainable employee engagement. Depending on the organisational type and industry, sometimes extensive changes and restructuring may be needed. The listed considerations should be perceived as the main considerations and/or a starting point to ground conversations about sustainability.

The considerations for sustainable employee engagement could serve as the enablers of *communication about sustainability* which emphasise the importance of employees' collaboration and co-creation to realise the organisation's sustainability goals. Such a view supports an integrative approach to sustainability communication whereby sustainability communication is broader than its origin of environmental communication (Fischer et al., 2016) and includes a focus on economic and socio-economic issues. A collaborative approach, proposed by the sustainable employee engagement considerations, could contribute to building a broader organisational mindset of societal well-being alongside an active awareness and desire among employees to contribute to the 2030 UN Sustainable Development Goals.

3.2 Communication **About** *Sustainability*

This section transitions from ways that organisations can engage in sustainable employee engagement to a focus on organisational communication about sustainability. Arguably, the considerations explained above (and ideally putting those considerations into action) serve as enablers for communication *about* sustainability. Communication about sustainability is a discourse-orientated approach implying an active awareness of and desire among employees to pursue the organisation's sustainability goals through debates and discussions. Arguably, not all considerations will suit all organisations. Therefore, we offer them here as avenues to shape or refine organisational processes to increase the likely effectiveness of communication about sustainability. This section further explores what is meant by communication about sustainability, specifically focusing on organisational contexts; we describe how the considerations previously mentioned may support an organisation's sustainability communication. Effective communication about sustainability within an organisational context is important for several reasons including the many benefits for employees and organisations from such an endeavour. Overall, we argue that communication for sustainability, through a sustainable employee engagement process, encourages communication about sustainability in the organisation. We end this section with a brief recognition that the opposite of this argument could also be true. Thus, from a broader sustainability perspective, "communication about sustainability" could also contribute to "communication for sustainability".

Sustainable employee engagement practices provide the necessary platform for conversations, debates, and shared solutions to achieve an organisation's sustainability goals. This is particularly important when contextualised within an innovation culture with clear and even ambitious sustainable or environmental goals or initiatives. The more ambitious the sustainable targets, arguably the more important it is to foster an innovation culture that allows discursive and debate-oriented communication about sustainability. Employees need to know that their inputs are being valued and supported. Employees who are engaged in discussion about sustainability initiatives and consulted for suggestions will likely feel a greater sense of ownership and engagement with the process of moving towards an organisation's sustainable goals. Because employees have had opportunities to contribute to organisational discourse on sustainability matters, should they be invited to enact sustainable practices in their work role, they may feel more compelled to respond positively. As Kataria et al. (2013) note, organisations have previously failed to engage employees in ways that recognise them as valuable stakeholders in relation to sustainability; employees have been neglected as a significant resource in realising organisational sustainability goals. Therefore, engaging employees in the process and discourse about sustainability matters may contribute to better sustainability outcomes. As Kataria et al. (2013) state "it is essential for them to engage all employees in sustainability initiatives to maximize value and use sustainability as a competitive advantage" (p. 50).

For organisations wishing to engage with or work towards the Sustainable Development Goals, the considerations described in this chapter may act as a framework or process to engage employees while sparking conversations about an organisation's wider sustainability initiatives. This also implies that communication about sustainability could feed into the communication for sustainability; conversations about sustainability in its ideal state need to lead to sustainability actions. The best sustainability vision means little unless it is backed by genuine and authentic action. Realising sustainability goals, as the SDGs underscore, requires action at all levels of society and at every level within organisations. From an organisational perspective, employees are a vital part of this process; if the intention is sustainability action, then this requires communication for sustainability. If done well, an organisation's sustainability initiatives may act as a springboard for employees to act more sustainably in their lives outside of work, in their community and more widely in society.

4 Conclusion

From the perspective that strategic communication is a "communicative cultivation process…within and beyond the organization's environment" (Weder, 2023: 229), this chapter highlights the significance of sustainable employee engagement to acculturate and co-create, serving as enablers of an organisation's sustainability messages and goals. Organisations seeking to create an innovation culture that is primed for communication about sustainability, may well consider taking a CCO approach to communication, engage in reflectivity and radical leadership, utilise the tenets of holacracy and aim for an innovation culture that is driven by curiosity while seeking interdisciplinary responses to complex and sometimes wicked problems. As we have argued, such considerations may serve as precursors to communication about sustainability. Genuine communication about sustainability should interlock with an organisation's wider vision or purpose whereby that purpose is understood and supported by an organisational workforce. This process does not only focus on reaching the organisation's sustainability goals; rather, it aims to create a true desire within employees to extend these goals beyond the borders of the organisation, to ultimately, contribute to the betterment of the economy, environment and society.

References

Adams, R., & Soudien, C. (2022). Introduction: Radical Reason. *South African Journal of Science, 118*, 1–3. https://doi.org/10.17159/sajs.2022/15000

Allen, M. (2016). *Strategic communication for sustainable organizations. CSR, sustainability, ethics & governance*. Springer. https://doi.org/10.1007/978-3-319-18005-2_6

Arif, S., Johnston, K. A., Lane, K. A., & Beatson, A. (2023). A strategic employee attribute scale: Mediating role of internal communication and employee engagement. *Public Relations Review, 49*(2), 102320. https://doi.org/10.1016/j.pubrev.2023.102320

Auh, S., Menguc, B., Spyropoulou, S., & Wang, F. (2016). Service employee burnout and engagement: The moderating role of power distance orientation. *Journal of the Academy of Marketing Science, 44*(6), 726–745.

Barendsen, W., Muβ, A. C., & Silvius, G. (2021). Exploring team members' perceptions of internal sustainability communication in sustainable project management. *Project Leadership and Society, 2*, 1–13. https://doi.org/10.1016/j.plas.2021.100015

Basque, J., Bencherki, N., & Kuhn, T. (2022). *The Routledge handbook of the communicative constitution of the organization*. Routledge.

Benecke, R., & Phumo, T. (2021). A road less travelled: Re-charting future strategic communication in southern Africa. In D. R. Benecke, S. Verwey, & T. Phumo (Eds.), *Strategic communication: South African perspectives* (pp. 471–480). Oxford.

Botan, C. H. (2018). *Strategic communication theory and practice: The cocreational model*. Wiley Blackwell.

Brummans, B. H. J. M., Cooren, F., Robichaud, D., & Taylor, J. R. (2014). Approaches to the communicative constitution of organizations. In L. L. Putnam & D. K. Mumby (Eds.), *The SAGE handbook of organizational communication* (pp. 173–194). SAGE.

Brunton, M., Eweje, G., & Taskin, N. (2017). Communicating corporate social responsibility to internal stakeholders: Walking the walk or just talking the talk? *Business Strategy and the Environment, 26*(1), 31–48. https://doi.org/10.1002/bse.1889

Cardona, P., & Rey, C. (2022). *Management by missions: Connecting people to strategy through purpose*. Palgrave Macmillan.

Chen, H., Richard, O. C., Boncoeur, O. D., Ford, J. R., & D.I. (2020). Work engagement, emotional exhaustion and counterproductive work behaviour. *Journal of Business Research, 11*, 30–41.

Churchman, C. W. (1967). Guest editorial: Wicked problems. *Management Science, 14*(4), 141–142.
Coombs, T., & Holladay, S. J. (2017). Social issue qua wicked problems: The role of strategic communication in social issues management. *Journal of Communication Management, 22*(1), 79–95.
Cooren, F., Taylor, J. R., & Van Every, E. J. (2013). *Communication as organizing: Empirical and theoretical explorations in the dynamic of text and conversation*. Routledge.
Cooren, F., & Seidl, D. (2022). The theoretical roots of CCO. In J. Basque, N. Bencherki, & T. Kuhn (Eds.), *The Routledge handbook of the communicative constitution of the organization* (pp. 27–46). Routledge.
Cooren, F., & Martine, T. (2016). Communicative constitution of organizations. *The international encyclopedia of communication theory and philosophy*. Wiley. https://doi.org/10.1002/9781118766804.wbiect097
Cornelissen, J., Durand, R., Fiss, P., Lammers, J. C., & Vaara, E. (2015). Putting communication front and center in institutional theory and analysis. *Academy of Management Review, 40*(1), 10–27.
Dhanesh, G. S. (2017). Putting engagement in its Proper place: State of the field, definition and model of engagement in public relations. *Public Relations Review, 43*(5), 925–933.
Dobni, C. B. (2008). Measuring innovation culture in organizations: The development of a generalized innovation culture construct using exploratory factor analysis. *European Journal of Innovation Management, 11*(4), 539–559. https://doi.org/10.1108/14601060810911156
European Communication Monitor: Looking back and ahead: 15 years of research on strategic communication. (2023). A. Zerfass, R. Tench, D. Verčič, A. Moreno, A. Buhmann, & J. Hagelstein (Eds).
Falkheimer, J., & Heide, M. (2022). Introduction: The emergent field of strategic communication. In J. Falkheimer & M. Heide (Eds.), *Research handbook of strategic communication* (pp. 1–12). Edward Elgar.
Fischer, D., Lüdecke, G., Godemann, J., Michelsen, G., Newig, J., Rieckmann, M., & Schulz, D. (2016). Sustainability communication. In H. Heinrichs, A. Wiek, P. Martens, & G. Michelsen (Eds.), *Sustainability science. An introduction* (pp. 139–148). Springer.
Golob, U., Podnar, K., & Zabkar, V. (2023). Sustainability communication. *International Journal of Advertising, 42*(1), 42–51.
Godeman, & Michels. (2011). Sustainability Communication: Interdisciplinary Perspectives and Theoretical Foundations. In J. Godeman, & G. Michels (Eds.)., Springer Science, London. https://doi.org/10.1007/978-94-007-1697
Gregory, A., & Halff, G. (2017). Understanding public relations in the 'sharing economy'. *Public Relations Review, 43*(1), 4–13. https://doi.org/10.1016/j.pubrev.2016.10.008
Heide, M., von Platen, S., Simonsson, C., & Falkheimer, J. (2018). Expanding the scope of strategic communication: Towards a holistic understanding of organizational complexity. *International Journal of Strategic Communication, 12*, 452–468. https://doi.org/10.1080/1553118X.2018.1456434
Henriksson, H., & Grunewald, E. W. (2020). *Sustainability leadership: A Swedish approach to transforming your company, your industry and the world*. Palgrave Macmillan.
Holtzhausen, D. R., Fullerton, J. A., Lewis, B. K., & Shipka, D. (2021). *Principles of strategic communication*. Routledge.
Holtzhausen, D. R., & Zerfass, A. (2015). Strategic communication: Opportunities and challenges of the research area. In D. R. Holtzhausen & A. Zerfass (Eds.), *The Routledge handbook of strategic communication* (pp. 3–17). Routledge.
Johnston, K. A., & Taylor, M. (2018). *The handbook of communication engagement*. Wiley Blackwell.
Kahn, W. A. (1990). Psychological conditions of personal engagement and disengagement at work. *Academy of Management Journal, 33*(4), 692–724. https://doi.org/10.5465/256287
Kataria, A., Kataria, A., & Garg, R. (2013). Effective internal communication: A way towards sustainability. *International Journal of Business Insights and Transformation, 6*, 46–52.

Kumar, S. V., & Mukherjee, S. (2018). Holacracy – the future of organizing? The case of Zappos. *Human Resource Management International Digest, 26*(7), 12–15. https://doi.org/10.1108/HRMID-08-2018-0161

Lee, T., O'Mahony, L., & Lebeck, P. (2023). *Creativity and innovation: Everyday dynamics and practice*. Palgrave.

McGreavy, B., Haynal, K., Smith-Mayo, J., Reilly-Moman, J., Kinnison, M. T., Ranco, D., & Leslie, H. M. (2022). How does strategic communication shape transdisciplinary collaboration? A focus on definitions, audience, expertise, and ethical praxis. *Frontiers in Communication, 7*(831727), 1–14. https://doi.org/10.3389/fcomm.2022.831727

Men, L. R. (2021). Evolving research and practices in internal communication. In L. R. Men & A. Tkalac Verčič (Eds.), *Current trends and issues in internal communication. New perspectives in organizational communication* (pp. 1–18). Palgrave Macmillan. https://doi.org/10.1007/978-3-030-78213-9_1

Morehouse, J. & Saffer, A. J. (2023). Putting stakeholders' engagement in the equation: Proposing the integrated network engagement model. *Public Relations Review, 49*(1), 102291. https://doi.org/10.1016/j.pubrev.2023.102291

Mumby, D. K. (2018). *Organizational communication: A critical introduction* (2nd ed., international student edition). SAGE.

Newig, J., Schulz, D., Fischer, D., Hetze, K., Laws, N., Lüdecke, G., & Rieckmann, M. (2013). Communication regarding sustainability: Conceptual perspectives and exploration of societal subsystems. *Sustainability, 5*(7), 2976–2990. https://doi.org/10.3390/su5072976

Overton-de Klerk, N. (2023). Reflection on the current state and future of strategic communication as paradigm and practice. *Communicare, 42*(1), 1–21. https://doi.org/10.36615/jcsa.v42i1.2610

Rafi, T., & World Economic Forum. (2022). *ESG: Why sustainability is crucial for corporate strategy*. Retrieved January 11, 2024, from https://www.weforum.org/agenda/2022/06/why-sustainability-is-crucial-for-corporate-strategy/

Rittel, H. W. J., & Webber, M. M. (1973a). Dilemmas in a general theory of planning. *Policy Sciences, 4*(2), 155–169.

Rittel, H. W., & Webber, M. M. (1973b). 2.3 planning problems are wicked. *Polity, 4*(4), 155–169.

Ruck, K., & Welch, M. (2012). Valuing internal communication: Management and employee perspective. *Public Relations Review, 38*(2), 294–302.

Saks, A. M. (2006). Antecedents and consequences of employee engagement. *Journal of Managerial Psychology, 21*(7), 600–619.

Saifudin, M. S. M. (2022). Sustainability communication: West vs Asia. In M. S. M. Saifudin, N. A. A. Rahman, & S. A. A. Kasuma (Eds.), *Sustainability communication across Asia: Fundamental principles, digital strategies and community engagement* (1st ed., pp. 1–14). Routledge.

Snyder, K., Hedlund, C., Ingelsson, P., & Bäckström. (2017). Storytelling: A co-creative process to support value-based leadership. *International Journal of Quality and Service Sciences, 9*(3/4), 484–497.

Van Ruler, B., & Verčič, D. (2005). Reflective communication management, future ways for public relations research. *Annals of the International Communication Association., 29*, 239–274. https://doi.org/10.1080/23808985.2005.11679049

Verwey, R., & Benecke, R. (2021). Paradigms, perspectives, and paradoxes. In S. Verwey, R. Benecke, & T. Phumo (Eds.), *Strategic communication: SA perspectives* (pp. 5–27). Oxford University Press.

Villaluz, V. C., & Hechanova, M. R. M. (2019). Ownership and leadership in building an innovation culture. *Leadership & Organization Development Journal, 40*(2), 138–150. https://doi.org/10.1108/LODJ-05-2018-0184

Tkalac Verčič, A., & Men, L. R. (2023). Redefining the link between internal communication and employee engagement. *Public Relations Review, 49*(1), 102279. https://doi.org/10.1016/j.pubrev.2022.102279

Tkalac Verčič, A., Galić, Z., & Žnidar, K. (2023). The relationship of internal communication satisfaction with employee engagement and employer attractiveness: Testing the joint mediating effect of the social exchange quality indicators. *International Journal of Business Communication, 60*(4), 1313–1340. https://doi.org/10.1177/23294884211053839

Tkalac Verčič, A. (2021). The impact of employee engagement, organisational support and employer branding on internal communication satisfaction. *Public Relations Review, 41*(1), 102009. https://doi.org/10.1016/j.pubrev.2021.102009

UNDP. (2023a). *Sustainable Development Goals: Background on the goals*. Retrieved January 25, 2024, from https://www.undp.org/sdg-accelerator/background-goals#:~:text=The%20Sustainable%20Development%20Goals%20(SDGs,economic%20challenges%20facing%20our%20world

UNDP. (2023b). *Sustainable Development Goals: Business and the SDGs*. Retrieved January 25, 2024, from https://www.undp.org/sdg-accelerator/business-and-sdgs#:~:text=In%20collaboration%20with%20partners%2C%20the,delivering%20on%20the%20SDG%20targets

Vásquez, C., Bencherki, N., Cooren, F., & Sergi, V. (2018). From 'matters of concern' to 'matters of authority': Studying the performativity of strategy from a communicative constitution of organization (CCO) approach. *Long Range Planning, 51*(3), 417–435.

Verčič, A. T., & Vokić, N. P. (2017). Engaging employees through internal communication. *Public Relations Review, 43*(5), 885–893. https://doi.org/10.1016/j.pubrev.2017.04.005

Weder, F. (2023). The evolution of the sustainability story: Strategic sustainability communication as niche construction. *International Journal of Strategic Communication, 17*(3), 228–244. https://doi.org/10.1080/1553118X.2023.2229304

Weder, F., Karmasin, M., Krainer, L., & Voci, D. (2021). Sustainability communication as critical perspective in media and communication studies—an introduction. In F. Weder, M. Karmasin, L. Krainer, & D. Vo (Eds.), *The sustainability communication reader* (pp. 1–12). Springer VS.

Zerfass, A., & Link, J. (2022). Communication management: Structures, processes and business models for value creation through corporate communications. In J. Falkheimer & M. Heide (Eds.), *Research handbook of strategic communication* (pp. 237–258). Edward Elgar.

Further Reading

Botha, Y. (2021). Extending polyphonic strategic communication to stakeholder engagement: Exploring deliberate-emergent and radical-emergent approaches. *Management Dynamics, 30*(2), 1–14.

Lee, T., O'Mahony, L., & Lebeck, P. (2023). *Creativity and innovation: Everyday dynamics and practice*. Palgrave.

Yolandi Botha is a Senior Lecturer of Strategic Communication in the School of Media and Communication at Murdoch University in Western Australia. She has 6 years' experience in Public Relations and Strategic Communication in financial and banking environments and 14 years of experience in higher education teaching, supervision, and curriculum development in both online and blended modes of tuition. She is an NRF-rated researcher in South Africa with a multidisciplinary research focus in communication and management sciences. Yolandi has published various articles, papers, and book chapters on strategic and organisational communication, stakeholder engagement, stakeholder inclusivity, crisis communication, and change communication.

Lauren O'Mahony is a Senior Lecturer in Communications at Murdoch University, Western Australia. Her research focuses on Australian women's literature as well as media analysis, media audiences, and creativity. Her research has been published in *The Journal of Intercultural Studies, The Journal of Postcolonial Writing*, The *Journal of Popular Romance Studies*, *Communication Research and Practice*, and *Text* Journal as well as the edited books *The Routledge Research Companion to Popular Romance Fiction* (2021), *Interrogating Boundaries of the Nonhuman* (2022) and *Difficult Death, Dying and the Dead in Media and Culture* (2024). In 2023, *Creativity and Innovation: Everyday Dynamics and Practice* (co-authored with Terence Lee and Pia Lebeck) was published by Springer Books.

The Role of Greenwashing Suspicion in Advertising Effectiveness

Ann-Elen Waldeier, Stefan Weinacht, and Christopher Morasch

Abstract Sustainability communication as an increasingly circumscribable research field grounds in communication, management and marketing concepts. This chapter focuses on communication of sustainability, sustainable consumption and green claiming and discusses the effects of non-transparent sustainability communication and greenwashing. It specifically presents a study on how the suspicion of greenwashing in green advertising affects attitudes towards the ad and the brand, as well as the purchase intention. The affect transfer hypothesis is used as a model to understand advertising effects in this context. Additionally, this study delves into potential influencing factors on suspicion of greenwashing, such as the type of green advertising and the "lifestyle of health and sustainability" (LOHAS). The research question is addressed through a quantitative experimental online survey. The results demonstrate that the suspicion of greenwashing has a negative effect on attitude towards the ad and on attitude towards the brand, as well as on purchase intention. The chapter therefore offers new insights into one of the critical aspects of strategic sustainable communication and lays the groundwork for future research on "washing" (green, pink, rainbow, etc.) and the need for re-framing sustainability in business-consumer discourses.

1 Introduction

In response to the increasing relevance of sustainability in everyday life (Mishra & Mishra, 2020) and with consumers demanding ethically responsible behaviour, particularly in the form of active climate protection (Chen & Chang, 2013), expectations for companies have shifted. To meet these expectations, companies use strategic sustainable communication—here defined as intentional and purposeful

A.-E. Waldeier (✉) · S. Weinacht · C. Morasch
Westfalian University of Applied Sciences, Institute for Journalism and Public Relations, Gelsenkirchen, Germany
e-mail: stefan.weinacht@w-hs.de; christopher.morasch@w-hs.de

F. Weder (ed.), *Strategic Sustainability Communication*, CSR, Sustainability, Ethics & Governance, https://doi.org/10.1007/978-3-031-89486-2_8

communication of sustainability that is aimed at achieving corporate goals based on Holtzhausen & Zerfaß (2013). Sustainable communication can be classified in the following three categories as described by Weder (2023, p. 22) in "the communication OF sustainability (pragmatic, functional understanding of communication, informing and educating people about certain "sustainable products"), communication ABOUT sustainability (conversations about more or less sustainable choices, political debates) and communication FOR sustainability (participatory approaches, questioning non-regenerative principles and imbalances in human-nature relationships, critique, sense and meaning-making, social learning)". This chapter is taking a marketing approach on the topic of strategic sustainable communication and therefore focuses on the communication OF sustainability. In this context companies sometimes portray themselves and their products as more environmentally friendly than they are (TerraChoice, 2010). This is known as greenwashing, which will be further defined in chapter "The In(de)finition of Sustainability: Strategic Communication for Sustainability as Conversational Contestation". Parallel to the literature on CSR, CSR communication and sustainability communication that considered non-transparent communication and greenwashing since the beginning (Elving et al., 2015), there is an increasing number of studies exploring specific cases of greenwashing, bringing in reflections on sustainable consumption or critical questions regarding the quality and specifically trustworthiness of communication (Amin & Tarun, 2021; Hui-Wen Chuah et al., 2020).

Next to the research on greenwashing as a strategic communication phenomenon, CSR and sustainability communication also seeks to better understand the effects of greenwashing on specific stakeholder groups, particularly consumers. So far, research has shown that greenwashing has a rather negative impact on advertising effectiveness (Chen & Chang, 2013; Schmuck et al., 2018; Szabo & Webster, 2020). However, consumers often lack the expertise to figure out whether they are encountering greenwashing or genuine green advertising (Ottman et al., 2006). Therefore, this study focuses on the influence of the suspicion of greenwashing and asks:

How does the suspicion of greenwashing in green advertising affect consumers' attitudes towards the ad, consumers' attitudes towards the brand and consumers' purchase intention?

This study is linked to advertising effectiveness research. It is contributing to a better understanding of communication OF sustainability by highlighting risks and benefits of green advertising and is proposing a model of the effects of greenwashing suspicion, which provides a framework for understanding the factors contributing to consumer scepticism towards green advertising.

2 Sustainability Marketing and the Challenges Around Greenwashing

2.1 *Green Advertising*

As described earlier, strategic sustainable communication includes the communication OF sustainability, which is part of sustainability marketing and "aims to satisfy consumers' needs or wants and creates a favourable position for the business in the marketplace by communicating how the business addresses environmental, social and economic concerns" (Villarino & Font, 2015). Sustainability marketing involves creating and communicating products, services, and corporate practices in an environmentally friendly way (Belz & Peattie, 2012). When it comes to communicating the connection between a company and sustainability, green advertising is often used and defined as advertising that targets consumers who are concerned about the environment. It establishes a connection between the advertised product/service and the environment, promotes a green lifestyle, and/or presents a "green" corporate image (Banerjee et al., 1995; Minton et al., 2012; Zinkhan & Carlson, 1995).

Green advertising often contains so-called green claims, which emphasise the connection to the environment (TerraChoice, 2010). It also includes executional elements (Segev et al., 2016), which refer to how the ad is presented visually (Stanton, 1998). In the context of green advertising, these visuals are linked to sustainability and include the most used colours, which are white, green, and blue, as well as imagery of vegetation, the sky, renewable energy, landscapes, or animals (Segev et al., 2016). Parguel et al. (2015) also mention the use of background images showing mountains or the ocean.

A reoccurring topic in the research regarding green advertising is the effectiveness of green influencers (Cooke et al., 2022; Buvár et al., 2023; König & Maier, 2024; Yıldırım, 2021; Jalali & Khalid, 2021) who aim to convince and educate their followers to lead a more sustainable lifestyle (Yıldırım, 2021). Recent research concerning green advertising mainly focuses on the effectiveness of eco-labels (Kolović et al., 2023; Sun et al., 2020) and the effect of consumers (non-) trust in green advertising messages (Amin & Tarun, 2021; Hui-Wen Chuah et al., 2020).

2.2 *Greenwashing*

In this study, greenwashing is defined as the act of misleading consumers regarding the actual sustainability of the company/product/service by spreading false information about their sustainability performance, taking and communicating only symbolic steps to improve environmental performance, or highlighting positive sustainability achievements in communication while hiding negative sustainability performances (Baum, 2012; Guo et al., 2014; Delmas & Burbano, 2011).

By looking at the different elements of green advertising described earlier, greenwashing can be separated into two types as well, regarding the way in which a green message is expressed: Claim greenwashing and executional greenwashing. Claim greenwashing applies when a claim contains a green message that is vague, untrue, not verifiable or irrelevant to consumers (TerraChoice, 2010; Bengtson et al., 2023; de Freitas Netto et al., 2020; Munir & Mohan, 2022; Mütze, 2022). According to a study by the European Commission (2023), 53.3% of the 150 environmental claims examined contained vague, misleading, or unfounded statements. To protect consumers, the EU took measures against greenwashing by introducing the Green Claim Directive (GCD), which only allows companies to profit of their environmental claims if they have been verified and substantiated (Ragonnaud, 2024). These measures and the growing scepticism of consumers has led some companies to be strategically silent around the topic of sustainability. This strategy is known as greenhushing (Ginder et al., 2021). Another way of avoiding these restrictions is to use executional greenwashing and therefore not include a green claim but instead only use visual sustainability elements in green advertising (Parguel et al., 2015). In research, executional greenwashing has only played a minor role so far. In a study by Parguel et al. (2015), an "executional greenwashing effect" was observed for the first time. Schmuck et al. (2018, p. 129) did examine the impact of using visual elements of sustainability on advertising success, not independently, but only in combination with green claims.

To fill this research gap, this paper will compare the effects of claim greenwashing and executional greenwashing on advertising success.

2.3 Green Consumers and LOHAS

As described earlier, green advertising and therefore greenwashing as well does target consumers who are concerned about the environment. Those are often named green consumers and described as "individuals looking to protect themselves and their world through the power of their purchasing decisions. In their efforts to protect themselves and their world, they are scrutinising products for environmental safety" (Ottman et al., 2006, p. 3). Therefore, they approach purchasing decisions with particular care and scrutiny. These consumers cannot be clearly identified by demographic characteristics (Wonneberger et al., 2016) because they are more defined by endogenous, exogenous, and structural factors (Jordan et al., 2015).

One way to categorise green consumers based on such factors is the so-called Lifestyle of Health and Sustainability (LOHAS), which was first described by Ray and Anderson (2000).

People who lead a lifestyle of health and sustainability (LOHAS) often engage with environmental issues and strive to live sustainable by purchasing products that are as sustainable as possible (Grimm & Malschinger, 2021). An Allensbach study (2023) compares the characteristics of individuals leading a LOHAS with the rest of the German population. It is notable that LOHAS individuals are significantly more interested in nature and environmental protection (LOHAS individuals: 95.6%; German population: 75.4%) (Grimm & Malschinger, 2021). Furthermore, LOHAS individuals are more willing to spend more money on sustainable products (LOHAS individuals: 77.9%; German population: 34%) (ibid.) and consider the sustainability of a product to be very important in their purchasing decisions (LOHAS individuals: 58.2%; German population: 14.4%) (ibid.).

In summation greenwashing is a form of sustainability communication that can be found in green advertising and is misleading the consumer regarding the actual sustainability of their products, services or practices. The use of greenwashing does have a negative effect on consumers which in return can have a negative effect on the company in form of a shrinking purchase intention (Szabo & Webster, 2020). This will be further explored later. While greenwashing has been relevant since the start of research on CSR communication (Elving et al., 2015, etc.), the topic is still important today as restrictions regarding greenwashing just recently changed (Ragonnaud, 2024), which does cause new forms of sustainability communication like greenhushing to form and it increases the importance of executional greenwashing, which is further explored in this study.

3 Theoretical Foundation

To understand advertising effects in this study the affect transfer hypothesis has been used. It postulates a direct influence of the attitude towards the ad on the attitude towards the brand, which then influences purchase intention. We therefore propose;

H_1: The more positive the attitude towards advertising, the more positive the attitude towards the brand.

H_2: The more positive the attitude towards the brand, the more likely it is that a purchase intention will be formed.

These hypotheses serve to test the transferability of the affect transfer hypothesis onto the data collected for the study.

3.1 Suspicion of Greenwashing in Green Advertising

A study by Chen and Chang (2013) analyses the effects of perceived greenwashing. Perceived greenwashing is defined as the ability of consumers to recognise greenwashing tactics as such (ibid.). It has been found that greenwashing that is unmasked confuses consumers and reduces trust in the company (ibid.). This study has been followed by others that also investigate the unmasking of greenwashing, such as Schmuck et al. (2018) and Szabo and Webster (2020).

Results show a more negative attitude towards the brand when greenwashing tactics are used and recognised by the recipient. In addition, a negative impact on the formation of purchase intention was observed (Szabo & Webster, 2020).

The studies mentioned all used stimuli that were self-created and categorised as greenwashing beforehand (Chen & Chang, 2013; Schmuck et al., 2018; Szabo & Webster, 2020). However, in addition to actual greenwashing, greenwashing is also sometimes suspected in sincere sustainability campaigns (Jahdi & Acikdilli, 2009; Mütze, 2022). This is because consumers often lack the expertise or ability to distinguish between actual sustainable initiatives and greenwashing, which in turn leads to confusion and scepticism towards green advertising (Ottman et al., 2006). Based on the Thomas theorem ("If men define situations as real, they are real in their consequences"; Thomas & Thomas, 1928, p. 572), the fundamental assumption of this paper is that the suspicion of greenwashing is already sufficient to impair advertising success. Thus, we propose:

H_3: The more greenwashing is suspected in green advertising, the more negative the consumers' attitude towards advertising.

H_4: The more greenwashing is suspected in green advertising, the more negative the consumers' attitude towards the brand.

H_5: The more greenwashing is suspected in green advertising, the lower the consumers' purchase intention.

3.2 Differences in the Suspicion of Greenwashing

As explained earlier, greenwashing is generally divided into two categories: claim greenwashing and executional greenwashing. As a greenwashing effect has already been observed in both forms (Parguel et al., 2015; Schmuck et al., 2018), this paper assumes that greenwashing is also suspected in both green advertising with a green claim and in green advertising with exclusively visual elements of nature. This is to be verified by comparing it with the suspicion of greenwashing in neutral advertising, i.e. advertising that makes no reference to sustainability. Parguel et al. (2015) describe the influence of visual elements of nature on the perceived sustainability of the advertised company as subtle. It is therefore assumed that green claims convey sustainability messages more obviously than visual elements of nature and that greenwashing is more likely to be suspected there. We therefore propose:

H_6: The suspicion of greenwashing:

(a) *is less pronounced in neutral advertising compared to green advertising with green claims,*
(b) *is less pronounced in neutral advertising compared to green advertising with only visual elements of nature, and*
(c) *is more pronounced in green advertising with green claims than in green advertising with only visual elements of nature.*

As described above, it is assumed that the suspicion of greenwashing has a negative impact on the attitude towards the ad and brand as well as on the purchase intention. It is also assumed that the type of green advertising (claim, executional) moderates this influence. The following hypothesis is therefore added:

H_7: Effects assumed in H_3, H_4 and H_5 *for green advertising are stronger for green claims than for green visuals.*

Green consumers are sceptical against green advertising and tend to find it untrustworthy (Wonneberger et al., 2016). In addition, people with a strong lifestyle of health and sustainability (LOHAS) are generally critical and scrutinise their purchasing decisions, as described earlier. Therefore, we assume that people with high level of LOHAS suspect greenwashing in green advertising more than people with low levels of LOHAS. This assumption is tested by the following hypothesis:

H_8: The more intense their lifestyle of health and sustainability (LOHAS), the more consumers are to suspect greenwashing in green advertising.

4 Research Design

The research question was addressed through a quantitative experimental online survey with a non-probability quota sample of n=345 German respondents. The stratified sample is representative of the German population in terms of gender (56% female and 42% male) and almost representative in terms of age, with younger people being slightly overrepresented.

Following a pretest, the data was conducted between June 27 and July 23 of 2023 via social media and the recruiting of missing segments by personal contacts. Respondents were divided into three experimental groups and exposed to different stimuli, as seen in Table 1.

Table 1 Test groups

Test group 1—"Claim"	Test group 2—"Executional"	Test group 3—"Neutral"
Green advertising from Vio with a green claim	Green advertising from Vio with exclusively visual natural elements	Neutral advertising from Vio
Green advertising from Volvic with a green claim	Green advertising from Volvic with exclusively visual natural elements	Neutral advertising from Volvic
Green advertising from Gerolsteiner with a green claim	Green advertising from Gerolsteiner with exclusively pictorial nature elements	Neutral advertising from Gerolsteiner

4.1 Variables and Operationalisation

The following variables were addressed in the derived hypotheses: *Attitude towards Advertising*, *Attitude towards the Brand*, *Purchase Intention*, *Suspicion of Greenwashing* and *LOHAS*. All variables are based on several items that were aggregated into index variables using scale averages. Additionally, the questionnaire also asked about age and gender.

4.1.1 The Attitude Towards Advertising and the Brand as Well as the Purchase Intention

To measure the variables relevant to the affect transfer hypothesis (*Attitude towards Advertising*, *Attitude towards the Brand* and *Purchase Intention*), the existing operationalisations according to MacKenzie et al. (1986) were translated into German and implemented. *Attitude towards Advertising* was measured with two items (favourable/unfavourable; interesting/boring), *Attitude towards the Brand* with three items (favourable/unfavourable; good/bad; reasonable/unreasonable) and *Purchase Intention* with two items (probable/unprobable; possible/impossible). MacKenzie et al. (1986) originally envisaged three items for the operationalisation of *Purchase Intention.* However, two of them (likely/unlikely; probable/improbable) had the same meaning after translation into German and were therefore combined into one item (likely/unlikely). A semantic differential with seven levels was used for all three variables. A low number stood for the negative adjective and a high number for the positive adjective.

4.1.2 Suspicion of Greenwashing

Chen and Chang (2013) created five items for their study on greenwashing that measure whether consumers recognise greenwashing intentions in advertisements. This operationalisation was extended by Schmuck et al. (2018) by two items and used under the same premise. As this study is not concerned with whether greenwashing is detected but whether, if it is suspected, the wordings of the items were adapted. For example, the first item, "This advertisement misleads with its words about the sustainability features of the product" (Schmuck et al., 2018, p. 143), was reformulated as "In my opinion, this advertisement misleads with its words about the sustainability features of the product". According to this principle, all seven items were adapted and added to the questionnaire on a seven-point Likert scale ranging from "I strongly disagree" (1) to "I strongly agree" (7). The question was: "To what extent do the following statements apply? Please rate the statements based on the advert you have just watched".

4.1.3 LOHAS

In a study by Choi and Feinberg (2021), an operationalisation was created for the "lifestyle of health and sustainability" (LOHAS). To this end, the number of 70 possible items intended to measure membership of the LOHAS was reduced to 28 using factor analysis (ibid.). The final 28 items can be divided into the categories of physical fitness (5), mental health (3), emotional health (4), spiritual health (3), environmental protection (10) and social awareness (3) (ibid.). The items were applied to the questionnaire and collected on a seven-point Likert scale ranging from "I strongly disagree" (1) to "I strongly agree" (7).

5 Analysis

To test H_1, H_2 H_3, H_4 and H_5 regression analysis was used. Results approve of *Attitude towards the Ad* affecting *Attitude towards the Brand* (B=0.664; SEM=0.04; β=0.673; Adj. R^2=0.451) and *Attitude towards the Brand* affecting *Purchase Intention* (B=0.767; SEM=0.043; β=0.694; Adj. R^2=0.48) supporting the affect transfer hypothesis. Thus, H_1 and H_2 must be regarded as true.

Additionally, results demonstrate that the *Suspicion of Greenwashing* has a negative effect on *Attitudes towards the Ad* (B=−0.348; SEM=0.042; β=−0.410; Adj. R^2=0.165), on *Attitudes towards the Brand* (B=−0.365; SEM=0.041; β=−0.435; Adj. R^2=0.187) and on *Purchase Intention* (B=−0.345; SEM=0.046; β=−0.372; Adj. R^2=0.136). The stronger respondents suspected greenwashing the lower they rated the ad and the brand or had the intention to buy the product. Therefore, H_3, H_4 and H_5 must be regarded as true.

H_6 is concerned with neutral advertising and green advertising consisting of green claims and green visuals. Results indicate that Greenwashing is more strongly suspected in green advertising with green claims (Mean=4.391) than in neutral advertising (Mean=3.686). Thus, $H_{6.(a)}$ must be regarded as true. Greenwashing is also more strongly suspected in green advertising which only contain visual elements of nature (Mean=4.249) than in neutral advertising (Mean=3.686). Therefore, $H_{6.(b)}$ must be regarded as true. $H_{6.(c)}$ proposed a higher suspicion of greenwashing in green advertising with a green claim than in green advertising that only contains visual elements of nature. This was not supported by the data, as the t-test did not find a significant difference in greenwashing suspicion between the two forms of green advertising. Thus, $H_{6.(c)}$ must be regarded as untrue. In total, H_6 can be regarded as widely true.

To test for moderator effects of green claim vs green executional as assumed in H_7, the PROCESS macro for SPSS (V. 4.1) by Andrew F. Hayes (2022) was used. No moderator effect of *Form of Advertising* was found in the relationship between *Attitude towards the Brand* and *Suspicion of Greenwashing.* Likewise, no moderator effect of *Form of Advertising* was found in the relationship between *Suspicion of Greenwashing* and *Purchase Intention.*

However, results indicate a moderator effect of *Form of Green Advertising* in the ANOVA model of *Attitude towards the Ad* (predictor) and *Suspicion of Greenwashing* (dependent variable) with the p-Value (0.085) of the interaction term being significant at the 10% level; At low and medium levels of *Suspicion of Greenwashing,* the claim has a stronger negative effect on *Attitude towards the Ad* than the executional. However, at a high level of *Suspicion of Greenwashing,* the executional affects *Attitude towards the Ad* slightly stronger negatively than the claim, as can be seen in Fig. 1. Results suggest that green visuals lead to less negative impact of *Suspicion of Greenwashing on Attitude towards the Ad.* However, with consumer's suspicion of greenwashing getting stronger, the form of green advertising (claim or executional) is becoming increasingly irrelevant.

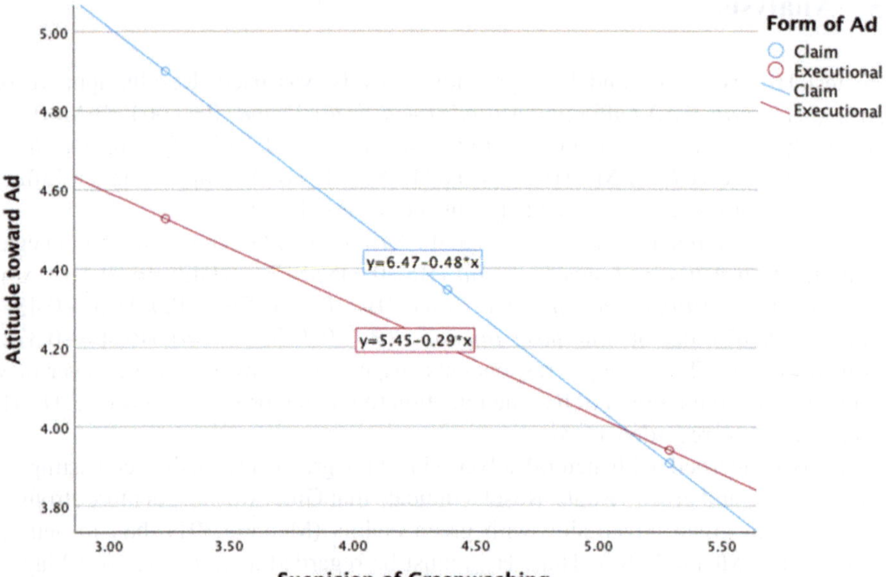

Fig. 1 Probing of interactions; moderator effect of Form of Advertising in the relationship of Suspicion of Greenwashing and Attitude towards the Advert

Attention should be paid to the relatively high p-Value, resulting in a probability of error of almost 9%. This could be due to a relatively small sample size. It is furthermore important to categorise this effect in the superordinate ANOVA model of *Suspicion of Greenwashing* and *Attitude towards the Ad* being negatively correlated. Thus, we must regard H_7 as widely untrue.

Regression analysis was used to test for H_8 which proposes a high level of *LOHAS intensity* resulting in a high level of *Suspicion of Greenwashing.* This is fully supported by our data: *Lifestyle of Health and Sustainability (LOHAS) Intensity* has a highly significant positive impact on Suspicion of Greenwashing. Therefore, we must regard H_8 as true.

However, results suggest this effect being weak, given a relatively small coefficient ($\beta=-0.158$) and a relatively low explanatory power (Adj. R2=0.022) of *Suspicion of Greenwashing*.

Grimm and Malschinger (2021) argue that the demand for green products and sustainable behaviour from companies is already so widespread that almost all consumers can be considered green consumers. Therefore, only a small group cannot be persuaded to choose green products. These considerations could explain why, even though *LOHAS intensity* and *Suspicion of Greenwashing* are positively correlated, this may not always result in LOHAS consumers being very suspicious of greenwashing.

6 Conclusion

The investigation of our research question, "How does the suspicion of greenwashing in green advertising affect consumers' attitudes towards the ad, consumers' attitudes towards the brand and consumers' purchase intention?" was intended to determine the role of the suspicion of greenwashing in the advertising impact process. Our results suggest that the suspicion of greenwashing negatively affects consumers' attitudes towards the ad, consumers' attitudes towards the brand and consumers' purchase intention. Marketers and communicators wishing to use green advertising should therefore be aware that this poses a major risk for their brands; sustainability marketing messages that seem implausible, due to suspicion of greenwashing are bound to trigger negative repercussions. Claims that are perceived as greenwashing from the consumer's point of view are likely to trigger more negative repercussions towards adverts than sustainability visuals that look like greenwashing to consumers. Therefore, communicators should be especially vigilant when using green claims that could be mistaken for greenwashing in their advertising. Regardless of the type of green advertising, the suspicion of greenwashing can lead to reputational and monetary losses for brands.

The effects found in this study are summarised in Fig. 2.

This study did bring a better understanding of consumers interpretation of green advertising and the effects caused by this interpretation by showing that the

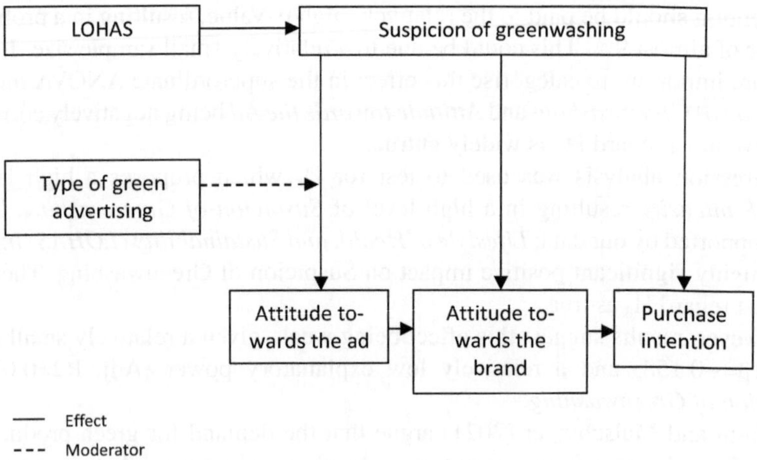

Fig. 2 The role of suspected greenwashing in the advertising impact process

suspicion of greenwashing in green advertising alone is enough to lead towards reputational damage and monetary losses.

Overall, our research results can be considered as robust, but it would be useful to carry out similar studies with larger sample sizes in the future.

As mentioned before, research on using only visuals linked to sustainability and no green claim in green advertising has only begun recently (Parguel et al., 2015) but is becoming more and more important as the usage of green claims becomes more restricted by the EU (Ragonnaud, 2024). It is plausible that the number of green advertisements that only include these visuals and no green claim will increase, which would be interesting to observe by research.

Results did indicate that green advertising that only uses visuals linked to sustainability and no green claim does in fact cause suspicion of greenwashing. Therefore the field of sustainability communication should expand research done on visuals linked to sustainability. For example, further research could examine which visuals exactly lead consumers to suspect greenwashing. This could help companies avoid this suspicion in the future and would bring a better understanding of green advertising to the research field.

References

Amin, S., & Tarun, M. T. (2021). Effect of consumption values on customers' green purchase intention: A mediating role of green trust. *Social Responsibility Journal, 17*(8), 1320–1336. https://doi.org/10.1108/SRJ-05-2020-0191

Banerjee, S., Gulas, C. S., & Iyer, E. (1995). Shades of green: A multidimensional analysis of environmental advertising. *Journal of Advertising, 24*(2), 21–31. https://doi.org/10.1080/00913367.1995.10673473

Baum, L. M. (2012). It's not easy being green … or is it? A content analysis of environmental claims in magazine advertisements from the United States and United Kingdom. *Environmental Communication, 6*, 423–440. https://doi.org/10.1080/17524032.2012.724022

Belz, F. M., & Peattie, K. (2012). *Sustainability marketing: A global perspective*. Wiley.

Bengtson, E., Mossberg, O., Finlayson, A., Martin, J., & Phillips, K. R. (2023). From the sins of greenwashing to the virtues of green marketing. In *The virtues of green marketing: A constructive take on corporate rhetoric* (pp. 51–64). Springer Nature. https://doi.org/10.1007/978-3-031-32979-1_6

Buvár, A., Zsila, A., & Orosz, G. (2023). Non-green influencers promoting sustainable consumption: Dynamic norms enhance the credibility of authentic pro-environmental posts. *Frontiers in Psychology*.

Chen, Y.-S., & Chang, C.-H. (2013). Greenwash and green trust: The mediation effects of green consumer confusion and green perceived risk. *Journal of Business Ethics, 114*(3), 489. https://doi.org/10.1007/s10551-012-1360-0

Choi, S., & Feinberg, R. A. (2021). The LOHAS (Lifestyle of Health and Sustainability) scale development and validation. *Sustainability, 13*(1598). https://doi.org/10.3390/su13041598

Cooke, P., Nunes, S., Oliva, S., & Lazzeretti, L. (2022). Open innovation, soft branding and green influencers: Critiquing 'fast fashion' and 'overtourism'. *Journal of Open Innovation: Technology, Market, and Complexity, 8*, 52.

de Freitas Netto, S., Sobral, M. F. F., Bezerra Ribeiro, A. R., & Da Luz Soares, G. R. (2020). Concepts and forms of greenwashing: A systematic review. *Environmental Sciences Europe, 32*(1). https://doi.org/10.1186/s12302-020-0300-3

Delmas, M. A., & Burbano, V. C. (2011). The drivers of greenwashing. *California Management Review, 54*(1), 64–87. https://doi.org/10.1525/cmr.2011.54.1.64

Elving, W. J. L., Golob, U., Podnar, K., Ellerup-Nielsen, A., & Thomson, C. (2015). The bad, the ugly and the good: New challenges for CSR communication. *Corporate Communications: An International Journal, 20*(2), 118–127. https://doi.org/10.1108/CCIJ-02-2015-0006

European Commission. (2023). *Environment – Green claims*. European Commission. https://environment.ec.europa.eu/topics/circular-economy/green-claims_en

Ginder, W., Kwon, W. S., & Byun, S. E. (2021). Effects of internal–external congruence-based CSR positioning: An attribution theory approach. *Journal of Business Ethics, 169*, 355–369. https://doi.org/10.1007/s10551-019-04282-w

Grimm, A., & Malschinger, A. (2021). *Green Marketing 4.0: Ein Marketing-Guide für Green Davids und Greening Goliaths* (1st). Springer Fachmedien Wiesbaden. https://doi.org/10.1007/978-3-658-03698-0

Guo, R., Tao, L., Yan, L., & Gao, P. (2014). The effect path of greenwashing brand trust in Chinese microbiological industry from decoupling view. *BioTechnology: An Indian Journal, 10*(7), 1827–1831.

Hayes, A. F. (2022). *Introduction to mediation, moderation, and conditional process analysis: A regression-based approach*. Guilford Press.

Holtzhausen, D. R., & Zerfass, A. (2013). Strategic communication – pillars and perspectives of an alternative paradigm. In A. Zerfaß, L. Rademacher, & S. Wehmeier (Eds.), *Organisationskommunikation und Public Relations* (pp. 73–94). Springer VS.

Hui-Wen Chuah, S., El-Manstrly, D., Tseng, M. L., & Ramayah, T. (2020). Sustaining customer engagement behavior through corporate social responsibility: The roles of environmental concern and green trust. *Journal of Cleaner Production, 262*.

Jahdi, K. S., & Acikdilli, G. (2009). Marketing communications and corporate social responsibility (csr): Marriage of convenience or shotgun wedding? *Journal of Business Ethics, 88*(1), 103–113.

Jalali, S. S., & Khalid, B. H. (2021). The influence of instagram influencers' activity on green consumption behavior. *Business Management and Strategy, 12*, 1.

Jordan, J., Sachdeva, S., Mazar, N., & SOM, O. (2015). Green consumerism: Moral motivations to a sustainable future. *Current Opinion in Psychology, 6*, 60–65. https://doi.org/10.1016/j.copsyc.2015.03.029

König, S., & Maier, E. (2024). Preaching to the choir: Do green influencers make a difference? *Journal of Cleaner Production, 447*. https://doi.org/10.1016/j.jclepro.2024.141449

Kolović, T., Vlastelica, T., & Krstić, J. (2023). Consumers' perception of green advertising and eco-labels: The effect on purchasing intentions. *Marketing, 54*(1), 54–66. https://doi.org/10.5937/mkng2301054K

MacKenzie, S. B., Lutz, R. J., & Belch, G. E. (1986). The role of attitude toward the ad as a mediator of advertising effectiveness: A test of competing explanations. *Journal of Marketing Research (JMR), 23*(2), 130–143. https://doi.org/10.2307/3151660

Minton, E., Lee, C., Orth, U., Kim, C.-H., & Kahle, L. (2012). Sustainable marketing and social media. *Journal of Advertising, 41*(4), 69–84. https://doi.org/10.1080/00913367.2012.10672458

Mishra, D., & Mishra, A. (2020). Sustainability inclusion in informatics curriculum development. *Sustainability, 12*(14), 5769. https://doi.org/10.3390/su12145769

Munir, S., & Mohan, V. (2022). Consumer perceptions of greenwashing: Lessons learned from the fashion sector in the UAE. *Asian Journal of Business Ethics, 11*(1), 1–44. https://doi.org/10.1007/s13520-021-00140-z

Mütze, S. (2022). Corporate Social Responsibility und die Effekte nachhaltiger Unternehmenspraktiken und unterschiedlicher Greenwashing-Methoden auf die Unternehmensreputation. *Junior Management Science, 2022*. https://doi.org/10.5282/jums/v7i3pp826-873

Ottman, J. A., Stafford, E. R., & Hartman, C. L. (2006). Avoiding green marketing myopia: Ways to improve consumer appeal for environmentally preferable products. *Environment: Science and Policy for Sustainable Development, 48*(5), 22–36.

Parguel, B., Benoit-Moreau, F., & Russell, C. A. (2015). Can evoking nature in advertising mislead consumers? The power of "executional greenwashing". *International Journal of Advertising, 34*(1), 107–134. https://doi.org/10.1080/02650487.2014.996116

Ragonnaud, G. (2024). *Green Claims Directive. Protecting consumers from greenwashing*. Briefing. EU Legislation in Progress. EPRS | European Parliamentary Research Service. PE753.958. https://www.europarl.europa.eu/ReData/etudes/BRIE/2023/753958/EPRS_BRI(2023)753958_EN.pdf.

Ray, P. H., & Anderson, S. R. (2000). *The cultural creatives: How 50 million people are changing the world.*

Schmuck, D., Matthes, J., & Naderer, B. (2018). Misleading consumers with green advertising? An affekt–reason–involvement account of greenwashing effects in environmental advertising. *Journal of Advertising, 47*, 127–145. https://doi.org/10.1080/00913367.2018.1452652

Segev, S., Fernandes, J., & Hong, C. (2016). Is your product really green? A content analysis to reassess green advertising. *Journal of Advertising, 45*(1), 85–93. https://doi.org/10.1080/00913367.2015.1083918

Stanton, J. L. (1998). Comparative effectiveness of executional elements in TV advertising: 15- versus 30-second commercials. *Journal of Advertising Research, 38*(6), 7–8. https://dialnet.unirioja.es/servlet/articulo?codigo=453238

Sun, Y., Luo, B., Wang, S., & Fang, W. (2020). What you see is meaningful: Does green advertising change the intentions of consumers to purchase eco-labeled products? *Business Strategy and the Environment, 30*(4). https://doi.org/10.1002/bse.2648

Szabo, S., & Webster, J. (2020). Perceived greenwashing: The effects of green marketing on environmental and product perceptions. *Journal of Business Ethics, 171*(4), 719–739. https://doi.org/10.1007/s10551-020-04461-0

TerraChoice. (2010). *The sins of greenwashing*. TerraChoice Environmental Marketing. https://www.twosides.info/wp content/uploads/2018/05/Terrachoice_The_Sins_of_Greenwashing_-_Homc_and_Family_Edition_2010.pdf
Thomas, W. I., & Thomas, D. S. (1928). The methodology of behavior study. In *The child in America: Behavior problems and programs* (pp. 553–576). A. A. Knopf.
Villarino, J., & Font, X. (2015). Sustainability marketing myopia: The lack of persuasiveness in sustainability communication. *Journal of Vacation Marketing, 21*(4), 326–335. https://doi.org/10.1177/1356766715589428
Weder, F. (2023). The evolution of the sustainability story: Strategic sustainability communication as niche construction. *International Journal of Strategic Communication, 17*(3), 228–244. https://doi.org/10.1080/1553118X.2023.2229304
Wonneberger, A., Matthes, J., Siegert, G., Wirth, W., Weber, P., & Lischka, J. A. (2016). Grüne Werbung : Inhalte und Wirkungen ökologischer Werbebotschaften. In *Handbuch Werbeforschung* (pp. 741–760). Springer Fachmedien. https://doi.org/10.1007/978-3-531-18916-1_32
Yıldırım, S. (2021). Do green women influencers spur sustainable consumption patterns? Descriptive evidences from social media influencers. *Ecofeminism and Climate Change, 2*(4), 198–210.
Zinkhan, G. M., & Carlson, L. (1995). Green advertising and the reluctant consumer. *Journal of Advertising, 24*(2), 1–6. https://doi.org/10.1080/00913367.1995.10673471

Further Reading

Greenwash and Green Trust: The Mediation Effects of Green Consumer Confusion and Green Perceived Risk

Can evoking nature in advertising mislead consumers? By Parguel, B., Benoit-Moreau, F., & Russell, C.A.

Ann-Elen Waldeier is a master's graduate in Communication Management from the Westphalian University of Applied Sciences, Germany. She is interested in consumer behaviour and advertising effectiveness, particularly in the case of products marketed as sustainable. She researched the suspicion of greenwashing in her master's thesis and now works in social and market research.

Stefan Weinacht is Professor of Communication Studies at the Institute of Journalism and Public Relations, Westphalian University of Applied Sciences, Gelsenkirchen, Germany. His research interests include the dependences between journalism and pr, data journalism and lately sustainability communication and sustainable communication shown in projects with the local soccer club Schalke 04.

Christopher Morasch is Professor of Public Relations at the Institute for Journalism and Public Relations, Westphalian University of Applied Sciences. His focus lies in digital communication management, in particular, online market research. He is the founder and CEO of digitell.me, a software company focused on online surveys.

Sustainability Communication Beyond the Business Case Frame: A Study of Investors Making Sense of Paradoxical Tensions

Johan Florén

Abstract Strategic sustainability communication in organizations managed by financial objectives is characterized by paradoxical tensions that arise within the organization when sustainability goals are added. Yet the dominant perspective on sustainability among practitioners and researchers is the Business Case Frame (BCF) which is ill-suited to address these tensions.

This chapter is built on a qualitative study of sustainable investors' approach to paradoxical tensions in their business. The interviewees are Swedish members of the Climate Action 100+ project, an investor-led engagement initiative on climate change.

The theoretical base is formative approaches to communication that regard it a sensemaking process where sustainability talk and practice constitute each other and meaning is continuously negotiated. Formative approaches to sustainability communication have come as a reaction to the dominance of instrumental, functionalist research perspectives.

The results indicate that investors have no objection to a BCF, although they confirm a potential conflict between sustainability and financial goals that are sometimes realized. Still, a BCF limits the ability of sustainable investors and companies to operate under the complex circumstances and contributes to the problems rather than solving them.

I propose an alternative to the dominant BCF and develop knowledge about formative sustainability communication by using the theory in a new field as it studies investors rather than companies in an inter-organizational rather than organization-centred context.

J. Florén (✉)
Department of Communication, Lund University, Lund, Sweden
e-mail: Johan.Floren@ap7.se; johan@floren.net

F. Weder (ed.), *Strategic Sustainability Communication*, CSR, Sustainability, Ethics & Governance, https://doi.org/10.1007/978-3-031-89486-2_9

1 Introduction

A public conversation about sustainability has been going on for decades. Complexity and uncertainty play an essential role in this debate (Godemann & Michelsen, 2011). Clear answers on how to reduce the subject's complexity and translate insights into solutions to problems are still missing. A manager with more than 15 years of experience in a sustainable investment organization succinctly captured the paradoxical tensions that pervaded public discourse:

> Either sustainability has to be the solution to everything *and* create the best return, or sustainability costs money. It's one or the other. It still is. Or, from the beginning it was that it always costs money, and now suddenly it's going to fix everything.

Sustainability communication is a fairly new concept. Ziemann (2011) sees it as a global social process consisting of a recursive sequence of contributions and debates for a better ecological, economic and social life. In organizations with financial drivers, there is a fundamental tension between financial goals and sustainability goals that affects communication on many levels. Complexity, ambiguity and contradictions challenge organizational sensemaking and strategic sustainability communication (Onkila et al., 2018), which tends to land in one-dimensional perspectives.

This chapter is based on a study of sustainable investors' approach to the paradoxical tensions that characterize their operations and communication. The purpose of the study is to develop knowledge and deeper understanding of challenges related to the dominant perspective, the *Business Case Frame* (Hahn et al., 2015), and to propose an alternative frame for strategic sustainability communication based on both theory and empirical material.

The premise is a constitutive, ritual view of communication and the empirical material consists of a qualitative study of Swedish investors participating in a global initiative to combat climate change, Climate Action 100+ (CA100). Communication research based on a constitutive perspective has in recent years been developed as "formative perspectives" (Schoeneborn et al., 2020) mainly focused on companies, and there are many examples of research that recognizes complexity and paradoxes as central prerequisites for sustainability communication, but empirical studies of investor organizations are rare. This study fills that gap and contributes to the development of knowledge about paradoxical dimensions of formative sustainability communication. Investors allocate resources and govern companies, but they are also a powerful force in the public conversation that shapes society's perception of what sustainability is.

The CA100 initiative consist of financial organizations that have a clear commitment to sustainability in both words and actions, and therefore face paradoxical tensions in their daily operations. To highlight the internal complexity related to sustainability in the organizations, a paradox lens is applied in the interviews. A paradoxical perspective can free sustainability research from the hegemony of the Business Case Frame (BCF) and a purely economically driven paradigm (Hahn et al., 2018). Paradoxes are triggers for sensemaking processes, leading members of organizations to rethink their practices and innovate new ways of reconciling what might at first have seemed irreconcilable (Child, 2020). The main research questions of the study are: *How do financial organizations make sense of the paradoxical tensions that comes with sustainability practices? What could an alternative to the BCF look like? What are the communication implications?*

Although investor engagement with companies is a central dimension of the CA100 initiative, the study is limited to investors' sensemaking around paradoxical tensions in this context. The study has a synthesizing approach when it comes to terminology. Strategic sustainability communication is consistently used as an umbrella concept, although some sources use other related prefixes such as CSR (Corporate Social Responsibility), CR (Corporate Responsibility), CS (Corporate Sustainability) and ESG (Environmental, Social, Governance) for example. The chapter begins with a problematization of the BCF and continue with a section about formative perspectives on sustainability communication and paradoxical frames. The third part presents the empirical study, and the last part consists of discussion and conclusions.

1.1 An Instrumental Frame

In organizations that are guided by financial objectives, sustainability is primarily handled within a conventional BCF that reflects an economic conceptual world where sustainability only is considered within the scope of business interests (Hahn et al., 2018). Sustainability is addressed at the organizational level, and social and environmental aspects are merely relevant if they improve the organization's financial performance. Other social and environmental benefits are ignored. In this way tensions are eliminated.

Vast scholarly attention has been devoted to instrumental economic frames and Dzhengiz and Hockerts (2022) point out that the excessive attention to this type of frames has depleted the potential for further research. The central paradigm of research on CSR, corporate communication, marketing literature and non-financial reporting is also that communication should ultimately contribute to profit maximization (Crane & Glozer, 2016; Verk et al., 2021). Focus is on exploring the ways in which organizations can profit from addressing societal challenges. This instrumental perspective does not sit well with the conditions of strategic sustainability communication for several reasons.

To begin with, sustainability *practices with negative financial effects* is altogether incompatible with a BCF. An organization that pursues sustainability for strictly financial reasons should immediately end it when the opposite proves to yield better returns, be it fossil energy, pollution, child labour or the like. But a consistent BCF can also be problematic for an organization that gets *positive financial results*. The challenge then is to explain how the business differs from conventional business or investments. Positive financial results are what everyone strives for. For example, investors who can increase risk-adjusted returns by taking sustainability into account are essentially committing malpractice if they don't seize the opportunity, whether they claim to be sustainable or not. The label "sustainable" in these contexts becomes a redundant addition that can even be perceived as greenwashing if it is combined with claims of social benefit beyond that which all companies and investors make (De Vries et al., 2015).

Furthermore, activities that do not generate economic benefits cannot be explained by profit-seeking (Schaltegger & Hörisch, 2017). With a strict instrumental view of sustainability, commitments that have *no positive effects on financial performance* are essentially a distraction that can even harm returns. An example of this concern is the oil and gas company BP which in 2024 came under pressure from activist hedge fund investors who claimed they were destroying shareholder value by moving away from hydrocarbons faster than society. The hedge fund wanted the company to abandon clean energy promises and the "irrational strategy" they believe has devalued BP's share price (Wilson, 2024).

All and all, one wonders why organizations that apparently believe they need to have an ambitious sustainability programme stick with the dysfunctional BCF. The obvious risk is that the organization's claims are not convincing either in terms of social or economic benefit (Chang et al., 2012; Revelli & Viviani, 2015; Billio et al., 2021).

2 Formative Sustainability Communication

The concept of sustainable development was widely spread with the so-called Brundtland report (World Commission on Environment and Development, 1987). It refers to a worldwide societal process aimed at tackling problems often characterized by high complexity, where communication is an essential part (Weder et al., 2021). Definitions of sustainability mostly relate back to the Brundtland report which states that we must meet the needs of today without compromising the ability of future generations to meet their needs. In line with the definition of strategic communication as "the purposeful use of communication by an organization or other entity to engage in conversations of strategic importance to its goals" (Zerfass et al., 2018, s. 493), strategic sustainability communication is an organizational communication practice of importance related to sustainability.

After a decade of increased attention paid to Corporate Social Responsibility (CSR) and the social licence to operate, Weder (2022) states that sustainability communication is slowly emerging as a new field of research at the intersection of Public Relations, Marketing, Consumer Research, Science and Environmental communication. The functional, instrumental understanding of communication dominates, which is confirmed by an increasing number of studies focusing on sustainability reporting, media channels for sustainability communication or related dissemination strategies. This instrumental "transmission view" of strategic sustainability communication reduces it to a tool for reaching financial goals and ignores the formative function of communication to constitute, change and maintain businesses (Schoeneborn & Trittin, 2013).

The instrumental transmission model has dominated the view of organizational communication during the twentieth century, but alongside it there is a communication model that sees communication as a process that constitutes society (Craig, 1999). Social order is created by common languages and norms, which are produced and reproduced through communication (Ziemann, 2011). Dhanesh (2022) identifies constitutive research paradigms as a promising way to generate insights into processes of sensegiving and sensemaking in sustainability communication.

Researchers with a strategic-*instrumental* orientation are primarily interested in how sustainability communication can be used as a means to achieve strategic ends such as reputation and profitability (Schoeneborn et al., 2022). In contrast, formative approaches to communication see sustainability practice as constructed by sustainability communication. Unlike the business case perspective, a formative perspective need not deny paradoxical tensions. Research with a formative perspective includes different ways of interpreting sustainability through sensemaking, framing or rhetoric that influence how sustainability is practiced (Schoeneborn et al., 2020). From this point of view, strategic sustainability communication is not only an instrument to achieve strategic goals, but a continuous process in which different societal actors explore, construct, negotiate and modify what it means to be a sustainable organization (Christensen & Cheney, 2011).

Based on the phrase "walk the talk", Schoeneborn et al. (2020) drafts three types of formative sustainability communication where the first type assumes that walk precedes talk. In the second type, the order between words and deeds is the reverse, the sustainability practice is shaped by the sustainability communication in the form of aspirational talk (Christensen et al., 2013). The third category is more complex and characterized by simultaneity rather than a sequence where one follows the other. Walk and talk are two sides of the same coin and the very dichotomy is erased. Time is always now in the sense that the past and future are always viewed through the present and are part of the present. Christensen et al. (2020) refer to this more complex time perspective as reflexive time.

Communication models are templates, sometimes unconscious, that guide concrete processes and human interaction (Carey, 2008). They not only describe human behaviour but also shape it. Therefore, an instrumental view not only makes it difficult for us to understand complex conditions but also contributes to the problems. From a formative perspective, an organization's sustainability function is instead stabilized and maintained by a number of communicative paradoxes that constitute it (Hoffmann, 2018). The implementation of sustainability practices is based on the continuous activation of these paradoxes to recreate sustainability as a communicatively constituted institution. Hoffmann concludes that organizations that are proactive about paradoxes and communicate them fare better because organizational tensions cannot be managed without verbalizing them. Although this "naming" may sound like a simple task, his analysis points to the amount of sophisticated communication effort devoted to paradox denial.

2.1 The Paradox Lens

To make sense of the complexity of sustainability issues, organizations need cognitive frames that enable them to assimilate relevant information and function effectively. Researchers such as Hoffmann (2018) point out that most sustainability discourses still avoid paradoxical tensions and argue for a communication theory that contributes not to denial but to the constitution of sustainability paradoxes. Smith and Lewis (2011) suggest that sensemaking can be facilitated if paradoxical tensions are accepted in an organization. Sensemaking is the continuous process where cues are interpreted using frames. "Frames tend to be past moments of socialization and cues tend to be present moments of experience" (Weick, 1995, p. 111). Paradoxical frames are mental templates that recognize and accept the coexistence of conflicting forces (Hahn et al., 2015).

The dynamic equilibrium model Smith and Lewis present (2011) has had considerable influence in academic research. It regards the conflicting forces of paradox within complex systems as dynamic rather than static and enduring rather than transitory. It also sees the forces as beneficial and powerful rather than harmful and threatening (Luo et al., 2020). Long-term sustainability requires continuous efforts to meet competing demands (Argento et al., 2022). Situations arise repeatedly where the organization must simultaneously achieve several desirable but conflicting economic, environmental and social outcomes at the company and societal level that operate within different time frames and follow different logics (Hahn et al., 2015). Instead of suppressing tensions, it is better to embrace them, raise awareness of them and analyse their characteristics (Heide et al., 2020).

Scherer et al. (2013) argue that organizations should take a paradoxical approach to preserve their legitimacy. In a heterogeneous and complex world with many conflicting expectations, organizations must be able to conduct moral reasoning simultaneously with traditional methods such as isomorphic adaptation to external stakeholders and strategic manipulation of expectations. Complexity in organizational practice is needed rather than a one-best-way approach. The authors perceive that legitimacy challenges are what primarily motivate organizations to adopt sustainable ways of working and engage in debates about sustainability issues. A study by Vanvik (2024) of the Norwegian oil company Equinor supports the view of legitimacy as a main driver, while showing that a paradoxical approach can be challenging.

For a researcher studying complexity, the BCF is a constraint since it only addresses sustainability at the organizational level and tensions are eliminated by defining sustainability as a contribution to the organization's financial goals. Environmental and social considerations are given no value in themselves (Hahn et al., 2018). A paradox lens is helpful in deepening the understanding of complex phenomena by revealing underlying assumptions (Sparr et al., 2022). Instead of choosing between competing demands, a researcher with a paradox lens can explore how to engage contradictions simultaneously.

3 Sustainable Investors Make Sense of Paradoxical Tensions

The empirical material in this chapter consists of a qualitative interview study of Swedish members of the Climate Action 100+ (CA100) project, a global investor-led initiative to persuade the world's largest corporate greenhouse gas (GHG) emitters to take action against climate change (Climate Action 100+, 2024).

Launched in December 2017, CA100 is today made up of 600 global investors across 33 markets, seeking to persuade companies with large GHG emissions to improve climate impact governance, reduce emissions and strengthen climate reporting. The 170 emitting companies targeted by CA100 account for 80% of global emissions from industry and are thus key to the transformation of this part of society. The signatories believe that engagement and collaboration with companies, which ensures better transparency of their climate risks and strategies to reduce emissions, is needed to achieve the goals of the Paris Agreement.

The initiative was formed in the wake of the 2015 Paris Agreement, which aims to keep the increase in global average temperatures below two degrees compared to pre-industrial levels. Initially, it was launched as a 5-year initiative (2018–2022), but in 2023, it announced a second phase that would run from 2023 to 2030. The new phase focus on a global scale-up in active ownership, shifting the focus from disclosure to the implementation of climate transition plans.

3.1 *Sample and Methodology*

Of the 700 global investors in CA100, 23 come from Sweden and consist of 14 asset owners, 8 asset managers and 1 service provider. Asset owners are, for example, pension funds, insurance companies and foundations—organizations that represent holders of long-term pension savings, insurance and other assets. Asset managers are organizations that manage assets on behalf of clients, who can be both institutional investors and private investors.

The empirical part of the study consists of semi-structured, qualitative interviews conducted 2022–2023 with Swedish representatives from both asset owners' organizations and asset management organizations. The interviews began with an introductory part about the initiative itself and the investors' perception of it before delving into the topic of paradoxical tensions. Prior to the interviews, a list of paradoxical tensions was compiled based on both literature and my experiences in the industry (see below). From this list, seven sustainability-related tensions relevant to the investors were selected and presented for the investors to comment and reflect on. In the analysis of the answers, the results were grouped into three paradox themes presented in this chapter.

The interviewees are primarily managers and specialists with responsibility for sustainability, corporate governance and communication who have been involved in CA100. They manage sustainability practices and communications and make sense of paradoxical tensions in financial organizations, on a daily basis. I have conducted interviews with 16 people in 13 different organizations, one of which was a background interview with an expert who was deeply involved in the project. The interviews took between 50 and 80 min and were conducted in physical meetings or via Teams. All interviews were conducted in Swedish and translated, except for one that was conducted in English.

The study's methodological approach is reflexive. The empirical part of the study consists of information gathering in a less explored area, but it is also a resource for developing ideas and problematizing existing theoretical frameworks. The empirical material is not regarded as a "reflection of reality" that can falsify or verify theories but as arguments for or against theoretical ideas and ways of understanding the world. Using a metaphor, one can say that empirical material conducts a critical dialogue with theory and argues for and against various positions (Alvesson & Sköldberg, 2018). This approach is qualitative and does not collect large amounts of quantitative data to draw general conclusions. Also, this study is limited to the internal aspect of the complex conditions of strategic sustainability communication. External complexities are an important topic for other studies.

I have been working at a pension fund for many years and, before that, been active in a human rights organization. These experiences have inevitably influenced my understanding of the subject as well as the interviews and interpretations of the answers. The experience as a practitioner can lead to more initiated interviews and a deeper understanding of the empirical material. At the same time, the same experiences entail a risk of uncritical acceptance of implicit assumptions and claims from participants. A problematizing and critical theoretical approach mitigates this risk to some extent.

3.2 *Findings*

Paradoxical tensions are an example of the complex conditions that strategic sustainability communication needs to handle. The results from the interviews are presented in three themes: (1) *Financial goals and sustainability goals*, which capture the fundamental tension; (2) *To own or not to own*, perhaps the most debated tension within sustainable investing, and (3) *Between greenwash and greenhush*, a predicament many sustainability communicators testify to.

3.2.1 Financial Goals and Sustainability Goals

The participants in the CA100 investor initiative are organizations with financial performance as their core business, while being committed to contributing to sustainable development. The commitment is arguably beyond mere sustainability statements and there is no immediate financial return, although each participant can choose the level of engagement. None of them consider sustainability incompatible with the objectives of the investment business. Nevertheless, they confirm that there is a potential conflict between sustainability and financial goals that are sometimes realized:

> "I would probably say that we have very clearly experienced that conflict in the last two years considering how bad it has been from a short-term perspective not to own fossil fuels."... "We think that it is a short-term effect, that of course may affect some of our customers depending on when they need to go in and out, but in the long run we still believe that this is a temporary situation and that it will not look like this for the next 5, 10, 15, 20 years".

The investors think this paradoxical tension is resolved in a longer perspective. They have a belief that unsustainable companies carry a risk that makes them poor long-term investments. With this approach, the tension between financial goals and sustainability goals is actually a tension between the short and long term:

> "We are a long-term investor; I think most people understand that. You can certainly make money in the short term from a lot of crap, but in the longer term we, and many others, believe that we get better returns if our holdings behave in a good way".

When asked directly, many downplay the challenge posed by tensions between sustainability and financial goals, but they emerge spontaneously in various places in the interviews, also depending on external factors. The lack of political decisions that adapt society's economic structures to a low-carbon future is repeated in several interviews:

> "To some degree, many of these companies can do quite radical things, which will bring about change, and do it in a profitable way. Lead some innovation and so on. But at some point it stops being profitable unless policy catches up and makes it profitable. So that's where we might be very soon, or maybe that's where we are now. These companies have built their capabilities and insights, but it is still not profitable to make the investments that will get them there, because the necessary conditions are missing in the market. That's the big difficulty".

3.2.2 To Own or Not to Own

> "That's probably the main tension of public interest".

A paradoxical tension that receives a lot of attention in the public debate is the choice between owning and not owning, that is, owning and influencing companies to become more sustainable or simply excluding companies deemed unsustainable from investment portfolios. In the individual case, it is impossible to do both. Ownership is required to vote, submit shareholder resolutions and collaborate with other owners.

> "One problematic thing is that the worst companies have been selected and since everyone focus on divesting to make the portfolio look good, we are not as many responsible owners as we could have been, who are actually involved and influencing".

Nevertheless, having a consistent stance proves difficult, if not impossible. While the investors have chosen to participate in the CA100 because they believe in engagement and active ownership, they invariably also exclude companies for various purposes. One common reason is that certain industries are considered unethical, either by the organization itself or by stakeholders. In this case, the motive is neither the investor's ability to influence the company nor financial reasons:

> "It's based on values, that you think that you shouldn't be invested in certain things".

At portfolio level, it is clear that nothing prevents investors from using both exclusion and active ownership as different tools in sustainability work. Exclusions can even be used as a means of influence. For example, it can be a final step that increases the effect of previous steps in an escalation process. An imminent exclusion from an investor portfolio tends to increase the engagement of many companies in a dialogue about solutions to sustainability problems.

The central question in the "to own or not to own" debate is what does the most good, but in reality it is often reputational risk and commercial considerations that determine final decisions. Even for asset owners who believe in and practice active ownership, exclusions are part of the business. For asset managers, who by definition manage other people's assets, it is inevitable:

> "We want to work actively and get involved. We don't want to exclude. Then we must get the clients on board. If we get the clients on board, we'd rather engage than exclude. But it's a huge job to get people on that train. And until then, you have to exclude, and engage in the companies you can" ... "But you can't run a fund today, or we can't, where we don't exclude. Because then we have no investors in the fund".

3.2.3 Between Greenwash and Greenhush

The external communication brings tensions which, among other things, materialize in the daily interaction within the organization. External communication challenges stakeholders to scrutinize the business. If the communication goes too far in the ambition to reach out, it risks being perceived as greenwashing.

> "There can be some tension internally between us who do the work and those who want to manage the communication. 'This is going to be too difficult'. That's the most common. 'This is too complicated. We need to simplify'. Or 'This is too long; we have to shorten it.' 'This was getting too boring. Can't we write like this instead?' And that's when it goes wrong. So, I would say that the internal conflict is quite tangible".
>
> "It's a tension, absolutely. Because it becomes a tension when they want to simplify so much that it ends up being wrong. And with that error, it becomes greenwash".

This risk, in turn, can lead to its opposite, "greenhushing", which also risks damaging reputation. Without communication, the outside world will never know what good work is being done:

> "For our part, we are more nervous. Better that we act without being seen than be seen without acting. So, we are probably more on that side of the court. It would be much worse to be perceived as greenwashing than if we do something good that nobody knows about".

The investors' external communication of CA100 mainly consists of sustainability reporting, governance reports and corporate websites. It is clear that participation in the specific project is primarily neither a marketing nor a legitimacy strategy. The conclusion that legitimacy has limited importance is confirmed by the generally low interest from stakeholders in the initiative:

> "We probably haven't put a lot of effort into marketing specifically linked to Climate Action compared to quite a few other engagement initiatives that we've done."... "I called it a hygiene factor before and I would say that it is a part of that legitimacy puzzle. But I have never seen that any stakeholders have asked for this particular thing or raised a lot of eyebrows for this particular initiative".

Part of the explanation for the low interest of Swedish stakeholders is likely that sustainability engagement is clearly related to mass media reporting, where global investor collaboration rarely reaches the top of the Swedish agenda. Investors need to create awareness of the sustainability work can also conflict with the collaborative effort itself because individual initiatives create better opportunities to profile their own organization.

4 Discussion

4.1 Paradoxes of the Business Case

This study challenges the dominant business case framing of sustainability in organizations driven by financial goals and explores how investors make sense of the paradoxical tensions that arise when sustainability goals are added. It turns out that the BCF is something of an unexamined presumption among investors. None of them have any objections to a BCF. Nor do any of the investors use a paradoxical framework for the integration of sustainability in the business. This is interesting since the BCF itself creates additional paradoxical tensions when it simultaneously tries to claim that the organization's motive is a sincere social commitment, and that the activity is driven solely by financial goals.

Furthermore, part of the daily business is dealing with recurring contradictions. But the BCF leaves no room for contradictions, since the organizations' sustainability practices are instrumental in relation to the financial goals, which means that contradictions must be denied. From the management perspective, paradoxical tensions are rejected to create uniformity and harmony in the organization (Hoffmann, 2018). Perhaps this to some part explain participants' responses in line with Child's (2020) observation that many practitioners, unlike researchers and critics, do not see sustainable investment and entrepreneurship as problematic. Tensions exist, but the consequences are downplayed. The answers can also be interpreted as examples of the acceptance that the dynamic equilibrium model advocates as a path to a "virtuous circle" (Smith & Lewis, 2011). This attitude indicates that the instrumental view of sustainability has a weaker position in practice than in research. To some extent, the participants in CA100 adopt a paradoxical perspective, although not explicitly. How is it then that they still uphold the BCF?

Several motives are possible. An obvious reason to keep this frame is precisely that it is the most established one. They just follow convention. Deviating from conventional beliefs is a risk in a field where there are many strong opinions and feelings, and public criticism can be stinging. Another reason to defend the sustainability contribution to the financial goals, may be that it's strongly linked to the legitimacy of the organization. The mere suspicion that returns do not come first is an existential threat to the business, illustrated by the aforementioned BP case (Wilson, 2024). External communication can also be a contributing factor to the dominance of the BCF. What is the alternative? A paradoxical perspective is hardly useful for explaining the organization's activities, which rather risk being perceived as having problems. Straight questions require straight answers.

A paradox lens is a powerful research tool to uncover complexities that characterize sustainability communication. It's a key to unlock deeper understanding of important preconditions. But judging from the study, a pure paradox frame is not a viable alternative to the BCF for practitioners. A straightforward explanation is that practitioners prefer to solve problems, not construct them. Although latent paradoxes abound, practitioners will avoid turning them into salient paradoxes if they can. And as Hahn and Knight (2021) argues, paradoxes have no standing until they are recognized by those affected. The paradox perspective is valuable, but to break the BCF's dominance, an alternative way of framing is needed that does not just problematize and uncover complexity.

With the dominant view in research and society at large, the instrumental frame may still appear to simplify, but in fact it hampers organizations' sensemaking, operations and external communication. As stakeholder theory has long made clear, a company is a "multi-purpose entity" (Crane et al., 2014). Dealing with paradoxical tensions requires first accepting that they exist (Smith & Lewis, 2011; Vangen, 2017; Hoffmann, 2018; Vallaster et al., 2021). A fundamental problem with the BCF is that it only addresses sustainability at the organizational level (Hahn et al., 2018) while sustainable development is a global social process for dealing with problems that pose an existential threat to both people and nature, that no one can handle on their own.

In all parts of society today, there is an ongoing conversation about how to deal with sustainability problems. Companies and investor organizations need to be part of that conversation, both for their own sake and for society's sake. They need to participate because they cannot decide what to do on their own. And they must deal with conditions that change at breakneck speed. Sustainability is an umbrella concept for a range of different problems, approaches, strategies and practices that are constantly negotiated and renegotiated. No one owns the concept, and everyone has partial insight into how the problems should be solved. For that reason, the organizations also need to participate in the conversation to put forward their beliefs. Scherer et al. (2013) concluded that it is in their self-interest not only to conform but also to contribute their views and shape the society. This is true, but it is not just a need for legitimacy. The Brundtland report was based on the realization that environmental issues and economic development cannot be treated separately (Fegler & Unemo, 2000). Organizations are part of society, and the problems cannot be solved without their participation. CA100 can be seen as a result of the need to participate.

4.2 An Alternative Frame

Frames are organizing principles of communication that are persistent over time and meaningfully structure the social world (Weder, 2023). The BCF restrains the ability of sustainable investors and companies to operate under present circumstances and contributes to the problems rather than solving them. A formative communication perspective offers a way out of this impasse. An alternative to the BCF begins with the idea that equally legitimate elements of organizations can nevertheless constitute a contradiction. With a formative frame, strategic sustainability communication constitutes a flow where practice and communication continuously constitute each other and meaning is negotiated between different societal actors who explore, construct and modify what it means to be a sustainable organization (Schoeneborn et al., 2020). This process gives the organization the opportunity to develop business structures and activities in an area characterized by complexity, contradictions and constant change. We can call an alternative frame based on this approach a *Sustainability Agent Frame (SAF).*

As this study indicates a paradox perspective can be valuable, but a pure paradox frame might provide too much complexity and too little meaning. At the other end of the scale, a frame without complexity is conceivable as well. Such a frame might dictate, for example, that profit and returns are the sole concern of companies and investors (Friedman, 1962), or conversely, that capitalism must be abandoned entirely if a sustainable society is to be achieved (Alexander, 2020). However, such a "dogmatic" frame is in principle, incompatible with sustainable investments and sustainable businesses, as the basic premise is that the combination of sustainability goals and economic goals is a zero-sum game (Dzhengiz & Hockerts, 2022). One can only be achieved at the cost of the other. So, in this case, dogmatic frames have no relevance, see Table 1.

The BCF is a low complexity frame. It does not consider business or investment operations incompatible with sustainability, but it denies the paradoxical tensions and minimizes the complexity by instrumentalizing sustainability as a means to achieve financial ends. To do this, the BCF must exclude dimensions of

Table 1 Comparison of frames

Frame	Business case	Sustainability agent
Level of complexity	Low	High
Paradoxical tensions	Deny	Acknowledge
Relation sustainability/financial objectives	Means-Ends	Field polarity
Sustainability focus	Organization	Society
Communication theory	Transmission	Constitutive

sustainability outside the organization, which in itself is paradoxical as the claim is that it contributes to a greater good. In line with a transmission perspective, communication focuses on reporting performed sustainability activities, with the paradoxical consequence that organizations are drawn to understated communication and reduced transparency to avoid accusations of greenwashing (Morsing et al., 2008). This in turn confirm the perspective of critical stakeholders who acknowledge the existence of tensions but attribute them to a discrepancy between rhetoric and practice and do not interpret the difference as a paradox but rather as a form of organizational hypocrisy (Hoffmann, 2018).

The SAF is a high-complexity frame that recognizes paradoxical tensions as part of the dual ambition to meet both financial and sustainability goals. Since the two objectives have an equal standing, a field polarity of continuous meaning negotiation is formed where economics and sustainability practices shape each other, and the organization's sustainability function is maintained and stabilized. (Hoffmann, 2018). By accepting tensions within the organization, the SAF has a degree of complexity needed to integrate sustainability aspects outside the organization. Only with a formative perspective on communication can paradoxical tensions be managed, and a constantly evolving practice constituted. The practice become "talked into being" through sustainability communication.

5 Conclusion

Strategic sustainability communication in organizations driven by financial goals will be problematic as long as the starting point is the BCF that only addresses sustainability at the organizational level. Since sustainability is merely seen as a contribution to the organization's financial goals, environmental and social considerations are given no value in themselves. This frame not only denies paradoxical tensions that exist in organizations but is also the source of new paradoxes that impair the organization's ability to achieve its goals. A BCF lowers the credibility of claims about both economic and sustainability benefit and may increase rather than decrease stakeholders' suspicions of greenwashing.

If organizations want to stop being accused of greenwashing, communication cannot be seen only as an instrument to spread information about their good deeds (Cooren, 2020). But judging by the study's results, the pure paradoxical frame is not an alternative for organizations to the BCF. A paradoxical lens helps researchers study tensions in the operations of organizations and can of course be applied with the same purpose by organizations, but a pure paradox frame is insufficient for organizations that need to establish and explain their positions and decisions. It highlights tensions and reveals complexity, but it does not provide the meaning and visions needed for sustainability communication and practice.

This chapter outlines an alternative to the BCF based on a formative communication model that sees communication as a sensemaking process that constitutes, changes and sustains companies. With this SAF, the organization is an actor that participates in an interactive process to continuously create and recreate the meaning of sustainability in an applied context (van Ruler, 2018). The organization is an active participant in the conversation but has limited control over its development and thus also over its own application of the concept. Thus, the SAF gives organizations driven by financial goals a way to relate to sustainability without minimizing the complexity of either the organization or the societal problem.

For practitioners, this does not mean that planning and reporting of sustainability practices is abandoned, but rather that it becomes a background structure to a more emergent strategy (King, 2009). The process is guided by ongoing internal and external conversations maintained by the organization's specialists, communicators and spokespersons. The conversation is complex, fluid and requires knowledge of the subject as well as rhetorical skills. The organization needs preparedness to be able to mix proactive and reactive participation in a conversation about both principles and pragmatics, and to go back and forth between arguing, explaining and listening, to both influence and adapt to the environment. Too much detailed planning and rigid management by objectives impair this necessary preparedness.

Traditional communication campaigns are not a vital component, which participation in various collaborative forums in contrast are. They provide the organization with opportunities to interact with actors who have similar as well as divergent points of view. Aspirational communication (Christensen et al., 2013) is required to drive progress through commitments to future action and to balance the risk aversion and historical focus of conventional reporting. The conversation will lead the way and practices must be adapted to what has been communicated. The words spoken will determine what deeds are done.

References

Alexander, S. (2020). Postcapitalism by design not disaster. *The Ecological Citizen, 3*, 13–21.

Alvesson, M., & Sköldberg, K. (2018). *Reflexive methodology: New vistas for qualitative research* (3rd ed.). Sage.

Argento, D., Broccardo, L., & Truant, E. (2022). The facets of the sustainability paradox. *Meditari Accountancy Research, 30*(7), 26–48.

Billio, M., Costola, M., Hristova, I., Latino, C., & Pelizzon, L. (2021). Inside the ESG ratings:(Dis) agreement and performance. *Corporate Social Responsibility and Environmental Management, 28*(5), 1426–1445.

Carey, J. W. (2008). *Communication as culture, revised edition*: Essays on media and society. Routledge.

Chang, C. E., Nelson, W. A., & Doug Witte, H. (2012). Do green mutual funds perform well? *Management Research Review, 35*(8), 693–708.

Child, C. (2020). Whence paradox? Framing away the potential challenges of doing well by doing good in social enterprise organizations. *Organization Studies, 41*(8), 1147–1167.

Climate Action 100+. (2024). *Progress Update 2023*. Retrieved February 20, 2024, from https://www.climateaction100.org/

Cooren, F. (2020). A communicative constitutive perspective on corporate social responsibility: Ventriloquism, undecidability, and surprisability. *Business & Society, 59*(1), 175–197.

Craig, R. T. (1999). Communication theory as a field. *Communication Theory, 9*, 119–161.

Crane, A., & Glozer, S. (2016). Researching corporate social responsibility communication: Themes, opportunities and challenges. *Journal of Management Studies, 53*, 1223–1252.
Crane, A., Palazzo, G., Spence, L. J., & Matten, D. (2014). Contesting the value of "creating shared value". *California Management Review, 56*(2), 130–153.
Christensen, L. T., & Cheney, G. (2011). Interrogating the communicative dimensions of corporate social responsibility. In *The handbook of communication and corporate social responsibility* (pp. 491–504).
Christensen, L. T., Morsing, M., & Thyssen, O. (2013). CSR as aspirational talk. *Organization, 20*(3), 372–393.
Christensen, L. T., Morsing, M., & Thyssen, O. (2020). Timely hypocrisy? Hypocrisy temporalities in CSR communication. *Journal of Business Research, 114*, 327–335.
De Vries, G., Terwel, B. W., Ellemers, N., & Daamen, D. D. L. (2015). Sustainability or profitability? How communicated motives for environmental policy affect public perceptions of corporate greenwashing. *Corporate Social Responsibility and Environmental Management, 22*, 142–154.
Dhanesh, G. S. (2022). 22. Corporate responsibility and strategic communication. *Research Handbook on Strategic Communication, 319.*
Dzhengiz, T., & Hockerts, K. (2022). Dogmatic, instrumental and paradoxical frames: A pragmatic research framework for studying organizational sustainability. *International Journal of Management Reviews*, 1–34.
Fegler, C., & Unemo, L. (Eds.). (2000). *Långtidsutredningen 1999/2000: [LU]. Bilaga 7 Vad är hållbar utveckling?* Fritzes offentliga publikationer.
Friedman, M. (1962). *Capitalism and freedom.* University of Chicago Press.
Godemann, J., & Michelsen, G. (2011). *Sustainability communication–an introduction* (pp. 3–11). Springer Netherlands.
Hahn, T., & Knight, E. (2021). The ontology of organizational paradox: A quantum approach. *The Academy of Management Review, 46*(2), 362–384.
Hahn, T., Preuss, L., Pinkse, J., & Figge, F. (2015). Cognitive Frames in corporate sustainability: Managerial sensemaking with paradoxical and business case frames. *The Academy of Management Review, 39*, 463–487.
Hahn, T., Figge, F., Pinkse, J., & Preuss, L. (2018). A paradox perspective on corporate sustainability: Descriptive, instrumental, and normative aspects. *Journal of Business Ethics, 148*, 235–248.
Heide, M., von Platen, S., Simonsson, C., & Falkheimer, J. (2020). Expanding the scope of strategic communication: Towards a holistic understanding of organizational complexity. In *Future directions of strategic communication* (pp. 124–140). Routledge.
Hoffmann, J. (2018). Talking into (non)existence: Denying or constituting paradoxes of corporate social responsibility. *Human Relations, 71*(5), 668–691.
King, C. L. (2009). Emergent communication strategies. *International Journal of Strategic Communication, 4*(1), 19–38.
Luo, B. N., Tang, Y., Chen, E. W., Li, S., & Luo, D. (2020). Corporate sustainability paradox management: A systematic review and future agenda. *Frontiers in Psychology, 11*, 1–15.
Morsing, M., Schultz, M., & Nielsen, K. U. (2008). The 'Catch 22'of communicating CSR: Findings from a Danish study. *Journal of Marketing Communications, 14*(2), 97–111.
Onkila, T., Mäkelä, M., & Järvenpää, M. (2018). Employee sensemaking on the importance of sustainability reporting in sustainability identity change. *Sustainable Development, 26*, 217–228.
Revelli, C., & Viviani, J. L. (2015). Financial performance of socially responsible investing (SRI): What have we learned? A meta-analysis. *Business Ethics: A European Review, 24*(2), 158–185.
Schaltegger, S., & Hörisch, J. (2017). In search of the dominant rationale in sustainability management: Legitimacy- or profit-seeking? *Journal of Business Ethics, 145*, 259–276.
Scherer, A. G., Palazzo, G., & Seidl, D. (2013). Managing legitimacy in complex and heterogeneous environments: Sustainable development in a globalized world. *Journal of Management Studies, 50*(2), 259–284.

Schoeneborn, D., & Trittin, H. (2013). Transcending transmission: Towards a constitutive perspective on CSR communication. *Corporate Communications, 18*(2), 193–211.
Schoeneborn, D., Morsing, M., & Crane, A. (2020). Formative perspectives on the relation between CSR communication and CSR practices: Pathways for walking, talking, and t (w) alking. *Business & Society, 59*(1), 5–33.
Schoeneborn, D., Glozer, S., & Trittin-Ulbrich, H. (2022). Constitutive views on CSR communication: The communicative constitution of responsible organization, organizing, and organizationality. In *The Routledge handbook of corporate social responsibility communication* (pp. 73–84). Routledge.
Smith, W., & Lewis, M. (2011). Toward a theory of paradox: A dynamic equilibrium model of organizing. *The Academy of Management Review, 36*.
Sparr, J. L., Miron-Spektor, E., Lewis, M. W., & Smith, W. K. (2022). From a label to a metatheory of paradox: If we change the way we look at things, the things we look at change. *Academy of Management Collections, 1*(2), 16–34.
Vallaster, C., Maon, F., Lindgreen, A., & Vanhamme, J. (2021). Serving Multiple Masters: The role of micro-foundations of dynamic capabilities in addressing tensions in for-profit hybrid organizations. *Organization Studies, 42*(6), 911–947.
Vangen, S. (2017). Developing practice-oriented theory on collaboration: A paradox lens. *Public Admin Rev, 77*, 263–272.
Vanvik, E. C. (2024). The paradoxes in corporate sustainability communication–a critical. *Strategic Communication–Contemporary Perspectives, 8153*.
van Ruler, B. (2018). Communication theory: An underrated pillar on which strategic communication rests. *International Journal of Strategic Communication, 12*(4), 367–381.
Verk, N., Golob, U., & Podnar, K. (2021). A dynamic review of the emergence of corporate social responsibility communication. *Journal of Business Ethics, 168*, 491–515.
Weder, F. (2022). Strategic problematization of sustainability reframing dissent in strategic communication for transformation. *Public Relations Inquiry, 11*(3), 337–360.
Weder, F. (2023). The evolution of the sustainability story: Strategic sustainability communication as niche construction. *International Journal of Strategic Communication, 17*(3), 228–244.
Weder, F., Karmasin, M., Krainer, L., & Voci, D. (2021). Sustainability communication as critical perspective in media and communication studies—an introduction. In F. Weder, L. Krainer, & M. Karmasin (Eds.), *The sustainability communication reader*. Springer.
Weick, K. E. (1995). *Sensemaking in organizations*. Sage.
Wilson, T. (2024, January 29). BP faces activist investor pressure to ditch clean energy pledges. *Financial Times*. Retrieved January 30, 2024, from https://www.ft.com/content/626e344c-e266-4a70-b947-f122aa2fe65b
World Commission on Environment and Development. (1987). *Our common future*. Oxford University Press.
Zerfass, A., Verčič, D., Nothhaft, H., & Werder, K. P. (2018). Strategic communication: Defining the field and its contribution to research and practice. *International Journal of Strategic Communication, 12*(4), 487–505.
Ziemann, A. (2011). Communication theory and sustainability discourse. In J. Godemann & G. Michelsen (Eds.), *Sustainability communication*. Springer.

Further Reading

Hahn, T., & Knight, E. (2021). The ontology of organizational paradox: A quantum approach. *Academy of Management Review, 46*(2), 362–384.
Kölbel, J. F., Heeb, F., Paetzold, F., & Busch, T. (2020). Can sustainable investing save the world? Reviewing the mechanisms of investor impact. *Organization & Environment, 33*(4), 554–574.

Johan Florén is a PhD student at the Department of Communication at Lund University. He has a bachelor's degree in philosophy and a bachelor's degree in marketing communication from Uppsala and Stockholm universities. He currently works as Chief ESG and Communication Officer at the Swedish pension fund AP7 and has a private sector background working in communications since the 1990s. Johan Florén is a member of the Taskforce on Nature-related Financial Disclosures (TNFD) and has previously been chairman of Sweden's forum for sustainable investments (SWESIF) and chairman of the Amnesty Business Group in Sweden. https://portal.research.lu.se/sv/persons/johan-flor%C3%A9n; https://www.linkedin.com/in/johan-flor%C3%A9n-78638b4/

Introducing "Corporate Environmental Efficacy" Beliefs: A Stakeholder-Centric Perspective on Strategic Sustainability Communication

Alexandra Schwinges and Markos Mpadanes

Abstract One of the fundamental questions in strategic sustainability communication is how to foster sustainable transformation utilizing purposeful communication. As stakeholders act in the context of their own perceptions of the capability and effectiveness of corporations to tackle the climate crisis, it is vital to approach strategic communication from a perceptual perspective in addition to functional, managerial, and others. This theoretical contribution proposes "corporate environmental efficacy" as a set of beliefs of the capability (corporate self-efficacy, corporate efficacy, and corporate responsiveness efficacy) and effectiveness (corporate response efficacy) of corporate environmental action. Standing on firm interdisciplinary legs, we propose an integrative framework to understand stakeholders' confidence in sustainable transformation. Key takeaways include a nuanced understanding and operationalization of corporate environmental efficacy, as well as a communication strategy for corporate efficacy. We suggest that further development of an empirical corporate environmental efficacy construct may offer scholars and, not least, practitioners a stakeholder-centric framework to constitute, organize, and evaluate strategic communication for sustainability. In doing so, corporations can effectively engage with stakeholders and reciprocally drive sustainable transformation—moving beyond strategic communication of sustainability to acknowledge that corporate environmental efficacy has the potential to initiate a new cycle of stakeholder-driven and, therefore, more resonating communication.

A. Schwinges (✉)
Amsterdam School of Communication Research, University of Amsterdam, Amsterdam, Netherlands
e-mail: a.schwinges@uva.nl

M. Mpadanes
Department of Communication and Media Research, University of Zürich, Zürich, Switzerland
e-mail: m.mpadanes@ikmz.uzh.ch

F. Weder (ed.), *Strategic Sustainability Communication*, CSR, Sustainability, Ethics & Governance, https://doi.org/10.1007/978-3-031-89486-2_10

1 A Stakeholder-Centric Perspective on Strategic Communication for Sustainability

In an era defined by pressing environmental challenges and growing societal demands for corporate responsibility, the role of communication in driving sustainable transformation has never been more vital. The human-made climate crisis has become the severest threat to all living creatures and biodiversity. Multinational corporations, in particular, are responsible for a vast amount of carbon emissions (Steenbergen & Saurav, 2023) and, thus, bear a special responsibility regarding several Environmental, Social, and Governance (ESG) aspects. Corporations' role in a sustainable transformation process hinges on their stakeholders' trust in their ability to act efficiently. The term "efficacy" captures this belief of being capable of engaging in sustainable action and of doing so effectively. Efficacy beliefs have long been recognized in social psychological research on environmental behavior (Bandura, 1977) but have barely found their way into strategic communication research. In this chapter, we introduce "corporate environmental efficacy" as a set of beliefs offering an integrated framework to understand and evaluate how corporations can strategically communicate on ESG issues; and, by acknowledging the role of stakeholders and their feedback, ultimately foster sustainable transformation for the better of societies. We define corporate environmental efficacy as:

> [...] a stakeholder's set of beliefs relating to a corporation and its capability to substantially tackle environmental challenges and respond to stakeholder demands (efficacy), and its effectiveness in achieving desired pro-environmental outcomes (response efficacy).

As stakeholders increasingly scrutinize the corporate role in sustainability and demand transparency, corporate sustainability communication emerges as a powerful tool for fostering climate understanding, engagement, and action (Weder, 2023). To address these demands, corporations must strategically communicate their efforts to stimulate and meet societal needs and expectations. Mainly on a voluntary basis, some corporations have already been making disclosures on environmental, social, and government (ESG) aspects. But to tackle the climate crisis more wide-ranging and substantially, the binding corporate sustainability reporting directive (CSRD) was introduced and received final approval from the Council of the European Union. The CSRD mandates that since the fiscal year 2024, businesses have to publish extensive sustainability information annually. As part of the "European Green Deal," large business entities and listed companies are held more accountable, different sustainability standards will be avoided, and the shift to a sustainable economy will be made easier (European Commission, 2023a).

Sustainability and ESG reporting, as a form of corporate sustainability communication, belong to the realm of strategic communications, a field of practice aiming at informing and influencing public media discourse, policy, and perceptions of corporations (Zerfass et al., 2018). Strategic communication *for* sustainability goes beyond simply reporting sustainability-related corporate activities; it aims to catalyze sustainable transformation and cultivate fresh perspectives on sustainability.

Our perception of anthropogenic climate change has seen a significant shift over the years, leading to different approaches to climate action compared to a few years ago. This is where strategic sustainability communication and its ability to inspire meaningful action and corporations' skills and effectiveness in addressing Environmental, Social, and Governance (ESG) challenges comes in—not only with outward-facing messages but also internal communication strategies to engage employees with sustainability objectives.

Within strategic sustainability communication, our chapter introduces "corporate environmental efficacy" beliefs. This innovative notion offers a stakeholder-centric perspective and bridges the realms of environmental efficacy and corporate engagement. Upfront, it is important to stress one careful caveat. Empirical corporate capability and efficiency are essential, but so are people's perceptions of these. While efficacy typically denotes the ability to produce a desired outcome, our framework shifts focus to the subjective perceptions thereof. In essence, it is about the *beliefs* of stakeholders, whether accurate or not, about a corporation's capability and effectiveness in tackling ESG issues. The transformative nature of communication cultivates new norms like sustainability (Weder & Erikson, 2023) that, intertwined with efficacy beliefs, become powerful drivers of sustainable change (Barth et al., 2016).

Our stakeholder-centric approach to sustainability communication offers several vital contributions. Firstly, we discuss corporate environmental efficacy as a *constitutive* principle. Through our participative and organizational perspective, we recognize corporate efficacy beliefs as pivotal in building the capacity for sustainable change. By acknowledging individuals as active contributors to the transformative process (Weder & Erikson, 2023), we empower them to perceive themselves as capable of leading effective, meaningful change. Adopting a stakeholder-centric approach enables organizations to tailor messages and practical applications to diverse stakeholder groups, thereby stimulating action for impactful change. Secondly, we offer corporate environmental efficacy as an *organizing* principle. Our concept is rooted in established theoretical frameworks from related disciplines (Bandura, 1977; Groening et al., 2018; Tam et al., 2021). By leveraging strong parallels between fields, we aim to integrate corporations as key players into existing efficacy typologies (see, e.g., Meijers et al., 2023). This translation introduces corporate efficacy beliefs as vital in aligning corporate action with sustainability goals and calls for further organization along interdisciplinary lines. Thirdly, we propose corporate environmental efficacy as an *evaluative* principle and a structured framework to evaluate how corporations actively shape sustainability transformations. This facilitates the development of measurable indicators for corporate environmental efficacy, ensuring effective communication of these metrics as required by, for example, the CSRD mandate. Robust measurement and assessment tools are pertinent for researchers and practitioners alike.

In the following sections, we delve into the conceptual foundation of corporate environmental efficacy, explore its components and dimensions, and discuss how organizations can use it as a principle for strategic sustainability communication.

2 Corporate Environmental Efficacy as a Driver of Sustainable Change

2.1 The Evolution of Environmental Efficacy Research

To demonstrate the firm theoretical legs that corporate environmental efficacy stands on, we offer a brief theoretical detour into environmental psychology and critical aspects of environmental efficacy research. Central to fostering sustainable transformation is the belief of being capable of engaging in environmental action (efficacy) and the confidence in achieving desired outcomes by engaging in these (response efficacy; Bandura, 1977). Enhancing efficacy beliefs has shown to be an effective strategy to stimulate sustainable action (Hart & Feldman, 2016; Jugert et al., 2016): The more individuals believe they can engage in environmental action and that it will have a meaningful impact, the more likely they are to do so. While there is widespread acknowledgment of climate change and its severe effects on all beings, particularly in Europe (European Commission, 2023b), efficacy beliefs remain crucial for driving behavior change aimed at mitigating environmental threats (Chu & Yang, 2020). Consequently, the focus should shift from debating the need for sustainability to devising capable and effective strategies for sustainable transformation.

Many models of sustainable behavior change center around the individual (Bandura, 1977; Ajzen, 1991). However, the main factors contributing to and solving environmental issues primarily lie within groups such as society, organizations, and governments rather than with individuals. Climate change is not solvable by the individual alone but necessitates collective and organized efforts. Hence, research has increasingly considered the importance of collective efficacy beliefs, which refer to a social group's capability to address climate change (Barth et al., 2016; Jugert et al., 2016). A substantial body of literature speaks to the vital importance of efficacy on both individual level—the belief that the *individual* is capable and effective (Tabernero & Hernández, 2011; Huang, 2016; Faraz et al., 2021)— and collective level—the belief that *people* as part of a social group are capable and effective (Barth et al., 2016; Jugert et al., 2016; Meijers et al., 2023)—for stimulating environmental action. Meijers et al. (2023) synthesized a few decades' worth of environmental efficacy research into a Personal-Collective-Governmental typology (PGG), encapsulating seven sub-dimensions of efficacy beliefs related to pro-environmental action (see Fig. 1).

While environmental efficacy beliefs, crucial for intentional climate action, have long been acknowledged in social psychological research (Bandura, 1977), their integration into strategic sustainability communication remains unexplored. Therefore, this chapter introduces corporate environmental efficacy, offering a highly interdisciplinary perspective.

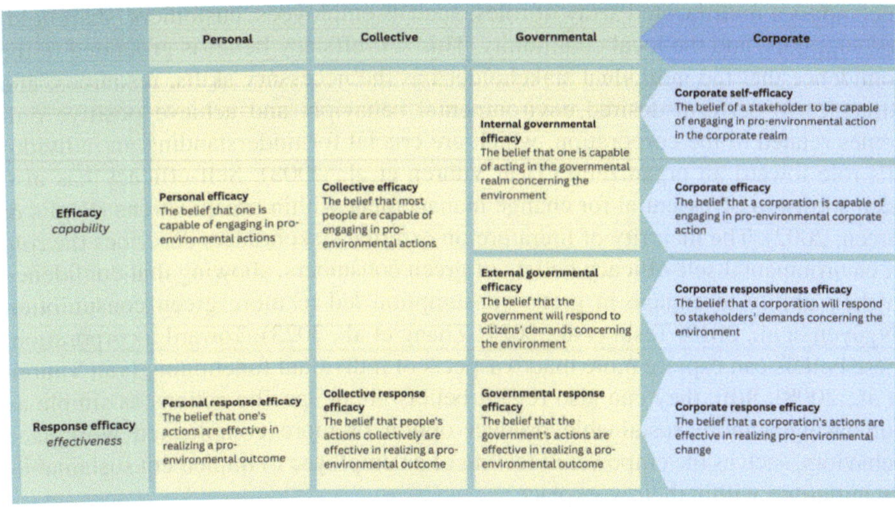

Fig. 1 The corporate-personal-collective-governmental (CPCG) typology. *Note. This figure illustrates the concept of 'corporate environmental efficacy' developed in this chapter, complementing Meijers et al.' (2023) Personal-Collective-Governmental (PCG) typology. The four corporate environmental efficacy beliefs are clustered into two dimensions: From top to bottom, the first three beliefs in the corporate column are classified as "efficacy," (perceived capability). The last belief is classified as "response efficacy" (perceived effectiveness).*

2.2 *Introducing Corporate Environmental Efficacy*

Based on environmental psychology research, we define corporate environmental efficacy as the belief in a stakeholder's and an organization's ability to effectively address environmental challenges (efficacy dimension; capability) and achieve desired environmental outcomes (response efficacy dimension; effectiveness). Thereby, we remain consistent with previous conceptualizations of efficacy and response efficacy as individual perceptions that influence environmental behavior (Bandura, 1977). This chapter complements Meijers et al.' (2023) typology with a corporate level. Organizational efficacy builds on a strong research base (van Vuuren et al., 2008; Fearon et al., 2013) and can be applied to corporations and sustainability. Our framework (see Fig. 1) differentiates four sub-dimensions of corporate environmental efficacy: corporate self-efficacy, corporate efficacy, corporate responsiveness efficacy, and corporate response efficacy. In the following, we will describe each conceptualized dimension in depth and ground them in existing literature.

Corporate self-efficacy closely relates to Bandura's concept of self-efficacy (1977) and, in this instance, describes the belief that oneself is capable of engaging in pro-environmental action as a stakeholder of or within the organization. This concept requires a nuanced interpretation of corporate stakeholders, defined here as any person who is affected by or has an interest, concern, or investment in a corporation's activities, decisions, and outcomes of a corporation. Stakeholders encompass a multifaceted array of roles, such as employees, customers, shareholders, suppliers, and the local community. This self-efficacy belief is grounded in the confidence that the individual stakeholder has the necessary skills, resources, and strategies to perform desired environmental behaviors and achieve positive outcomes related to the corporation, which are crucial for understanding an individual's role toward an organization (van Vuuren et al., 2008). Self-efficacy has also been shown to be essential for change management within organizations (Paglis & Green, 2002). The majority of literature on external stakeholders describes the role of environmental self-efficacy beliefs of green consumers, showing that confidence in their ability to engage in green consumption led to more green consumption (Nguyen et al., 2019; Testa et al., 2021; Zhang et al., 2023). Toward a corporation, these beliefs can capture more macro aspects of individual functioning (van Vuuren et al., 2008). Still, they can also regard beliefs about specific actions as simple as purchasing a more sustainable product option or more complicated workplace behaviors, such as the empowered feeling of an employee to implement sustainability initiatives within their workplace.

Corporate efficacy describes the stakeholder's belief that the corporation, as a whole, is capable of engaging in corporate environmental action. It denotes the individual's confidence in the corporation's ability to influence environmental issues and enact significant sustainable transformation. Here, we borrow and translate a concept from organizational studies. Within this field, organizational efficacy is defined as "a generative capacity within an organization to cope effectively with the demands, challenges, stressors, and opportunities it encounters within the business environment. It exists as an aggregated judgment of an organization's individual members about their (1) sense of collective capacities, (2) sense of mission or purpose, and (3) a sense of resilience" (Bohn, 2002, p. 65). Again, the concept can denote macro beliefs about corporate functioning (van Vuuren et al., 2008) and overall capabilities and organizational goals (Fearon, McLaughlin and Morris, 2013) or more specific corporate actions. Also taking a stakeholder-centric approach, Lee (2020) proposed the related concept of perceived sustainability, denoting stakeholder's perceptions of the "strength" (stakeholder beliefs about a firm's expertise in sustainability), "magnitude" (an individual's belief in a firm's sustainability), and "mobilizability" (the perceived ability of an organization to motivate stakeholders to engage in environment-protective behavior). Similar to efficacy beliefs, this construct would positively affect consumer behavior, such as "brand resonance" and "purchase intention." Aiding the assessment of a corporation's capability on a more general level, also "perceived corporate ability," i.e., perceptions of a company's competence or ability to produce an intended output (Lin et al., 2011), has shown to be an important antecedent of "product purchase intention." Inspired by these concepts, our conceptualization captures the belief in a corporation's capability to meet its objectives, a shared sense of collective capability in engaging in corporate sustainability. These corporate actions can range from investing in renewable energy sources to reduce its carbon footprint, implementing and actually adhering to green policies and practices, to creating new technologies to mitigate environmental impact. It requires corporations to not only set ambitious sustainability goals but also demonstrate their capacity to achieve them through concrete actions and resilient strategies.

Corporate responsiveness efficacy labels the belief that a corporation will respond to both the general public's as well as specific stakeholder demands concerning the environment. We see parallels in strategic CSR management, where the fulfillment of consumer expectations toward CSR and perceived corporate ability has been shown to bring strategic advantage (Lin et al., 2011; Kuokkanen & Sun, 2020). In today's business landscape, there is a noticeable shift toward prioritizing sustainability practices, driven partly by heightened consumer awareness and expectations. Consumers increasingly view the protection of stakeholder interests, including environmental concerns, as integral factors when making purchasing decisions, as noted by Hur et al. (2016). Stakeholder theory shows that corporations need to strategically and interdependently engage with stakeholder interests to create value (Freeman et al., 2010). Overall, corporate responsiveness efficacy underscores the increasing interconnectedness between stakeholders and corporations—also fostered by technology—in addressing environmental concerns and advancing sustainability goals. By prioritizing stakeholder responsiveness, corporations can foster sustainable relationships necessary for the corporation's long-term survival. In practice, these beliefs are, among other things, potentially related to stakeholders' beliefs that they can form an opinion on the environmental matter, that the corporation pays attention to these opinions, and can thus influence corporate change.

Corporate response efficacy is the belief that a corporation's actions are effective in realizing pro-environmental change. Put simply, while corporate responsiveness efficacy relates to stakeholder beliefs on a corporation being capable of—and actually—responding to demands, corporate response efficacy relates to beliefs on the effectiveness of the response; this sub-dimension hinges on the stakeholder's goal to positively impact the environment through their actions within the corporate sphere. The primary focus of previous research positions stakeholders as consumers, exploring communicative interventions to promote sustainable product consumption (Young et al., 2010; Cojuharenco et al., 2016). Termed perceived consumer effectiveness (PCE) (Antil, 1984), consumers' beliefs about the impact of their product choice moderate green consumption (Nguyen et al., 2019): The stronger their belief in their choice's effectiveness, the more likely they are to consume green products. Confidence in the ability to contribute to environmental improvement through eco-friendly choices aligns consumers' actions with their intentions. Consumers' collective response efficacy beliefs—the belief that all together, the masses of consumers can make a difference—are associated with environmentally conscious purchasing behavior (Cojuharenco et al., 2016). The belief that a corporate product itself will make a difference is captured in another related concept. Drawing from business research, the concept of product efficacy pertains to inferences about a product's functional performance and quality (Zeithaml, 1988). Conversely, Young et al. (2010) demonstrated that when information about a product's performance is lacking, consumers are less inclined to engage in green consumption. This underscores the importance of effectively communicating about the effectiveness of corporate action for sustainability.

2.3 *Examples of Audience Engagement in Corporate Sustainability Through Corporate Efficacy*

Building on this theoretical foundation, we offer an exemplary exploration of various stakeholder groups and their corporate environmental efficacy beliefs to start an academic debate to further develop analytical criteria and operationalizable dimensions for empirical work in the future.

Employees may experience a high sense of corporate self-efficacy when they perceive their capacity to actively participate in office environmental campaigns, such as reducing waste or implementing energy-saving measures. However, they might question whether their efforts will elicit a response from the company and lead to meaningful changes in organizational management practices, exemplifying low corporate responsiveness efficacy beliefs. Nonetheless, employees might recognize that these actions can significantly contribute to improving the company's overall environmental footprint (indicating high "corporate response efficacy"). Some corporations have recognized the power that employees hold and have developed so-called corporate green teams. A green team consists of a group of employees who are engaged in advancing sustainability within an organization. Consulting partners like *Planetgroups*, a not-for-profit consulting, support employees in climate action at the workplace. They strategically appeal to individuals' "responsibility in their own workplace to build truly sustainable, successful and value-creating companies", thereby stressing how a company's response efficacy depends on its employees' efficacy.

Meanwhile, a *consumer* may feel empowered to choose the sustainable version of a product over the less sustainable one, demonstrating high corporate self-efficacy. In doing so, they might perceive that this choice actively contributes to the environment, thus driving change and indicating high "personal response efficacy." Alternatively, an impassioned consumer might express dissatisfaction by leaving critical feedback in the product review section, believing that such action can influence sustainable transformation as they anticipate the company's responsiveness to their concerns (highlighting "corporate responsiveness efficacy"). For example, recent years have seen a surge in plant-based meat substitutes, often advertised using clear efficacy appeals. Companies like *Beyond Meat* follow the mission to „[…] feed a better future […]" as "[…] the positive choices we make every day—no matter how small—can have a great impact on our world"). These appeals shall motivate consumers' response efficacy by highlighting how even the smallest actions can drive meaningful impact.

The dimension of "corporate self-efficacy" for *investors/shareholders* could be seen in their belief in their own ability to influence environmental outcomes through their investment decisions. This includes their confidence in choosing environmentally responsible companies or divesting from those with poor environmental practices. Investors with high "corporate self-efficacy" may actively seek out investment opportunities that align with their environmental values and advocate for sustainable business practices within their investment portfolios. High "corporate responsiveness efficacy" involves their confidence that their engagement, whether through shareholder activism, proxy voting, or direct communication with corporate management, can influence corporate behavior toward more environmentally sustainable practices. Investors with high "corporate responsiveness efficacy" may engage in dialogues with companies, submit shareholder proposals related to environmental issues, or support resolutions that call for improved environmental performance. Shareholders with high "corporate response efficacy" believe that a corporation's environmental initiatives, such as setting emissions reduction targets or implementing sustainable supply chain practices, are effective in realizing sustainable transformation. An example of sustainability-focused investing is exchange-traded funds (ETF), which allow investors to diversify their portfolios by investing in a collection of stocks or bonds that meet specific ESG criteria. This type of investment not only contributes to the global movement toward sustainable development but also offers investors a way to contribute to positive social and environmental outcomes. At the macro level, the European NGO *Finance Watch* advocates reform of financial regulation to give society, and thus individual investors, a strong voice. In addition, organizations such as *Sustainalytics* evaluate the ESG statements of listed companies to help investors make more informed investment decisions. What both examples have in common is that they increase the knowledge and confidence of financial actors, ideally helping them to make environmentally sound decisions.

The digital transformation of media production led *journalists* to develop new forms of journalistic reporting. For example, the current trend toward more *solutions-oriented journalism* embraces evidence of a solution's success or solution. While alarming news about climate change dominates the news, solutions-oriented journalism shifts focus on how corporations aid sustainable transformation. By showcasing successful corporate initiatives or strategies, they help imagine a desirable, sustainable future led by organizations (Atanasova, 2019). This can build confidence among stakeholders—both within and outside the organization—that the corporation has the capability, commitment, and effectiveness to achieve sustainable transformation (Thier & Namkoong, 2023). This article by the Guardian provides its readers with "reasons to be hopeful" by presenting corporate developments and initiatives in key sectors as part of the solution to tackle the climate crisis. This form of constructive style of journalism is increasingly being institutionalized. For example, the US-based nonprofit Solutions Journalism Network aims to inspire readers to engage in pro-environmental engagement by supplying its readers with stories that have led to meaningful change, showcasing the capability and effectiveness of corporate action.

There are also opportunities to reach stakeholders with important information about corporate activities or planned measures outside of tried-and-tested or new forms of journalistic storytelling. *Corporate influencers* and *corporate ambassadors* (Hudders et al., 2021) skillfully blend authenticity, intimacy, and knowledge to forge robust bonds with their audience (Enke & Borchers, 2019). They influence followers' pro-environmental and sustainable behavior (Breves & Liebers, 2022), for example, sustainable consumption (Kapoor et al., 2023) or environmental activism (Knupfer et al., 2023). According to the latest report from the Intergovernmental Panel on Climate Change (IPCC), scientists affirm that "social influencers and thought leaders have the potential to enhance the uptake of low-carbon technologies, behaviors, and lifestyles." Corporate influencers, for example, under the successful #ecotok, do so by modeling and normalizing sustainable action for the individual, demonstrating that they can eat, buy, or travel sustainably to #savetheworld—claiming that so can everyone else. Notably, corporate influencers are also crucial in internal communication since they co-create corporate culture and a brand's meaning (Hesse et al., 2021).

To summarize, these are the first examples of five key stakeholders and a brief application of the corporate environmental efficacy concept as first thoughts to start academic debates. Journalists provide assessed and processed information, resources, and platforms for engagement, empowering stakeholders in advancing sustainable practices and driving positive impact at both organizational and societal levels. In the realm of corporations and sustainability issues, research on the persuasiveness of online influencers toward their audiences is conclusive. What is largely missing, however, is what kind of information influencers receive, how they receive it, how they evaluate it, and how information is produced for influencers in the first place. Therefore, more research is needed on corporate strategic environmental communication, including the distinct professional communication practices of such large organizations, such as media relations or community-building activities.

As a first step, studies could analyze the content of strategic communications explicitly aimed at opinion leaders—also in comparison to content targeting the general public or specific groups such as investors. Another conceivable line of research could be to examine the individual recipients of media-based corporate communication, such as consumers. Future studies could focus on how the cognitive process of information evaluation works using, for example, eye-tracking to capture (sub)conscious attention and information processing. It would be worthwhile to analyze which communication tools and channels lead to an increase or maintenance of the level of corporate environmental efficacy and how. To this end, we propose longitudinal studies that link corporate communication activities to survey measurement points.

Ultimately, these are rudimentary first ideas for future research projects and an attempt to evaluate strategic communication efforts. This is a challenging task, given that it has been, and probably will continue to be, an ongoing challenge for practitioners for decades. All the more reason, we believe, to consider not only the dimensions of producers and content but also the cognitive processes of individuals as recipients or intermediaries of strategic environmental corporate communication.

3 Communication Is Key: How Strategic Communication for Sustainability Becomes Reality

The aim of this chapter was to bring a stakeholder or a media reception perspective to the corporate lens that is represented in most of the chapters in this handbook. We have discussed why all the scientific evidence points to the need to think about communication and sustainable change in a holistic way. Therefore, we have derived the theoretical concept of corporate environmental efficacy beliefs from decades of interdisciplinary research in media psychology and the latest developments in the field of environmental communication (see Meijers et al., 2023) and adapted it to the corporate context. Yet the various dimensions of corporate environmental efficacy are by no means to be understood as mere addenda. Only in its entirety can the concept deliver its full value for societies.

Sustainable transformation emerges through sustainable actions. The necessity of this transformation process being a joint effort by both organizations and their stakeholders becomes apparent when considering the complexity of the rapid fight against the climate crisis. Communication is key here, but it is also a reciprocal process of co-creating reality, with its accompanying challenges. A cultural change in corporations is crucial, as they emit vast amounts of greenhouse gases. It would, therefore, be easy if they simply reduced or even abolished emissions. At the same time, it often seems as if these multinational business conglomerates would only be responding to external pressure. Looking back more than two decades, the public's demands for ecological accountability led to the first corporate sustainability reports in Europe. Today's climate movement was one of the catalysts for the new CSDR EU directive on mandatory extensive sustainability reporting. The question now is, what impact will its implementation have on individual and collective stakeholders? After all, it would be most desirable if strategic communication of Environmental, Social, and Governance (ESG) criteria led to informed, sustainable stakeholder decisions. However, it remains to be seen whether this will stay a pipe dream or become a reality.

We are convinced that the concept of Corporate Environmental Efficacy will play a decisive role in the scientific and practical accompaniment of the sustainable transformation of the economic sector at large and initiatives such as the European Green Deal. As a *constitutive* principle, all those involved in strategic sustainability communication—be it corporate employees, legislators, or researchers, among others—must finally direct their attention to the micro-level psychological perspective. It is an *organizing* principle that makes it possible to combine approaches, theories, models, and methods from different academic fields. Because, ultimately, there will not be the one and only right solution; to systematically gain knowledge, here as well, communication will be key. And lastly, as an *evaluative* principle to monitor the implementation and impact of, for example, the new CSDR directive. Future research must first (1) empirically operationalize and validate a corporate environmental efficacy measure. Subsequently, cross-sectional and then longitudinal survey studies must be carried out to (2) assess whether sustainability reports as strategic corporate communication of sustainability lead to a formation and increase of individual and collective corporate environmental efficacy belief scores. Next, it must be investigated whether these scores associated with sustainability reports (3) are followed by sustainable behavior (changes). This triad of corporate environmental efficacy then has the potential for strategic communication *of* sustainability morphing to strategic communication *for* sustainability—because companies will react to the sustainable behavior of their stakeholders with further sustainability measures, starting a new loop of this reciprocal communicative process.

Recommendations for Future Reading:

1. Industry and the Green Deal
2. The World Economic Forum (weforum.org)

References

Ajzen, I. (1991). The theory of planned behavior. *Organizational Behavior and Human Decision Processes, 50*(2), 179–211. https://doi.org/10.1016/0749-5978(91)90020-T

Antil, J. H. (1984). Socially responsible consumers: Profile and implications for public policy. *Journal of Macromarketing, 4*(2), 18–39. https://doi.org/10.1177/027614678400400203

Atanasova, D. (2019). Moving society to a sustainable future: The framing of sustainability in a constructive media outlet. *Environmental Communication, 13*(5), 700–711. https://doi.org/10.1080/17524032.2019.1583262

Bandura, A. (1977). Self-efficacy: Toward a unifying theory of behavioral change. *Psychological Review, 84*(2).

Barth, M., Jugert, P., & Fritsche, I. (2016). Still underdetected – Social norms and collective efficacy predict the acceptance of electric vehicles in Germany. *Transportation Research Part F: Traffic Psychology and Behaviour, 37*, 64–77. https://doi.org/10.1016/j.trf.2015.11.011

Bohn, J. G. (2002). The relationship of perceived leadership behaviors to organizational efficacy. *Journal of Leadership & Organizational Studies, 9*(2), 65–79. https://doi.org/10.1177/107179190200900206

Breves, P., & Liebers, N. (2022). #Greenfluencing. The Impact of Parasocial Relationships with Social Media Influencers on Advertising Effectiveness and Followers' Pro-environmental Intentions. *Environmental Communication, 16*(6), 773–787. https://doi.org/10.1080/17524032.2022.2109708

Chu, H., & Yang, J. Z. (2020). Risk or efficacy? How psychological distance influences climate change engagement. *Risk Analysis, 40*(4), 758–770. https://doi.org/10.1111/risa.13446

Cojuharenco, I., Cornelissen, G., & Karelaia, N. (2016). Yes, I can: Feeling connected to others increases perceived effectiveness and socially responsible behavior. *Journal of Environmental Psychology, 48*, 75–86. https://doi.org/10.1016/j.jenvp.2016.09.002

Enke, N., & Borchers, N. S. (2019). Social media influencers in strategic communication: A conceptual framework for strategic social media influencer communication. *International Journal of Strategic Communication, 13*(4), 261–277. https://doi.org/10.1080/1553118X.2019.1620234

European Commission. (2023a). *Financial markets: Corporate sustainability reporting*. Accessed March 12, 2024, from https://finance.ec.europa.eu/capital-markets-union-and-financial-markets/company-reporting-and-auditing/company-reporting/corporate-sustainability-reporting_en

European Commission. (2023b). *Special Eurobarometer 538 Climate Change - Report 538.* European Union. Accessed March 12, 2024, from https://doi.org/10.2834/653431

Faraz, N. A., et al. (2021). The interplay of green servant leadership, self-efficacy, and intrinsic motivation in predicting employees' pro-environmental behavior. *Corporate Social Responsibility and Environmental Management, 28*(4), 1171–1184. https://doi.org/10.1002/csr.2115

Fearon, C., McLaughlin, H., & Morris, L. (2013). Conceptualising work engagement: An individual, collective and organisational efficacy perspective. *European Journal of Training and Development, 37*(3), 244–256. https://doi.org/10.1108/03090591311312723

Freeman, R. E., et al. (2010). *Stakeholder theory: The state of the art.* Cambridge University Press. https://doi.org/10.1017/CBO9780511815768

Groening, C., Sarkis, J., & Zhu, Q. (2018). Green marketing consumer-level theory review: A compendium of applied theories and further research directions. *Journal of Cleaner Production, 172*, 1848–1866. https://doi.org/10.1016/j.jclepro.2017.12.002

Hart, P. S., & Feldman, L. (2016). The influence of climate change efficacy messages and efficacy beliefs on intended political participation. *PLoS One, 11*(8), e0157658. https://doi.org/10.1371/journal.pone.0157658

Hesse, A., Schmidt, H. J., & Baumgarth, C. (2021). How a corporate influencer co-creates brand meaning: The case of Pawel Dillinger from Deutsche Telekom. *Corporate Reputation Review, 24*(4), 191–204. https://doi.org/10.1057/s41299-020-00103-3

Huang, H. (2016). Media use, environmental beliefs, self-efficacy, and pro-environmental behavior. *Journal of Business Research, 69*(6), 2206–2212. https://doi.org/10.1016/j.jbusres.2015.12.031

Hudders, L., De Jans, S., & De Veirman, M. (2021). The commercialization of social media stars: A literature review and conceptual framework on the strategic use of social media influencers. *International Journal of Advertising, 40*(3), 327–375. https://doi.org/10.1080/02650487.2020.1836925

Hur, W.-M., Kim, H., & Jang, J. H. (2016). The role of gender differences in the impact of CSR perceptions on corporate marketing outcomes. *Corporate Social Responsibility and Environmental Management, 23*(6), 345–357. https://doi.org/10.1002/csr.1380

Jugert, P., et al. (2016). Collective efficacy increases pro-environmental intentions through increasing self-efficacy. *Journal of Environmental Psychology, 48*, 12–23. https://doi.org/10.1016/j.jenvp.2016.08.003

Kapoor, P. S., Balaji, M. S., & Jiang, Y. (2023). Greenfluencers as agents of social change: The effectiveness of sponsored messages in driving sustainable consumption. *European Journal of Marketing, 57*(2), 533–561. https://doi.org/10.1108/EJM-10-2021-0776

Knupfer, H., Neureiter, A., & Jörg Matthes. (2023). From social media diet to public riot? Engagement with "greenfluencers" and young social media users' environmental activism. *Computers in Human Behavior, 139*, 107527. https://doi.org/10.1016/j.chb.2022.107527

Kuokkanen, H., & Sun, W. (2020). Companies, meet ethical consumers: Strategic CSR management to impact consumer choice. *Journal of Business Ethics, 166*(2), 403–423. https://doi.org/10.1007/s10551-019-04145-4

Lee, Y.-C. (2020). Communicating sustainable development: Effects of stakeholder-centric perceived sustainability. *Corporate Social Responsibility and Environmental Management, 27*(4), 1540–1551. https://doi.org/10.1002/csr.1900

Lin, C.-P., et al. (2011). Understanding purchase intention during product-harm crises: Moderating effects of perceived corporate ability and corporate social responsibility. *Journal of Business Ethics, 102*(3), 455–471. https://doi.org/10.1007/s10551-011-0824-y

Meijers, M. H. C., et al. (2023). Introducing and testing the personal-collective-governmental efficacy typology: How personal, collective, and governmental efficacy subtypes are associated with differential environmental actions. *Journal of Environmental Psychology, 85*, 101915. https://doi.org/10.1016/j.jenvp.2022.101915

Neureiter, A., & Jörg Matthes. (2023). From social media diet to public riot? Engagement with "greenfluencers" and young social media users' environmental activism. *Computers in Human Behavior, 139*, 107527. https://doi.org/10.1016/j.chb.2022.107527

Nguyen, H. V., Nguyen, C. H., & Hoang, T. T. B. (2019). Green consumption: Closing the intention-behavior gap. *Sustainable Development, 27*(1), 118–129. https://doi.org/10.1002/sd.1875

Paglis, L. L., & Green, S. G. (2002). Leadership self-efficacy and managers' motivation for leading change. *Journal of Organizational Behavior, 23*(2), 215–235. https://doi.org/10.1002/job.137

Steenbergen, V., & Saurav, A. (2023). *The effect of multinational enterprises on climate change: Supply chain emissions, green technology transfers, and corporate commitments*. World Bank. https://doi.org/10.1596/978-1-4648-1994-0

Tabernero, C., & Hernández, B. (2011). Self-efficacy and intrinsic motivation guiding environmental behavior. *Environment and Behavior, 43*(5), 658–675. https://doi.org/10.1177/0013916510379759

Tam, L., Choi, S., & Kim, J.-N. (2021). Reconceptualizing the self-efficacy construct in public relations research: The case of sojourners and their communicative behaviors. *Public Relations Review, 47*(4), 102090. https://doi.org/10.1016/j.pubrev.2021.102090

Testa, F., et al. (2021). Drivers to green consumption: A systematic review. *Environment, Development and Sustainability, 23*(4), 4826–4880. https://doi.org/10.1007/s10668-020-00844-5

Thier, K., & Namkoong, K. (2023). Identifying major components of solutions-oriented journalism: A review to guide future research. *Journalism Studies, 24*(12), 1557–1574. https://doi.org/10.1080/1461670X.2023.2230314

van Vuuren, M., de Jong, M. D. T., & Seydel, E. R. (2008). Contributions of self and organisational efficacy expectations to commitment: A fourfold typology. *Employee Relations, 30*(2), 142–155. https://doi.org/10.1108/01425450810843339

Weder, F. (2023). The evolution of the sustainability story: Strategic sustainability communication as niche construction. *International Journal of Strategic Communication, 17*(3), 228–244. https://doi.org/10.1080/1553118X.2023.2229304

Weder, F., & Erikson, M. B. (2023). *CSR communication and cultures of sustainability*. The University of Queensland. https://doi.org/10.14264/428d81b

Young, W., et al. (2010). Sustainable consumption: Green consumer behaviour when purchasing products. *Sustainable Development, 18*(1), 20–31. https://doi.org/10.1002/sd.394

Zeithaml, V. A. (1988). Consumer perceptions of price, quality, and value: A means-end model and synthesis of evidence. *Journal of Marketing, 52*(3), 2–22. https://doi.org/10.2307/1251446

Zerfass, A., et al. (2018). Strategic communication: Defining the field and its contribution to research and practice. *International Journal of Strategic Communication, 12*(4), 487–505. https://doi.org/10.1080/1553118X.2018.1493485

Zhang, R., Ma, Q., & Guan, D. (2023). The impact of financial scarcity on green consumption: Sequential mediating effects of anxiety and self-efficacy. *Psychology & Marketing, 40*(6), 1162–1178. https://doi.org/10.1002/mar.21801

Alexandra Schwinges is an Assistant Professor in Corporate Communication at the Amsterdam School of Communication Research, University of Amsterdam, the Netherlands. Her research and teaching activities center around issue communication, corporate digital responsibilities, and digital sustainability. Proficient in diverse research methods, she employs, among others, quantitative and qualitative content analysis, interviews, surveys, and experimental research design.

Markos Mpadanes is a Research and Teaching Assistant in Strategic Communication and Media Management at the Department of Communication and Media Research (IKMZ), University of Zurich, Switzerland. He is also a PhD Candidate in the excellence program of the Digital Society Initiative at the University of Zurich. His empirical research focuses on "Strategic CEO Communication" and "Socio-Political Positioning of Companies and Executives." Methodologically, Markos Mpadanes employs a mixed-method approach integrating quantitative, qualitative, and computational social science methods.

Framing and Argumentative Strategies in Modern Slavery Statements: Debating Sustainability or Avoiding Responsibility?

Rudi Palmieri and Ekaterina Balabanova

Abstract This chapter develops an analysis of the linguistic content of modern slavery statements by UK-based businesses, which represent a genre of sustainability communication that has emerged in recent years. We built a dataset of all modern slavery reports filed in 2019 by FTSE100 firms. Our analysis followed a multi-step procedure including qualitative and quantitative discourse-analytic methods. First, a small sample was screened to find claims and supporting arguments. Subsequently, a corpus-assisted analysis of keywords-in-context was undertaken to identify micro-level framing strategies. Our results provide evidence of a very minimalistic argumentative engagement. Rhetorical appeals to ethos prevail over logos and pathos appeals, to construct trust-building and responsibility-shifting frames while avoiding critical discussion. The main rhetorical strategy that emerges from these frames consists in (a) companies portraying themselves as reliable and compliant agents of social sustainability and (b) companies putting themselves in the expedient position of a benevolent intermediary between governments and stakeholders towards whom ultimate responsibility is shifted. Our findings advance our understanding of how companies communicate rhetorically to comply with regulatory demands of accountability while promoting the desired image of ethics and sustainability in front of markets and society. Our study is the first to offer a systematic analysis of the content of modern slavery statements from a discourse perspective. It contributes to strategic communication research on framing with a focus on its linguistic aspects. The findings have implications also for policymakers and their attempt to design a more effective disclosure framework to enhance transparency and accountability on modern slavery and other social sustainability issues.

R. Palmieri (✉) · E. Balabanova
Department of Communication and Media, University of Liverpool, Liverpool, UK
e-mail: Rudi.Palmieri@liverpool.ac.uk; E.Balabanova@liverpool.ac.uk

F. Weder (ed.), *Strategic Sustainability Communication*, CSR, Sustainability, Ethics & Governance, https://doi.org/10.1007/978-3-031-89486-2_11

1 Introduction

Environmental, social and governance (ESG) issues are nowadays an essential ingredient of a company's strategic communication efforts to promote an image of sustainability in front of both its financial and non-financial stakeholders. Alongside environmental issues (net-zero, climate change, wasting and recycling, etc.), social issues regarding, in particular, the working conditions of employees, represent a delicate object of discussion. This is especially the case for large global corporations who delocalise their production process to countries where the respect of basic workers' rights—and human rights more generally—cannot be taken for granted. Within the vast and interdisciplinary body of academic research on sustainability communication, a growing interest in the linguistic and argumentative analysis of public controversies on social sustainability issues has emerged, for instance, around the fast fashion industry (e.g., Greco & De Cock, 2021; Balabanova et al., 2023). In this chapter, we contribute to this research area by investigating the micro-linguistic strategies of a recently emerged genre of sustainability communication in the UK: modern slavery statements.

Modern slavery refers to "the exploitation of a person who is deprived of individual liberty anywhere along the supply chain, from raw material extraction to the final customer, for the purpose of service provision or production" (Gold et al., 2015, p. 487). In recent years, it has become an increasingly important area of corporate social responsibility (Caruana et al., 2021), urging firms to communicate more transparently about these issues.

In the UK, the Modern Slavery Act 2015 (MSA) requires businesses to publish a "slavery and human trafficking statement" (MSA statement) every year. Similar to annual reports and Corporate Social Responsibility (CSR) reports, the modern slavery statement constitutes a genre of business discourse that has emerged in response to a new rhetorical situation (Miller & Kelly, 2016), which reflects a new strategic exigence from a corporate viewpoint. As many accounting and business communication scholars have highlighted, any genre of corporate reporting simultaneously fulfils an *informative* function and a *promotional/rhetorical* function (Rutherford, 2005; Merkl-Davies & Brennan, 2007; Beattie et al., 2008; Malavasi, 2010). The former relates to stakeholders' demand for information that allows public scrutiny and accountability, requiring firms to disclose accurate and transparent information. The latter reflects the firm's interests in promoting corporate goals and avoiding undesired inferences which lead corporate managers to advance and argumentatively defend a claim.

While MSA statements should enable stakeholders to hold big businesses accountable for their engagement with modern slavery risks in their supply chains, many experts and scholars lament a lack of serious engagement by companies (Christ & Burritt, 2018; Monciardini et al., 2021). Some of the main criticisms highlight the templating and ticking-the-box approach, while a serious attempt to develop convincing explanations and provide audiences with rationale and evidence is lacking (Rogerson et al., 2020, p. 1515).

Although Section 54 of the MSA has been revised to counter such a ticking-the-box approach and offer guidance on how to write statements, research has critically pointed to the vagueness and ambiguity that dominate these reports (Hsin et al., 2021; Meehan & Pinnington, 2021). Through a qualitative analysis of MS statements by government suppliers, Meehan and Pinnington (2021) identified three ambiguous techniques (defensive reassurance, transfer of responsibility and scope reduction) through which companies try to defend the status quo, reduce accountability and delay action against modern slavery within supply chains.

Overall, existing research suggests that (a) MS statements display a strategic use of language, resulting in a highly rhetorical genre of business communication; (b) these statements are of disappointing quality in relation to the purpose for which they have been mandated by law. There is, however, little evidence of the linguistic-discursive strategies featured within MS statements, which ultimately reflects the scarcity of discourse-analytic studies on this genre of strategic sustainability communication. Our study addresses this gap with an analysis of MS statements published by FTSE100 firms aimed at detecting argumentation and framing strategies by which companies try to defend or promote their image of sustainability while fulfilling the requirements of MS regulation.

Our main research questions are:

(Q1) What themes are tackled in MS statements and to what extent are these themes problematised (i.e. presented as argumentative issues)?
(Q2) How do companies (strategically) frame modern slavery and human rights issues, events and situations?
(Q3) To what extent do companies engage in argumentative discussions rather than limiting themselves to report basic information and/or self-promotional messages?

By combining corpus-assisted discourse analysis, argumentation analysis and trust discourse theories, two main findings have emerged: (1) firms deploy an *argument-avoidance strategy* based on trust-building signals that put them in a less risky position regarding an issue which they do not seem confident enough to publicly discuss; (2) firms *frame themselves* (and their role in the fight against modern slavery) as *benevolent intermediaries* between regulators and business partners, towards whom ultimate responsibility is shifted.

The chapter is organised as follows: in the next section, the theoretical framework, methods, and data adopted in this study will be explained. Subsequently, the main findings will be presented. The concluding section will discuss the practical implications of this study as well as its limitations that open up questions for future research.

2 Theory, Data, Methods

2.1 Argumentation as Reasonable Persuasion

Argumentation is an inherent feature of corporate communications (Palmieri, 2014). It is a means to build legitimacy (Green, 2004), justify organisational change (Sillince, 2002), influence opinions about performance (Brennan et al., 2010) and, more generally, gain support from multiple stakeholders (Palmieri & Mazzali-Lurati, 2021). The use of argumentation in corporate texts (letters to shareholders, sustainability reports, strategy documents, etc.) reflects the simultaneous aims of reasonableness and effectiveness (van Eemeren, 2010). Thus, argumentation can be understood as the rhetorical communication of reasons to achieve reasonable persuasion, or as "a discourse aimed at convincing a reasonable critic to accept a standpoint with reasons that justify the standpoint" (van Eemeren & Grootendorst, 2004, p.1).

A commitment to reasoned justification is implied in most corporate disclosure regulations, including those related to modern slavery. The spirit of the Modern Slavery Act is to make companies more directly responsible towards MS issues and more transparent in demonstrating actual engagement. Accordingly, one would expect companies to (a) recognise the presence of actual or potential modern slavery issues, (b) discuss them by advancing claims regarding their actions and performances, and (c) defend such claims with convincing reasons and anticipate or respond to criticisms raised by stakeholders. From a rhetorical perspective, we would expect companies to frame these arguments strategically in order to promote favourable views about themselves (image/reputation) and the actions they have undertaken to comply with the law.

A common approach to understand and analyse rhetorical arguments in organisational contexts is the Ethos-Logos-Pathos framework (Green, 2004). Rooted in Aristotle's Rhetoric, this framework conceptualises persuasion along three dimensions, one being about reasoning and justification (logos), the other two about the arguer's credibility (ethos) and the audience's emotions and perspective (pathos). At the discourse-analytic level, logos are represented by the set of claims and supporting arguments expressed in the text, whereas ethos and pathos appeals typically manifest themselves through different types of framing strategies. Ethos is frequently signalled through self-referenced statements usually activated by first-person pronouns ("we" and "our"), while pathos appeals can be realised through the lexicalisation of emotions and audience framing expressions, activated, for instance, by second-person pronouns ("you" and "your") (see Palmieri et al., 2022). Table 1 gives an overview of the main forms of ethos, logos and pathos appeals that were searched for in this study.

Table 1 Ethos, logos and pathos appeals: forms of textual manifestation and examples

Rhetorical appeal	Manifestation form	Example
Ethos (source credibility)	Self-referenced statements with first-person pronouns ("We"; "our")	"We have taken a systematic approach to reviewing the risk of modern slavery in our businesses and supply chains" (BP 2019 Modern slavery statement, p.4)
Logos (argumentation)	Claims justified by reasons, possibly signalled by argumentative indicators ("because", "as", etc.)	"At HSBC UK we understand the important role banks play in a thriving society and we care about our impact. This is why we work every day to meet the needs of our customers, colleagues and the communities we serve" (HSBC 2019 CSR report, p.2)
Pathos (audience emotions)	Lexicalisation of emotions and audience framing expressions, activated for instance by second person pronouns	"When you get the emails or the thank you cards or the videos from the beneficiaries saying, "You've completely changed how we can help our communities' and there are tears of joy, you can't help but feel absolutely amazing about the decision you've made to support that organisation" (HSBC 2019 CSR report, p.8)

2.2 *Framing and Its Argumentative Functions*

In general, framing refers to how events and situations are put into a particular perspective so that some elements gain saliency over others (Entman, 1993). Goffman suggests frames are "principles of organization which govern events—at least social ones—and our subjective involvement in them" (1974, pp. 10–11). Frames make sense of situations by selecting "players, positions and places that are relevant in a controversy" (Musi & Aakhus, 2019).

The argumentative implications of framing have been discussed by several scholars who highlighted how the choice of a particular frame may activate and foreground preferred lines of reasoning (e.g., Rocci, 2009; Greco Morasso, 2012; Musi, 2024). Some framing devices (e.g., personal pronouns) help put the speaker/writer under a certain perspective, thus building ethos; or elicit the audience's emotions to appeal to pathos.

In this study, we focus on lexical frames, which represent a class of framing devices that exploit the semantic and pragmatic properties of some (key)words, and their ability to highlight a particular perspective and interpretation of the underlying event (Fillmore, 1976). Our frame analysis is inspired by Congruity Theory (Rigotti & Rocci, 2001; Rigotti et al., 2006), in which lexical frames are reconstructed as predicate-argument structures characterised by a set of presuppositions and entailments. Presuppositions are implicit, taken-for-granted, and shared conditions of meaningfulness of a predicate. Each predicate-argument structure involves a specific number of entities[1] each one holding specific properties that are required by the predicate itself. For example, the verb "to give" presupposes three entities (x_1, x_2 and x_3), being the giver, the given object, and the receiver. Each of these three entities has some requirements without which the use of the predicate would not make sense, like the fact that the giver (x_1) and not the receiver (x_3) must hold the object (x_2) before the act of giving takes place. Entailments are the conditions that should be realised if the predicate is affirmed, i.e. if the action the predicate refers to takes place. That the object x_2 is held by the receiver (x_3) and is no longer possessed by the giver (x_1) is the main entailment of the predicate "to give".

Presuppositions and entailments work argumentatively as *virtual standpoints* (Van Eemeren et al., 1993). Their truth is an implicit opinion the speaker is committed to and which can be called out by the audience (Mohammed, 2019). If I say that this morning John has given his car keys to Emily, I have also said (implicitly) that John has a car, that John was holding his car keys this morning, that at some point the same car keys were held by Emily, and similar presuppositions and entailments. Also, further lines of reasoning are warranted, like the inference that Emily was driving the car in the afternoon and, therefore, she must be held responsible for the penalty ticket received at that time of the day.

Arguers often make lexical choices strategically, exploiting the presuppositions and entailments of a predicate to convey preferred arguments. Back in 2009, when Kraft attempted a hostile acquisition of Cadbury, the directors of Cadbury solicited shareholders to reject the bid by telling them: "Don't let Kraft steal your company" (Palmieri, 2014). The choice of the verb "to steal", rather than others like "to buy" has clear argumentative implications. The predicate "to buy" presupposes four semantic entities: the buyer (x_1), money (x_2), a good (x_3), the seller (x_4), where x_1 legally owns x_2 and x_3 legally owns x_4. The main entailment is that the legal ownership of x_2 and x_3 are shifted: x_1 becomes the new owner of x_3 while x_4 becomes the new owner of x_2. The semantics of "to steal" features some marked differences that have important argumentative implications. First, "to steal" has only three entities (x_1, x_2, x_3), as no money is involved. Second, no legal transfer of ownership is entailed. X_1 holds the good (x_2) but does not own it. The stealer (x_1) takes possession of x_2 without the permission of x_3, and without paying x_3 for it. By framing Kraft's financial proposal as a stealing act, Cadbury manages to emphasise the inadequate level of offer price, which was indeed the main argument for rejection (Palmieri, 2014).

Besides presuppositions and entailments, which work as virtual argumentative features, lexical frames activate arguments by working as cultural keywords that are grounded in common opinions and values, known as *endoxa* (Rigotti & Rocci, 2005; Bigi & Greco, 2012). An attempt to steal is generally judged as bad both

[1] In semantics, these entities are called 'arguments' or 'argument places'. In this article we refer to them as 'entities' to avoid confusion with the argumentative concept of 'argument' as a justification reason.

legally and morally, signalling therefore the presence of a malevolent intention. The "steal" frame chosen by Cadbury appeals to this endoxon to instil a negative view of Kraft in the eyes of shareholders who are, thereby, encouraged to distrust the takeover proposal.

2.3 Dataset and Procedure of Analysis

We collected the 2019 MS statements published by FTSE100 firms. We downloaded them in pdf format and converted them into txt to make them compatible with the analytic tools used for this study. The corpus included 96 statements with a total of 377,353 words.

Our study involved two main steps. The first consisted of an explorative analysis of a small sample of statements (n=20) to familiarise ourselves with the genre and gain an initial appreciation of its argumentative and rhetorical features. Using UAM Corpus Tool (O'Donnell, 2008), we annotated the topic of each document section to build the thematic macro-structure of the report. We also intended to annotate each sentence of the document according to its argumentative function, e.g., type of claim, type of reason, etc. However, we decided not to complete this level of analysis as it quickly became evident that companies avoid using an argumentative style, as we will explain more in detail in Sect. 3.

The findings of the first analytic step informed the subsequent framing analysis which was based on a corpus-assisted discourse analysis (Gillings et al., 2023). We combined well-established quantitative techniques in corpus linguistics with a qualitative semantic analysis of the prevailing predicates using the Congruity Theory framework explained in Sect. 2.2 (for a similar approach, see for instance Rigotti et al., 2006; Palmieri & Rigotti, 2014). The results of the first screening of the documents revealed the neat predominance of ethos-signalling expressions activated by "we + verbal phrase" statements. We decided, accordingly, to focus on these statements and their concordance lines, which allowed us to investigate the linguistic environment (or co-text) of these expressions (Sinclair, 2003; Baker, 2006). Using the *AntConc* tool (Anthony, 2018), we performed a collocation analysis of "we" to identify the most frequent verbs associated with it and the most frequent objects that are predicated by these verbs. The Congruity Theory-based semantic analysis of these collocates allowed us to identify the main framing strategies deployed by companies. As Gillings, Mautner and Baker explain, "a collocation analysis can be useful in spotting hidden or subtle meanings which can signal the author's stance" (2023, p.30). They also connect the use of collocation to the concept of *semantic prosody*, which, similar to framing, refers to the strategic association of apparently neutral words to convey evaluations and opinions that reflect the author's rhetorical goals.

Table 2 Macro-structure of MS statements

Report sections	# reports	% reports
Introduction	73	75
Chairman/CEO letter	13	13
Organisational structure/Business model	90	93
Approach to MS (incl. policies)	89	92
Training	76	78
Risk assessment/Due Diligence	85	88
Performance review	50	52
Specific initiatives	63	65
Case studies	29	30

3 Findings

Our analysis revealed four key findings that will be exposed in the next subsections: (1) a templated reporting genre which largely reflects regulatory prescriptions; (2) a non-argumentative-rhetorical style, in which ethos appeals dominate over logos and pathos; (3) a systematic deployment of trust-building expressions to discourage critical discussion; (4) a framing strategy aimed at shifting responsibility.

3.1 A Templated Reporting Genre

In line with prior research, we found that companies largely follow a templating approach in composing their MS statement.

As Table 2 shows, the titles of each section tend to mirror the list of areas that the UK government recommends covering in its guidance (Organisation structure and supply chains; Policies in relation to slavery and human trafficking; Due diligence processes; Risk assessment and management; Key performance indicators to measure effectiveness of steps being taken; Training on modern slavery and trafficking).[2] The Chairperson's introductory letter and the presentation of case studies, which have the potential to elaborate on corporate performance beyond the basic requirements, are the least frequent sections in the whole dataset. This suggests companies have opted for a minimalistic reporting approach, which prioritises box-ticking compliance over engagement with relevant issues and critical conversations.

[2] https://www.gov.uk/guidance/publish-an-annual-modern-slavery-statement#demonstrate-compliance-with-the-minimum-legal-requirements

3.2 *Mitigated Argumentation*

A closer inspection of the content of each section revealed a rather surprising finding, if compared to results obtained in prior corporate discourse research: arguments in support of asserted claims are nearly non-existent. Each statement in the report tends to express unopinionated facts or, if a claim is advanced, it refers to very generic and hardly disputable principles:

(1) "Diageo has a zero-tolerance approach to modern slavery. This statement describes what we are doing to prevent modern slavery in our business operations and our value chain" (HSBC, 2019, p.2)
(2) "We believe every worker deserves to be treated with dignity and respect. People are at the very heart of our business and whether they work directly for us, or within our supply chain, we take care to treat people well" (ABF, 2019, p.2).

This rhetorical style, which seems to avoid making arguments that can resolve doubts or promote corporate actions, is in stark contrast to other similar genres of sustainability communication, such as CSR reports. For example, we can compare the 2019 MS of HSBC with the CSR report published by the same corporation in the same year (HSBC, 2019).

The latter contains several argumentative features that support a clearly rhetorical/promotional aim, which the modern slavery statement does not exhibit at all, such as explicit logical connectors (3); contrastive connectives to convey need-based reasons for actions (4); quotations of arguments made by third parties like customers (5); headlines phrased as sub-claims that support the overarching sustainability-related claim promoted by the report (6):

(3) "At HSBC UK we understand the important role banks play in a thriving society and we care about our impact. This is why we work every day to meet the needs of our customers, colleagues and the communities we serve" (p.2).
(4) "Accessing the financial system is vital for everyone, but for some people in society this can be incredibly hard. At HSBC UK, we have been working to support some of the UK's most vulnerable people, giving them access to the financial system, which can change lives" (p.6).
(5) "These events are definitely worthwhile as they give us a chance to give back to the community which, in my eyes, is very important" (p.19).
(6) Headlines as sub-claims: "We are North" (p.26), "Supporting our communities" (p.28), "Developing the employability and financial skills people need to thrive" (p.12).

In contrast, as example (7) further highlights, MS slavery statements do not use headlines or paragraphs to put forward a claim (e.g., "Training" rather than "Our training is effective" or "Outstanding training support, etc.):

(7) "Training

We provide training on our code of conduct, which underpins and reinforces elements of our human rights commitments. We train bp employees about how to apply the code in their daily work. Training is available in 18 languages and in 2019 around 39,000 employees completed it in full" (BP, 2019 modern slavery statement)

In this example, the style of which is representative of the typical writing style in the whole dataset, a fully-fledged argument can be potentially identified only after a highly elaborated and context-intensive reconstruction: the reported information that training is available in 18 languages etc. could be taken as a reason that justifies the claim that training is effective or, at least, compliant with the regulations. However, companies choose to avoid making this logical link explicit and prefer not to advance any self-promotional claim in this regard.

Similarly, actions and decisions are causally explained, rather than argumentatively justified and promoted:

(8) "We know that genuinely committing to address modern slavery requires concrete action. Accordingly, we continue to set clear expectations that our people and suppliers must follow, so they can be alert to potential involvement in modern slavery, take steps to prevent and address it" (Rio Tinto, 2019, p.1).

In example (8), the company's acknowledgement that concrete action is required is presented as *the cause* of the decision to continue to set clear expectations for people and suppliers, which in turn will *cause* them to be alert and take steps. By formulating such decision-making in an explanatory style, the questionable nature of the decision is overshadowed, favouring thereby the impression that what has been decided is the only possible course of action that can be taken.

3.3 Trust-Building Frames as an Argument-Avoidance Strategy

The first analytic screening of the dataset has shown unequivocally that companies refrain from advancing explicit and direct arguments and make very scarce use of promotional language that can engage and emotionally involve the reader. The persistent and systematic absence of these rhetorical elements, as shown in Sect. 3.2, suggest that modern slavery statements are an inherently ethotic genre: firms adopt a rhetorical style in which ethos appeals dominate over logos and pathos.

The prevailing feature of these statements is, indeed, the constant use of self-referential language to promote a particular image of trustworthiness (ethos), which can refer to the ability/intelligence, integrity or benevolence of the company (Mayer et al., 1995; Fuoli, 2018). We found an abundant use of assertions describing what the company has done or is doing which foregrounds the company itself through first-person pronouns, such as "we" or "our":

(9) "We have not identified any instances of modern slavery occurring in our supply chain and therefore it can be difficult to measure the effectiveness of our policies in reducing the risk of slavery and human trafficking. However, we recognise that there is not room for complacency and will continue to work to enhance our policies and procedures and measure awareness within our business" (Sage, 2019, p.3).

In example (9), rather than simply saying that there was no instance of modern slavery or that there is no room for complacency, the company uses the first-person plural to emphasise their awareness and control of the situation ("we have not identified..."; "we recognise that..."), thus conveying an image of ability-based trust.

As Table 3 shows, 83.62% of the sentences that refer to actions or outcomes are internally attributed, i.e., they thematise something that has or will be done/achieved by the company itself. Almost half of these sentences are grammatically constructed as a "we+verb" phrase (e.g., "we recognise that...") while 35% have an "our + company's attribute" form (e.g., "our approach to modern slavery...").

This evidence led us to focus the second analytic step—the frame analysis based on keywords in context—on first-person pronouns (We/our) and explore their most frequent verb collocates.

As Table 4 indicates, the most frequent verbs used in "we+verb" expressions signal trustworthiness at the level of intelligence/prudence (e.g., "to recognise" and "to believe") and honesty/integrity (e.g., "to be committed", "to expect", "to operate"). For reasons of space, in this sub-section, we will examine only the verb "to recognise", which signals ability-based trust, while other verbs signalling integrity or benevolence will be discussed in the next sub-section (Sect. 3.4).

The verb "to recognise" (x_1, x_2) activates a predicate which presupposes two semantic entities: x_1 is a human subject capable of intellectual scrutiny and critical judgement, while x_2 is an event/situation related to x_1's domain of activity. The entailment is that x_1 is aware of x_2 and acknowledges its relevance. As the following examples show, in those instances where x_1 is represented by the pronoun "we", the most frequent types of situations x_2 refers to are: risks (10), priorities (11), responsibilities, (12), pending tasks (13) and limits (14):

Table 3 Frequency of internally attributed actions in MS statements

Outcome/action attribution	**N=4299**	
Internal	3595	83.62%
External	704	16.38%
Internally attributed agent	**N=3595**	
We	1751	48.71%
Company's attribute (e.g. Our business, our approach)	1269	35.30%
Company's name	303	8.43%
Corporate members	272	7.57%

Table 4 Collocation analysis: most frequent verbs in "we+verb" sentences

Verb	Frequency
Recognise	127
Are committed	105
Expect	96
Believe	88
Operate	81
Use	57
Aim	52
Require	49
Provide	44
Know	37
Identify	30
Monitor	30
Encourage	29
Conduct	28
Consider	27
Understand	27
Engage	25
Support	23
Ensure	19

(10) We recognise the risks of forced labour modern slavery, human trafficking and other labour standards violations within our supply chains (Glencore, 2019 MS statement)

(11) "We recognise modern slavery and human trafficking as a growing global concern and an important issue for Johnson Matthey, our employees, customers and other stakeholders" (JMAT, 2019, p.2).

(12) "In situations where we identify we have caused or contributed to human rights harm, including modern slavery, in line with the UNGPs, we recognise our responsibility to provide for or cooperate in remediation (Rio Tinto, 2019, p. 9).

(13) "We recognise that additional work is still required on an ongoing basis to further develop our ability to correctly identify other potential signs of modern slavery that may be less evident (FRES, 2019, p. 12)

(14) "We recognise that it is not easy to quantify and measure some of the outcomes and impacts regarding our approach to human rights and efforts to tackle modern slavery as these are multi-dimensional issues" (Aviva, 2019, p.9)

Overall, these lexical choices realise a framing strategy by which the company portrays itself as a competent and honest actor who (a) knows and controls the situation and (b) acts accordingly with responsibility towards the law. The construction of a trustworthy image through frames that signal ability and integrity appears to be instrumental to an argument-avoidance strategy: the audiences/readers are invited to trust the company and refrain from casting doubts or raising criticisms. Following this reconstruction, the MS statement would be governed by an implicit overarching argument that can be informally put as follows: "We know what we are doing, we know what we should do, we have everything under control, therefore: don't worry and don't ask".

3.4 *Shifting Responsibility Frames*

We found extensive use of integrity-signalling verbs ("to expect" and "to welcome") and benevolence-signalling verbs ("to encourage" and "to support") by which companies frame their role as intermediaries between governments, on the one side, and suppliers or employees on the other side. By doing so, they try to place ultimate responsibility for the prevention of modern slavery onto these external actors.

3.4.1 Downward Responsibility Shifting

Examples (15) and (16) are representative of a frequent framing strategy performed mainly by three verbs:

(15) "**We expect** all suppliers to commit to responsible sourcing as part of their engagement with Anglo American and recognise the need to **support** suppliers to identify and manage sustainability risk" (AA, 2019, p. 9).

(16) "We are committed to respecting the human rights of our employees in the workplace. **We encourage** our suppliers and business partners to adopt the same or similarly high standards of ethical behaviour. This includes **giving guidance and support** to our directly contracted suppliers and business partners in taking appropriate steps to prevent slavery and human trafficking occurring in our business or supply chain" (BAE, 2019, p.2).

The three-place predicate "to expect" involves the company (x_1), a lower-end actor in the supply chain (x_2), typically partners and suppliers, and an action (x_3) that is required to fulfil MS laws. The entailment of "to expect" is that x_1 attributes to x_2 the responsibility and future accountability for x_3 and hopes x_2 will do x_3. The three-place predicate "to encourage" has a similar structure, but x_2 tends to refer to employees while the entailment is that x_1 solicits but does not force x_2 to perform x_3 attributing responsibility accordingly. The predicate "to support" targets again employees and entails that x_1 helps x_2 accomplish x_3 for which x_2 remains the main responsible agent. Companies (x_1) frame themselves as agents acting with integrity and goodwill, but the actions they expect, support or encourage are in the hands of other stakeholders.

3.4.2 Upward Responsibility Shifting

A similar framing strategy is deployed in relation to governmental and regulatory authorities, their aims and initiatives (underlined in the example below). These entities fill the semantic role of x_2 or x_3 in predicates such as "to welcome" (17) and "to support" (18 and 19), which signal corporate benevolence:

(17) **"We welcome** the increased focus of governments around the world on the issues of modern slavery and human trafficking. As a global firm, **we would especially welcome** international coherence and mutual recognition of different reporting requirements and standards, to help mobilise global efforts as well as to prevent good legislative intentions from being undermined by different reporting standards" (HSBC, 2019, p.1).
(18) **"We support** the aims of the Act and seek to ensure slavery and human trafficking does not feature in our business or supply chain" (Admiral, 2019, p.2).
(19) **"We support** the government's work to end modern slavery and aim to play our part as a good corporate citizen" (Hargreaves Lansdown, 2019, p.1).

The predicate "to welcome" (x_1, x_2) typically involves a new initiative/measure (x_2) not made by the company (x_1) and entails that x_1 limits itself to support it and include it in its sphere of action. The predicate "to support" (x_1, x_2, x_3) addresses government/international (x_2) organisations and a goal they have (x_3). The entailment frames the company (x_1) as an actor who accepts to join the cause of x_2 towards goal x_3 for which x_2 remains the main responsible agent.

4 Conclusion

This chapter focused on sustainability strategic communication through a linguistic-rhetorical analysis of modern slavery statements aimed at better understanding how companies respond to new disclosure obligations. Starting from the hypothesis that companies do make use of rhetorical arguments as they do in other genres of (sustainable) communication, we found instead evidence of minimally argumentative behaviour. Rather than engaging in critical and promotional argumentation, firms deploy framing strategies that fulfil an argument-avoidance strategy. These strategies have been identified through a linguistic analysis combining keyword-in-context quantitative analysis and qualitative frame semantic analysis. The main emergent finding is that companies frame themselves as a trustworthy intermediary in the fight against modern slavery between two parties that are framed as the ultimate responsible agents: on the one side, they call for, welcome and support regulatory and intergovernmental initiatives; on the other side, they expect partners and employees to comply with modern slavery legal requirements, helping them do so.

Rather than advancing fully-fledged arguments, firms privilege an expository style which, through a systematic preference for ethos signals over logos and pathos appeals, fulfils an argument-avoidance strategy. This lack of argumentative engagement is in stark contrast to the communicative behaviour that companies exhibit in other genres of strategic communication, such as takeovers (Brennan et al., 2010; Palmieri, 2014) and proxy fights (Uysal et al., 2018). Research in some of these genres of strategic communication seems to suggest that companies engage in more relevant and critical strategic conversations when (a) regulations explicitly require companies to justify/argue an opinion rather than simply reporting information (e.g., takeover rules) and/or (b) an influential stakeholder (e.g., an activist shareholder) raises an issue and challenges corporate directors to discuss it. We call for future research to systematically compare modern slavery statements with other genres of social sustainability communication (e.g., activist shareholder campaigns or crisis responses) where modern slavery is discussed. Such a comparative analysis could help better understand the limits of existing modern slavery disclosure rules and design more effective systems that are able to create both regulatory and market-driven incentives for relevant sustainability communication.

References

Anthony, L. (2018). *AntConc: A freeware corpus analysis toolkit for concordancing and text analysis*. http://www.laurenceanthony.net/software/antconc.

Baker, P. (2006). *Using corpora in discourse analysis*. Continuum.

Balabanova, E., Palmieri, R., & Liu, Z. (2023). Twitter (X), fast fashion and backlash: Argumentation and ethics on social media. *Business and Professional Communication Quarterly*, 23294906231208411.

Beattie, V., D'Hanane, A., & Jones, M. J. (2008). Investigating presentational change in UK annual reports: A longitudinal perspective. *The Journal of Business Communication, 45*(2), 181–222.

Bigi, S., & Morasso, S. G. (2012). Keywords, frames and the reconstruction of material starting points in argumentation. *Journal of Pragmatics, 44*(10), 1135–1149.

Brennan, N. M., Daly, C. A., & Harrington, C. S. (2010). Rhetoric, argument and impression management in hostile takeover defence documents. *The British Accounting Review, 42*(4), 253–268.

Caruana, R., Crane, A., Gold, S., & LeBaron, G. (2021). Modern slavery in business: The sad and sorry state of a non-field. *Business & Society, 60*(2), 251–287.

Christ, K. L., & Burritt, R. L. (2018). Current perceptions on the problem of modern slavery in business. *Business Strategy & Development, 1*(2), 103–114.

Entman, R. M. (1993). Framing: Toward clarification of a fractured paradigm. *Journal of Communication, 43*(4), 51–58.

Fillmore, C. J. (1976). Frame semantics and the nature of language. *Annals of the New York Academy of Sciences, 280*(1), 20–32.

Fuoli, M. (2018). Building a trustworthy corporate identity: A corpus-based analysis of stance in annual and corporate social responsibility reports. *Applied Linguistics, 39*(6), 846–885.

Gillings, M., Mautner, G., & Baker, P. (2023). *Corpus-assisted discourse studies*. Cambridge University Press.

Goffman, E. (1974). *Frame analysis: An essay on the organization of experience*. Harvard University Press.

Gold, S., Trautrims, A., & Trodd, Z. (2015). Modern slavery challenges to supply chain management. *Supply Chain Management: An International Journal, 20*(5), 485–494.

Greco, S., & De Cock, B. (2021). Argumentative misalignments in the controversy surrounding fashion sustainability. *Journal of Pragmatics, 174*, 55–67.

Greco Morasso, S. (2012). Contextual frames and their argumentative implications: A case study in media argumentation. *Discourse Studies, 14*(2), 197–216.

Green, S. E. (2004). A rhetorical theory of diffusion. *Academy of Management Review, 29*(4), 653–669.

HSBC (2019). *Environmental, Social and Governance Update 2019*. Available at: https://www.hsbc.com/-/files/hsbc/investors/hsbc-results/2019/annual/pdfs/hsbc-holdings-plc/200218-esg-update-2019.pdf

Hsin, L., New, S. J., Pietropaoli, I., & Smit, L. (2021). *Accountability, monitoring and the effectiveness of Section 54 of the Modern Slavery Act: Evidence and comparative analysis*. Modern Slavery Policy and Evidence Centre.

Malavasi, D. (2010). The multifaceted nature of banks' annual reports as informative, promotional and corporate communication practices. In P. Evangelisti Allori & G. E. Garzone (Eds.), *Discourse, identities and genres in corporate communication. Sponsorship, advertising and organizational communication* (pp. 211–233). Peter Lang.

Mayer, R. C., Davis, J. H., & Schoorman, F. D. (1995). An integrative model of organizational trust. *Academy of Management Review, 20*(3), 709–734.

Meehan, J., & Pinnington, B. D. (2021). Modern slavery in supply chains: Insights through strategic ambiguity. *International Journal of Operations & Production Management, 41*(2), 77–101.

Merkl-Davies, D. M., & Brennan, N. M. (2007). Discretionary disclosure strategies in corporate narratives: Incremental information or impression management? *Journal of Accounting Literature, 27*, 116–196.

Miller, C. R., & Kelly, A. R. (2016). Discourse genres. In A. Rocci & L. de Saussure (Eds.), *Verbal communication* (pp. 269–286). Mouton de Gruyter.

Mohammed, D. (2019). Standing standpoints and argumentative associates: What is at stake in a public political argument? *Argumentation, 33*(3), 307–322.

Monciardini, D., Bernaz, N., & Andhov, A. (2021). The organizational dynamics of compliance with the UK Modern Slavery Act in the food and tobacco sector. *Business & Society, 60*(2), 288–340.

Musi, E. (2024). Framing to make an argument: The case of the genocide hashtag in the Russia-Ukraine war. *Argumentation*, 1–20.

Musi, E., & Aakhus, M. (2019). Framing fracking: Semantic frames as meta-argumentative indicators for knowledge-driven argument mining of controversies. *Journal of Argumentation in Context, 8*(1), 112–135.

O'Donnell, M. (2008, April). The UAM CorpusTool: Software for corpus annotation and exploration. In *Proceedings of the XXVI Congreso de AESLA* (Vol. 3, p. 5). Almeria.

Palmieri, R. (2014). *Corporate argumentation in takeover bids*. John Benjamins.

Palmieri, R., & Mazzali-Lurati, S. (2021). Strategic communication with multiple audiences: Polyphony, text stakeholders, and argumentation. *International Journal of Strategic Communication, 15*(3), 159–176.

Palmieri, R., Mercuri, C., & Mazzali-Lurati, S. (2022). Persuasive reasons in crowdfunding campaigns: Comparing argumentative strategies in successful and unsuccessful projects on Kickstarter. *International Journal of Strategic Communication, 16*(2), 332–355.

Palmieri, R., & Rigotti, E. (2014). Suspicion as an argumentative move. *Journal of Argumentation in Context, 3*(3), 287–321.

Rigotti, E., & Rocci, A. (2001). Sens–non-sens–contresens. *Studies in Communication Sciences, 1*(2), 45–80.

Rigotti, E., & Rocci, A. (2005). From argument analysis to cultural keywords (and back again). In P. Houtlosser & F. van Eemeren (Eds.), *Argumentation in Practice* (pp. 125–142). John Benjamins.

Rigotti, E., Rocci, A., & Greco, S. (2006). The semantics of reasonableness. In P. Houtlosser & A. M. van Rees (Eds.), *Considering pragma-dialectics* (pp. 257–274). Routledge.

Rocci, A. (2009). Manoeuvring with voices. The polyphonic framing of arguments in an institutional advertisement. In F. H. van Eemeren (Ed.), *Examining argumentation in context. Fifteen studies on strategic maneuvering* (pp. 257–283). John Benjamins.

Rogerson, M., Crane, A., Soundararajan, V., Grosvold, J., & Cho, C. H. (2020). Organisational responses to mandatory modern slavery disclosure legislation: A failure of experimentalist governance? *Accounting, Auditing & Accountability Journal, 33*(7), 1505–1534.

Rutherford, B. A. (2005). Genre analysis of corporate annual report narratives: A corpus linguistics-based approach. *Journal of Business Communication, 42*(4), 349–378.

Sillince, J. A. (2002). A model of the strength and appropriateness of argumentation in organizational contexts. *Journal of Management Studies, 39*(5), 585–618.

Sinclair, J. (2003). *Reading concordances: An introduction*. Pearson Longman.

Uysal, N., Yang, A., & Taylor, M. (2018). Shareholder communication and issue salience: Corporate responses to 'social' shareholder activism. *Journal of Applied Communication Research, 46*(2), 179–201.

Van Eemeren, F. (2010). *Strategic maneuvering in argumentative discourse*. John Benjamins.

Van Eemeren, F. H., & Grootendorst, R. (2004). *A systematic theory of argumentation: The pragma-dialectical approach*. Cambridge University Press.

Van Eemeren, F. H., Grootendorst, R., Jackson, S., & Jacobs, S. (1993). *Reconstructing argumentative discourse*. University of Alabama Press.

Further Reading

Mercuri, C. (2023). Characterization frames constructing endoxa in activists' discourse about the public controversy surrounding fashion sustainability. *Topoi, 42*(2), 635–650.

Pinnington, B., & Meehan, J. (2023). Learning to see modern slavery in supply chains through paradoxical sensemaking. *Journal of Supply Chain Management, 59*(4), 22–41.

Rudi Palmieri is a Senior Lecturer in Strategic Communication at the University of Liverpool. His research focuses on the analysis of argumentation in different contexts of strategic communication, including financial communication, entrepreneurial rhetoric, and crisis communication. His work tackles important economic and societal issues with an argumentative-rhetorical perspective, such as the interplay between sustainability and business goals in contemporary financial and investor communications, the dynamics of trust repair in crisis communication and the configuration of persuasive strategies in complex and multiple audience situations in digited media.

Ekaterina Balabanova is Professor of Politics and Media at the University of Liverpool. Her research interests include public and policy debates on migration and refugees, human rights campaigning and communication of human rights issues by business, government and non-government organisations. She has led a research project that examined the backlash on social media over working conditions and environmental costs in the fashion industry, looking at conversations between businesses and consumers on social media.

Sustainability as a CEO Activism Trigger: The Portuguese Case

Alexandre Duarte and Patricia Dias

Abstract This study aimed to explore if and how the international growing trend of CEO activism is manifesting in the Portuguese context, by exploring the perspective of Portuguese CEOs on the topic, the course of action that they are taking, and how they evaluate it, namely within the sustainability framework.

Following Howard Bowen's legacy for responsible business practices and adding to that the actual growing pressure of societal problems, business leaders all around the world are being asked, if not demanded, to act and communicate as action persons who effectively contribute to a better common value creation.

To address this topic, a qualitative methodology was implemented based on interviews to 24 Portuguese CEOs and used thematic analysis to explore them. Our findings reveal that, although Portuguese CEOs are aware of this trend, they acknowledge both risks and opportunities, but their activism in still incipient, although they strongly believe that their activism actions must be aligned with the values and mission of the organization they represent. Finally, Portuguese CEOs are convict that their activism should be firstly implemented and communicated within the organization, which is where most of them focus their action. This work provides an original mapping of the CEO activism landscape in Portugal and an insightful discussion of the CEO's perspective on this growing phenomenon, looking at it by the power position that the leaders have, also in sustainability-related causes.

A. Duarte (✉)
ICNOVA, Faculty of Social and Human Sciences, Universidade Nova de Lisboa, Lisbon, Portugal

P. Dias
CECC, Faculty of Human Sciences, Universidade Católica Portuguesa, Lisbon, Portugal
e-mail: pdias@ucp.pt

F. Weder (ed.), *Strategic Sustainability Communication*, CSR, Sustainability, Ethics & Governance, https://doi.org/10.1007/978-3-031-89486-2_12

1 Introduction

Because of the persistent and growing pressure from citizens to incorporate sustainability concerns into their corporate agendas (Perrault & Clark, 2016), modern organizations seem to have adopted responsibility as their guiding principle. As civil society is increasingly demanding "social good" from all sectors, and consumers leverage their influence and purchasing power to demand tangible actions from organizations, many business leaders have recognized the need to respond to these societal expectations. Therefore, discourses on sustainability have become central in modern societies. Governments, activists, associations, NGOs, and citizens have brought this issue to the table, and now companies have felt their confidence undermined, as they have been identified as one of the main ones responsible for the catastrophic situation we are experiencing.

In this sense, Chief Executive Officers (CEOs) have felt the need to face this global requirement, responding to sociopolitical issues, namely sustainability, adapting their positions to discourses of sustainable development in order to show and persuade stakeholders that their actions are congruent with the societal demands (Sama et al., 2020; Pless et al., 2021; Arivdson & Sabelfeld, 2023).

Thus, having CEOs taking a stand on sociopolitical themes is the first phenomenon that has been gaining a big relevance in recent years and justifies our attention in this article. The second one is the fact that in the same way that societal discourse is fundamental to providing legitimacy to sustainable development (Newig et al., 2013), a company leader's discourse is also key for the alignment of the whole organization around its principles, values, and mission. Thirdly, activism has been recognized as a powerful tool to influence decision-making at different levels (Mäkelä & Olkkonen, 2021) and finally, as stated by Weder and Dobric (2021), there are still few studies on sustainability communication, and even fewer on specific countries or cultural settings (p. 59).

Most studies on this topic concentrate on the effects and outcomes of CEO activism rather than the "role" and self-identity of the CEO as an activist, and even less on specific countries or cultural environments.

This study aims, thus, to shed light on this theme. By interviewing 24 CEOs of companies operating in Portugal in areas ranging from Insurance to Public Relations, from Media to Health and Baby Care products, and from Marketing and Advertising to Technologic consultancy, among others, the authors reunite, for the first time, some insightful data on how Portuguese CEOs see and what's their position toward this topic, bringing some light also to the less researched areas on strategic sustainability communication such as the questions of power (Weder, 2023, p.240).

2 Theoretical Framework

In recent years, several social changes have profoundly influenced corporate governance. Duarte and Chambel (2023) highlighted some of the movements that boosted modernity, such as "the affirmation and empowerment of ethnic, cultural, religious minorities, but also of class, sexual orientation and gender, among others" (p. 205). These movements often arise from a social desire to address perceived injustices and challenge systemic issues but also contribute to positive social transformation, resulting in diverse forms of activism.

Activism, defined as the efforts, actions, or campaigns of individuals, groups, or organizations to promote social, political, economic, or environmental change, plays a critical role in shaping public discourse, challenging the status quo, and driving social progress. The increasing prevalence of these sociopolitical issues in everyday life is undeniable for both individuals and organizations, and this widespread involvement in promoting the common good suggests that the social good is becoming dominant in today's society (Vredenburg et al., 2020, p. 456).

This perspective is based on legitimacy theory, which posits that organizations have an implicit social contract with their communities and must therefore align with society's values and expectations (Shocker & Sethi, 1973). As social expectations evolve and awareness of social causes grows, as is the case with sustainability, the company's top leaders must strive to achieve social legitimacy, in line with both legitimacy theory and the theory of upper echelons (Velte, 2023b).

Pless et al. (2021) describe a responsible and sustainable business orientation as being accountable to others, including future generations, and ensuring the well-being of society and the environment. Sama et al. (2020) note that the recent integration of Corporate Social Responsibility (CSR) and the United Nations' Sustainable Development Goals (SDGs), promoted both academically and professionally, has enhanced ethical and responsible business practices, leading companies to act as agents of positive social change.

Furthermore, there has been a notable increase in the appointment of sustainability experts to boards of directors and the establishment of the Chief Sustainability Officer (CSO) position. While CSOs are crucial and positively impact corporate environmental strategies, the influence of traditional top management roles (e.g., CEOs or Chief Financial Officers—CFOs) on company decisions remains significant (Velte, 2023b).

The emphasis on responsibility and sustainability is particularly evident in the growing number of sustainability reports, which now feature extensive coverage of environmental, social, and governance (ESG) issues. Velte (2023a) observed that, following the 2008–2009 financial crisis, there was a rising demand from stakeholders for greater emphasis on environmental issues in corporate sustainability reports.

In a review of 250 CEO messages from companies in the international energy sector, Franziska Weder (2023) identified 62 topics in their CSR reports, with climate change, workplace conditions, and factory safety being the most frequently mentioned. Another study by Arivdsson and Sabelfeld (2023), which analyzed 120

CEO messages in sustainability reports from 2008 to 2019 across 10 Swedish multinational industrial companies, identified three distinct phases with key themes. In the latest phase, beginning in 2016, prominent themes included transformation/transition, climate action, climate crises/climate urgency, science, circular economy/models, diversity, reducing carbon emissions, attracting employees/focus on employees, and sustainable business efforts.

Sustainability, understood as a holistic concept, morality constructed by political, corporate, and non-governmental actors (Weder, 2022, p.352–353), necessarily involves communication (Newig et al., 2013). Joining these two constructs, we get to the sustainable communication concept, whose task, for Godemann and Michelsen (2011), is to critically evaluate and introduce an understanding of the human-environment relationship into social discourse.

In search of the legitimacy that organizations need to survive and thrive, many CEOs frame sustainability on two dimensions: (a) in line with responsible leadership (RL), which Maak et al. (2016), have defined as "a relational (and purpose-driven) influence process between leaders and stakeholders geared toward the establishment of accountability in matters about organizational value creation" (p. 464). (b) Within dynamic evolutionary discourses, represented by national and transnational policies, regulations, and sociopolitical events, to be perceived as conforming to institutionalized concepts and structures over time (Arivdson & Sabelfeld, 2023).

As strategic communication is "the purposeful use of communication by an entity to engage in conversations of strategic significance to its goals" (Zerfass et al., 2018, p.2), and 65% of consumers want CEOs to speak on major social issues (Hou & Poliquin, 2023), the authors posit that strategic sustainability communication is not only one of the CEOs main goals but one of its survival's requisites. In fact, business leaders are not just leading the discourse of profitable sustainability, they are also slowly adapting their sustainability talk to other discourses (Arivdsson & Sabelfeld, 2023). Furthermore, CEOs' extreme influence, a consequence of their huge public visibility and positional power (Branicki et al., 2021), makes their public stances and discourses attract a great mediatic interest and therefore social repercussions.

Activism and sustainability have been sharing the road for so long (Duarte & Chambel, 2023), as activists always pushed and pressured companies to make their operations more sustainable, but what's new nowadays is the fact that new actors such as shareholders and CEOs start to become more actively engaged in the phenomenon (Mäkelä & Olkkonen, 2021). Although these authors differentiate several kinds of sustainable activisms, such as environmental activism, shareholder activism, and social activism, we will assume the concept of sustainability according to Elkington's (1999) triple bottom line, i.e., based on economic prosperity, environmental quality, and social justice, and therefore analyze activism in a holistic approach.

Despite recent, in recent years, a plethora of publications on the effects of CEO activism seems to have flourished but mostly focused on the consequences of the public opinions of business leaders (Hambrick & Wowak, 2021; Rumstadt & Kanbach, 2022; Hou & Poliquin, 2023; Mkrtchyan et al., 2022). In this study, we focus on the perception of the Portuguese CEOs on this phenomenon. How they see it, what they believe it's the right way to explicitly talk, or not, when, where, and what themes they think are more relevant and pertinent to talk.

3 Methodology

The objective of this research is to delve into the involvement of Portuguese CEOs in activism, particularly focusing on sustainability and their corresponding communication strategies. In pursuit of this objective, we pose the following inquiries:

RQ1. What is the importance of activism for organizations and CEOs?
RQ2. What are the most important topics in corporate and CEO activism?
RQ3. Why are companies and CEOs engaging in activism?
RQ4. Does CEOs see the sustainability as a strategic form of communication?

To address these inquiries, we undertook an exploratory investigation, interviewing 24 CEOs, as outlined by Creswell and Poth (2017).

3.1 Data Collection Technique

Our data collection method involves semi-structured interviews chosen for their suitability in an exploratory study. These interviews follow a structured script that adapts to the natural flow of conversation between researchers and participants. Consequently, within the context of an exploratory investigation, the semi-structured approach allows for the emergence of novel topics not previously explored in the scientific literature (Brinkmann & Kvale, 2014).

Our interview script comprises six questions designed to explore the significance of activism for CEOs in the contemporary landscape, to map the engagement of Portuguese CEOs in activism, to identify exemplary practices in CEO activism, and to understand their commitment to sociopolitical topics.

The interviews took place in February 2023, facilitated through the Zoom videoconference platform. Prior to the interviews, informed consent was obtained, adhering to the General Regulation of Data Protection and ethical guidelines for qualitative research (Miller et al., 2012). Each interview session was recorded and subsequently transcribed automatically by the platform utilized. On average, the interviews lasted 28 min. While participants consented to the disclosure of their names, job roles, and company affiliations, we are committed to anonymizing the transcriptions.

3.2 Sampling Technique

We employed a purposive sampling approach, specifically targeting Portuguese CEOs operating within Portugal. Initially, we reached out to five Portuguese CEOs within our LinkedIn network. Subsequently, we utilized the snowballing technique, whereby these initial interviewees were asked to recommend other Portuguese CEOs exemplifying commendable CEO activism (Ray, 2012). Based on these referrals, we identified additional participants who met our criteria, once again making contact through LinkedIn.

In total, we reached out to 31 CEOs, of whom 24 graciously agreed to participate in our research endeavor.

3.3 Data Analysis Technique

For data analysis, we employed thematic analysis by corresponding the answers obtained with themes previously identified in our theoretical framework and adding afterwards the new topics that came in the interviews. So, we conducted two rounds of coding on the transcriptions. In the initial round, we focused on predefined coding topics corresponding to each of the six questions derived from our literature review. Subsequently, in the second round of coding, we identified emergent new topics (Braun & Clarke, 2021). To ensure intercoder validity, excerpts selected by two distinct researchers were compared for three randomly chosen topics (Vanover et al., 2021).

Table 1 outlines the initial codes utilized, alongside the emergent codes, corresponding to each question outlined in the interview script.

Table 1 Codes used in the thematic analysis

Script question	Initial codes	Theoretical grounding	Emergent codes
1. Which concerning risks or threats do you think your business will face in a near future?	Political instability Inflation Cybersecurity Supply chain management Social fragmentation/conflict Governmental regulation Climate change Natural disasters	Brown et al. (2020)	Attracting/retaining talent

(continued)

Table 1 (continued)

Script question	Initial codes	Theoretical grounding	Emergent codes
2. Which opportunities and risks do you identifying in a CEO taking a stand about a cause or topic, engaging in activism?	Opportunities Risks Both	Larcker et al. (2018); Chatterji and Toffel (2019); Shetty et al. (2019); Eyada (2020); Hou and Poliquin (2023)	O—Growing the business O—Contributing to well-being of all O—Putting purpose and values in practice R—Navigating turbulent social media R—Organizational constraints R—Losing support R—Crises R—Being irrelevant O/R—Separation between CEO and company B—Credibility and reputation B—Honesty vs. Opportunism B—Positioning vs. neutrality
3. Why do you think companies and CEOs are engaging in activism?	Commitment to a positive social change Alignment with company's values Improving reputation Request/demand from clients Request from collaborators/ shareholders Request/demand from administration Because other companies are doing it Increasing sales Pressure from activist groups	Chatterji and Toffel (2019); Hambrick and Wowak (2021)	Right thing to do Attracting talent

(continued)

Table 1 (continued)

Script question	Initial codes	Theoretical grounding	Emergent codes
4. Which topics and/or causes is your organization (and/or you) engaged with?	Ethnic equality Gender equality Climate change LGBT+ rights Health reform Salary equality Retirement policies Immigration Homeless Abortion	Brown et al. (2020); Pordata (2022a)	Education Minimum wage
5. Can you give us good examples of Portuguese CEO activists? Why do you think they are good?	Individual traits Message Action	Branicki et al. (2021); Rumstadt and Kanbach (2022); Hambrick and Wowak (2021)	I—Discretion I—Empathy I—Courage I—Conviction I—Good listener M—Simplicity M—Focus M—Consistency M—Honesty A—Positive impact in community A—People-driven A—Innovation A—Example A—Opportune A—Crises management M/A—Alignment with company purpose and values
6. How do you put your activism into practice? What are your strategies and tactics? Which media do you use?	Yes No Media Public	Mkrtchyan et al. (2022)	M—Informal conversations M—Social media M—Press M—Marketing/ Branding M—RP M—Podcasts M—Conferences/ academia M—Internal communication M—Action P—Collaborators P—Partners P—Community Work Mentorship

3.4 Sample Characterization

Table 2 provides a comprehensive overview of our sample, consisting of 24 Portuguese CEOs operating within Portugal. The interviewees are arranged chronologically based on the date of their respective interviews.

Of particular note is the gender distribution within our sample, revealing that merely 6 out of the 24 CEOs interviewed are female. Moreover, among the 26 CEOs identified as exemplary instances of CEO activism through our snowball sampling approach, only 4 were women. While our sample encompasses a diverse array of markets, it's essential to acknowledge that due to the snowball sampling method, certain sectors are underrepresented. In contrast, others, such as marketing, advertising, and business consultancy, are more represented.

Table 2 Characterization of the sample

	Name	Gender	Company	Market
1	João Ferreira	Male	Expo Group	Real estate
2	Daniel Redondo	Male	Licor Beirão	Alcoholic beverages
3	Diogo Teixeira	Male	Beta-i	Start-up hub
4	Fernando Figueiredo	Male	Fuel TV	Media
5	Luísa Oliveira	Female	Henkel	Consumer Products
6	Miguel Moreira Rato	Male	Adagietto	Public Relations
7	Pedro Janela	Male	Wy Group	Marketing and advertising
8	Nuno Campos	Male	Nearshore	Technology
9	Rodrigo Viana de Freitas	Male	Central de Informação	Public Relations
10	Tim Vieira	Male	Brave Generation Academy	Education
11	Vitor Vasques	Male	Pick a Buy	Storage
12	Pedro Rodrigues	Male	Desafio Global	Events
13	Vasco Falcão	Male	Konica Minolta	Technology
14	Nuno Santos	Male	JWT	Marketing and advertising
15	Pedro Grad Brito	Male	IMCI Group	Business consultancy
16	Sandra Alvarez	Female	PHD Media	Marketing and advertising
17	Carlos Freire	Male	AON	Insurance
18	Rudolph Gruner	Male	Observador	Media
19	Maria João Mergulhão	Female	MustSee Portugal	Tourism
20	João Nogueira Santos	Male	Learninghubz	Online Education
21	Elisabete Ferreira	Female	New Normal	Business consultancy
22	Maria Estarreja	Female	Core	Research
23	Nuno Conceição	Male	InovCorp Group	Business consultancy
24	Filipa Remígio	Female	Artsana	Health & Baby care

4 Findings

4.1 Activism as a Management Tool

Corporate and CEO activism is increasingly recognized as a strategic approach for companies to address challenges, respond to societal demands, and foster robust relationships with diverse stakeholders (Chatterji & Toffel, 2019; Hambrick & Wowak, 2021). This proactive stance allows organizations to cultivate strong reputations and demonstrate their commitment to social responsibility. By engaging in activism, companies can align their actions with the values and expectations of their stakeholders, including consumers, investors, employees, and communities. This approach helps build trust and enhances the company's resilience in the face of evolving market dynamics and public scrutiny.

The responses from the CEOs highlight several significant challenges and threats that they anticipate facing soon (Fig. 1). The most prominent concern among CEOs is related to the attraction and retention of talent, reflecting the growing importance of human capital in today's competitive business environment. Sixteen out of 24 CEOs in our sample (66.7%) expressed concerns about their companies' capability to attract and retain talent, as "a misalignment between what the organization says and does may harm its reputation, and it could lead to a talent drain." This concern underscores talent's critical role in driving organizational success and innovation.

Another major concern mentioned by a significant portion of the CEOs is inflation, which has been exacerbated by external factors such as geopolitical events like the invasion of Ukraine. The rise in inflation poses challenges for businesses in terms of managing costs, pricing strategies, and overall economic stability.

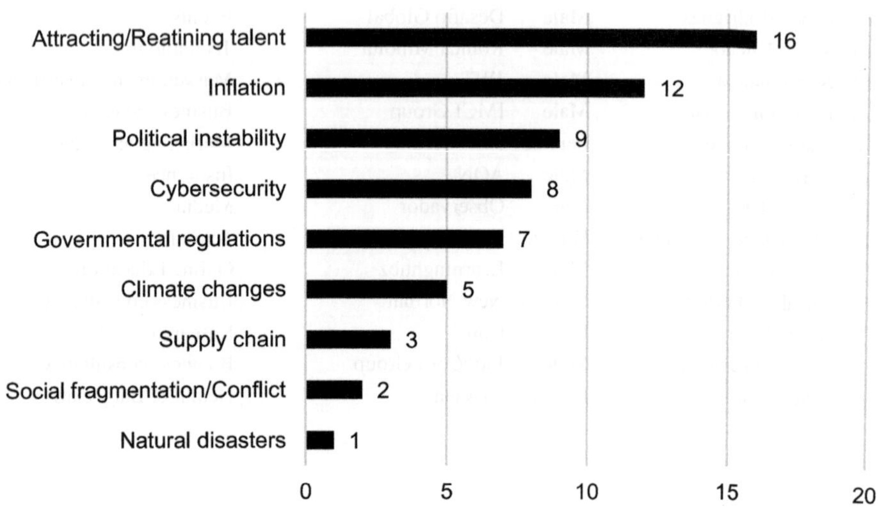

Fig. 1 Main challenges or threats that their businesses will face in the near future

Additionally, political instability, cybersecurity threats, and governmental regulations are identified as significant concerns. Political instability can create uncertainty and disrupt business operations, while cybersecurity threats pose risks to data security and operational continuity. Governmental regulations, on the other hand, can impact businesses in various ways, including compliance costs and operational restrictions.

Interestingly, climate change appears to be a lower priority concern among the CEOs, despite its growing significance globally. This may suggest that while environmental sustainability is recognized as an important issue, it is not perceived as an immediate threat compared to other pressing challenges facing businesses in Portugal.

Overall, the responses highlight the complex and multifaceted nature of the challenges and threats that CEOs anticipate facing soon, underscoring the need for proactive and strategic approaches to addressing these issues.

Nevertheless, when we delved into why our interviewees perceive corporate and/or CEO activism as crucial, we discovered that their perspectives were primarily rooted in ethics rather than solely business-driven motivations (Fig. 2). For 17 of our respondents, corporate and CEO activism hold significance because they serve as vehicles for embodying the purpose and values of organizations, thereby manifesting alignment between these principles and those of the CEO, as well as various stakeholders such as clients, fans, shareholders, partners, employees, and even the broader community. One of the interviewees said that "CEOs need to actively promote social and environmental progress by aligning their organization's actions and values accordingly." Additionally, 15 respondents emphasized that activism reflects a personal commitment, both from the CEO and the company, toward effecting positive social change. Although these motivations are predominantly altruistic and ethically grounded, an emerging theme was the notion that advocating for causes

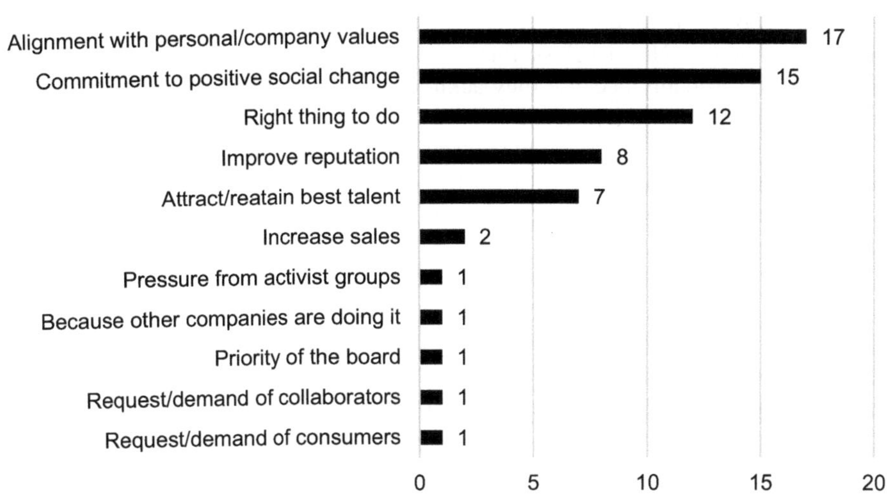

Fig. 2 Why corporate/CEO activism matters

one genuinely believes in is fundamental "the right thing to do," a sentiment echoed by 12 interviewees, as this one: "A company should behave in line with its principles. If the company's actions don't match its core values, it can harm its performance, brand image, and reputation."

When considering the potential benefits that activism can confer upon companies, eight participants concurred that it could bolster a positive reputation. Furthermore, aligning with the expressed concern regarding the future attraction and retention of talent, an emergent theme mentioned by seven respondents is the belief that corporate and CEO activism can serve as a strategy to mitigate this challenge.

It is worth noting, however, that in contrast to these perspectives, two CEOs acknowledged that activism could serve as another tool to enhance profitability.

The responses from our interviewees reveal diverse perspectives on CEO activism, with some CEOs viewing it as a fundamental responsibility rather than merely an optional strategic decision. This perspective underscores the belief that business leaders are more committed to advocating for positive change beyond purely economic interests, which is in line with the broader sustainability concept assumed in this article.

Interestingly, while there is a recognition of the importance of CEO activism, the responses also highlight a disparity between the perceived advantages and disadvantages. The fact that there were more risks (30) mentioned than opportunities (16) suggests that many CEOs perceive engaging in activism as a potentially risky endeavor. This may indicate hesitancy or caution among CEOs, despite recognizing the potential benefits.

It's worth noting that despite the acknowledgment of risks, a significant portion of the CEOs still consider themselves activists (20 out of 24), indicating a commitment to driving positive change. For some CEOs, activism "is not an opportunity, it is a necessity" or "it is a responsibility." This perspective is supported by quotes like "A businessperson is inherently an activist, as they always aim to improve something for the better." Nevertheless, this activism is shy (mostly internally) or insipid (few actions), probably because they acknowledge more risks associated with this course of action than advantages.

This discrepancy between the number of risks and opportunities mentioned suggests that there may be a need for further exploration and understanding of how CEOs perceive the balance between the potential benefits and drawbacks of activism.

Looking at the opportunities, several CEOs highlighted the potential for CEO activism to generate various benefits for companies. Three CEOs identified the opportunity to generate more business, whether through new sales, attracting new customers, or generating new ideas. Another CEO specifically mentioned that "activism can not only attract clients but also fans or followers," who are valuable for generating positive electronic word-of-mouth (eWOM).

Furthermore, one CEO pointed out that activism can present an opportunity for a company to diversify its business, while another CEO emphasized that it can be an effective way to differentiate from competitors. These opportunities are often grounded in ethical considerations. For instance, two CEOs believe that activism provides companies with an opportunity to fulfill their purpose and align their actions with their values and mission. Additionally, one CEO mentioned that "activism presents an opportunity to contribute to the greater good."

Reinforcing a good reputation emerged as another important opportunity acknowledged by three CEOs. While activism can pose a risk to reputation, particularly if not executed thoughtfully, these CEOs believe that it can ultimately enhance reputation.

Two CEOs highlighted the role of CEO activism in humanizing the CEO, which can positively impact the company's reputation. Finally, one CEO mentioned that activism can contribute to motivating collaborators, especially if there is alignment between their values and the public stance assumed by the CEO on behalf of the organization. This aligns with the concern about attracting and retaining talent, suggesting that CEO activism can play a role in addressing this challenge.

Regarding risks, the most cited concern, expressed by four CEOs, is the potential for activism to trigger crises within companies. Taking a firm stance on divisive issues risks alienating those with opposing views, potentially leading to negative word-of-mouth and reputational damage, alongside potential declines in sales, as seen in cases of boycotts. One CEO notes that "such crises typically arise when there is a disconnect between the values expressed through CEO activism and those held by consumers and other stakeholders."

Within this realm, three CEOs highlight the unpredictable nature of social media, which can serve as a breeding ground for crises. As a result, three others suggest that CEOs may engage in activism, provided their positions are not overly extreme or radical, enabling the company to continue serving a broad range of stakeholders. Similarly, one CEO emphasizes the importance of maintaining neutrality or impartiality regarding social causes. An additional three express concerns about losing customers or followers on social media due to controversial stances.

While one CEO argues that it "is not social media itself that is unpredictable but rather the volatile nature of societal issues," another highlights the risk of not engaging in activism aligned with the company's purpose and values. In their view, failing to do so could result in talent attrition, suggesting that not being an activist also carries risks.

Another significant concern, raised by the three CEOs, is the challenge of distinguishing between the CEO and the company. While the CEO represents the company, they are also entitled to personal opinions on social issues. Ideally, the CEO should clearly separate personal actions from those undertaken on behalf of the company. Failure to do so could damage the company's reputation, particularly if CEO activism becomes associated with a crisis (Duarte, 2023). However, one CEO notes that if a CEO's charisma and inspiring activism overshadow the company itself, it could lead to the company becoming overly dependent on the CEO, posing its own risks.

Another significant topic highlighted by CEOs is the importance of honesty, authenticity, transparency, and genuineness, which were mentioned by four different CEOs. One CEO emphasized that "CEOs shouldn't engage in activism solely for marketing purposes but should strive for a genuine positive impact." Similarly, two CEOs cautioned against "opportunistic activism" or "gold digger activism," warning against aligning with trending causes that do not reflect the company's core values, mission, or consistent actions over time. One CEO specifically denounced the practice of greenwashing, emphasizing the need for authenticity in corporate activism.

Conversely, one CEO expressed skepticism about the trend of overhyping activism, questioning the perceived benefits it brings to companies. However, another CEO argued that "companies risk irrelevance if they do not engage in activism." This diversity of perspectives underscores the complexity of navigating corporate activism and the need for companies to carefully consider their motivations and approaches to ensure genuine, impactful engagement.

5 Hot Topics

In Fig. 3, we present the "hot topics" in contemporary activism as identified by our interviewees. Notably, 16 out of 24 CEOs believe that gender equality is a pivotal issue in contemporary society that warrants attention through CEO activism. This aligns with the significant gender inequality prevalent in Portugal, where women earn salaries approximately 16% lower than men for similar positions and where women typically devote an average of 3 more hours per day to a combination of work, household chores, and family caregiving (Pordata, 2021, 2022b).

Education emerged as another prominent topic, mentioned by 14 respondents. This issue's importance is underscored by recent social unrest regarding the

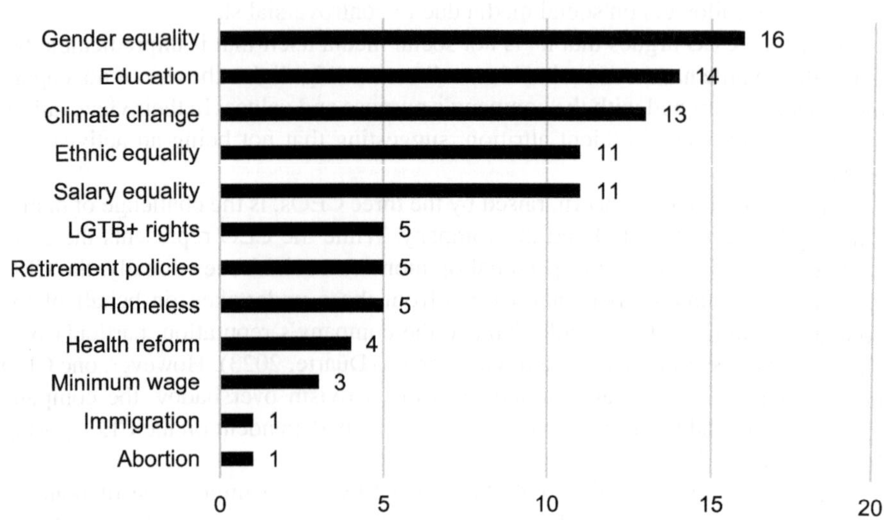

Fig. 3 Corporate/CEO activism "hot topics"

Portuguese education system, particularly in response to significant setbacks in children's and adolescents' learning and development during prolonged lockdowns amid the Covid-19 pandemic, as well as demands for improved working conditions from teachers.

Climate change is the third most cited "hot topic" that garnered attention from 13 Portuguese CEOs, reflecting the growing recognition of its significance as a pressing global concern (Perrault & Clark, 2016).

Ethnic equity was deemed important by 11 CEOs, although only 5 felt that equality in terms of sexual orientation and gender diversity should be addressed through corporate or CEO activism. Many CEOs cited the contentious and polarizing nature of the latter issue as a reason for its exclusion from their activism agendas. Additionally, several CEOs emphasized the importance of avoiding radical or extreme positions to ensure broad customer identification with the CEO and the company/brand.

Another emergent topic, albeit mentioned by only three respondents, is the minimum wage. This issue is gaining increasing relevance considering declining purchasing power among the Portuguese population due to rising inflation (Pordata, 2022c).

6 Best Practices

Among the CEOs interviewed, four assert that they are not activists. Their rationale for this decision typically revolves around the belief that activism does not align with the purpose or market positioning of their companies, coupled with concerns about the perceived risks associated with engaging in activism.

Conversely, among the 20 CEOs who identify as activists on behalf of their organizations, they employ a variety of approaches. Firstly, we explored the audiences with whom they communicate or act within the scope of their activism, followed by their methods of advocacy.

For 11 CEOs, the most crucial audience is their organization's employees. They emphasize the importance of fostering internal activism by embodying the causes they advocate. Two CEOs specifically mention sustainability initiatives, such as energy-saving and recycling programs, while two others focus on promoting ethics, citing examples such as implementing non-discrimination policies, diversity promotion, and fostering inclusiveness. Informal conversations within the organization's daily operations are highlighted by five CEOs as the primary method of implementing internal activism. Others utilize formal communication channels such as meetings, internal communication platforms, internal TV channels, and corporate events. Additionally, two CEOs integrate activism into their companies' daily operations by exclusively accepting projects that align with the organization's purpose and values, thereby considering their organizations intrinsically activist.

Another significant audience is the community, as mentioned by only three CEOs, indicating that corporate and CEO activism is still relatively nascent in

Portugal. Within this context, one CEO mentions corporate social responsibility and solidarity initiatives as forms of activism, albeit acknowledging the distinct categorization outlined in the scientific literature. Engagement with the community is often intertwined with partnerships, as noted by three other CEOs. One CEO cites a long-standing partnership with a solidarity organization, while another mention selecting business partners aligned with the causes they advocate, such as sustainability.

It is intriguing that none of the CEOs directly referenced consumers, customers, fans, or followers as target audiences for their activism efforts. However, seven CEOs assert their commitment to communicating about the causes and issues they advocate on social media platforms, particularly Instagram and LinkedIn. Among them, four CEOs utilize their personal profiles, prominently identifying themselves as CEOs of their respective companies, while three leverage their companies' social media channels for advocacy purposes. Additionally, seven CEOs express their activism through engagement with the press, employing public relations tactics such as sending press releases, participating in interviews, and penning opinion articles. One CEO elaborates on a strategic approach involving market research to inform press releases and agenda-setting efforts. This strategy is augmented by partnerships with universities and research centers, a strategy adopted by two CEOs, who engage in academic conferences and support research initiatives. Other methods of advocacy include utilizing company-owned media channels such as websites, newsletters, internal magazines, flyers, and corporate events, as mentioned by two CEOs. Furthermore, one CEO hosts regular podcasts focused on important causes and topics. Only one CEO mentions advocating causes through the marketing and branding of the company. Finally, one CEO adopts a unique strategy of mentoring young entrepreneurs, believing that this endeavor contributes to fostering future activism among business leaders. In summary, CEO activism is encapsulated by the idea that "activism is expressed through actions, much more than through words."

When solicited for examples of CEO activism and their justifications, CEOs provided insights organized into three categories: individual traits, characteristics of the advocated topic or cause, and their actions.

Regarding individual traits, 10 CEOs identified 11 different traits deemed crucial. Empathy emerged as a prominent trait, highlighted by two CEOs. Additionally, transparency, honesty, and genuineness were underscored, seen as interconnected attributes. Conviction and courage were also lauded, along with a keen sense of opportunity. One CEO stressed the importance of active listening, while another emphasized discretion, believing that activism need not always be publicized for reputational gain.

Concerning the message, 12 CEOs referenced eight topics as key considerations. Five CEOs emphasized alignment with the organization's purpose and values. Another CEO stressed the importance of genuine activism, devoid of opportunism or ulterior business motives. Three CEOs emphasized consistency in communication and actions, ensuring alignment with company values and public interests. Another CEO advocated for coherence in this regard. Content-wise, CEOs cautioned against adopting extreme or radical stances that could alienate the public or incite crises. Conversely, one CEO advocated for topics that foster unity and appeal to humanity's shared values. In terms of form, simplicity, focus, and the use of metaphors and examples were deemed effective and engaging by respective CEOs.

Regarding action, six CEOs proposed six different ideas. One CEO emphasized the necessity for activist CEOs to have a positive impact on society. Another CEO suggested achieving this impact through corporate social responsibility. A third CEO endorsed innovation as the most effective means of effecting positive change. For another CEO, activism is only meaningful if the CEO and the company exemplify the causes they advocate, practicing what they preach. Another CEO stressed the importance of people-focused activism over business-driven initiatives. Lastly, a CEO emphasized the importance of having a crisis management and communication strategy in place for CEOs engaging in activism.

7 Conclusions

This study represents the initial mapping of CEO activism in Portugal, as perceived by the CEOs, looking at the relationship between CEO activism and sustainability communication within the context of Portuguese corporate leadership. Compared to the global landscape, our findings indicate that CEO activism is still nascent in Portugal. While Portuguese CEOs acknowledge this trend, they may not fully comprehend its significance in the eyes of consumers globally. Research on consumer behavior suggests a rising emphasis on ethical considerations, with consumers increasingly demanding ethical actions from companies as they become more concerned about contemporary global sustainability issues such as climate change and social equality (Benoit et al., 2022). In a marketplace inundated with similar offerings, consumers are showing a preference for brands that demonstrate honesty, diversity, inclusivity, transparency, and sustainability. Companies, and their CEOs, that go beyond mere incorporation of sustainability, solidarity, and transparency into their daily operations, and actively engage in advocating for causes unrelated to their core business, are gaining respect, bolstering their reputation, and fostering loyalty (Mena et al., 2019).

Nevertheless, in Portugal, while CEOs recognize the significance of activism, particularly in the realm of sustainability, their engagement remains cautious and aligned primarily with internal organizational values and missions. This approach reflects a broader trend where sustainability serves as both a trigger and a framework for CEO activism, suggesting that corporate leaders view activism not merely as a response to external pressures but as a strategic component of their communication and leadership practices.

Previous research highlights the nuanced role of sustainability in shaping CEO activism, where the integration of sustainable practices into corporate strategies is seen as a pathway to legitimacy and social responsibility. However, the activism displayed by Portuguese CEOs is characterized by a strong alignment with corporate values and a cautious approach to public advocacy, indicating a preference for internal rather than external communication strategies. This more conservative stance contrasts with the broader perspective adopted by international CEOs. Moreover, Portuguese CEOs tend to focus internally, striving to imbue their organizational values and beliefs into daily operations and fostering strong relationships with their employees. They perceive an opportunity in CEO activism for talent attraction and retention, with many committed to causes directly related to talent and labor conditions in Portugal, such as supporting education and advocating for salary increases. However, only a fraction of them extends their activism beyond internal initiatives to take a public stance on pertinent sociopolitical issues. Fear of reputational crises and consumer alienation are cited as primary reasons for their reluctance to adopt a more assertive position on controversial topics.

While Portuguese CEOs acknowledge the importance of aligning their actions with the causes they advocate within their organizations, there appears to be a missed opportunity to deepen connections with stakeholders and contribute meaningfully to important societal issues. Further research is warranted to juxtapose the perspectives of CEOs with those of employees, consumers, and other corporate stakeholders, to gain a comprehensive understanding of this emerging phenomenon.

Overall, this study contributes to the understanding of how personal beliefs, such as sustainability, among others, concerns influence CEO activism, emphasizing the importance of strategic communication in modern corporate governance. The insights gained from the Portuguese context provide a valuable perspective on the evolving role of CEOs as activists and the potential for sustainability to serve as a critical driver of corporate communication and leadership.

8 Implications and Recommendations for Strategic Communication Scholars

In Portugal, due to fear of possible consequences, many CEOs opt for a more discreet and careful stance on sociopolitical topics. Although the vast majority of those interviewed in our study consider themselves activists, the truth is that, on the one hand, their actions and public stances are incipient, and on the other, there seems to be little clarification of concepts. For example, the fact that many CEOs cannot differentiate between corporate activism and traditional CSR opens the door to a misinterpretation of their role as organizational leaders. The fact that CSR activities are typically seen as beneficial to the whole society (Vredenburg et al., 2020), while CEO activism can offend or at least contribute to alienating customers, employees, or other stakeholders who do not identify with the causes defended, makes a big difference in the actions of CEOs.

In 2009, Fassin and Rossem conducted interviews with 41 prominent Belgian experts and CEOs both from the economy and civil society, and in 2011, Godos-Díez et al. gathered data from 149 CEOs in Spain. Both studies found that the influence of a business leader's profile on CSR practices is primarily shaped by their perception of the outcomes associated with ethical and social commitment. These conclusions, corroborated by our study and supported by Mäkelä and Olkkonen (2021), who proved that activism is a powerful tool to influence decision-making at different levels, can (should?) be used by various stakeholders to pressure CEOs toward a relevant, positive, sustainable social change by their organizations.

A comparison between the perceptions, commitment, and behavior of CEOs on activist issues from different countries, regions of the world, and cultural origins could be an interesting field for future research. Also, the way they communicate, which channels they choose to do so, when, and which topics most concern them and make them get involved, will certainly be relevant topics to understand how different contexts shape and influence their attitudes from the point of view of strategic communication.

References

Arivdsson, S., & Sabelfeld, S. (2023). Adaptive framing of sustainability in CEO letters. *Accounting, Auditing & Accountability Journal, 36*(9), 161–199. https://doi.org/10.1108/aaaj-11-2019-4274

Benoit, S., Baker, T., Bolton, L. E., Gruber, T., & Kandampully, J. (2022). Consumer demand for ethical action: A meta-analysis and future research directions. *Journal of Management, 48*(2), 461–489. https://doi.org/10.1177/0149206321997044

Branicki, L., Brammer, S., Pullen, A., & Rhodes, C. (2021). The morality of "new" CEO activism. *Journal of Business Ethics, 170*, 269–285. https://doi.org/10.1007/s10551-020-04656-5

Braun, V. and Clarke, V., 2021. Thematic analysis: A practical guide. .

Brinkmann, S., & Kvale, S. (2014). *InterViews: Learning the craft of qualitative research interviewing*. Sage.

Brown, L. W., Manegold, J. G., & Marquardt, D. J. (2020). The effects of CEO activism on employees' person-organization ideological misfit: A conceptual model and research agenda. *Business and Society Review, 125*(1), 119–141. https://doi.org/10.1111/basr.12196

Chatterji, A. K., & Toffel, M. W. (2019). Assessing the impact of CEO activism. *Organization & Environment, 32*(2), 159–185. https://doi.org/10.1177/1086026619848144

Creswell, J. W., & Poth, C. N. (2017). *Qualitative inquiry and research design: Choosing among five approaches*. Sage.

Duarte, A. (2023). The impact of CEO activism on brand reputation: The PROZIS case. In: *The normative imperative sociopolitical challenges of strategic and organizational communication* (pp. 161–168). LabCom Books. https://doi.org/10.25768/654-941-1

Duarte, A., & Chambel, S. (2023). Brand activertising: From profit-purpose to social positioning. In *The role of brands in an era of over-information* (pp. 198–215). IGI Global. https://doi.org/10.4018/978-1-6684-8351-0.ch009

Eyada, B. (2020). Brand activism, the relation and impact on consumer perception: A case study on Nike advertising. *International Journal of Marketing Studies, 12*(4), 30–42. https://doi.org/10.5539/ijms.v12n4p30

Elkington, J. (1999). Triple bottom-line reporting: Looking for balance. *Australian CPA, 69*, 18–21.

Godemann, J., & Michelsen, G. (2011). Sustainability communication – an introduction. In *Sustainability Communication* (pp. 3–11). Springer Netherlands. https://doi.org/10.1007/978-94-007-1697-1_1

Hambrick, D. C., & Wowak, A. J. (2021). CEO sociopolitical activism: A stakeholder alignment model. *Academy of Management Review, 46*(1), 33–59. https://doi.org/10.5465/amr.2018.0084

Hou, Y., & Poliquin, C. W. (2023). The effects of CEO activism: Partisan consumer behavior and its duration. *Strategic Management Journal, 44*(3), 672–703. https://doi.org/10.1002/smj.3451

Larcker, D. F., Miles, S., Tayan, B., & Wright-Violich, K. (2018). The double-edged sword of CEO activism. Rock Center for Corporate Governance at Stanford University Closer Look Series:

Topics, Issues and Controversies in Corporate Governance No. CGRP-74, Stanford University Graduate School of Business Research Paper, (19–5).

Maak, T., Pless, N. M., & Voegtlin, C. (2016). Business statesman or shareholder advocate? CEO responsible leadership styles and the micro-foundations of political CSR. *Journal of Management Studies, 53*(3), 463–493. https://doi.org/10.1111/joms.12195

Mäkelä, M., & Olkkonen, L. (2021). Sustainability activism – a review of the state of the art. In *Research handbook of sustainability agency* (pp. 139–156). Edward Elgar. https://doi.org/10.4337/9781789906035.00015

Mena, S., Rintamäki, T., & Fleming, P. (2019). CEO activism: A review and research agenda. *Academy of Management Annals, 13*(1), 193–226. https://doi.org/10.5465/annals.2017.0144

Miller, T., Birch, M., Mauthner, M., & Jessop, J. (2012). *Ethics in qualitative research*. Sage. https://doi.org/10.4135/9781473913912

Newig, J., Schulz, D., Fischer, D., Hetze, K., Laws, N., Lüdecke, G., & Rieckmann, M. (2013). Communication Regarding Sustainability: Conceptual Perspectives and Exploration of Societal Subsystems. *Sustainability, 5*(7), 2976–2990. https://doi.org/10.3390/su5072976

Mkrtchyan, A., Sandvik, J., & Zhu, V. (2022). *CEO activism and firm value*. Available at SSRN 3699082. https://bit.ly/3yIu4qV

Perrault, E., & Clark, C. (2016). Environmental shareholder activism: Considering status and reputation in firm responsiveness. *Organization & Environment, 29*(2), 194–211. https://doi.org/10.1177/1086026615571939

Pless, N. M., Sengupta, A., Wheeler, M. A., & Maak, T. (2021). Responsible leadership and the reflective CEO: Resolving stakeholder conflict by imagining what could be done. *Journal of Business Ethics*, 1–25. https://doi.org/10.1007/s10551-021-04865-6

Pordata. (2021). *Disparidade de género no ganho médio mensal dos trabalhadores 2021*. https://bit.ly/3JpKgDg

Pordata. (2022a). *Salário médio mensal dos trabalhadores 2022*. https://bit.ly/3mEFuZR

Pordata. (2022b). *Duração média semanal do trabalho efetivo dos trabalhadores, por sexo 2022*. https://bit.ly/3JpNBT3

Pordata. (2022c). *Poder de compra per capita 2022*. https://bit.ly/3Zxus7n

Ray, A. (2012). *The methodology of sampling and purposive sampling*. Green Verlag.

Rumstadt, F., & Kanbach, D. (2022). CEO activism. What do we know? What don't we know? A systematic literature review. *Society and Business Review, 17*(2), 307–330. https://doi.org/10.1108/sbr-10-2021-0194

Sama, L. M., Stefanidis, A., & Horak, S. (2020). Business ethics for a global society: Howard Bowen's legacy and the foundations of United Nations' sustainable development goals. *International Studies of Management & Organization, 50*(3), 201–208. https://doi.org/10.1080/00208825.2020.1811526

Shetty, A. S., Venkataramaiah, N. B., & Anand, K. (2019). Brand activism and millennials: An empirical investigation into the perception of millennials towards brand activism. *Problems and Perspectives in Management, 17*(4), 163–175. https://doi.org/10.21511/ppm.17(4).2019.14

Shocker, A. D., & Sethi, S. P. (1973). An Approach to Incorporating Societal Preferences in Developing Corporate Action Strategies. *California Management Review, 15*(4), 97–105. https://doi.org/10.2307/41164466

Vanover, C., Mihas, P. and Saldana, J., 2021. Analyzing and interpreting qualitative research: After the interview.

Velte, P. (2023a). Sustainable board governance and sustainable supply chain reporting: European evidence. *Journal of Strategy and Management, 17*(2), 222–243. https://doi.org/10.1108/jsma-07-2023-0189

Velte, P. (2023b). Chief sustainability officer expertise, sustainability-related executive compensation and corporate biodiversity disclosure: Empirical evidence for the European capital market. *Journal of Global Responsibility, 14*(2), 241–253. https://doi.org/10.1108/jgr-06-2022-0055

Vredenburg, J., Kapitan, S., Spry, A., & Kemper, J. A. (2020). Brands Taking a Stand: Authentic Brand Activism or Woke Washing? *Journal of Public Policy & Marketing, 39*(4), 444–460. https://doi.org/10.1177/0743915620947359

Weder, F., & Dobrić, N. (2021). Conceptualisation of sustainability. *Colloquium: New Philologies, 6*(2), 46–64. https://doi.org/10.23963/cnp.2021.6.2.3

Weder, F. (2022). Strategic problematization of sustainability reframing dissent in strategic communication for transformation. *Public Relations Inquiry, 11*(3), 337–360. https://doi.org/10.1177/2046147x211026857

Weder, F. (2023). The evolution of the sustainability story: Strategic sustainability communication as niche construction. *International Journal of Strategic Communication, 17*(3), 228–244. https://doi.org/10.1080/1553118x.2023.2229304

Zerfass, A., Verčič, D., Nothhaft, H., & Werder, K. H. (2018). Strategic communication: Defining the field and its contribution to research and practice. *International Journal of Strategic Communication, 12*(4), 487–505. https://doi.org/10.1080/1553118x.2018.149

Further Reading

Adae, E. K. (2023). Doing well to do good: A quick map of the field of corporate sociopolitical involvement. *CEOs on a Mission* (*Communicating Responsible Diversity, Equity, and Inclusion*) (pp. 37–48). Emerald Publishing. https://doi.org/10.1108/978-1-80382-215-020231002

Wang, T., Fu, Y., Rui, O., & De Castro, J. (2024). Catch up with the good and stay away from the Bad: CEO decisions on the appointment of chief sustainability officers. *Journal of Management Studies, 61*(4), 1295–1326. https://doi.org/10.1111/joms.12926

Alexandre Duarte, PhD in Communication Sciences (*specialization in Organizational and Strategic communications)* by Universidade do Minho, is a Professor at Universidade Nova de Lisboa, a Researcher at ICNOVA/UNL, and an Invited Professor at Universidade Católica Portuguesa, where coordinates the Postgraduate Courses in *Communication and Advertising Creativity*, and *Service Design*.
Having worked in several multinational advertising agencies for more than 20 years as a Senior Creative Copywriter, he was a Tutor of the *Young Marketers Academy* of the EUROBEST Festival, CEO of RESTART—*Institute of Arts and Creativity* and advisor to SUPERBRANDS Portugal.
Since 2018, is a member of the Board of EDCOM—*European Institute for Commercial Communications Education,* and from 2024, assumed the Coordination of the Advertising Working Group of the Portuguese Association of Communication Sciences—SOPCOM.

Patrícia Dias, PhD in Communication Sciences by Universidade Católica Portuguesa, coordinates the Postgraduate course Social Brands—Communication and Marketing in a digital environment. In 2019, she was a visiting researcher at the Arctic University of Norway. Researcher at CECC and CRC-W currently coordinates the project "0–3 DigiKids—Use of touch screen devices by children up to 3 years old" and leads the Portuguese team of the project "Use of digital technologies by children and adolescents during the COVID-19 pandemic." coordinated by the JRC-European Commission. Her main research interests are the use of digital media by children and also within organizational communication, marketing and branding. She is the author of several publications about the social impact of digital media and the use of smartphones.

Strategic Communication of and for Sustainability in Higher Education Institutions: Conceptional Considerations and Results of a Case Study

Mira Rochyadi-Reetz and Jens Wolling

Abstract In this chapter, conceptual considerations on strategic communication *of* and *for* sustainability in higher education institutions are presented and discussed in light of the results of a case study. The chapter defines key aspects such as sustainability in higher education institutions, sustainability communication, strategic communication, and the importance of a "whole institution approach" for sustainability transformation in Higher education institutions (HEIs). The case study analyzed the first years of a transformative research project at a HEI, revealing that effective strategic sustainability communication should be flexible, opportunity-driven, and often interpersonal rather than public. It emphasizes the importance of aligning strategic communication with the values and interests of stakeholders.

1 Sustainability in Higher Education Institutions

Higher education institutions (HEIs) play a key role in the transformation process to a more sustainable society (UNEP, 2021). Therefore, these institutions should not only address the SDGs as topics in their regular research and teaching; they should go beyond that and integrate the philosophy and didactic approach of Education for Sustainable Development (ESD) in all study programs. In Europe, many universities have already started to transform their curricula to increase students' competence in sustainability (Albareda-Tiana et al., 2018; European University Association, 2021). However, as institutions that have an important role in society, universities should also comply with the sustainability principles beyond their teaching and research activities. They should follow a holistic approach and include the various aspects of sustainability in all areas of the institution, such as

M. Rochyadi-Reetz (✉) · J. Wolling
Department of Economic Sciences and Media, Group for Media Research and Political Communication, Technical University of Ilmenau, Ilmenau, Germany
e-mail: Mira.Rochyadi-Reetz@tu-ilmenau.de; Jens.Wolling@tu-ilmenau.de

F. Weder (ed.), *Strategic Sustainability Communication*, CSR, Sustainability, Ethics & Governance, https://doi.org/10.1007/978-3-031-89486-2_13

administration, governance, and procurement, as well as the management of buildings, infrastructure, and green spaces on campus. In the literature, this is called the whole institution approach (Kohl et al., 2022; Schopp et al., 2020). Within this framework, HEIs should make sustainability a central priority and establish it in their strategic goals.

As there are different understandings of the concept of sustainability, it is necessary to define how the term will be used in this chapter. This is particularly important because the term "sustainable" is often used in the sense of permanent, which is a problematic oversimplification of the concept. Furthermore, it is necessary to explain the difference between sustainability and sustainable development. We argue that sustainability highlights the aspect of stability and protection, while sustainable development emphasizes change. Both perspectives are essential to understanding the concept properly.

Concerning the substance of its content, we believe that it is reasonable to build on the concept of Sustainable Development Goals (SDGs) adopted by the UN in 2015.[1] To overcome the problem of missing prioritization of the goals, we apply the idea of differentiating between the three basic dimensions of sustainability, which we label *satisfaction of economic needs*, *social cohesion*, and *environmental protection*. This differentiation allows criteria to be developed for prioritization if in concrete situations certain goals are in conflict. All 17 SDGs are related to all three principal dimensions, albeit in different intensities. This relative position can help to prioritize between the goals. Therefore, we propose a hierarchy model of the three dimensions. This model is developed based on the planetary boundary concept (Rockström et al., 2009), which stresses the importance of the natural or biophysical system of the planet to be able to fulfill humans' needs. Even if such hierarchization is criticized as neo-colonialism, there are good arguments for following this approach (Hoppe & Wolling, 2017).

As sustainability includes both *preservation* and *change*, we propose two pyramids: a *pyramid of preservation* and a *pyramid of change* (Fig. 1). As all interventions in the ecosystem pose a major risk to the natural balance, the environment is the base of the pyramid of preservation. Concerning society, a careful process of change is required. From history, we can learn that social experiments, which

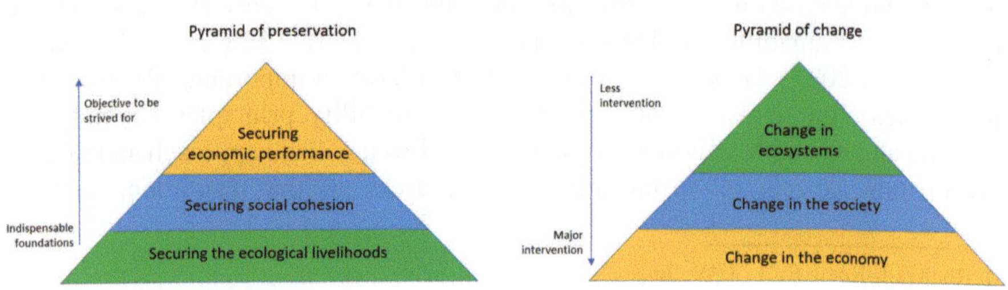

Fig. 1 The pyramids of preservation and change

[1] https://sdgs.un.org/goals

attempted to radically change people and society, had consistently devastating consequences. However, the existing inequalities, polarization, and injustices demand social change. With respect to the economy, fundamental changes are required. A circular economy and the reduction of consumption, for those who currently consume more than needed, are indispensable components of sustainable development.

In this context, technology plays a major role. Technologies are tools that foster the economy; they can be helpful or a threat to society, and they are frequently a risk for the environment. Even technologies that are designed to reduce environmental problems, e.g., wind turbines for climate protection, have always had a negative impact on the environment. Accordingly, while the use of technological innovations may be necessary for economic development, its potential impact on society must be carefully considered and its impact on nature should be avoided as far as possible. In this context, HEIs play an important role as places where new technologies and innovative solutions for a sustainable future are investigated in research and teaching and are transferred to society.

In order to implement sustainability as a strategic goal, universities have to mobilize, inspire, and empower students, faculties, and staff to act for sustainability. This requires targeted internal communication. Furthermore, universities' external communication is needed in order to engage with society and to fulfill their role as transformation agents. Accordingly, strategic sustainability communication is very important for addressing both target groups within the organization and the wider public.

This chapter deals with the question of whether and how such strategic communication can advance the sustainability transformation at HEIs in Germany. Since HEIs fulfill special tasks for society and therefore have to fulfill particular roles, we first explain what is meant by strategic communication of universities. We then explain what we mean by strategic sustainability communication of universities. Finally, we show how we implement these considerations in a case study and discuss the success of the chosen approach.

2 Strategic Communication of Higher Education Institutions

In the first publication of the *International Journal of Strategic Communication*, Hallahan et al. (2007, p. 3) define strategic communication as "the purposeful use of communication by an organization to fulfill its mission." Holtzhausen and Zerfass (2013) developed the concept further and defined strategic communication as follows: "the practice of *deliberative* and purposive communication that a *communication agent* enacts in the *public sphere* on behalf of a *communicative entity* to reach set goals" (p. 74). Unlike the previous one, this definition recognizes that organizations (communicative units) cannot communicate themselves. Rather, there are actors or agents who speak on behalf of the organization. Furthermore, this definition emphasizes that strategic communication is public communication. However, based on our understanding and the results of our research, we do not believe that strategic communication must always be public. Most importantly, by introducing the concept of deliberation, the definition shows that strategic communication is not limited to persuasive communication. The definition confirms that strategic communication is clearly interest-driven but also recognizes that in many cases discourse can be more effective in achieving the organization's goals than a unidirectional approach.

Strategic communication is done by very different organizations, each pursuing specific goals. While most companies primarily pursue economic goals, universities have other objectives. It is important for them to address these objectives because the legitimacy of an organization depends on the perception of whether their actions are in line with their mission and societal expectations and norms (Raupp, 2017). Higher education institutions should fulfill three central tasks for society: teaching, research, and the adoption of responsibility for social development (third mission). The fulfillment of these tasks determines HEIs' organizational legitimacy in society (Entradas et al., 2023). Accordingly, HEIs should fulfill and communicate these requirements in order to gain legitimacy. Relevant target groups of the communication are policymakers, industry, and society, as well as their internal audience, such as faculty staff, researchers, and students.

Raupp (2017) argues that the official strategic communication of HEIs has two different types of goals: organizational objectives, which relate to the (positive) self-presentation of the organization, and social objectives, such as the communication of scientific content or the significance of scientific results.

Accordingly, one aspect of strategic university communication is strategic science communication, which is defined as the intentional, inward, and outward communication of science organizations that serves to maintain and expand organizational legitimacy (Raupp, 2017). This communication is mainly about the results of scientific research and can be realized by various agents, including official PR of the HEIs, institutes, departments, or even single scientists, who can pursue very different individual goals (Friedrichsmeier et al., 2015).

Another aspect of strategic university communication is referred to as university PR and marketing. Its role is not limited to disseminating or creating a public understanding of science but focuses mainly on increasing the visibility and attractiveness of the organizations in order to attract students and funding (Volk et al., 2023). Entradas et al. (2023, p. 4) refer to such orientations as "self-interested rationales." Recent empirical research in European countries shows that most university PR officers focus more on self-interested rationales than on creating a societal dialog about science, which is seen as a public good orientation (Entradas et al., 2023).

However, this differentiation might not be as strict as it seems. One could argue that whenever a HEI communicates about scientific content and the relevance of the results obtained by its researchers, it does so not only for social motives but also because the PR department believes that such communication enhances the organization's reputation.

For the purpose of this paper, we are focusing on all sustainability-related activities of the PR department and on all sustainability-related activities that are realized jointly or in coordination with the PR department. In addition, we will address all communicative activities within the HEI (internal communication) that aim to promote organizational change in order to foster a culture of sustainability within the organization and create the structural conditions for this. Due to capacity reasons, we cannot consider sustainability-related communication by individual scientists.

3 Sustainability Communication

Sustainability communication can be divided into three forms: communication *about*, *of*, and *for* sustainability (Weder et al., 2021). Communication *about* sustainability refers to journalistic content that presents sustainability issues from the perspective of an observer whose primary intention is to inform recipients about the topic. This communication has primarily cognitive goals. Communication *of* sustainability, meanwhile, refers to the intentional communication of an actor about its own activities in the field of sustainability. The aim of this communication is usually to create a positive impression of the actor on the recipient. Public relations of companies that show their efforts to reach sustainability are typical examples of this type of communication. It has primarily *emotional* goals. Thus, when a HEI utilizes communication *of* sustainability, it is based on the assumption that by demonstrating its sustainability activities, the HEI will create a *positive image* of the organization. Additionally, communication *for* sustainability can be summarized as forms of intentional campaign communication or advertising, with the primary aim of ensuring that the target groups react to the communication with certain sustainable actions. Accordingly, these are *behavioral* goals.

Based on the definition of strategic communication presented in the previous section, it becomes clear that communication *of* sustainability and communication *for* sustainability are two types of strategic sustainability communication (SSC), while communication *about* sustainability follows another rationale. Only a few studies investigate communication *about* sustainability. A reason for this might be that in journalistic publications, the term "sustainability" is not frequently used. Journalists frequently address topics related to sustainability and the SDGs without even mentioning the term "sustainability." However, in the case of strategic communication, the opposite appears to be true. Communicators talk about sustainability even if the activities they are promoting are just loosely connected to sustainability. Therefore, it is remarkable that the vast majority of sustainable communication research is on SSC (Hoppe & Wolling, 2017).

Communication *of* sustainability is what an organization such as a HEI normally does when it wants to address sustainability, while communication *for* sustainability is what NGOs normally do when they want to mobilize people to act in line with or in favor of the SDGs. Nevertheless, in this chapter we will argue and demonstrate that communication *for* sustainability is also a necessary activity for a HEI if it wants to drive the transformation of the organization to a more sustainable one.

4 Strategic Sustainability Communication of HEIs

Combining the concepts of sustainability communication and strategic communication of HEIs, we understand the strategic sustainability communication (SSC) of HEIs as any communication of a responsible agent of a HEI that aims to reach organizational goals related to sustainability. Normally, this responsible agent is the communication or PR department of the HEI. There are several possible ways the agent can communicate the HEI's activities related to sustainability. An important way to do so is through the website. Research on the representation of sustainability on websites has mainly been conducted in America. A study conducted in 2009 demonstrates that at that time in the United States only one in five of the HEIs communicated on sustainability on a special subpage of the HEI's website (Dade & Hassenzahl, 2013). Nine years later, a study by Amey et al. (2020) on the situation in Canada revealed that two out of three HEIs communicated on sustainability at least to some degree on their website. An analysis of Fernández-Vázquez's (2021) study on Latin American HEIs showed that almost 60% of the HEIs addressed environmental aspects of sustainability on their homepages.

Besides the official websites of the HEI, there are several other ways a HEI can communicate about their activities related to sustainability. The communication strategy can include various communicative practices, mediated and non-mediated (Kannengießer, 2021). The HEI's agents can publish sustainability reports; communicate on social media; organize events, such as science nights, conferences, or symposia; organize press conferences; or publish press releases to create awareness on sustainability in and beyond the HEI. These activities are of high importance because the lack of awareness of sustainability issues is among the most relevant barriers to the transformation process in HEIs (Verhulst & Lambrechts, 2015). To overcome these barriers, effective strategic communication *of* sustainability is considered one of the most important factors (Sacchi et al., 2023). Interviews conducted by Carpenter et al. (2016) with sustainability managers and leaders in several universities in the United States show that dialogic communication strategies are often used to recruit active participants and build stakeholder relations across campuses. However, such a communication strategy needs adequate time and financial resources. The question is, which agents in the HEI can do that, and how can they communicate their sustainability messages?

Usually, the communication or PR departments of HEIs do the official organizational communication *of* sustainability, but they typically rely on other agents for necessary input, such as the administrative staff, scientists, lecturers, and students in the university. However, these communication activities are not only about informational input; they also depend on the implementation of real activities related to sustainability. Someone has to conduct real activities such as the installation of solar panels on the roof of the HEI's building, the planting of a blooming meadow, research, and lectures related to sustainability, or the implementation of sustainable procurement criteria. All these activities have to be implemented before the sustainability-related aspects can be communicated. Some of these activities can be ordered by the HEIs' presidium; however, most of the activities for sustainability in HEIs depend on the voluntary ventures of the members of the HEI. Thus, for the university to transition toward greater sustainability, its members must be motivated to actively engage in the process. The support of the HEI's presidium is crucial to creating favorable conditions for success; without it, change is nearly impossible. If support is lacking, pressure from university members is necessary to compel the presidium to initiate the transformation.

5 A Case Study on Transformative Research: Strategic Communication for Sustainability in a HEI in Germany

In this section, we describe our experiences in planning and implementing strategic sustainability communication at Technische Universität Ilmenau (TU Ilmenau, Germany) to drive sustainability transformation in the organization. The central idea of the project is the assumption that strengthening strategic sustainability communication is a necessary and effective means to promote organizational change.

With around 4,500 enrolled students, TU Ilmenau is a small university, compared to other technical universities in Germany, such as TU Berlin with more than 30,000 students or TU Munich with more than 50,000 enrolled students. However, TU Ilmenau is the only technical university in the federal state of Thuringia and the largest employer in the district. With regard to research, the university ranks eighth among all universities in Germany in terms of patent applications. Many small technology companies located in the vicinity of the university were founded by professors, employees, and graduates of the university, which demonstrates the importance of TU Ilmenau's transfer activities. As mentioned in the introduction, technology plays a major role in the transformation process toward sustainability, which we conceptualized in the second pyramid above.

In the following paragraphs, we will present our activities to promote sustainability throughout the whole organization. These activities were conducted as part of the Klima-N project, which started at the end of 2022 and is funded by the Federal Ministry of Education and Research (BMBF, FKZ: 01UN2204D, https://klima-n.de/, or see QR code in the last section for further information). The project consists of a consortium of four HEIs, which are all located in Thuringia. The aim of the project is to push forward the sustainability transformation process in the respective HEIs. The following description of the strategic communication *for* sustainability during the first years of the transformation process is based on observations, meeting minutes, qualitative content analysis, and interviews with relevant stakeholders and key agents involved in the transformation process.

The project was conceptualized as a transformative research project, which means that the role of the researchers in this project is twofold. On the one hand, they take on the classical role of scientists who observe reality and collect data by means of observation, interviews, and surveys in order to analyze how and why reality changes or remains unchanged. On the other hand, in this approach, the researchers also take on the role of activists who initiate measures to transform the university into a more sustainable place (Horcea-Milcu et al., 2024). Accordingly, as communication scientists, we applied strategic communication *for* sustainability to initiate, promote, and support the transformation of the university. Following the whole institution approach, the target groups of this strategic communication are all members of the university, including academics, staff, students, and especially the presidium.

The aim of the project—to drive forward the process of sustainability transformation at the HEI—is very broad and unspecific, and therefore it was clear that the measures required to achieve this goal could not be determined from the outset. Accordingly, it was impossible to develop a clear roadmap with well-defined milestones, which only had to be processed step by step. On the contrary, it was necessary to apply a flexible approach that considers the actual situation, looks for opportunities, and identifies milestones that appear to be reachable in a manageable time period.

The process of strategic communication started even before the project was accepted by the funding organization. In order to secure the project, the applicants had to analyze the situation at the HEI and identify possible partners who already shared similar values, ideas, and motivations concerning the importance of a transformation of the HEI to a more sustainable organization. In the first step—about 18 months before the KLIMA-N-project started—one of the authors of this chapter wrote a working paper on the need for and usefulness of a stronger focus on sustainability at the HEI and distributed this paper to relevant people in the organization. Furthermore, in a bachelor's thesis about sustainability communication, a student (Hermeyer, 2023) conducted interviews with all deans of the university to obtain information about their understanding of and perspectives on sustainability in teaching, research, administration, and campus management.

The feedback on the working paper helped to identify relevant partners and refine the argumentation, while the results of the bachelor's thesis provided valuable insights into the perspectives of the faculties on the topic. These inputs helped to reach the *first milestone*, the official support of the presidium for the project. This support was necessary because the funding organization asked for an official commitment from the presidium.

To get full support from the presidium, the project idea was discussed with two university vice presidents who were identified as having similar values to the applicants and who recognized the potential benefits of the transformation. Based on this discussion, a presentation about the project was prepared and presented to the president and the other members of the presidium. To obtain their support, we addressed the biggest problem of the organization, the decreasing number of students in the engineering study programs. We argued that by addressing sustainability in the study programs and by implementing the principles in the whole university culture, the university might be successful in attracting new students. In other words, we framed the sustainability transformations of the university as a solution to an essential problem faced by the university. The strategic communication at this phase was successful, as the office of the president agreed to make an official commitment to BMBF that they fully support the project and that the university plans to continue the project even after funding has ended. The statement made by the vice president for research in an interview confirms the success of the strategy. When asked how he assessed the importance of the topic of sustainability for the presidium, he said,

> I have the feeling that there has been a very positive response and that this initiative has been very well received. […] Our biggest concern is the [decreasing] number of students. And we have to somehow manage to address our topics in such a way that they also appeal to young people. I would say it's a good mixture of pragmatic motivation, but also the awareness that we need to pay more attention to sustainability in our society as a whole. That would be my assessment. And I'm personally very pleased that, in my view, we've clearly succeeded in putting the issue on the agenda. In other words, there is now hardly a single meeting of the Executive Board of the presidential office that doesn't address topics of sustainability.

A similar assessment was shared by the vice president of international affairs. He explained in an interview that the topic of sustainability is increasingly prominent in presidium meetings and discussions. Now more than ever before, the presidium members recognize the significance of sustainability in the university's daily operations and are striving to enhance sustainability efforts on campus.

After both the declaration of support from the presidium and the funding approval from BMBF were received, it was of central importance to achieve organizational integration of the project and to ensure that the necessary and desired target groups of the university got involved. At this step, we worked on reaching the *second milestone*, namely the installation of the Sustainability Task Group, which consists of members from different organizational parts of the university. To reach this goal, we implemented a communication strategy that aimed to (1) convince the presidium to officially install the group and (2) to motivate agents who represent different organizational parts of the university to join the Sustainability Task Group. Interpersonal communication via email, telephone call, or face-to-face interaction with the prospective candidates was chosen at this step.

The strategy was successful as the Sustainability Task Group was founded and nine people joined the group. The members of the working group were selected based on the principles of the whole institutional approach (HOCH-N, 2020). Accordingly, the group consists of actors from the areas of teaching, research, campus operations, staff representatives, university marketing, and university administration. The group also consists of student representatives and two members of the Presidium. The establishment of the task group marked an important milestone in the sustainability transformation process.

The *third milestone* was creating awareness on the importance of the sustainability topic among the members of the TUIL and demonstrating to people inside and outside the organization that the TUIL was already working on the topic. To achieve this objective, the authors of this chapter, together with the members of the task group developed an integrated communication strategy. In the first step, a logo and a slogan stating "TU Ilmenau on its way towards a sustainable community" were created. In the second step, a new webpage with a wide variety of content on the university's activities related to sustainability was developed, and a link to this webpage was placed on the top level of the university website. This important step could be implemented very quickly because one of the members of the task group works for the communication department of the university.

In a third step, the task group prepared an email that was sent in the name of the president to all members of the university, asking them to submit ideas that could be implemented in the university to improve sustainability. Within a few weeks, more than 100 people sent ideas on how to make the university more sustainable. The topics submitted included a wide range of aspects, including waste reduction, carsharing mobility, installation of renewable energy resources, campus gardening, biodiversity, increasing the quantity of vegetarian menu at the university's canteen, conducting curricula reform to include more sustainability topics in teaching, and creating a sustainability certificate for students. This high rate of response to the call demonstrates that there was already a certain level of awareness of sustainability-related issues and problems that should be addressed.

At the same time, the vice president for research declared sustainability as the main focus of the Year of Science 2023. The authors of the chapter were invited to give the opening lecture of the scientific year, and several scientific talks and discussion panels on various sustainability topics were organized throughout the year. These activities raised awareness on the topic throughout the university and beyond.

Furthermore, several other practical projects were started. These projects had two objectives: On the one hand, we wanted to initiate real changes in the organization because we believe that real changes are the most convincing arguments for obtaining support from university members. On the other hand, we planned to utilize strategic communication of these activities and projects in order to motivate and mobilize university members to join the existing sustainability activities or to start their own sustainability projects on campus. A description of activities and agents leading the activities can be seen in Table 1.

Strategic communication plays an important role for all activities listed in Table 1. Strategic communication *for* sustainability is used to motivate and convince university members to participate in these activities. Intensive interpersonal communication is used in this process to connect and stimulate collaboration between possible future agents. Meanwhile, strategic communication *of* sustainability is used to inform the different target groups within the university and also external audiences about these activities. Accordingly, several communication channels are used simultaneously, such as writing and disseminating press releases to the local media and reporting the event on the university's website and social media channels and in the student radio program. For most of the activities, such as the sustainability days, all university members and, frequently, the general public are invited. For these activities, promotional materials, such as rollups, flyers, posters, and cubes illustrating the 17 sustainability goals, are prepared. All these materials aim to attract attention, explain the vision of the university, and demonstrate the steps already taken by the university on its way to a sustainable community.

Table 1 Sustainability activities and agents involved at Technische Universität Ilmenau[a]

Sector	Activities	Status	Agents
Teaching	Interdisciplinary teaching module on circular economy for students from all study programs at the HEI	Realized in Summer Semester 2023 and 2024	University professors from several faculties
	Development of a teaching module on "Communication for a better future" for Master's study program Media and Communication Science	Realized in Winter Semester 2023/2024	Lecturer at Institute of Media and Communication Science
Research	Survey of university students on their attitudes and perceptions of sustainability	Realized in Spring 2023/ publication in progress	Klima-N
	Theme year 2023: Research on sustainability	Realized in 2023	Presidium and several researchers
	Sustainability was a key topic at the Night of Science, the university's central science communication event for the public	Realized in Summer 2023	Presidium and researchers from all faculties
	Classification of all funded research projects of the university regarding their connection to sustainability	Realized in Spring 2023	Department for research support and transfer
Campus management	Initiative to install solar panels on buildings of the HEI	In progress	Presidium and campus management department
	Blooming meadows for biodiversity in campus	Started in Spring 2023 and planned to continue in next years	Campus management department
	New tree planting in campus area with climate-resilient trees	Implemented in Spring 2024 and planned to continue in next years	Campus management department
	Car-sharing initiative	In process since Winter Semester 2023, actualized in summer 2025	Presidium, professors, and staff members of HEI and the city administration
	Free water dispenser in canteen	Implemented in Spring Semester 2024	Canteen management, health department, and student council

(continued)

Table 1 (continued)

Sector	Activities	Status	Agents
Students' initiatives	Sustainability days at HEI. Exhibitions on existing initiatives of the university and the city and discussions on sustainability topics at the HEI	Started in Summer Semester 2023 and will occur once per semester	Student council and staff members of the HEI
	Clothes swap	Started in Summer Semester 2023, occurs once a week	Student council
	Radio Program "Sustainability Corner"	Started in Winter Semester 2023, aired once in every 2 weeks	Student council, student radio, and staff of HEI

[a]Examples on communication of sustainability of various actors at TU Ilmenau can be found through the QR code provided after the conclusion section of this paper

The *fourth milestone* involved creating a binding concept where the university commits to greater sustainability and outlines specific measures. The authors proposed this concept as an appendix to the university's overall strategy. However, since the university strategy was due for revision, it was decided to integrate sustainability into the updated strategy. To ensure this goal was appropriately implemented and prominently featured, the authors vigorously communicated their ideas and arguments to the task group, the presidium, and to some university senate members. The authors also suggested various improvements based on the initial concept paper. Following an intensive communication process spanning several months, the senate approved the new university strategy, which now clearly commits to sustainability as an institutional goal across the entire university.

No systematic evaluation of transformation communication has been undertaken to date. One reason for this is that in most cases it is almost impossible to quantify whether a goal has been achieved in a meaningful way, as the goals relate to the entire institution and there is no benchmark for comparison. For example, it would be possible to count the number of participants in the events or the frequency with which the topic of sustainability appears in the university's strategy paper. However, as we have only conducted one case study so far, such figures would not be very meaningful and would be difficult to interpret because it is not possible to compare them with the results from other institutions. For this reason, we can only describe our approach and results qualitatively and name possible success factors without being able to conclusively decide whether our interpretations are correct. We can only state that we have achieved certain results with the respective approach. We cannot say whether we would have achieved the same, better, or worse results with other strategies.

In the second half of the project, we will intensify our efforts to systematically evaluate our approach and make a comparative classification of our findings. Within the university, a quantitative survey of the various groups (students, teaching staff, administrative staff, etc.) will be conducted. This survey should provide information on the extent to which the communication activities have succeeded in reaching the various target groups, how the target groups assess the activities and communication, and what prevents people from becoming more involved in the activities. In order to compare the results achieved at TU Ilmenau with the situation in other locations, we will use the network of our Klima-N research project to conduct interviews with the relevant people responsible for sustainability transformation at the three partner universities. The aim of the interviews will be to find out what the initial situation was, what goals were sought to be achieved there, what results were achieved, and by what means these results were achieved. Only through comparison with the other institutions will it be possible to find out how the results presented in this article can be assessed.

6 Conclusion

In this chapter, we explained our understanding of sustainability in the context of HEIs, strategic communication of HEIs, sustainability communication, strategic communication *of* and *for* sustainability, and the important role of sustainability for HEIs, which is expressed in the "whole institution approach." Furthermore, we would argue that HEIs, even when they communicate *about* sustainability, will do that in their own organizational interest. Accordingly, a university's sustainability communication will be predominantly strategic communication *of* their sustainability activities and commitment. However, especially during the transformation process of a HEI, strategic communication *for* sustainability is also needed. This assessment is based on the experiences and observations we made in our case study.

One conclusion from our research is that communication *for* sustainability in the context of a HEI's transformation process seems to be different from what is normally understood under strategic communication. While strategic communication is often defined in the literature as public communication (Holtzhausen & Zerfass, 2013), we found that much of the strategic sustainability communication in our university's transformation process involved interpersonal communication with key agents rather than focusing on communication with the public. The interpersonal communication aimed to motivate and encourage individuals to take an active role in the transformation process or to use their influence to guide the process.

Another key insight we gained from the process is the necessity of being pragmatic and flexible and aligning our sustainability transformation goals with the existing objectives of the HEI. To secure support from the university's presidium, we argued that the sustainability transformation will address current challenges faced by the university, especially decreasing student enrollments. Attracting new students is not a task that has anything to do with sustainability. However, we used this argument to gain the necessary support for our actual goals. This example illustrates that it is necessary to look pragmatically for an alliance.

However, it is crucial to ensure that this alignment of interests is not reduced to superficial adjustments or even pure strategic communication activities. If the activities are perceived as a greenwashing campaign merely aimed at enhancing the university's image for prospective students, there is a high risk that even the opposite effects will emerge. To avoid this situation, we used the slogan "TU Ilmenau *on its way* to a sustainable community" in order to highlight that there is still much to be done.

The communication must be underpinned by real changes. For the success of the strategy, it is crucial that the university is genuinely committed to becoming more sustainable in all aspects. To make this happen, it is highly important that all members of the university, including students and staff, actively contribute to and participate in the transformation process. Here, too, it is necessary to look for possible allies who actually have other tasks but are willing to implement certain sub-projects due to their value orientations. We found such allies in the campus management department, health department, and student council, among others. These allies implemented projects such as flowering meadows, drinking water fountains, and the sustainability days.

Before we started the process of strategic sustainability communication, we analyzed the situation in the HEI and came to the conclusion that we needed to implement a transformative research project. Therefore, we started the process of strategic communication with the conscious decision to conduct communication activities to implement the idea of sustainability as an important value and a central goal of the organization. In other words, we started with communication *for* sustainability (see Fig. 2).

Our initial communication activities targeted several key people at the university whom we assumed had similar values in relation to sustainability. The goal of this first step was to initiate some activities to drive concrete action on sustainability and real change in the organization. The outcome of this first step led us to the second cycle in Fig. 2, namely communication *of* sustainability. In this step, the organization communicates its activities related to sustainability, including the actions that were stimulated by the communication *for* sustainability. This communication is aimed at both the internal and external public in the form of press releases, websites, and social media content. These communication activities can in turn be used in a next step to promote further sustainability activities and communication for sustainability within the university. In this way, the two forms of strategic communication can function as a cycle, each enhancing the effectiveness of the other.

A second possible output of communication of sustainability could be reports and news *about* the university's activities that are published in the mass media. Sharing content from the university's social media or the creation of new posts by stakeholders based on the communication *of* sustainability activities are also important forms of communication *about* sustainability (Fig. 2). This communication *about* the university's sustainability activities can have a strong influence on internal decision-making processes and on the university's further communication. An external response that shows that these activities are noticed and recognized by external stakeholders, and the public can encourage actors to drive the sustainability transformation process and also the university's communication *for* sustainability and *of* sustainability. In our case, we were successful in creating media attention on

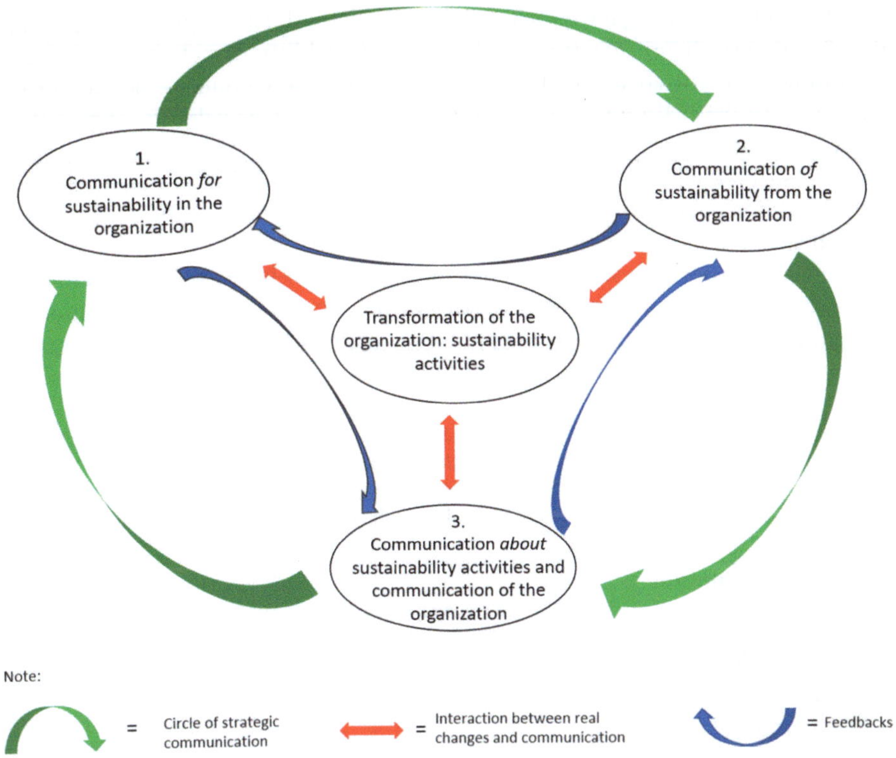

Fig. 2 Cycles of strategic communication for sustainability

some of the activities. For example, the local newspaper reported on the opening lecture of the scientific year on sustainability held by one of the authors.

As with other case studies, the results cannot be generalized. Therefore, future research is needed to examine whether the strategic sustainability communication process we identified is applicable in other universities. As explained above, we plan to conduct some comparative research with the partner universities within the project. These results will give some additional insights into how much our findings are specific.

Furthermore, we assume that the applicability of our findings strongly depends on the context and the specific situation of the HEI. In the case of the university studied, for example, there were very few student activities to promote sustainability, and there was no pressure from students to move the university in the direction of sustainability. This situation is different in other universities. Therefore, it is reasonable to expect that the relevant agents might be different in other HEIs. However, based on our case study, we were able to create a model that could be used to analyze the transformation process at other universities. A study using a process-tracing approach (George & Bennett, 2005) would be an ideal method to find out whether the process we propose in Fig. 2 can be generalized to other universities and perhaps even beyond to other organizations such as companies. It seems plausible to assume that for almost all organizations, fostering sustainability is not the central goal. Accordingly, it will always be the challenge to somehow connect the idea of sustainability to the main aims of the organization. It is reasonable to expect that a successful transformation process depends on the question of whether the primary targets can be connected in a meaningful way to sustainability and whether the transformation agents can successfully communicate this idea to the decision makers and also to all members of the organization.

For such a study, a systematic comparative approach is needed, and an interdisciplinary research project between scholars from the fields of communication and organizational change could be an appropriate approach to gain a holistic perspective on strategic sustainability communication.

Acknowledgments The German Federal Ministry of Education and Research (BMBF) is funding the project KLIMA-N within the framework of the Strategy „Research for Sustainability" (FONA) www.fona.de/en/ as part of its Social-Ecological Research funding priority, funding no 01UN2204D. Responsibility for the content of this publication lies with the authors.

Further Information

	Link to strategic communication of sustainability at TU Ilmenau
	Link to website Klima-N project (only in German)

References

Albareda-Tiana, S., Vidal-Raméntol, S., & Fernández-Morilla, M. (2018). Implementing the sustainable development goals at university level. *International Journal of Sustainability in Higher Education, 19*(3), 473–497. https://doi.org/10.1108/IJSHE-05-2017-0069

Amey, L., Plummer, R., & Pickering, G. (2020). Website communications for campus sustainability: An analysis of Canadian universities. *International Journal of Sustainability in Higher Education, 21*(3), 531–556. https://doi.org/10.1108/IJSHE-04-2019-0137

Carpenter, S., Takahashi, B., Lertpratchya, A. P., & Cunningham, C. (2016). Greening the campus: A theoretical extension of the dialogic communication approach. *International Journal of Sustainability in Higher Education, 17*(4), 520–539. https://doi.org/10.1108/IJSHE-02-2015-0036

Dade, A., & Hassenzahl, D. M. (2013). Communicating sustainability. *International Journal of Sustainability in Higher Education, 14*(3), 254–263. https://doi.org/10.1108/IJSHE-08-2011-0053

Entradas, M., Marcinkowski, F., Bauer, M. W., & Pellegrini, G. (2023). University central offices are moving away from doing towards facilitating science communication: A European cross-comparison. *PLoS One, 18*(10), e0290504. https://doi.org/10.1371/journal.pone.0290504

European University Association. (2021). *Networks on greening and sustainability: EUA survey on greening in European higher education institutions*. European University Association. https://www.eua.eu/publications/reports/greening-in-european-higher-education-institutions.html

Fernández-Vázquez, J.-S. (2021). Measuring environmental website communications in Latin American universities: Multimodal and ecolinguistic benchmarking. *International Journal of Sustainability in Higher Education, 22*(3), 599–614. https://doi.org/10.1108/IJSHE-07-2020-0248

Friedrichsmeier, A., Lautkötter, E., & Marcinkowski, F. (2015). Hochschul-PR als Restgröße: Wie Hochschulen in die Medien kommen und was ihre Pressestellen dazu beitragen. In M. S. Schäfer, S. Kristiansen, & H. Bonfadelli (Eds.), *Wissenschaftskommunikation im Wandel* (pp. 128–152). Köln.

George, A. L., & Bennett, A. (2005). *Case studies and theory development in the social sciences*. The MIT Press.

Hallahan, K., Holtzhausen, D., van Ruler, B., Verčič, D., & Sriramesh, K. (2007). Defining strategic communication. *International Journal of Strategic Communication, 1*(1), 3–35. https://doi.org/10.1080/15531180701285244

Hermeyer, L. (2023). *Nachhaltigkeitskommunikation an der TU Ilmenau: Eine Untersuchung der Fakultäten auf Nachhaltigkeit in Forschung und Lehre*. Bachelor's thesis. Technishe Universität Ilmenau.

HOCH-N. (2020) *Die Deutsche Gesellschaft für Nachhaltigkeit an Hochschulen*. Positionspapier HOCH-N. https://wiki.dg-hochn.de/images/c/c5/Positionspapier-Roadmap.pdf

Holtzhausen, D., & Zerfass, A. (2013). Strategic communication: Pillars and perspectives of an alternative paradigm. In K. Sriramesh, A. Zerfass, & J.-N. Kim (Eds.), *Public relations and communication management* (pp. 283–302). SAGE.

Hoppe, I., & Wolling, J. (2017). Nachhaltigkeitskommunikation. In H. Bonfadelli (Ed.), *Forschungsfeld wissenschaftskommunikation* (pp. 339–354). Springer VS.

Horcea-Milcu, A.-I., Dorresteijn, I., Leventon, J., Stojanovic, M., Lam, D. P., Lang, D. J., Moriggi, A., Raymond, C. M., Stålhammar, S., Weiser, A., & Zimmermann, S. (2024). Transformative research for sustainability: Characteristics, tensions, and moving forward. *Global Sustainability, 7*, e14. https://doi.org/10.1017/sus.2024.12

Kannengießer, S. (2021). Sozial-ökologische Transformationsforschung: Desiderata der Kommunikations- und Medienwissenschaft. In N. S. Borchers, S. Güney, U. Krüger, & K. Schamgerger (Eds.), *Transformation der Medien – Medien der Transformation* (pp. 245–258). Westend Verlag. https://doi.org/10.53291/QNCO5200

Kohl, K., Hopkins, C., Barth, M., Michelsen, G., Dlouhá, J., Razak, D. A., Abidin Bin Sanusi, Z., & Toman, I. (2022). A whole-institution approach towards sustainability: A crucial aspect of higher education's individual and collective engagement with the SDGs and beyond. *International Journal of Sustainability in Higher Education, 23*(2), 218–236. https://doi.org/10.1108/IJSHE-10-2020-0398

Raupp, J. (2017). Strategische Wissenschaftskommunikation. In H. Bonfadelli (Ed.), *Forschungsfeld wissenschaftskommunikation* (pp. 143–163). Springer VS.

Rockström, J., et al. (2009). Planetary boundaries: Exploring the safe operating space for humanity. *Ecology and Society, 14*(2).

Sacchi, A., Molino, M., Dansero, E., Rossi, A. A., & Ghislieri, C. (2023). How sustainable is the governance for sustainability in higher education? Insights from an Italian case study. *International Journal of Sustainability in Higher Education, 24*(8), 1970–1990. https://doi.org/10.1108/IJSHE-08-2022-0254

Schopp, K., Bornemann, M., & Potthast, T. (2020). The whole-institution approach at the University of Tübingen: Sustainable development set in practice. *Sustainability, 12*(3), 861. https://doi.org/10.3390/su12030861

UNEP. (2021). *UNEP's sustainable university framework.* UNEP. https://wedocs.unep.org/bitstream/handle/20.500.11822/36341/USUF.pdf

Verhulst, E., & Lambrechts, W. (2015). Fostering the incorporation of sustainable development in higher education. Lessons learned from a change management perspective. *Journal of Cleaner Production, 106*, 189–204. https://doi.org/10.1016/j.jclepro.2014.09.049

Volk, S. C., Vogler, D., Fürst, S., Schäfer, M. S., & Sörensen, I. (2023). Role conceptions of university communicators: A segmentation analysis of communication practitioners in higher education institutions. *Public Relations Review, 49*(4), 102339.

Weder, F., Karmasin, M., Krainer, L., & Voci, D. (2021). Sustainability communication as critical perspective in media and communication studies—an introduction. In F. Weder, L. Krainer, & M. Karmasin (Eds.), *The sustainability communication reader* (pp. 1–12). Springer.

Further Reading

Kohl, K., Hopkins, C., Barth, M., Michelsen, G., Dlouhá, J., Razak, D. A., Sanusi, A. B., & Z. and Toman, I. (2022). A whole-institution approach towards sustainability: A crucial aspect of higher education's individual and collective engagement with the SDGs and beyond. *International Journal of Sustainability in Higher Education, 23*(2), 218–236. https://doi.org/10.1108/IJSHE-10-2020-0398

Volk, S. C., Vogler, D., Fürst, S., Schäfer, M. S., & Sörensen, I. (2023). Role conceptions of university communicators: A segmentation analysis of communication practitioners in higher education institutions. *Public Relations Review, 49*(4), 102339.

Mira Rochyadi-Reetz is a Researcher and Lecturer at the Institute of Media and Communication Science Technische University Ilmenau. Her research focus is on environmental and climate communication.

Jens Wolling is Professor and Chair of the research group "Empirical Media Research and Political Communication" of the Technische Universität Ilmenau. His research focus is, among others, on environmental and political communication.

Strategic Sustainability Communication—Local Perspectives

From Global to Local: Communication of, About, and for Sustainability in the Local Are(n)as

Beatrice Dernbach

Abstract Climate change and the resulting necessary transformation require a change in ecological, economic, and social terms, not only from a global or national perspective but also from a regional viewpoint. Even if there are many similarities, each region must define its own challenges and understand the transformation process as a holistic cultural change. The focus of this chapter is on local initiatives and their communication strategies of, about, and for the idea of sustainability.

1 Globalization Versus Regionalization

First of all, it should be clarified why it is so important to differentiate between communication strategies *for* sustainability at the *global and regional levels*. The first documents of sustainability do not focus on the region as a geographical and cultural sphere. The Brundtland Commission defined sustainable development in its final report "Our Common Future" as follows: "Humanity has the ability to make development sustainable to ensure that it meets the needs of the present without compromising the ability of future generations to meet their own needs" (WCED, 1987, p.15). This presented the idea of sustainability to a *broad and global public* for the first time. The report can be seen as the basis for many other efforts in the context of sustainability at the *international level*. Since the conference in Rio de Janeiro in 1992, the motto for sustainable action in a political, economic, and social context has been "Think global, act local." Both directives are implemented in the 2030 Agenda for Sustainable Development, which was adopted by the UN in 2015. The sustainable development goals (SDGs) contained therein are considered a milestone for global development toward sustainability (see Pufé, 2017, p. 55). Balancing the three dimensions of sustainability—ecological, economic, and social—is the core aspiration of the paradigm. To make the global Agenda 2030 work, it needs to

B. Dernbach (✉)
Technische Hochschule Nürnberg, Georg Simon Ohm, Nürnberg, Germany
e-mail: beatrice.dernbach@th-nuernberg.de

F. Weder (ed.), *Strategic Sustainability Communication*, CSR, Sustainability, Ethics & Governance, https://doi.org/10.1007/978-3-031-89486-2_14

be broken down to the *national level* and further down to the *regional* and at least *local level.*

The German Advisory Council on Global Change (WBGU), which was founded in 1992 in the context of the climate conference in Rio de Janeiro, advocates an all-encompassing transformation toward sustainability and therefore speaks of a *Great Transformation* (GT), similar to the Neolithic and Industrial Revolutions (WBGU, 2011, p. 87). The GT is intended to bring about an "epochal upheaval" (ibid., p. 89). The aim is to initiate processes of change, which not only include technological innovations but also a change in society, culture, and politics with the aim of guaranteeing "prosperity, stability and democracy within the boundaries of the Earth system" (ibid., p.87). The Great Transformation is a process that must be negotiated by society, in which a wide variety of sectors and a large number of different actors are involved. Based on this report, Schneidewind (2019) developed the "Concept of Future Art." The art of the future takes place in *seven turns*, which are considered necessary for sustainable development and thus form the core of the Great Transformation: the *energy transition*, the *resource transition*, and the *prosperity/consumption transition* mark the three cornerstones of a civil society acting within planetary boundaries. In addition, there are four other necessary turnarounds that together cover the majority of global sustainability challenges: the *mobility turnaround*, the *urban turnaround*, the *food turnaround*, and the *industrial turnaround* (ibid., p.169).

At the same time as *globalization*, the trend toward sustainability is moving toward *regionalization*. There is even talk of a *renaissance of regionality* (Sinz, 2018, p. 1978). Regions offer clear areas for identification and action against tendencies toward centralization and alienation in politics and society (ibid.). Regionality is described as the motor and driving force of innovation, which is justified by the uncomplicated everyday practical networking of actors and the local dissemination of knowledge (Felgenhauer, 2013, p.48). If a problem is identified in a specific area, a regional network of different municipalities, organizations, civil society, and community-led initiatives can work out a solution (Bernhard, 2014, p.84; Ecolise, 2019).

As a result, regions are defined and constituted mainly by networks rather than by administrative borders (Bernhard, 2014, p.87). This *network-based understanding of a region* includes, for example, urban and rural associations, voluntary municipal associations, and also cross-border cooperation areas, and regional development concepts (Sinz, 2018, p.1982). Characterized by informal planning, personal contacts, and close networking, such regions result from a shared vision of the further development of the spatial section (Felgenhauer, 2013, p.48). The high level of complexity associated with the Great Transformation can be reduced at the regional level since local problems can be dealt with in smaller sub-areas. National sustainability strategies focus on the big picture and therefore take fewer regional specifics into account (Brinkmann, 2021, p. 69). A regional approach thus offers a municipality the advantage of a regionally tailored specialization, rather than having to address all transformation fields equally (ibid, p.233).

"Supersystem risks (e.g., pandemics, environmental hazards, and financial crises) manifest locally, and tend to fall unfairly on communities in areas where geographical remoteness, social inequity, and under-representation in political decision-making structures create an extreme environment of vulnerability" (Shannon et al., 2023, p. e850). In the course of the COVID-19 pandemic, the effects of the globally spreading life-threatening virus on people's social environment, national and regional political decisions, and the consequences for economic processes at international, national, and local levels have been analyzed (Delhey et al., 2021).

It is not about a juxtaposition but about the *links between global, national, and regional processes*. These trends are discussed as "translocalism" or "translocality" (Shannon et al., 2023, p. e851) or also as "metropolization" (Moini & Pizzo, 2017, p.4). It should be briefly noted at this point that there is a difference between the terms *local* and *regional*, but this will not be elaborated on further. Both refer to *geographic scales*, but they denote different levels of scope and context. *Local* pertains to smaller, more specific areas and immediate community-focused concerns, while *regional* covers a broader geographic scope, addressing wider issues that affect multiple local areas within a larger defined boundary (Salder, 2023). Being a "local" means being rooted in a particular area, where the surroundings are deeply familiar and comfortable. We invest in a locality in various ways—socially, economically, culturally, and psychologically (Hess & Waller, 2017, p.6).

The main questions are about the understanding of governance in transformation processes and the role of *community-based organizations* (CBO; Shannon et al., 2023, p. e850). "Globally, community-based organisations (CBOs) working in these settings must navigate interlocking supersystem challenges. (…) Studies of successful sustainability initiatives have found that these initiatives are often translocal (ie, rooted in place-based solutions), but also connected to global knowledge and practice networks; examples include energy cooperatives, transition towns, indigenous guardian networks, and sustainable urban food networks." (ibid.) As described by Moini and Pizzo (2017, p.5), they "reconstruct the relationships between economic, political, and social actors" and "they influence the metropolization process." Who drives these "metropolization dynamics" how and in whose interests?

What must be pointed out is the high importance of the *social environment* and the *social capital* for the Great Transformation, defined as *trust*, *norms*, and *networks*. In the local are(n)as, social capital "is embedded in the community, encompasses the relationships and communication patterns of actors, and is manifested in the way, the community works together in practical activities and projects" (Dlouhá et al., 2022, p.3). Further essential factors of local changes are *strategic communication*, raising awareness, and providing and distributing data and facts to educate, as well as establishing and promoting this reliable knowledge in networks and through the exchange, motivating people to participate (Dernbach et al., 2022; Dernbach, 2021; Weder, 2022; Genç, 2017).

2 Sustainable Development Goals Go Local

The UN sustainable development goals are *very abstract* in their global idea and don't resonate with people. Various studies in recent years have shown that only a *few people are familiar* with sustainability concepts and goals. Unfortunately, this ignorance is proving to be relatively stable (Focus2030). As in all comparative empirical studies on complex socio-political issues, there are major challenges regarding the *validity and reliability* of the measurement. The results cannot be considered without looking at the client and the research institute, without putting the overall population and their political attitudes, etc. in relation to each other. For example, there is a global study on the perception of the SDGs in the context of the World Economic Forum (WEF, 2021) and one focusing on America, Great Britain, France, and Germany (Focus2023). While this second study reveals that only 8–12% of respondents are aware of the SDGs, the WEF study does not go into this but instead presents the ranking of the SDGs in comparison. According to this, SDGs 2 (no hunger), 1 (no poverty), and 3 (health) are the most frequently mentioned. Citizens attest that their governments are doing too little with regard to sustainability. The Global Survey (2020), funded by the Government of the Federal Republic of Germany, sees the global average level of awareness of the SDGs at well under 50%. Prioritization differs depending on demographic characteristics such as region and age; but the SDGs *Climate Action* (SDG 13) is the most frequently cited SDG of personal interest, followed by *Good Health* and *Well-being* (SDG 3) and *Quality Education* (SDG 4). No Poverty (SDG 1) and Zero Hunger (SDG 2) are more urgent in Africa than in Europe.

Focusing on the topic of climate change, the authors of the book "Global Warming in Local Discourses" (Brüggemann & Rödder, 2020) discuss that the major debate transcends national borders and is conducted transnationally, primarily based on the reports of the *Intergovernmental Panel on Climate Change* (IPCC). This reveals similarities between world regions. However, the debates conducted at the national and regional levels are *embedded* in the specific political, economic, and social *contexts of each region.* In these contexts, abstract scientific findings often play a minor role, while events that people can personally and directly experience take precedence.

Even if there are many commonalities and similarities between the members of the global community, each country, each region, and at least each city must define its own challenges and understand the transformation process as a *holistic cultural change*. Countries in the Global South in particular are confronted with the special requirements of the major transformation in the face of regional challenges (such as droughts and floods, crises, and wars) (Abdourahamane dit Baffa, 2021). Countries in the West, such as the USA and Europe, have to cope with different economic, ecological, and social *challenges* than those in India as well. It is the same at a regional level: not every region in Germany, the USA, Ghana, or India can develop under the same conditions, they can differ in terms of economic factors, political, and social structures. In view of the fact that up to 70% of the world's population will live in *cities* by 2050, they are of particular importance (UN DESA). The proportion of *megacities* will increase, especially in Africa and Asia. Many of them have joined together in large networks in order to find similar solutions to similar problems (such as energy supply and waste disposal etc.) (Dernbach, 2021, p.526). Many (mega) cities have already set out to become *eco-cities*. Five criteria (Yedla, 2015) could be helpful in order to describe them: "Therefore, the cities have to develop a combination of policies to address in tandem and realize ecologically efficient production, ecologically efficient landscaping, ecologically friendly culture, ecologically efficient and secured services provision, and ecologically efficient sanitation system."

The German Bertelsmann Foundation has developed the platform *SDG Indicators for German municipalities*.[1] Its purpose is to identify suitable criteria for *implementing the SDGs locally* and to provide the associated data on the portal. In the first phase of portal development, the SDG sub-goals were examined for their relevance to German municipalities. In the second phase, the selected indicators were compared with additional catalogs and databases and examined based on their validity, comprehensibility, data availability, data quality, and functionality. Finally, they were extensively described in profiles. In the third phase, the indicator values were collected and analyzed in cities, municipalities, and districts with over 5000 inhabitants.[2] The portal features concrete examples of measures taken by German municipalities and also allows for comparisons between them.

The only platform found under the tag *SDGs go local* leads to Nuremberg in Germany. The association Bluepingu, which has been promoting the topic in the Bavarian metropolitan region for years, wants to invite everyone to participate. "Civil society, business, science, politics, and culture are working together on implementing the ecological, economic, social, and cultural facets of sustainability. Citizens contribute in their various roles with their abilities and skills, forming working groups and making the Sustainable Development Goals (SDGs) individually tangible, thereby anchoring the SDGs in the consciousness of the population. We focus on knowledge dissemination, empowerment, and cooperation with actors in urban and rural areas." (Bluepingu[3]).

3 Communicating Sustainability on a Local Level

In the SDG framework, *communication is not explicitly mentioned*, but implicitly, it is *embedded in every goal* (Christensen et al., 2017; Commission for Economic Policy, 2022). The one and only model will not lead to the desired success,

---

[1] https://www.bertelsmann-stiftung.de/fileadmin/files/Projekte/Monitor_Nachhaltige_Kommune/SDG_Indicators_for_Municipalities_2020.pdf

[2] See the SDG portal https://sdg-portal.de/de/

[3] See the Website in English https://www.betterplace.org/en/organisations/3619-bluepingu-e-v

especially with regard to *Goal 11* "Make cities and human settlements inclusive, safe, resilient, and sustainable." Both the sustainability concept and the corresponding underlying communication concept must analyze and integrate *location-specific conditions*. The UN has set up the *Local 2030 platform* to raise awareness of the sustainable development goals at a local level in particular. The aim is: "Localizing the SDGs is a network and platform that supports the on-the-ground delivery of the SDGs, with a focus on those furthest behind. It is a convergence point between local and regional governments and their associations, national governments, businesses, community-based organizations and other local actors, and the United Nations system. Local2030 supports local leaders in collaboratively incubating and sharing solutions, unlocking bottlenecks and implementing strategies that advance the SDGs at the local level."

In June 2022, the United Nations released *Communication Guidelines* (UNSDG, 2022). They build up on *UN Country Teams* (UNCTs) and *UN Communications Groups* (UNCGs) "to better leverage the power of joint communications to make people care and take action for the Sustainable Development Goals (SDGs)." The core are *five guiding principles*:

1. *Emphasize* the shared values, norms, and standards of the United Nations system.
2. *Adapt* the guidance to the country context.
3. Consistent and coherent *messaging* is a shared responsibility of all UNCT members.
4. Agency-specific messages must be consistent with agreed *common positions*.
5. *Coordination and sharing* of timely information among UNCT members are essential.

These guidelines are formulated in a very general sense. But combined with the tools on the Local 2030[4] website, they could indeed provide communicators at the regional and local levels with some inspiration. The basic understanding of *localization* is: "Localizing development means taking into account subnational contexts in the achievement of the 2030 Agenda, from the setting of goals and targets, to determining the means of implementation and using indicators to measure and monitor progress. It is also putting the territories and their peoples' priorities, needs and resources at the center of sustainable development. There should be sustained exchanges between the global, national and local facets."

Before starting to use the tools, every actor in the field of local sustainability communication should take four steps:

1. *Identify* the stage of the initiative and the current actor's role
2. *Explore* the list of tools and choose the most relevant
3. *Implement and share* your insights, experiences, and questions
4. *Contribute*: implementation and monitoring of the SDGs at the local level (Local 2030)

[4] Local 2030 is a global initiative where stakeholders, especially from cities and municipalities, support each other and exchange knowledge and experience (https://www.local2030.org/).

In general, the concept of *sustainability communication at the regional level* follows basic rules that must be taken into account in many fields when planning *communication campaigns*: first, the situation must be analyzed, and then the goals, strategies, target groups, responsibilities, budget, and individual measures must be defined. These elements need to be evaluated multiple times and regularly compared with the goals and strategies (Local 2030; Weder, 2023; Weder & Eriksen, 2023).

The requirements for regional sustainability communication can undoubtedly be described as particularly *challenging*. This is due to *several factors*: first, the global and transnational nature of the challenges involved; second, the necessity to align with scientific insights from various disciplines; and third, the aspect that sustainable processes are not necessarily directly observable in everyday life. Consequently, the hurdle for people to change their lifestyles and demand corresponding frameworks from politics and the economy is quite high. So, the whole social change is challenging in every way.

4 Communication of, About, and for Sustainability

In order to understand communication in the context of sustainability and model it effectively, we need to take a fundamental look at it. The simplest one-way model is Lasswell's (1948): who says what in which channel to whom with what effect? This is a very pragmatic and functional understanding of the meaning of *one-way communication*. In other more complex models, other elements are added, such as time and duration. In a wider sense, communication enables individuals, groups, communities, organizations, and last but not least the whole society to create topics, ideas, concepts, and narratives to re-construct the world around them. In the social environment of citizens, political, economic, cultural, and many other actors communicate, as individuals (micro level), at the level of organizations and institutions (meso level), and at the level of the respective system (macro level; for example, all local political bodies of the legislative and executive branches). This respective socio-cultural context is essential for the success of the Great Transformation described above. The corresponding goal of sustainability communication is to initiate *communication processes that enable new cultural patterns*, techniques, and practices or "to create a cultural anchoring of the principle of sustainability through, via and in social communication processes" (Weder, 2022, pp.153–154). Weder combines the approaches of strategic organizational communication and corporate society with those of environmental, climate, science, and risk communication. Genç (2017) also highlights the importance of sustainability as both a theme and a concept for action and communication from the perspective of strategic organizational communication. She distinguishes between *three levels of strategic organizational communication*: the organizational level, the societal level, and the individual level. At the organizational level, it involves how companies integrate and communicate sustainability as part of their corporate strategy. This may include implementing environmentally friendly practices in the production or promoting socially responsible initiatives. At the societal level, strategic organizational communication refers to how companies communicate their sustainability efforts in relation to the needs and expectations of society. This may include transparency about environmental impacts, engaging stakeholders in the decision-making process, or supporting nonprofit organizations. Finally, the level of each single organization refers to how companies use sustainability as part of their *brand communication* to increase consumer awareness and loyalty. This can be achieved by emphasizing eco-friendly products and services, communicating corporate values, or involving customers in sustainable initiatives.

The three dimensions of sustainability communication according to Genç (2017: 514–516) can be summarized as follows:

Communication of sustainability (CoS) is primarily instrumental or managerial in nature. It involves a unidirectional, sender-receiver flow of communication, where the sender pursues a specific objective, often aligned with strategic or organizational goals.

Communication about sustainability (CaS) refers to the dynamic processes wherein information, interpretations, and opinions related to sustainability issues are actively exchanged and discussed. It emphasizes dialogue and mutual understanding.

Communication for sustainability (CfS) emphasizes the normative aspect of sustainable development. It goes beyond merely disseminating information and raising awareness about sustainability. The ultimate goal of CfS is to foster societal transformation in alignment with the normative objectives of sustainable development.

Weder (2022) adds to this model the dimension of *sustainable transformative communication (STC)*. It differs in direction and mode from communication *for* transformation: instead of participative, self-organized, many-to-one with a focus on social change, the STC is co-constructive, collaborative, and at the time polarizing with the function of negotiation, transformation, and cultivation (Weder, 2022, p.154). At the level of individual organizations and companies, models can certainly be identified that expand *transfer-oriented communication* of sustainability from the sender to the recipient (company–customer) to include dialogical and participatory communication elements for sustainability (ibid., p.152). However, scaling this model to the level of society as a whole is a highly ethical standpoint, as it considers sustainability as a paramount value for all social actions, particularly for communication, and media activities.

A study about *Sustainability Communication in the Nuremberg Metropolitan Region (MRN)* (Dernbach, 2021:531–535) showed among other results that the communication direction of the stakeholders differed distinctly and a tendency toward communication *of* sustainability. Civil society actors stand as an exception, as their communication is predominantly *about* and *for* sustainability, i.e., it contains a more or less strong call to action. They mainly use—due to lower costs and higher reach—digital channels and social media. Political and economic stakeholders display a higher diversity of communication tools and target groups. However, the analysis has not found any significant interconnection or networking activities of the stakeholders, rather practical cooperation and a generally low diversity. Or to put it more clearly: there are many individual groups in the local area that are working for more sustainability alongside but not with each other. The results from the sub-studies can be summarized in one term: *networking*. This would not only be essential for joint communication but also helpful for achieving the joint goal of greater sustainability in everyday life.

5 Importance of Local Players

Networks within and between stakeholder groups from politics, business, culture, sport, educational institutions, church communities, media, and others are created at a local level, where many people know each other personally. Therefore, this chapter takes a special look at the sustainability actors at the local level.

Dlouhá et al. (2022) studied the "Role of Actors in the Processes of Sustainable Development at Local Level." Several researchers from the Czech Republic have been running a 3-year project in non-metropolitan areas in cooperation with *local action groups (LAG)*. Their main questions were about the role of LAG representatives attributed to actors from different levels and sectors in different regions related to sustainable development. How are they involved in the processes of transition and SDG implementation? The sustainable development aspects of these themes were analyzed by assessing current actor involvement and future prospects, such as goal-setting, implementation, and innovation. This aligns with the research objectives to explore sustainable development opportunities and constraints at the local level, focusing on actor engagement and bottom-up processes (Dlouhá et al., 2022, p.6).

The results demonstrate the *important role of the LAGs* in terms of understanding and implementing the sustainability concept as a whole and some SDGs in particular—especially when there are quickly visible successes in social and economic areas. However, this does not apply to progress in the ecological area, particularly with regard to climate change. Nevertheless, the LAGs fulfill their function of supporting, inspiring, and coordinating the actions of community members and thus promoting their well-being (ibid., p.10).

Another explorative study was done in six European countries (Austria, Germany, Italy, Netherlands, Romania, and Spain) by Hauck et al. (2020) to understand *actor roles in sustainability initiatives*. An exploratory approach was taken due to the "lack of clear definitions of actors, and their roles" (ibid., p.1). The researchers selected two initiatives in each of the six countries and defined *six types of actor roles* (Hauck et al., 2020, p. 9, see Table 1).

In the table of actors, five different characteristics are shown: actors from the *private sector* (e.g., companies), *civil society* (e.g., initiatives), *governance* (e.g., parties and mayors), *media, education, and knowledge* (ibid.: 12). *Media* is an actor in all cases and plays an important role in Germany and Spain. For the Spanish

Table 1 Actor roles (Adapted from Hauck et al., 2020)

Actor role	Description of actor role
Catalyst	Actor accelerating and supporting certain initial developments and individuals (e.g., frontrunners) with knowledge, experience, and/or money
Opponent	Actor plays the role of an antagonist of an initiative
Intermediary	Actor who builds and maintains networks, connecting different actors
Frontrunner	Pioneers acting in niches, starting up to implement innovative ideas
Driver	Actors who are very active and keep initiatives alive over time by providing continuous support
Visionary	Actors with visions, who are opinion leaders and who, due to their spirit, idealism, authenticity, and charisma, influence the initiative beyond their niche in a positive way

initiatives, "the mass media are seen as an opportunity to disseminate ideas about sustainability among a wider audience." On the other hand, the Austrian initiatives do not use connections to the media but depend largely "on informal talks, meetings, and encounters that take place outside the realm of media." According to the qualitative data, the *significance of media* lies mainly in their flexibility across space, time, content, and application. Media actors within these initiatives support various functions, including internal and semi-public communication, networking, meetings, collaboration, exchange, and production (ibid.). The *function of media* depends on the age of the actors, the structure of the initiative, the attributed level of media influence, and the awareness of the advantage to reach many people at minimal costs. Some of the project groups appreciate the variety of media channels enabling independence and autonomy, particularly when they use or run alternative Internet platforms. "Hence, media here take on the role of a frontrunner." (ibid., p.14).

Many other studies also show that the media play a special role at the local level: they set topics, identify actors, and ideally contribute to social discourse. However, they are under particular pressure from digitalization (Hess & Waller, 2017). Newspapers grapple with steeply declining circulation, while visual and audiovisual media face competition from countless private sector publishers now able to communicate via *social media platforms*. The aspects of accessibility for both producers and audiences, coupled with wide reach and low costs, contribute to this shift. The media's role as a *multiplier* appears to diminish in significance and visibility. They could play an important role in communicating sustainability—but journalists have always been rather critical of that. The word sustainability is too unwieldy, the subject area too complex, and it cannot be reduced to a few buzzwords (Dernbach, 2021: 529–530). This prompts local sustainable initiatives to seek channels for not just one-way communication, but two-way dialogue, aiming to inspire action among people.

6 Looking at Community Initiatives

One common result of the studies cited and others is a large number of sometimes small, sometimes specialized *local action groups* and projects in the field of sustainability. Most of them don't have money but plenty of highly committed members with creative ideas. Their communication activities are often reduced to *self-made social media channels*. Networks or *community-led initiatives (CLI)* can be very helpful to spread the same idea the same way to more people in different local areas. The term CLI is not yet visible in scientific literature clearly related to terms such as community-based or -driven, grassroots movement, and others alike. They arise and exist across Europe, take different forms, and focus on a wide range of issues. They try to build up communities and motivate people to take action on climate change and sustainability worldwide. Their work complements the large national and global agenda in the context of *climate change and planetary boundaries at the local level*, where the main focus is on *recruiting volunteers*.

There is a very broad variety of CLIs in the world, like, for instance, *Transition Network*, *Fairtrade International*, and other initiatives in the field of agriculture, community gardens, etc. They aim to combine not only the three dimensions of sustainability—economic, ecology, and social—but also include technological choices and governance methods for society-wide transitions. Further important criteria are the use of scientific data and the communication of evident facts.

The European-wide network of community-led sustainability initiatives *ECOLISE*, founded in 2014 by 43 member organizations in 21 European countries, is a prime example of this approach. Their *core activities* are catalyzing social, economic, and ecological transformation. The First Status Report on Community-led Action on Sustainability and Climate Change in Europe (2019) named two main targets:

1. To provide a comprehensive and scientifically rigorous account of the documented extent, nature, impacts and potential of community-led initiatives (CLIs) in Europe, incorporating relevant academic studies, grey literature, practitioner records and informal and experiential knowledge, to inform policy, advocacy, practice, and future research.
2. To establish an active, self-organising knowledge co-creation community that collaborates on an inclusive, open source basis to maintain, update, extend and deepen the knowledge base on which the Status Report draws, in a way that also makes this knowledge base available for other uses. (Ecolise, 2019, p.2).

The main *communicative challenges* of these initiatives are:

- To *reduce the complexity* of climate change and transformation processes
- To *attract attention* also to uncomfortable aspects such as lifestyle changes
- To take care of the continuous, reliable, and factually *correct information* of the citizens
- To motivate and encourage people that they can make a difference and bring about *change*

7 Communicative Potentials and Ideas for Sustainability Communication in the Local

The *transition* toward sustainability demands a comprehensive, radical *transformation* of society and its societal subsystems. *Transition theory* (Pel et al., 2016) can help to describe this transformation, which will not proceed linearly. The theory is fed by four different perspectives: the multilevel perspective, strategic niche management, technological innovation systems, and transition management (Pel et al., 2016, p. 451). The reorganization of the whole system also needs the reorganization of the technological system, for example, with regard to energy. The dynamics of this necessary technological progress will vary across different areas (such as the energy sector, mobility, etc.)—depending on the extent of influence from specific actor groups such as politicians, scientists, and special interest groups. Engel (2022, pp. 17–38) has identified three main pillars in her analysis of *social innovations in transformation processes* in the context of the municipal energy transition, in order to measure functions, barriers, limits, and success factors.: the *multi-phase concept* to assess the speed, direction, and magnitude of the transformation, the *multi-level perspective* (MLP) to measure the functional relationships between actors, and the *multi-pattern concept* to depict the interaction forms among the actors.

All this leads to the key question: *how can local initiatives motivate people to act sustainably?*

This question has been regularly and frequently asked in recent years in various scientific disciplines (including psychology, medicine, and communication science) as well as in communication practice. In media research, for example, media coverage, communication strategies of companies in fields such as energy, or political sustainability communication in selected countries are analyzed (see, for example, Weder et al., 2019). In communication practice, recommendation lists are available, which mostly give concrete instruments for communication in general. Let's consider the different conditions of global, national, and regional/local sustainability communication. Key aspects include *visibility*, *personal relevance*, and the *promise of positive impacts*. These criteria apply to different individual knowledge bases, values, expertise, attitudes, behaviors, beliefs, motivations, and also material conditions such as income, social status, wealth, etc.

To avoid getting stuck on either the strongly individualistic or the highly abstract perspective, approaches such as *transition theory and social innovation* are helpful. Toya Engel (2022) examined the energy transition in a small German town against this theoretical background. Using the *multi-level perspective*, she identifies the micro-level (actions and interactions of individual actors), the meso-level (institutions and social structures), and the macro-level (societal structures and phenomena). At this point, her comprehensive study cannot be fully presented. Engel impressively explores the *functions and limits of social innovations in (local) transformation processes*, theoretically and empirically, across four fields: ecology, socio-culture, organization and economy, and politics (Engel, 2022, p.363).

The functions identified are *participation*, *education* and *reflection*, *collective formation*, *market actor relations*, and *mediation* and *communication*. The *limitations* lie in provision, mediation, and communication, as well as regulation and control (ibid., pp.365–367). Engel primarily sees the communication and mediation function in the field of politics, as opening a new path for legitimacy, by acquiring competence and accepting responsibility, and the communicative development in sensitive issues. Limitations may arise from the lack of majority in committees and the risk of political dynamics in discussion and decision-making processes (ibid., p.367). One of her central insights, if not the central insight, is this: to successfully transform society, it requires exchange and permeability between functional levels, a strong sense of responsibility among all actors, and the openness and willingness of the population to participate in participatory opportunities (Engel, 2022, pp. 360–361).

On both theoretical and empirical grounds, a *communication model for community-led initiatives* could be built. The following *central aspects* must be considered in the analysis and conception of this model:

- *Integration and participation* of the civil society: prioritize integrating and engaging the members of the local community, employing *co-creation* as a key tool in sustainability processes and communication strategies.
- Align with global and local objectives: ensure initiatives are *framed within the context of both global sustainability goals and their local impacts*, considering actors across politics, economy, culture, and civil society.
- *Empower individuals* through *personalized benefits*: analyze the micro-level by highlighting how sustainable practices enhance individual quality of life, emphasizing financial savings, improved health from local food sources, and enhanced environmental conditions.
- Appeal to *shared values and relatability*: connect sustainability messages with audience values and experiences through *storytelling*, illustrating relatable examples of ordinary people embracing sustainable living.
- *Provide practical guidance and overcome barriers*: offer actionable steps to reduce environmental impact, addressing common barriers like cost concerns and convenience issues with eco-friendly solutions.
- Nurture community *support* and *collaboration*: foster collaboration among like-minded individuals through platforms for sharing experiences and learning, building a sense of community around sustainable living.
- *Optimistic messaging* and incentivized *action: frame messages positively*, highlighting the benefits of sustainability and offering incentives for adopting eco-friendly behaviors, such as discounts and recognition.
- *Educate and empower* for informed choices: provide educational resources and opportunities to empower individuals with the knowledge and skills needed to make sustainable choices, leading by example to inspire change.

8 At Last…

This chapter outlines why the global sustainable development goals (SDGs) need to be transferred and communicated *locally* to be implemented and drive *societal transformation*. However, localizing the SDGs requires analysis and structuring into a tailored *local agenda*. It is important to recognize that not all local actors and initiatives can address every goal equally. There is no hierarchy in the framework, but *SDG 11: Sustainable cities and communities* should be prioritized on the local level.

When successfully shaping sustainable processes at the local level, many aspects should be taken into account in order to develop their potential for change in people's social living environment. In this respect, changes will not be achieved in the short or medium term. Looking at the political changes in Europe, the gate for the Great Transformation is slowly closing again, as wars and crises are increasingly capturing people's attention. Climate change is no longer at the top of the sociopolitical agenda, where it has been replaced by the topic of migration and people are worried about their economic future (YouGov, 2024).

Alongside many other actors from politics, economy, culture, and civil society, the media as the main communication channels (could) play a central role in this process. However, as traditional media, especially local newspapers, have come under economic pressure due to digitalization, local sustainability actors are creating and utilizing *new social media channels*.

The essential factor for the success of local *communication for sustainability* is the *networking of local actors*. These local networks, in turn, share knowledge and experiences on platforms such as Local2030 among communities worldwide. Creating a local agenda hinges on understanding the needs, values, and behaviors of social groups and individuals, respecting their diversity, and acting responsibly. *Action plans* should involve compromises, and information about communication goals, strategies, and tools. One objective remains constant above all: *communication for sustainability* must integrate and motivate all participants. The most promising approach to success is local initiatives coming together. The formats for this can vary greatly, such as citizens' councils, fairs, and festivals. Although we are running out of time due to climate change, local stakeholders still need *time* to network, learn about sustainable measures, exchange ideas, and jointly develop and implement a smart plan for communicating sustainability. More bottom-up than top-down.

References

Abdourahamane dit Baffa, K. (2021). From Accountability and Learning to evidence based evaluation: A roadmap to SDGs for Developing Countries. *Academia Letters*, Article 2936. https://doi.org/10.20935/AL2936.

Bernhard, C. (2014). Region ungleich Region. Vom normativen Regionsbegriff zur interpretativen Regionalität. *REPORT - Zeitschrift für Weiterbildungsforschung, 4*, 83–95.

Bluepingu. *SDGs go local*. https://sdgs-go-local.bluepingu.de/ueber/das-projekt
Brinkmann, R. (2021). *Introduction to sustainability*.
Brüggemann, M., & Rödder, S. (Eds.). (2020). *Global warming in local discourses: How communities around the world make sense of climate change*. Open Book Publishers. https://doi.org/10.11647/OBP.0212
Christensen, L. T., Morsing, M., & Thyssen, O. (2017). License to critique: A communication perspective on sustainability standards. *Business Ethics Quarterly, 27*, 239–262.
Commission for Economic Policy. (2022). *Synergies between the Sustainable Development Goals and the national recovery and resilience plans – Best practices from local and regional authorities*. https://doi.org/10.2863/518460
Delhey, J., Steckermeier, L. C., Boehnke, K., Deutsch, F., Eichhorn, J., Kühnen, U., & Welzel, C. (2021). *A virus of distrust? Existential insecurity and trust during the coronavirus pandemic*. https://www.researchgate.net/publication/354779571_A_Virus_of_Distrust_Existential_Insecurity_and_Trust_during_the_Coronavirus_Pandemic
Dernbach, B. (2021). Sustainability in the local area. In: Weder et al. (2021): *The sustainability communication reader* (pp. 525–542). https://doi.org/10.1007/978-3-658-31883-3
Dernbach, B., Hufksy, J., & Kumar, A. (2022). Regional communication for the great transformation towards sustainability. *Global Media Journal-Indian Edition, 14*(2). https://gmj.manipal.edu/issues/December2022/P%205.pdf
Dlouhá, J., Vávra, J., Pospíšilová, M., & Dvořáková Líšková, Z. (2022). Role of actors in the processes of sustainable development at local level—Experiences from the Czech Republic. *Frontiers in Sustainability, 3*, 888406. https://doi.org/10.3389/frsus.2022.888406
Ecolise. (2019). *Reshaping the future: How local communities are catalysing social, economic and ecological transformation in Europe*. https://www.ecolise.eu/wp-content/uploads/2016/02/Status-Report-on-Community-led-Action-on-Sustainability-Climate-Change-in-Europe-2019.pdf
Engel, T. (2022). *Soziale Innovationen in Transformationsprozessen*. oekom.
Felgenhauer, T. (2013). Regionalität als Rationalität. Die argumentative Konstruktion von Regionen. *Europa Regional, 21*(1–2), 47–59.
Genç, R. (2017). The importance of communication in sustainability & sustainable strategies. *Procedia Manufacturing, 8*, 511–516. https://doi.org/10.1016/j.promfg.2017.02.065
Global Survey. (2020). *Report of Results Global Survey on Sustainability and the SDGs*. https://www.globalsurvey-sdgs.com/wp-content/uploads/2020/01/20200205_SC_Global_Survey_Result-Report_english_final.pdf
Hauck, J., et al. (2020). *Understanding actor roles in sustainability initiatives: An exploratory study in five European countries*. (UFZ Discussion Papers, 2/2020). Leipzig. https://nbn-resolving.org/urn:nbn:de:0168-; https://doi.org/10.20935/AL2936.
Hess, K., & Waller, L. (2017). *Local journalism in a digital world*. Palgrave.
Lasswell, H. D. (1948). The structure and function of communication in society. In L. Bryson (Ed.), *The communication of ideas A series of addresses* (pp. 32–51). Harper & Brs..
Local2030. *Discover tools*. https://www.local2030.org/discover-tools
Moini, G., & Pizzo, B. (2017). Riding roughshod over people. Reading politics, economy and civil society in metropolization. A case study in Rome. In T. Reilly (Ed.), *The governance of local communities. Global perspectives and challenges* (pp. 3–20). Nova Publishers.
Pel, B., Avelino, F. R., & Jhagroe, S. S. (2016). Critical approaches to transitions theory. In H. Brauch, Ú. Oswald Spring, J. Grin, & J. Scheffran (Eds.), *Handbook on sustainability transition and sustainable peace. Hexagon series on human and environmental security and peace* (Vol. 10). Springer. https://doi.org/10.1007/978-3-319-43884-9_19

Pufé, I. (2017). *Nachhaltigkeit*. UVK.

Salder, J. (2023). Reading the relational local economy: Regional governance and the rescaling of local dynamics. *Geoforum, 144*. https://doi.org/10.1016/j.geoforum.2023.103785

Schneidewind, U. (2019). *Die große Transformation. Eine Einführung in die Kunst gesellschaftlichen Wandels*. Fischer.

Shannon, G., Basu, P., et al. (2023). Think global, act local: Using a translocal approach to understand community-based organisations' responses to planetary health crises during COVID-19. *Lancet Planet Health, 7*, e850–e858. https://www.thelancet.com/action/showPdf?pii=S2542-5196%2823%2900193-6

Sinz, M. (2018). Region. In: ARL – Akademie für Raumforschung und Landesplanung (Hrsg.), *Handwörterbuch der Stadt- und Raumentwicklung* (S. 1975–1984).

United Nations Sustainable Development Group (UNSDG). (2022). *Communicating together. Guidelines for UN Country Team communications on the SDGs*. https://unsdg.un.org/sites/default/files/2022-06/UNSDG_Report_Communicating__REV__.pdf

WBGU. (2011). *World in transition. A social contract for sustainability*. https://www.wbgu.de/fileadmin/user_upload/wbgu/publikationen/hauptgutachten/hg2011/pdf/wbgu_jg2011_en.pdf

WCED. (1987). *Report of the World Commission on Environment and Development: Our common future*. https://sustainabledevelopment.un.org/content/documents/5987our-common-future.pdf

Weder, F. (2022). *Nachhaltigkeit kultivieren. Communicatio Socialis, 55*(2), 146–159.

Weder, F. (2023). Paradigms in strategic communication. In F. Weder & M. Eriksen (Eds.), *CSR communication and cultures of sustainability* (pp. 42–47). The University of Queensland. https://doi.org/10.14264/428d81b

Weder, F., & Eriksen, M. (2023). *CSR communication and cultures of sustainability*. The University of Queensland. https://doi.org/10.14264/428d81b

Weder, F., Lemke, S., & Tungarat, A. (2019). (Re) storying sustainability: The use of story cubes in narrative inquiries to understand individual perceptions of sustainability. *Sustainability, 11*(19), 5264.

World Economic Forum (WEF). (2021). *Ending hunger and poverty are the top priorities for global public: Forum, Ipsos poll*. https://www.weforum.org/agenda/2021/06/hunger-poverty-improve-health-survey/

Yedla, S. (2015). Chapter 1. Cities and sustainability dimensions. In S. M. Dev & S. Yedla (Eds.), *Cities and sustainability* (pp. 1–21). Springer India.

YouGov. (2024). *The most important issues facing the country full*. https://yougov.co.uk/topics/society/trackers/the-most-important-issues-facing-the-country-full

Further Reading

Ecolise. (2019). *Reshaping the Future: How local communities are catalysing social, economic and ecological transformation in Europe*. https://www.ecolise.eu/wp-content/uploads/2016/02/Status-Report-on-Community-led-Action-on-Sustainability-Climate-Change-in-Europe-2019.pdf

Hauck, J., et al. (2020). *Understanding actor roles in sustainability initiatives: An exploratory study in five European countries*. (UFZ Discussion Papers, 2/2020). Leipzig. https://nbn-resolving.org/urn:nbn:de:0168-; https://doi.org/10.20935/AL2936.

Beatrice Dernbach has been the holder of the Research Professorship of Sustainability and Science Communication since October 2021 at Technische Hochschule Nürnberg; since March 2014, she has been a professor of "Practice of Journalism." From October 1999 to February 2014, she taught and pursued research in the field of journalism, media, and communication at the University of Applied Sciences Bremen. She has published books and scientific contributions on journalism and sustainability, sustainability and science communication, specialized journalism, journalism education, and the relation between journalism and public relations. Before starting her academic career, she was working as an editorial journalist at a newspaper.

The Strategic Making of Sustainable Venice

Monica Porzionato

Abstract Drawing on a constitutive understanding of strategic communication [Falkheimer and Heide (Strategic communication: An introduction. Routledge, 2018), (Strategic communication: An introduction to theory and global practice. Routledge, 2022); Winkler and Schoeneborn (The research handbook of strategic communication. Edward Elgar, 2023)], this chapter examines how the municipality of Venice's strategic sustainability communication not only represents the city's sustainability practices but also constitutes what a sustainable Venice means in the first place. In particular, it analyses a communication campaign to inform tourists about flooding events as well as an interview with a local politician. The findings show that the making of a sustainable Venice is the result of a negotiation of the meaning of sustainability between the specificities of the local urban setting and the meaning found in the global discourse. Extending analyses of the strategic use of language to communicate about sustainability, this chapter shows how strategic sustainability communication is a process that involves both crafted and emergent instances of strategic communication in the constitution of sustainable places. As such, it helps to circumvent some of the limitations of representational understandings of strategic sustainability communication by showing how such practice is inherently political, influencing the transformation of urban places in the age of climate change.

1 Introduction

Over the past century, the steady burning of fossil fuels to generate energy for human activities has increased the concentration of carbon dioxide in the atmosphere, leading to a warming of the Earth's average surface temperature (nasa.gov).

M. Porzionato (✉)
Department of Communication, Lund University, Lund, Sweden
e-mail: monica.porzionato@iko.lu.se

F. Weder (ed.), *Strategic Sustainability Communication*, CSR, Sustainability, Ethics & Governance, https://doi.org/10.1007/978-3-031-89486-2_15

As the temperature of land and water began to rise at an unprecedented rate, there has been an increase in the frequency and severity of extreme weather events since the mid-twentieth century as well as the melting of ice sheets at the Earth's poles and mountain glaciers, causing global sea levels to rise exponentially (nasa.gov). In this scenario, many of the world's coastal areas will become climate-vulnerable, in the sense that they will have a high propensity to be adversely affected by climate change-related impacts such as sea-level rise (IPCC 2022). In these settings, beyond geographical and economic differences, one factor strongly influences the implementation of timely and effective sustainability strategies: the degree of importance devoted to the issue of climate change, the way it is communicated and understood, and the resulting understanding of what it means to make these areas sustainable for the approaching future. In this regard, studies have shown that because climate change is most often framed as a problem of the future (Adams, 2022), there are varying degrees of salience in today's global discourse on the issue, with decision-makers deferring sustainability interventions in favour of more immediate and visible social difficulties, such as poverty, crime or migration flows (Roper et al., 2016). The corporate sector, and the fossil fuel industry in particular, has also been repeatedly found to deliberately amplify doubts about climate change in order to deflect warnings to take action against this global crisis and to keep their business model intact (Greenberg et al., 2011; Ihlen & Roper, 2014). It is clear, then, that by framing climate change in a particular way, highlighting some aspects and obscuring others, communication can and does influence how this crisis is understood and how it is ultimately addressed (Nerlich et al., 2010). As a result, sustainability in climate-vulnerable urban areas is not simply a set of pre-existing best practices and agendas but a highly politicised and contested issue, where sustainability solutions and responses are strongly influenced by the current socio-political landscape and the way strategic actors communicate about climate change to urban publics (Potter, 2020). In other words, in these climate-vulnerable areas, economic and political interests may lead to the climate crisis being strategically framed as a less serious problem than it is, with the result that sustainability solutions are postponed or disregarded in order to maintain urban life as usual.

The Italian city of Venice is an interesting case in this respect. Indeed, the city is one of the most famous and iconic tourist destinations in the world, with an estimated 23 million visitors walking its streets every year (Bertocchi & Visentin, 2019). The importance of tourism interests in recent decades has slowly but inexorably eroded the social fabric of the city, where overcrowding, high housing costs, loss of quality of life and cultural heritage have driven residents away, reducing the population to a third of what it was in the early 1960s (Salerno, 2020). In the same vein, recent local administrations have reduced funding for awareness-raising initiatives, environmental education and climate change research. In 2015, for example, they closed down the Lagoon and Territory Observatory, an important municipal service that had been collecting scientific and environmental data on the Venice Lagoon for years. Similarly, a series of seminars and educational activities aimed at informing the public about the lagoon environment and climate change in Venice were discontinued (www.irpimedia.irpi.eu). These policies are in stark contrast to scientific data and recommendations, which instead point to the vulnerability of this urban area, where periodic flooding, usually referred to as *acqua alta* and considered an inescapable part of urban life, has increased significantly in the second half of the twentieth century due to climate change (Umgiesser, 2020). As an icon of popular culture, the prototype of the romantic city, the fate of Venice in the era of climate change is now under close global scrutiny. Indeed, widespread interest in the future of the city on water is fuelling public discourse, with national and international media speculating on its ability to stay afloat in the crucial years ahead. In this context, the local government has recently challenged such doomsday scenarios and apocalyptic narratives, which have a tangible impact on tourist flows and international investment, threatening the city's economic survival. In order to protect the city of Venice from the relentless rise of sea level, while at the same time preserving it as the world-famous and attractive tourist destination that the world knows it to be, Venice Municipality has launched a communication campaign in the summer of 2022 to explain to national and international visitors what floods are and how the city is immediately protecting itself from them (Venezia Unica 2022). Special emphasis is placed on the Experimental Electromechanical Module (MOSE) system, a system of 78 mobile gates positioned in the three inlets of the lagoon, whose function is to isolate the lagoon from the sea during exceptional floods.

In this context, the present study analyses the strategic sustainability communication of Venice Municipality as a way to understand how Venice is constituted as a sustainable city in the face of exponentially more frequent and dangerous flooding events due to climate change. The chapter proceeds as follows: it briefly presents previous studies on strategic sustainability communication and the communication models underlying them. It then introduces a constitutive approach to communication through which strategic sustainability communication analyses can be rethought and extended. Next, it uses such a constitutive approach to analyse a series of communicative events to show how they contribute to the constitution of a climate-vulnerable city into a sustainable place. The chapter concludes with a discussion of the findings and a presentation of the contributions to strategic sustainability communication research.

2 Past Studies on Strategic Sustainability Communication

As a consequence of the fact that institutions and organisations are increasingly expected to follow sustainability standards and to openly inform customers and stakeholders about them, scholars recognised the importance of understanding how to better communicate sustainability practices and agendas to the public and stakeholders (Christensen et al., 2013, 2017; Genç, 2017). In this context, sustainability communication analyses that adopt a strategic communication perspective began to flourish (Weder et al., 2019; Weder & Eriksen, 2023). Most studies of strategic sustainability communication apply an instrumental view of strategic communication, where they tend to focus on who says what, through what medium, for what purpose and how this message is received (Lasswell, 1948). As such, studies most often examine how organisations communicate strategically about their sustainability practices (i.e. CSR) through an analysis of the use of metaphors, images, slogans and frames, especially in sustainability reports. For example, Jaworska (2018) argues that the communication strategies used by oil companies to position themselves within the climate discourse in the early 2000s, through the use of verbs such as 'combat', 'tackle' or 'address', were 'predominantly formulated as future goals' and as 'references to other stakeholders' (207), in this way preventing radical changes to be performed and allowing these companies to continue 'to a large extent, with business as usual' (208). As a result, these analyses tend to understand sustainability as a fixed state of affairs that an organisation can simply represent in and through communication, in a strategic way, according to calculated sets of objectives (Schoeneborn & Tritting, 2013; Schoeneborn et al., 2020). As such, these analyses fall short of providing a deeper and more holistic understanding of the role of this practice in creating sustainable organisations and places. In other words, because current studies of strategic sustainability communication see it merely as a purposeful act of informing about sustainability in order to achieve certain kinds of effects, these studies inevitably limit what sustainability is and how it is practiced to the rational intention of strategic actors. However, they overlook the fact that strategic sustainability communication, even when strategically crafted, rarely achieves only what each sender intends. This is because, while strategic communication certainly implies strategic intent, '[t]his, however, does not preclude the possibility that communication is dialogical, polyphonic or even subversive, and that receivers are able to play powerful roles in shaping the outcomes of communication' (Christensen & Christensen, 2023: 38).

An analysis of strategic sustainability communication should therefore not see it merely as a purposive act of transmitting information about sustainability, as this would only understand it as an unlocated practice stripped of its socio-cultural understandings. On the contrary, it should be understood as a dialogical process that is highly dependent on the context in which it is received and understood, as an intentional act of communication with unintended consequences. In this sense, what sustainability and sustainable places are depends not only on intentional acts of communication but also on broader and more complex communicative dynamics. That is, the interplay between strategic sustainability communication and the specificities of the context in which it takes place. Indeed, sustainability is not only a phenomenon to be communicated about but also an ideological, cultural and symbolic issue that takes on different contested meanings depending on the context of communication. Conversely, then, as Carvalho et al. (2017) suggest, there is a need 'for an analytical shift by…looking at communication as a constitutive practice, rather than as a matter of transmission of messages' (124). Strategic sustainability communication should also be seen as a practice that not only purposefully informs about sustainability practices and agendas but also constitutes what sustainability is and how it is practiced in particular socio-cultural settings. As Schoeneborn et al. (2020) argue, constitutive views of communication are those in which communication is attributed a role in 'creating, maintaining and transforming CSR practices'

(1). As such, strategic sustainability communication, such as CSR communication, does not simply inform about CSR practices, but 'in a recursive logic ... CSR communication also works back on the CSR practices that gave rise to the communication in the first place' (12). In other words, in contrast to a more linear understanding in which communication is seen merely as a way of talking about pre-existing sets of practices, constitutive analyses consider not only how sustainability is talked about but also how sustainability communication shapes and constitutes these sets of practices in the first place. The next section will elaborate on this theoretical proposition.

3 Strategic Sustainability Communication as a Constitutive Practice

To circumvent the perceived shortcomings of strategic sustainability communication analyses highlighted above, this chapter argues that strategic sustainability communication research should broaden its approach to communication from a representational to a constitutive one. A constitutive approach relies on an understanding of strategic communication as doing something, namely driving social practices and subsequently constituting specific states of affairs (Christensen et al., 2013, 2017; Christensen & Christensen, 2023; Falkheimer & Heide, 2018, 2022). Indeed, words are not simply 'just words', as they often imply the performance of specific practices in accordance with those words (e.g. 'Can you pass me the salt, please?' is a communication that hopefully gets someone else to pass you the salt in accordance with a pre-existing social convention). In this line of thinking, van Ruler (2018) suggests that strategic communication should be thought of as 'an ongoing process that develops as it occurs and cannot be predicted' (371), 'a continuous, reflective learning loop; (378). As a result, strategy and organisations are seen as outcomes of communication processes rather than pre-existing phenomena. On the one hand, strategy becomes something that organisations do rather than something they have, an 'amalgam of continuous communication processes in order to build, define, present, realize, and rebuild strategy' (378, emphasis erased). On the other hand, organisations are seen as happening simultaneously with communication, with communication being the process through which any form of organisation can be said to rest (Weick, 1995; Taylor & Van Every, 2000).

A constitutive understanding of strategic sustainability communication implies that communication continuously constitutes organisations as sustainable (Christensen et al., 2013, 2017). As such, different communicative events involving different organizational actors beyond management or the communication department can be seen as constituting what sustainability means in specific settings and thus maintaining organizations as sustainable. In the words of Winkler and Schoeneborn (2023):

> this [processual understanding of strategic communication] requires broadening the analytic perspective on mundane, mostly ambiguous, everyday communicative practices, various internal and external actors, and tools-in-use in order to understand the complex logics that ultimately define what counts as strategic communication in an organizational setting. (124)

Thus, what we count as strategic communication is a cultural and situated process that involves multiple communicative practices, both intentional in their framing of the meaning of climate change and unintentional but which can be valuable in their potential to become strategically relevant to these ends. Winkel and Schoeneborn (2023) refer to these as 'emergent instances' of strategic communication, namely those 'unintended, hence non-purposive, communication which, however, can become influential' (124) in sustaining organisational forms. Thus, for communication to be considered strategic, there does not necessarily need to be a planned and measured intention behind it. Indeed, as Christensen and Christensen (2023) argue '[t]here is, of course, more to strategic communication than what the sender intends. One thing is the overall purpose of such communication; another is what the communication actually accomplishes' (33). In other words, there are many other influences that make communication strategic outside and beyond strategic plans. This is not to say that strategic communication cannot be intentional, planned and measured. Sometimes, it will be. What it does mean is that strategic communication should be understood more as a process in which a wide range of intended but also unintended, informal and mundane communicative events can be said to contribute to the constitution and maintenance of particular organisational forms. Understood in this way, strategic communication becomes more of a continuous communicative process of strategis*ing* (Gulbrandsen & Just, 2016, 2023), which does not necessarily require only a specific initial intention, as it will always end up with unintended consequences.

Content analyses of strategic sustainability communication, such as the one presented above, tend to view strategic communication not as a constitutive practice but as a representational one, in the sense that they assume that strategic communication was planned to communicate about pre-existing sustainability values and practices in order to achieve some predetermined effects and that it either succeeded or failed in doing so. On the contrary, according to a constitutive approach to strategic communication, the focus of strategic sustainability communication analyses is not only on how organisations strategically use language to represent their sustainable practices but also on how communication as a dialogical practice is what stimulates and achieves the performance of sustainability practices in the first place. These analyses consider that the meaning of sustainability is shaped not only by the strategic intentions of actors but also by the recursive and unplanned communication of multiple organisational actors.

In what follows, this chapter adopts a constitutive approach to strategic sustainability communication (Weder & Eriksen, 2023) to examine how Venice, as a climate-vulnerable coastal city, is constituted as a sustainable place amidst exponentially more frequent and dangerous flooding events. The empirical material was collected during an ethnographic study as part of the author's doctoral research in

2021 and 2022. After collection, the analysis followed the method of discourse analysis in organisational communication research (Alvesson & Kärreman, 2000; Cooren, 2015), where the empirical materials were treated as textual instances of strategic sustainability communication with the potential to shed light on what sustainability means and how Venice is constituted as a sustainable place in it. Thus, in addition to identifying how sustainability is strategically communicated through the design of the communication campaign, the focus of the analytical endeavour was to identify how sustainability is also produced in and through various communication events by other members of the Venice municipality, such as a local politician in charge of the city's planning and environment.

4 Venetian Sustainability as a Specific Kind of Sustainability

I begin the analysis by examining the communication campaign launched by Venice Municipality to inform tourists about floods. The first excerpt I present here is a transcription of the narrator's words in the first of the six videos released:

> Venice is a city built on water, which allows its citizens and visitors to live in symbiosis with the natural flow of the tide. The cyclical movement of the water surface can cause some of the roads to flood, resulting in the famous *high water* [acqua alta].
> (*Acqua alta – What is High Water?,* Venezia Unica, 2022)

From the very beginning of the communication campaign, flooding is presented as being caused by 'the natural flow of the tide', which sometimes floods, giving rise to the phenomenon of '*acqua alta*'. Through the name '*acqua alta*', floods are presented as part of the normal tidal movements of Venice, which in turn are the cause of the fact that 'Venice is a city built on water' and therefore of the fact that there is an inevitable interdependence between water and land. In this way, floods are turned into a positive experience for tourists, since the experience of *acqua alta* allows them, like the citizens of Venice, to live 'in symbiosis' with the watery environment. This excerpt therefore shows how the communication campaign, from the very beginning, emphasised the positive aspects of living in harmony with nature. This harmonious relationship between humans and nature is similar to how the concept of sustainability is usually understood in the global discourse, namely as a set of practices, ideals and agendas aimed at sustaining a harmonious relationship between all forms of life in the midst of climate change (Weder, 2021). In other words, in this particular geohistorical era in which humans are said to have altered ecological equilibria, the concept of sustainability aims to restore a mindset in which human practices should only exist in harmony with the natural world.

But in Venice, where the harmonious relationship between water and land dates back to the city's birth ('Venice is a city built on water', says the narrator), the concept of sustainability takes on a new meaning, as the next excerpt from an interview with a local politician shows. When I ask him about Venice's sustainability plans for the coming future, he replies:

Politician: I am of the opinion that this city should continue... to do what it has done for 1600 years.
Monica: In harmony with nature.
Politician: In harmony with nature... no! In contrast... look, Venice opposes nature... the history of Venice, perhaps we forget it today, had always opposed nature because if it had let nature take its course, the lagoon would not exist!

Politician 1, Interview, Nov. 4, 2021

In this excerpt, while I am tempted to think of sustainability as implying harmony with nature, the politician urges me to reconsider such an assumption when applied to the context of Venice. In fact, in Venice, he says, harmony with nature is a constant endeavour to 'oppose' nature. In other words, unlike the way sustainability is understood in the global discourse as the search for a harmonious relationship between all forms of life on the planet, in Venice, sustainability means the opposite, because, as the politician argues: 'if Venice had let nature take its course', the lagoon would have been silted up. This misunderstanding is repeated later in the same interview when the politician argues:

Politician: From an environmental perspective, I must say that this is a city that has always, I must say, in 1600 years of history, left a culture of approaching the environment...
Monica: Yes, in relation
Politician: No, it's not a matter of relation, it's a matter of survival! Eight rivers used to flow into the lagoon... since the 1400s, six rivers flow into it because two rivers were diverted to the open sea, no longer into the lagoon. The big problem with rivers is the transport of sediments... it means that... as the seabed level rises, emerged lands start to appear. … It means that Venice would have been one with the mainland. Instead, the city's defensive system was to create, like a medieval castle, a moat with water... but six rivers still flow today...

Politician 1, Interview, Nov. 4, 2021

At the beginning of this conversation, the politician points out that sustainability in Venice, from an environmental perspective, is a specific type of relationship with the environment: that of survival. So, while I once again assume the harmony inherent in the usual sustainability discourse, here the local politician instead portrays nature (i.e. water) as an unruly element that stubbornly and persistently threatens Venice's survival. To explain this, he refers to past interventions to divert the lagoon's main tributaries out of the lagoon to prevent sedimentation, as proof that Venice needs the lagoon around it 'like a medieval castle' in order to 'defend itself'. In this way, over the centuries, Venice has protected itself from the water by means of technological interventions that have allowed the lagoon, and therefore Venice, to exist. This suggests that while the global sustainability discourse is known to involve a high degree of trust in technological solutions (Peeters et al., 2016), in Venice, according to my interviewee, this trust in technology has always existed, as technological interventions are here said to be the basis of Venice's very existence.

In relation to this latter point, the next excerpt shows how the politician responded once I asked him to talk about the MOSE system, namely the majestic engineering project designed to protect the lagoon against extreme flooding.

> **Politician**: ... the MOSE saves Venice, and the MOSE is one of those works that helps Venice to remain preserved, [but] I always say, be careful, the water issue should not be addressed in a single way. Venice, everyone thinks about seawater, the MOSE is a barrier against the sea, it raises its hands in front of the sea, and I remind you that there are six rivers behind that flow into the lagoon... we must not forget what the water carries.
>
> *Politician 1*, Interview, Nov. 4, 2021

When I ask him about the MOSE system, the politician argues that 'the MOSE saves Venice', but that this technology alone is not enough to solve what he calls 'the water issue'. Indeed, while 'everybody thinks of seawater' when it comes to Venice's sustainability problems, this politician stresses the need to remember that 'there are six rivers behind our back' that could bring in debris and possibly silt up the lagoon. As in the previous excerpt, water is an unruly element, stubbornly and persistently threatening Venice's survival from both the front (the sea) and the back (the rivers). Here, the MOSE system becomes the 'saviour' ('MOSE saves Venice'), turning water and the debries it carries into a controlled natural element. Interestingly, in the context of Venice, sustainability is a matter of survival, because it cannot be limited to a single issue, but is an ongoing, constant battle that Venice has fought on several fronts since day one against the chronic enemy of water. Therefore, in contrast to a global sustainability discourse that places faith in technological interventions with the specific aim of combating the effects of climate change, in Venice, technological interventions against water, such as the MOSE system, are seen as the logical continuation of a series of technological interventions carried out over the past centuries to keep Venice in a sustainable relationship with water. The technology of the MOSE system, in this interview, is not necessarily seen as an intervention against the current ecological challenges of climate change, against flooding due to rising sea levels, but against the chronic enemy of water. As such, in a continuation of this centennial story, the trust in technology is reaffirmed here in MOSE and in the future, as he calls for more investment in science and technology, given the magnitude of the battle ahead.

If we also look at the official campaign presented at the beginning of this analytical section, we can see how, by mentioning the MOSE system, the meaning of floods is strategically adapted to the current climate challenge. Indeed, in the fifth video, the narrator says:

> Flooding in the lower part of the city has become a recurrent phenomenon, especially in autumn and winter. Unfortunately, over the last 20 years, every two or three years, the high-water level has reached or exceeded 140 centimetres, thereby flooding at least 45% of the streets. The MOSE system is capable of blocking the temporary entry of sea water into the lagoon, consequently preventing *acqua alta* from occurring in the event of intense phenomena.
>
> (Acqua Alta – Mose, What Is It?, Venezia Unica, 2022)

Whereas in the previous video analysed above, the flooding of the city was strategically named *acqua alta* in order to portray Venice as being in harmony with the watery world around it, in this video, there is an explicit reference to 'floods' which are no longer framed as the result of 'the natural flow of the tide' but have instead 'unfortunately become a recurrent phenomenon'. Whereas in the previous video, the recursiveness of the tide was a positive sign of Venice living 'in symbiosis' with the watery world, this is no longer the case, and the recursive character of the tide is now understood as the act of an enemy that has become chronic. In fact, *acqua alta* is here an 'intense phenomenon' that requires the activation of the MOSE system in order to 'prevent it'. Here, then, water is turned into an enemy in the sense that MOSE can be appointed as a saviour while remaining the harmless and authentic phenomenon of *acqua alta* in all other cases. In other words, such a harmonious relationship is only possible by making *acqua alta* a phenomenon that is kept harmless only by the functioning of the MOSE system. In this way, the communication campaign is still able to present MOSE as a protection against the 'cyclical movement of the tide', which is, unfortunately, becoming more frequent and not against floods caused by climate change.

5 Discussion and Conclusion

Research on strategic sustainability communication most often adopts an instrumental view of communication to analyse this practice. Such an approach implies that strategic sustainability communication merely represents pre-existing sustainability meanings and practices adopted by organisations. This research focus has been criticised for leaving unexplored the formative potential of strategic sustainability communication (Weder & Eriksen, 2023), i.e. the potential of such a practice not only to inform about sustainability but also and above all to constitute what sustainability means and what sustainable practices are. In this context, the aim of this chapter was to propose a constitutive approach to strategic sustainability communication by looking at how Venice as a sustainable place is constituted in and through the strategic sustainability communication of Venice Municipality. Thus, by considering not only a designed communication campaign to strategically communicate about floods but also an interview with a local politician, the analysis showed that Venice municipality's communication does not only inform about what sustainability means in Venice but also constitutes a specific way in which Venice is made sustainable by negotiating the meaning of sustainability with the wider social sphere. In other words, by analysing both intentional and emergent instances of strategic communication (Winkler & Schoeneborn, 2023), the analysis showed that the meaning of sustainability in Venice is negotiated against the backdrop of a dominant view of sustainability proposed by the global discourse. Indeed, while the concept of sustainability usually implies the search for a harmonious balance between all forms of life on Earth, with the consequent downgrading of human use of natural and animal resources, a sustainable Venice is constituted through an opposing view of sustainability. In Venice, in fact, sustainability has been shown to mean continuing the quest to confront the chronic enemy of water. In Venice, sustainability is not a concept used specifically against climate change but against the recalcitrant element of water. However, precisely because of the existence of an alternative meaning of sustainability, sustainability in Venice turns out to imply the continuous control of water through hard forms of technological intervention. Contrary to a view of sustainability that sees technology as a promising tool for achieving a harmonious equilibrium between humans and nature, in Venice, these interventions are used to constantly contrast nature.

Referring only to an analysis of the communication campaign created by Venice Municipality to inform tourists about floods would have limited the understanding of its strategic sustainability communication. In fact, it would have missed the fact that sustainability and sustainable Venice emerge from the negotiation of such meaning against different, contrasting viewpoints. In this case, the global discourse of sustainability that I propose in the interview with the local politician. This negotiation wouldn't have been possible with a content analysis of the communication campaign alone. In fact, such an analysis would only have shown the way in which the municipality changed the meaning of floods from a harmless phenomenon (*acqua alta*) to a harmful one in order to communicate the existence of the MOSE system. However, it would not have been able to show that such a shift responded to a broader social field in which flooding phenomena are also understood as effects of climate change. Thus, a constitutive approach to strategic sustainability communication is useful to examine how the constitution of a city as sustainable is an inherently political endeavour to negotiate the specificity of the setting of Venice against the backdrop of widespread conceptualisations of sustainability and technological interventions.

References

Adams, P. C. (2022). 'Not a Big Climate Change Guy': Semiotic gradients and climate discourse. *Environmental Communication, 16*(1), 79–91.

Alvesson, M., & Kärreman, D. (2000). Varieties of discourse: On the study of organizations through discourse analysis. *Human Relations, 53*(9), 1125–1149.

Bertocchi, D., & Visentin, F. (2019). "The Overwhelmed City": Physical and social over-capacities of global tourism in Venice. *Sustainability, 11*, 6937.

Carvalho, A., van Wessel, M., & Maeseele, P. (2017). Communication practices and political engagement with climate change: A research agenda. *Environmental Communication, 11*(1), 122–135.

Christensen, L. T., Morsing, M., & Thyssen, O. (2013). CSR as aspirational talk. *Organization, 20*(3), 372–393.

Christensen, L. T., Morsing, M., & Thyssen, O. (2017). License to critique: A communication perspective on sustainability standards. *Business Ethics Quarterly, 27*, 239–262.

Christensen, E., & Christensen, L. T. (2023). The saying and the doing: When communication is strategic. In J. Falkheimer & M. Heide (Eds.), *The research handbook of strategic communication*. Edward Elgar.

Cooren, F. (2015). *Organizational discourse: Communication and constitution*. Polity Press.

Falkheimer, J., & Heide, M. (2018). *Strategic communication: An introduction*. Routledge.

Falkheimer, J., & Heide, M. (2022). What is strategic communication? In *Strategic communication: An introduction to theory and global practice*. Routledge.

Genç, R. (2017). The importance of communication in sustainability & sustainable strategies. *Procedia Manufacturing, 8*, 511–516.

Greenberg, J., Knight, G., & Westersund, E. (2011). Spinning climate change: Corporate and NGO public relations strategies in Canada and the United States. *International Communication Gazette, 73*(1–2), 65–82.

Gulbrandsen, I. T., & Just, S. N. (2016). *Strategizing communication: Theory and practice*. Samfundslitteratur.

Gulbrandsen, I. T., & Just, S. N. (2023). From strategy to strategizing. In J. Falkheimer & M. Heide (Eds.), *The research handbook of strategic communication*. Edward Elgar.

Ihlen, Ø., & Roper, J. (2014). Corporate reports on sustainability and sustainable development: 'We Have Arrived'. *Sustainable Development, 22*, 42–51.

Jaworska, S. (2018). Change but no climate change: Discourses of climate change in corporate social responsibility reporting in the oil industry. *International Journal of Business Communication, 55*(2), 194–219. https://doi.org/10.1177/2329488417753951

Lasswell, H. D. (1948). The structure and function of communication in society. In L. Bryson (Ed.), *The communication of ideas*. Harper and Brothers.

Nerlich, B., Koteyko, N., & Brown, B. (2010). Theory and language of climate change communication. *Wiley Interdisciplinary Reviews: Climate Change, 1*(1), 97–110.

Peeters, P., Higham, J., Kutzner, D., Cohen, S., & Gössling, S. (2016). Are technology myths stalling aviation climate policy? *Transportation Research Part D, 44*, 30–42.

Potter, E. (2020). Contesting imaginaries in the Australian city: Urban planning, public storytelling and the implications for climate change. *Urban Studies, 57*(7), 1536–1552.

Roper, J., Ganesh, S., & Zorn, T. E. (2016). Doubt, delay, and discourse: Skeptics' strategies to politicize climate change. *Science Communication, 38*(6), 776–799.

Salerno, G. M. (2020). *Per una critica dell'economia turistica. Venezia tra museificazione e mercificazione*. Quodlibet.

Schoeneborn, D., & Trittin, H. (2013). Transcending transmission. Towards a constitutive perspective on CSR communication. *Corporate Communications: An International Journal, 18*(2), 193–211.

Schoeneborn, D., Morsing, M., & Crane, A. (2020). Formative perspectives on the relation between CSR communication and CSR practices: Pathways for walking, talking, and t(w)alking. *Business and Society, 59*(1), 5–33.

Taylor, J. R., & Van Every, E. J. (2000). *The emergent organization: Communication as its site and surface*. Lawrence Erlbaum Associates.

Umgiesser, G. (2020). The impact of operating the mobile barriers in Venice (MOSE) under climate change. *Journal for Nature Conservation, 54*, 125783.

Van Ruler, B. (2018). Communication theory: An underrated pillar on which strategic communication rests. *International Journal of Strategic Communication, 12*(4), 367–381.

Venezia Unica (2022). High water campaign. Available at: https://www.veneziaunica.it/en/content/high-water

Weick, K. E. (1995). *Sensemaking in organizations*. Sage.

Weder, F., Lemke, S., & Tungarat, A. (2019). (Re)storying sustainability: The use of story cubes in narrative inquiries to understand individual perceptions of sustainability. *Sustainability, 11*(19), 5264.

Weder, F. (2021). Sustainability as master frame of the future? Potency and limits of sustainability as normative framework in corporate, political and NGO communication. In *The sustainability communication reader: A reflective compendium* (pp. 103–119).

Weder, F., & Eriksen, M. (2023). *CSR communication and cultures of sustainability*. The University of Queensland.

Winkler, P., & Schoeneborn, D. (2023). Understandings of organization in strategic communication research. In J. Falkheimer & M. Heide (Eds.), *The research handbook of strategic communication*. Edward Elgar.

Further Reading

Baldacci, C., Bassi, S., De Capitani, L., & Omodeo, P. D. (2020). *Venice and the Anthropocene. An ecocritical guide*. Wetlands Publishing.

Cavallo, F. L., Vallerani, F., & Visentin, F. (2021). Heart of wetness: Living, narrating, and representing ancient memories and new water rhythms in the Venetian Lagoon. *Shima Journal, 15*(1). https://doi.org/10.21463/shima.122

Monica Porzionato is a Researcher at the Department of Communication, Lund University, Sweden. She holds a BA in communication from the University of Padova, an MA in semiotics from the University of Bologna and an MA in women's and gender studies from the University of Utrecht. Her research interests lie at the intersection of communication and urban studies, specifically the communicative making of sustainable urban places. Her dissertation explores the possibility of an affective approach to the strategic communication of climate change in the tourist setting of Venice, investigating how the communication of flooding events, becoming more dangerous and frequent due to climate change, implies the strategic maintenance of affective atmospheres, thus of seemingly ordinary ways in which Venice should look and feel like in the midst of climate change.

Creating Lines to Land: The Country-of-Origin Effect from a Sustainability Communication Perspective

Florentina Höhs

Abstract Over the past decade, sustainability has received rising attention in politics, the business world and throughout other societal systems. Much more, today, sustainability-related policies and regulations standardize sustainability-related behavior and sustainability communication across all sectors. The emerging field of sustainability communication seeks to understand all areas of promotional culture, from public relations to marketing. However, while strategic communication of and about sustainability predominantly focuses on corporate reporting, marketing studies deal with green marketing (and related 'washing')—both fields without acknowledging cultural and local impacts on and in strategic sustainability communication. At the intersection between PR and marketing, the study at hand investigates customer perceptions of products related to specific locations. The country-of-origin theory states that the image people have about a certain country might affect their behavior and attitude toward products from that country and their perception of how 'sustainable' the product is. With a concrete case, delving into the perception and evaluation of Austrian white wine among customers in Sweden, the country-of-origin effect is explored in regard to its potential in sustainability marketing. The chapter also offers a country-of-origin-related example for creating local proximity and ideas for future strategic sustainability communication that takes strategies to create cultural and social local/regional proximity into account.

1 Introduction

Sustainability as a global concept has been established in society at the latest with the sustainability development goals (SDGs, UN, 2024). Also, it has become a corporate standard to apply environmentally sustainable governance (ESG, European

---

F. Höhs (✉)
Department of Business Communication, WU Vienna University of Business and Economics, Vienna, Austria
e-mail: Florentina.Hoehs@wu.ac.at

F. Weder (ed.), *Strategic Sustainability Communication*, CSR, Sustainability, Ethics & Governance, https://doi.org/10.1007/978-3-031-89486-2_16

Commission, 2024) criteria. Apart from the global standardization of sustainability, it is also increasingly addressed at a regional level. 'Sustainable cities' and 'climate neutral regions' are concepts applied across the globe leading to infrastructure projects, initiatives and projects in all areas and related cultural change. Sustainability-related policies and regulations, projects and initiatives and the mentioned cultural change are supported by 'strategic sustainability communication' which includes informing, contributing and debating all sustainability topics, whether this is environmental protection, climate changes or social acceptance of renewable energy sources and thus a large variety of issues, from fair traded fruits and vegetables to a new bike lane. Furthermore, sustainability communication aims to create a social and critical debate about sustainability (Newig et al., 2013; Voci & Karmasin, 2023; Weder et al., 2019) and to thereby enhance ecological, economic and social aspects of life (Ziemann, 2011).

From a strategic communication perspective, sustainability communication can be approached and conceptualized on three levels: an individual, organizational and societal level. The chapter at hand explores the dynamics of strategic sustainability communication on an organizational level, where sustainability communication seeks to cope with and impact the constantly changing organizational environment by cultivating new values in the organization and in the stakeholder relationships (Weder, 2023). As a discipline with unspecific boundaries, (strategic) sustainability communication does not take place solely in organizations but also between organizations and their environment (Weder, 2023; Weder et al., 2021).

Recent literature distinguishes between three dimensions of sustainability communication: communication of sustainability, communication about sustainability and communication for sustainability (Adomßent & Godemann, 2011; Newig et al., 2013; Weder, 2023). Communication of sustainability follows a pragmatic approach and thus includes a functional understanding of communication (Weder, 2023). Thereby, a one-way transmission of information related to specific sustainability issues or policies seeks to reach a certain (strategic) goal (Golob et al., 2023). This also includes 'informing and educating people about certain sustainable products' (Weder, 2023, p. 228).

Furthermore, communication about sustainability acknowledges the discursive character of communication and includes political debates and public forums such as conferences and aims at raising political awareness and initializing communication on sustainability between organizations and the societal subsystems. Communication for sustainability as the third dimension can be understood as a call to action including participative-involving approaches, initiating sense and meaning-making processes, engagement and empowerment in a specific stakeholder group or audience segment for example by initiating media initiatives with a sustainability focus or creating own solutions for societal problems (Golob et al., 2023; Newig et al., 2013; Weder, 2023).

That being said, sustainability communication is not only a research field within the discipline of communication or sustainability studies; much more, concepts and approaches to sustainability from consumer research, environmental communication, advertising and marketing need to be included to better understand strategic sustainability communication from an organizational perspective (Golob et al., 2023; Weder, 2023). The chapter at hand brings in a marketing perspective and discusses the potential of theoretical concepts from marketing to fill gaps in sustainability communication as an emerging research area. From a marketing perspective, it can be stated that the main goal of sustainability communication is to achieve the 'marketing goal' of selling sustainable products (Prexl, 2010). When sustainability communication and the marketing perspective are brought together, terms such as green advertising, green marketing, environmental marketing communications and sustainability marketing come into play and drive the conceptual thinking as well as research approaches, while the mentioned terms are often blurry, overlapping or used simultaneously. As a common ground, all marketing approaches contributing to sustainability communication deal predominantly with environmental and/or ecological and not so much with social issues (Golob et al., 2023). This might be related to the lack of clear conceptualizations and definitions, as the sustainability term is hard to grasp (Golob et al., 2023; Wang et al., 2020). Also, still there is a lack of generic theoretical concepts in the field of strategic sustainability communication (Wang et al., 2020). With an empirical study on a strategic sustainability marketing strategy, the chapter does not only complement existing work in the field of strategic sustainability communication by bringing in some marketing thinking; much more, it shows a specific gap in strategic sustainability research: the lack of recognition of cultural differences in understanding sustainability as well as the potential of linking sustainable projects, services or products to a specific region or locality and communicating sustainability by creating a connection and a notion of proximity between the customer and the product or project.

The chapter presents a study where the country-of-origin effect has been explored, which describes that a product produced in a certain country might be perceived in a certain way only based on the image customers have of that country. This perception then also leads to certain (purchase) attitudes (Koschate-Fischer et al., 2012; Roth & Diamantopoulos, 2009; Schooler, 1965; Usunier, 2006). The study explores at first the country-of-origin effect for Austrian white wine in Sweden. Second, the study examines the influence of the country of origin on product perception and consequently on the purchase decision of Austrian white wine among Swedish consumers (B2C). The contribution of this chapter lies in improving the understanding of the country of origin in consumer behavior, with a specific focus on the perception of the product from a strategic sustainability perspective; going beyond the specific case of Austrian wine, this study investigates the customer perception of *resources, which are related to specific locations*. Thereby, the question of whether resources are attached to certain regions and/or locations and the transfer of this connection to a social representation of sustainability is raised.

The chapter is structured as follows: first, an overview of the relevant country-of-origin research is given to then discuss its potential for strategic sustainability communication from a marketing perspective. Second, the study is introduced with a brief explanation of the methodology and findings. Third, in the actual discussion, implications and future research ideas are derived from the findings, and the country-of-origin study is reflected from a critical sustainability communication perspective.

2 Literature Review

2.1 *Country-of-Origin Effect*

To explain the theory of the country-of-origin effect properly, the country image must be looked at in the first place since the country of origin builds upon the country image (Kuzmany, 2008; Roth & Diamantopoulos, 2009). A country image is the total of all descriptive, inferential and informational beliefs one has about a particular country (Martin & Eroglu, 1993). In general, it refers to the idea the consumer has about products, countries, services or companies (Kroeber-Riel & Gröppel-Klein, 2013). It can also be described as an idea associated with a concept, which underlies certain emotional backgrounds and conceptual ideas (Nagashima, 1970). From a marketing perspective, the country image is explained as the beliefs, ideas and impressions a person has on an object (Kotler, 1991; Papadopoulos & Heslop, 2014). Importantly, each individual's perception is different, which means that the image an individual has about an object also varies from individual to individual (Papadopoulos & Heslop, 2014).

The psychological country-of-origin effect (COO effect) describes how attitudes, perceptions and purchasing decisions are influenced by the country of origin of a certain product (Johnson et al., 2016). Initially investigated by Schooler (1965), the country-of-origin effect developed over the years from the product country image to the image a person has of the country of origin, which then can have a relevant influence on the purchase decision (Bruhn & Homburg, 2004). The definition of the country of origin itself depends: the determination ranges from the country of origin as the country in which the product was manufactured (Liefeld, 2004) to the country of origin as the location of the corporate's headquarters responsible for marketing the product (Ozsomer et al., 1991). A major finding in the country-of-origin research is that the origin of a product acts as a quality attribute and thus influences consumers' purchase decisions (Han, 1989; Roth & Diamantopoulos, 2009). The country-of-origin information for assessing a product's quality seems to be even more important than price and any brand information. Especially, when customers don't have much product information, the country of origin is seen as a favorable decision tool (Wall et al., 1991).

The country-of-origin effect does not only depict a major theoretical concept but also influences parts of the typical customer journey, e.g. how a product is perceived, the customer's product evaluation, the purchase intention and the willingness to pay (Brookes, 1993; Felzensztein et al., 2004; Moon & Jain, 2002). Beyond being a quality cue, the country of origin also has a symbolic and emotional meaning for customers (Hong & Wyer Jr, 1989). For example, the country of origin can lead to the association of a product with characteristics such as authenticity, status and exoticness (Batra et al., 2000) and is also related to identity, pride and autobiographic aspects (Verlegh & Steenkamp, 1999). In addition, when a product's country of origin and national identity are in a relation, this can lead to emotional identification with a certain brand and/or product (Fournier, 1998). Also, the country of origin has a decisive influence on the expectation of a product for both experts and non-experts (Chiou, 2003). The country-of-origin's influence on consumer decision-making even increases when there is little to no other information about the product (Al-Sulaiti & Baker, 1998).

Apart from the country-of-origin's influence on product perception in general, research also confirmed that the country of origin influences consumers' product evaluation when they consider buying a product (Brookes, 1993; Felzensztein et al., 2004; Okechuku and Onyemah, 1999; Supanvanij & Amine, 2000). A perceived stronger country-of-origin image can lead to a more positive product evaluation and a more positive attitude toward the product (Carter, 2020). Also, the perceived product quality seems to be significantly dependent on the product's country of origin (Hui & Zhou, 2002), like products from developed countries are rated more positively (Lee et al., 2001). Therefore, it doesn't matter whether the consumers are in a developed or a developing country (Zain & Yasin, 1997). Also, income could be proven as a major attribute for how a product is evaluated: in countries with lower average incomes, inhabitants tend to be less educated and hence evaluate a product differently than consumers from countries with higher incomes per capita, who tend to be better educated on average (Insch & McBride, 2004).

To summarize, various pieces of information contribute to a customer's evaluation of a certain product—hereby, the country of origin is an extrinsic product criterium, such as price, brand name and retailer reputation (Teas & Agarwal, 2000; Thorelli et al., 1989). To better understand how the link between a product and individual associations of the customer are created, dimensions of these associations and therefore attitudes and evaluation of the product need to be clarified, which will be done in the following subchapter.

2.2 *Country-of-Origin Effect and Its Attitude Dimensions*

As long-term thought patterns, which are difficult to change, attitudes are highly relevant for marketing research (Argyriou & Melewar, 2011; Galan Ladero et al., 2015). This is because consumers do not re-evaluate a certain product every time but stick to this long-term thought pattern (Bruhn & Homburg, 2004). Attitudes are a complex psychological phenomena and have various dimensions. Here, the focus is on those that stand, according to research, in relation to the country-of-origin (effect) (Bruhn & Homburg, 2004; Iversen et al., 2022; Pisharodi & Parameswaran, 1992). Accordingly, the country-of-origin effect arises 'by a market country's people's cognition (…), affect (…) and conation (…) regarding the source country' (Pisharodi & Parameswaran, 1992). These categories are strongly related to each other and are partly overlapping during the consumer's process of decision-making. Moreover, the cognitive, affective and conative components of the country of origin have a significant influence on product evaluation and the willingness of potential buyers to buy foreign products and goods (Iversen et al., 2022).

Cognitive Aspects The cognitive dimension of the country image measures the beliefs consumers have about other countries (Maher & Carter, 2011). The whole concept of product-country image is based on people's cognitive beliefs of a certain product that has a certain quality only because the product is produced in a certain country (Baughn & Yaprak, 2014).

Affective Aspects The affective components of the country image measure the emotions and reactions a consumer has about other countries (Maher & Carter, 2011). For instance, affective components arise from direct experiences such as vacations or indirect experiences such as education and mass media consumption (Kock et al., 2019). In terms of the country-of-origin's affective product evaluation, research shows that positive country emotions increase product evaluations and negative emotions decrease product evaluations (Klein et al., 1998; Laroche et al., 2005).

Conative Aspects The conative component of attitude describes the way individuals act, which is in other words their behavior (Baldwin, 2023). More specifically, the conation is the level of interaction with the country's people in the source country (Pisharodi & Parameswaran, 1992). For example, the conative view can lead a person to the behavior of only buying domestically produced products but it can also lead to the avoidance of buying from a certain country a person disapproves (Laroche et al., 2005; Papadopoulos et al., 1990). Often it is a reflection of national pride, which can lead to patriotic behavior (Iversen et al., 2022).

To give an example of the various dimensions in which the connection between a specific product and a location (region, country and cultural area) is emphasized, the following subchapter explores the country-of-origin effect in the wine industry.

2.3 Country-of-Origin Effect on the Wine Industry

The country-of-origin effect hasn't only been examined in general, but there is also research with a focus on the wine industry. Many studies have shown that consumers use the origin of wine as a quality indicator (i.e. Lockshin & Timothy Rhodus, 1993). Consequently, information about the origin of the wine (grapes) is often the basis for making wine-buying decisions (Duhan et al., 1999); therefore, the literature states a country-of-origin effect on wine purchase decisions (Brookes, 1993; Felzensztein et al., 2004). This seems to be quite logical since wine comes mainly from one territory, i.e. mostly from one region or one country and has a short value chain. Particularly, in the case of wines from Europe, historically, the winegrower is usually also the producer and often also the seller of the wine (Balestrini & Gamble, 2006). Additionally, customer sensitivity toward the origin of wine is relatively high (Duhan et al., 1999), and mainly extrinsic cues such as price and country of origin influence the purchases as mentioned above (Lockshin & Timothy Rhodus, 1993). In one study, the wine's country of origin was even the most important cue before the price for Chinese customers (Balestrini & Gamble, 2006). Overall, a typical 'quality indicator' for a purchase decision is the origin of the wine, like the specific, unique vineyard location (Balestrini & Gamble, 2006). In *marketing*, the country of origin can be emphasized when selling wine from a well-known wine country. The country of origin is then, for example, added as visual or textual information on the labels or included in the corporate communication. This can serve as a purchase decision factor for wine consumers when deciding on a wine (Balestrini & Gamble, 2006). Classical wine countries have an advantage in this regard. However, due to globalization and emerging markets, wine-producing countries today can no longer rely on their automatic reputation. Emerging wine countries such as Australia or South America have created new competition (Beverland, 2001). Consequently, established wine countries, such as Italy and France, are at risk of having to increasingly defend their market position. The place of origin also plays a weighty role for these traditional wineries, as it reinforces the country-of-origin effect. It is precisely this more specific, place-based designation of origin that can be used as a competitive advantage through the use of marketing strategies (Felzensztein et al., 2004).

In summary, place-based marketing strategies seem to have the potential to not only directly influence purchase decisions but also stimulate associations and imaginaries of the country that the product is connected to. How places are communicated was explored with a study, which will be briefly introduced in the following, before the country-of-origin communication is discussed as a potential strategy in strategic sustainability communication.

3 Austrian White Wine: Creating Imaginaries and Associations

3.1 Methodological Considerations

While there is a large number of studies investigating the country-of-origin effect in Western countries (Ribberink & Washburn, 2017; Visbal et al., 2017), there is only a little research published in which Austria is investigated in the context of the country-of-origin effect (Diamantopoulos et al., 2020). Also, the country-of-origin effect concerning wine as a product has rarely been studied (Felzensztein et al., 2004). While there is some literature on the perception of established, large wine regions, such as Italy and France, these are not directly comparable to the wine from the little-production country, Austria, discussed in this chapter (McDougall & Lockshin, 2000).

Austrian white wine is a product that is becoming increasingly popular not only in Austria but also internationally. Numerous awards in international wine rankings prove this (Österreich Wein, 2022). The target country investigated is Sweden, which seems to be a suitable sales market for the export of Austrian wine for several reasons: first, Sweden can't grow its grapes over a wide area due to climatic conditions (Jones & Schultz, 2016; Ridder, 2023). Second, Sweden is dependent on the import of wine because they produce their wine only in a limited amount (Rauhut Kompaniets, 2022). Third, statistics show that wine consumption in Sweden has been rising in the last few years (Ridder, 2023).

Based on the literature review and the abovementioned considerations, the following *research questions* emerged:

1. What is the country-of-origin effect for Austrian white wine in Sweden?
2. To what extent does the country-of-origin effect of Austria have an impact on the product perception and evaluation of Austrian white wine among B2C consumers in Sweden?

Based on the research question, the *hypothesis* is developed, which is then empirically investigated.

H: Austria's image of people living in Sweden influences respondents' attitudes toward Austrian wine.

3.2 Conceptual Model

To conduct the study properly, a measurement model, basically composed of the hypothesis' dependent and independent variables, was established. The variables to be tested are retrieved from the constructed hypothesis 'Austria's image of people living in Sweden influences respondents' attitudes toward Austrian wine' and consist of the independent variable 'the image foreign consumers have of Austria' and the dependent variable, 'the participants attitudes toward Austrian wine'. The variables are split into several dimensions, based on the theoretical background to make them measurable (De Tavares Canto Guina & de Moura Engracia Giraldi, 2015; Khanal, 2018; Pisharodi & Parameswaran, 1992; Salam & Chowdhury, 2015). The following conceptual model (Fig. 1) illustrates the relationship between the variables and the respective dimensions.

The independent variable 'country image' is represented by seven dimensions, based on the classification of Pisharodi and Parameswaran (1992) and also applied in more recent studies such as De Tavares Canto Guina and de Moura Engracia Giraldi (2015). Accordingly, the dimensions belonging to the independent variable are the general image of Austrian people, the general image of Austrian products, the image of communication, distribution and differentiation of Austrian products, perceived similarity between Austria and Sweden, internationalization of Austria, the image of Austria in general and negative aspects of Austrian products (De

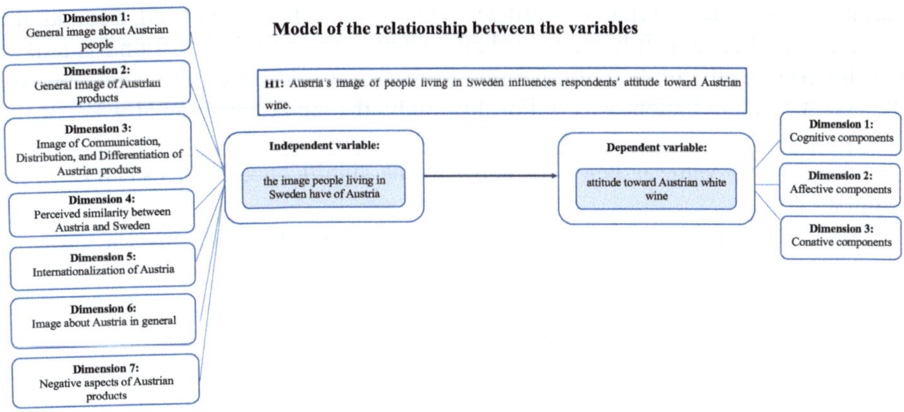

Fig. 1 Conceptual model of the variables

Tavares Canto Guina & de Moura Engracia Giraldi, 2015; Pisharodi & Parameswaran, 1992).

The dependent variable 'attitude' consists of three literature-based dimensions: the first dimension is the conative or behavioral dimension defined by Brengman (2002) and Khanal (2018). The other dimensions were also already used in earlier research, concretely the affective dimension by Watson et al. (1988) and the cognitive dimension by Khanal (2018).

3.3 *Data Analysis*

For conducting this study, a quantitative research method was applied. An online survey in *Qualtrics* was conducted to capture customers' perceptions, attitudes and beliefs. Participants had to be either Swedish or people living in Sweden. Due to money and time constraints, this study used a non-probability sample, which means that a subjective sample was used. In consequence, the sample does not represent the real population living in Sweden. Thus, no generalized conclusions can be drawn. The questionnaire was online from May 2023 to June 2023. In total, 261 responses were collected from which 147 responses could be used for statistical data analysis in *Jamovi.*

The above-explained conceptual model is the base for the construct to be measured. The data was collected with the questionnaire by asking the respondents about their agreement or disagreement with each dimension of both the independent and dependent variables. Thereby, each dimension was represented by several statements. A five-point Likert scale was used to measure the respondent's level of agreement on a scale of 1 (strongly disagree) to 5 (strongly agree).

In terms of the measurement analysis, the statistical test correlation matrix was used to determine whether and to what extent there is a correlation between the variables. First, the correlation coefficient (=Pearson's r) was analyzed. Second, the p-value is looked at as it helps to evaluate whether the correlation found is real or not. It describes the probability that the result also would have been found in case the correlation coefficient was 0. For this study, the significance level is set at the 5% commonly used in marketing research, whereby the result is statistically significant when the p-value is below or equal to the significance level (STHDA, n.d.; Minitab, 2023).

The evaluation of the demographic data has shown the following: in terms of age, around 26% of the participants were 18–30 years old, and 74% were 30 years or older. Regarding gender, the sample included 52.4% female and 46.9% male respondents, one person (0.7%) preferred not to say the gender. When it comes to education, most of the participants (46.3%) have a master's degree, 24.5% have a bachelor's degree, followed by 13.6% with a high school diploma, 10.2% have a PhD and 3.4% compulsory education. Regarding the respondents' country of origin, the answers covered a total of 18 different countries, with a majority from Germany (35.4%), Sweden (31.3%) and Austria (20.4%). Additionally, 2.7% are from Switzerland.

3.4 Findings

The findings show that most of the dimensions of the country image have a positive influence on the overall attitude:

1. The general image that people living in Sweden or Swedes have about Austrian people has a positive influence on their attitude toward Austrian white wine.
2. The general image that people living in Sweden or Swedes have about Austrian products has a positive influence on their attitude toward Austrian white wine.
3. The image of communication, distribution and differentiation that people living in Sweden or Swedes have of Austrian products has a positive influence on their attitude toward Austrian white wine.
4. The degree of internationalization that people living in Sweden or Swedes assume of Austria has a positive influence on their attitude toward Austrian white wine.
5. The image of Austria in general people living in Sweden or Swedes have has a positive influence on their attitude toward Austrian white wine.

For the negative dimension of country image, there is a negative coherence on the overall attitude: the negative aspects of Austrian products people living in Sweden or Swedes perceive have a negative influence on their attitude toward Austrian white wine.

Thus, the perceived similarity between Austria and Sweden that people living in Sweden or Swedes assume has a positive influence on their overall attitude toward Austrian white wine.

Therefore, it can be stated that Austria's image of people living in Sweden and Swedes influences respondents' affective, cognitive and conative dimensions of attitude toward Austrian white wine.

4 Discussion: Creating Lines to Land in Sustainability Communication

This study dealt with the country-of-origin effect and individual perceptions of a country influencing consumer behavior from a strategic communication perspective, especially following a marketing approach. With a study, investigating whether Austria's image of people living in Sweden has a significant and positive effect on respondents' attitudes toward Austrian wine, we are able to discuss the importance of creating the connection to land, to a specific region and thus 'borrowing' the countries' image and transferring it to the product—here, white wine. The findings of the study show that Austria's image of people living in Sweden does influence respondents' affective, cognitive and conative attitudes toward Austrian white wine. For example, the general image that people living in Sweden or Swedes have about Austrian people and products has a positive influence on their attitude toward Austrian white wine. Also, the image of communication, distribution and differentiation people living in Sweden or Swedes have of Austrian products, has a positive influence on their attitude toward Austrian white wine. Lastly, the degree of internationalization that people living in Sweden or Swedes assume of Austria has a positive influence on their attitude toward Austrian white wine. From these findings, the conclusion can be drawn that there does exist a country-of-origin effect that promotes Austrian white wine in Sweden. This aligns with the theory that the country-of-origin effect means that the consumer's image of a country is linked to the perception of the product (Johnson et al., 2016; Schooler, 1965).

A learning from the study for strategic sustainability communication thus is, that a country that is perceived as 'traditionally green' and 'sustainable'—again, like Austria (Pearlmutter et al., 2020), offers cognitive, affective and conative aspects of drivers that potentially influence individual decision marking for or against a product. Future research thus needs to delve deeper into the country images and map more and less 'green' countries. The study further contributes to future strategic sustainability communication from a marketing perspective by showing that sustainability communication in general should include regional and local interpretations of sustainability and in particular environmental and ecological impacts of products. The regionality of the investigated product Austrian wine likely creates sustainability by supporting local economies, creating a product high in quality and supporting cultural identity. That being emphasized, the country-of-origin effect provides one possible answer as to how regional/location-based proximity in sustainability communication can be created. This can also shed more light on the partly contradicting and inconsistent research (Golob et al., 2023).

A few considerations that might be examined in future research and thereby enhancing the strategic sustainability communication discussion are first, there might be a time-related aspect of regional/location-based proximity, which refers to historic events or traditions. Second, social activities such as community engagement, but also political values, which are targeted in local journalism, might as well indicate regional/location-based proximity. Third, the question arises of what it means for different cultures/cultural customs in different regions to establish a regional/location-based proximity in sustainability communication. One example, of incorporating several regional/location-based proximity facets (cultural, social and political) is renewable energy. This is a regional/location-based topic, since in different regions different energy sources are available and accepted, established and/or perceived as sustainable. Following that, future research should focus on attitudes and social representations of renewable energy (e.g. low-carbon energy technology) in a local and regional setting.

References

Adomßent, M., & Godemann, J. (2011). Sustainability communication: An integrative approach. In T. Godemann & G. Michelsen (Eds.), *Sustainability communication* (pp. 27–37). Springer.

Al-Sulaiti, K. I., & Baker, M. J. (1998). Country of origin effects: A literature review. *Marketing Intelligence & Planning, 16*(3), 150–199.

Argyriou, E., & Melewar, T. C. (2011). Consumer attitudes revisited: A review of attitude theory in marketing research. *International Journal of Management Reviews, 13*(4), 431–451.

Baldwin, T. (2023). *Attitude Types & Components in the Workplace*. Available at: https://study.com/learn/lesson/attitude-types-components.html#:~:text=Conative%2DThe%20conative%20component%20refers,avoid%20running%20into%20a%20snake (accessed at: June 25, 2023).

Balestrini, P., & Gamble, P. (2006). Country-of-origin effects on Chinese wine consumers. *British Food Journal, 108*(5), 396–412.

Batra, R., Ramaswamy, V., Alden, D. L., Steenkamp, J. B. E., & Ramachander, S. (2000). Effects of brand local and nonlocal origin on consumer attitudes in developing countries. *Journal of Consumer Psychology, 9*(2), 83–95.

Baughn, C. C., & Yaprak, A. (2014). Mapping country-of-origin research: Recent developments and emerging avenues. In *Product-country images: Impact and role in international marketing* (p. 89).

Brengman, M. (2002). *The impact of colour in the store environment: An environmental psychology approach*. Doctoral dissertation, Ghent University.

Brookes, G. (1993). Hungarian and Bulgarian wines prospects in the UK market. *British Food Journal, 95*(7), 33–36.

Bruhn, M., & Homburg, C. (2004). Gabler Lexikon Marketing.

Carter, L. L. (2020). The effects of country-of-origin image, consumer ethnocentrism, and animosity upon foreign product purchases. *Journal of Marketing Development & Competitiveness, 14*(2), 9–23.

Chiou, J. S. (2003). The impact of country of origin on pretrial and post-trial product evaluations: The moderating effect of consumer expertise. *Psychology & Marketing, 20*(10), 935–954.

De Tavares Canto Guina, F., & de Moura Engracia Giraldi, J. (2015). The country of origin effect on Brazilian beef trade in Europe: The moderating role of gender, age, and product involvement. *Journal of Food Products Marketing, 21*(2), 123–140.

Diamantopoulos, A., Arslanagic-Kalajdzic, M., & Moschik, N. (2020). Are consumers' minds or hearts guiding country of origin effects? Conditioning roles of need for cognition and need for affect. *Journal of Business Research, 108*, 487–495.
Duhan, D. F., Kiecker, P. L., Areni, C. S., & Guerrero, C. (1999). Origin information and retail sales of wine. *International Journal of Wine Marketing, 11*(3), 44–57.
European Commission. (2024). *Sustainable finance—environmental, social and governance ratings and sustainability risks in credit ratings.* Available at: https://ec.europa.eu/info/law/better-regulation/have-your-say/initiatives/13330-Sustainable-finance-environmental-social-and-governance-ratings-and-sustainability-risks-in-credit-ratings_en (Accessed at: 1st August 2024).
Felzensztein, C., Hibbert, S., & Vong, G. (2004). Is the country of origin the fifth element in the marketing mix of imported wine? A critical review of the literature. *Journal of Food Products Marketing, 10*(4), 73–84.
Fournier, S. (1998). Consumers and their brands: Developing relationship theory in consumer research. *Journal of Consumer Research, 24*(4), 343–373.
Galan Ladero, M. M., Galera Casquet, C., & Singh, J. (2015). Understanding factors influencing consumer attitudes toward cause-related marketing. *International Journal of Nonprofit and Voluntary Sector Marketing, 20*(1), 52–70.
Golob, U., Podnar, K., & Zabkar, V. (2023). Sustainability communication. *International Journal of Advertising, 42*(1), 42–51.
Han, C. M. (1989). Country image: Halo or summary construct? *Journal of Marketing Research, 26*(2), 222–229.
Hong, S. T., & Wyer, R. S., Jr. (1989). Effects of country-of-origin and product-attribute information on product evaluation: An information processing perspective. *Journal of Consumer Research, 16*(2), 175–187.
Hui, M. K., & Zhou, L. (2002). Linking product evaluations and purchase intention for country-of-origin effects. *Journal of Global Marketing, 15*(3–4), 95–116.
Insch, G. S., & McBride, J. B. (2004). The impact of country-of-origin cues on consumer perceptions of product quality: A binational test of the decomposed country-of-origin construct. *Journal of Business Research, 57*(3), 256–265.
Iversen, N. M., Hem, L. E., & Ulf, H. (2022). Willingness to buy US products in three Southeast European countries: The effects of cognitive, affective and conative components of country-of-origin image. *Journal for East European Management Studies, 27*(3), 487–518.
Johnson, Z. S., Tian, Y., & Lee, S. (2016). Country-of-origin fit: When does a discrepancy between brand origin and country of manufacture reduce consumers' product evaluations? *Journal of Brand Management, 23*, 403–418.
Jones, G. V., & Schultz, H. R. (2016). Climate change: Climate change and emerging cool climate wine regions. *Wine & Viticulture Journal, 31*(6), 51–53.
Khanal, J., (2018). Influence of affective, cognitive and behavioral intention on customer attitude toward coffee shops in Norway: Comparative study of local and international branded coffee shop. , .
Klein, J. G., Ettenson, R., & Morris, M. D. (1998). The animosity model of foreign product purchase: An empirical test in the People's Republic of China. *Journal of Marketing, 62*(1), 89–100.
Kock, F., Josiassen, A., & Assaf, A. G. (2019). Toward a universal account of country-induced predispositions: Integrative framework and measurement of country-of-origin images and country emotions. *Journal of International Marketing, 27*(3), 43–59.
Koschate-Fischer, N., Diamantopoulos, A., & Oldenkotte, K. (2012). Are consumers really willing to pay more for a favorable country image? A study of country-of-origin effects on willingness to pay. *Journal of International Marketing, 20*(1), 19–41.
Kotler, P. (1991). *Marketing management* (7th ed.). Prentice-Hall.
Kroeber-Riel, W., & Gröppel-Klein, A. (2013). *Konsumentenverhalten.* Vahlen.

Kuzmany, K. (2008). *Country-of-Origin Effekt und Länderimage Osteuropas. Auswirkungen auf die Produktbeurteilung institutioneller Käufer*. Doctoral dissertation, WU Vienna University of Economics and Business.

Laroche, M., Papadopoulos, N., Heslop, L. A., & Mourali, M. (2005). The influence of country image structure on consumer evaluations of foreign products. *International Marketing Review, 22*(1), 96–115.

Lee, C. W., Suh, Y., & Moon, B. J. (2001). Product-country images: The roles of country-of-origin and country-of-target in consumers' prototype product evaluations. *Journal of International Consumer Marketing, 13*(3), 47–62.

Liefeld, J. P. (2004). Consumer knowledge and use of country-of-origin information at the point of purchase. *Journal of Consumer Behaviour: An International Research Review, 4*(2), 85–87.

Lockshin, L. S., & Timothy Rhodus, W. (1993). The effect of price and oak flavor on perceived wine quality. *International Journal of Wine Marketing, 5*(2/3), 13–25.

Maher, A. A., & Carter, L. L. (2011). The affective and cognitive components of country image: Perceptions of American products in Kuwait. *International Marketing Review, 28*(6), 559–580.

McDougall, G., & Lockshin, L. (2000). Tanunda winery. *International Journal of Wine Marketing, 12*(3), 48–69.

Minitab Statistical Software. (2023). *Interpret the key results for Correlation*. Available at: https://support.minitab.com/en-us/minitab/21/help-and-how-to/statistics/basic-statistics/how-to/correlation/interpret-the-results/key-results/ (Accessed at: 10 June 2023).

Moon, B., & Jain, S. (2002). Consumer processing of foreign advertisements: Roles of country-of-origin perceptions, consumer ethnocentrism and country attitude. *International Business Review, 11*(2), 117–138.

Nagashima, A. (1970). A comparison of Japanese and US attitudes toward foreign products. *Journal of Marketing, 34*(1), 68–74.

Newig, J., Schulz, D., Fischer, D., Hetze, K., Laws, N., Lüdecke, G., & Rieckmann, M. (2013). Communication regarding sustainability: Conceptual perspectives and exploration of societal subsystems. *Sustainability, 5*(7), 2976–2990.

Österreich Wein Marketing GmbH. (2022): *Weltweit Auszeichnungen für Österreichs Weine.* Available at: https://www.oesterreichwein.at/presse-multimedia/pressetexte/news/article/weltweite-auszeichnungen-fuer-oesterreichs-weine-1 (Accessed at: 8 June 2023).

Ozsomer, A., Bodur, M., & Tamer Cavusgil, S. (1991). Marketing standardisation by multinationals in an emerging market. *European Journal of Marketing, 25*(12), 50–64.

Papadopoulos, N., & Heslop, L. A. (2014). *Product-country images: Impact and role in international marketing*. Routledge.

Papadopoulos, N., Heslop, L. A., & Beracs, J. (1990). National stereotypes and product evaluations in a socialist country. *International Marketing Review, 7*(1).

Pearlmutter, D., Theochari, D., Nehls, T., Pinho, P., Piro, P., Korolova, A., et al. (2020). Enhancing the circular economy with nature-based solutions in the built urban environment: Green building materials, systems and sites. *Blue-Green Systems, 2*(1), 46–72.

Pisharodi, R.M. / Parameswaran, R. (1992): Confirmatory factor analysis of a country-of-origin scale: Initial results. Advances in Consumer Research 19, pp. 706–714.

Prexl, A. (2010). *Nachhaltigkeit kommunizieren—nachhaltig kommunizieren.* VS Verlag für Sozialwissenschaften. https://doi.org/10.1007/978-3-531-92471-7

Rauhut Kompaniets, O. (2022). Sustainable competitive advantages for a nascent wine country: An example from southern Sweden. *Competitiveness Review: An International Business Journal, 32*(3), 376–390.

Ribberink, N., & Washburn, M. (2017). Estimating consumer demand for country of origin labelling: Cross country comparisons of german and us consumer home appliances. *International Journal of Scientific Management and Development, 5*(1), 1–8.

Ridder, M. (2023). *Annual consumption volume of wine in Sweden from 2010 to 2021.* Available at: https://www.statista.com/statistics/562628/consumption-volume-of-wine-in-sweden/ (Accessed at: 8 June 2023).

Roth, K. P., & Diamantopoulos, A. (2009). Advancing the country image construct. *Journal of Business Research, 62*(7), 726–740.

Salam, M., & Chowdhury, S. K. (2015). Predicting attitude based on cognitive, affective and conative components: An online shopping perspective. *Stamford Journal of Business Studies, 6*(7), 101–115.

Schooler, R. D. (1965). Product bias in the Central American common market. *Journal of Marketing Research, 2*(4), 394–397.

STHDA. (n.d.). Statistical tools for high-throughput data analysis: *Correlation test between two variables in R*. Available at: http://www.sthda.com/english/wiki/correlation-test-between-two-variables-in-r (Accessed: 10 June 2023).

Supanvanij, J., & Amine, L. S. (2000). Consumer perception of country-of-origin effect and brand effect. *Latin American Business Review, 1*(4), 47–60.

Teas, R. K., & Agarwal, S. (2000). The effects of extrinsic product cues on consumers' perceptions of quality, sacrifice, and value. *Journal of the Academy of Marketing Science, 28*(2), 278–290.

Thorelli, H. B., Lim, J. S., & Ye, J. (1989). Relative importance of country of origin, warranty, and retail store image on product evaluations. *International Marketing Review, 6*(1), 35–46.

United Nations. (2024). *Our common agenda and the sdgs*. Available at: https://www.un.org/en/common-agenda/sustainable-development-goals (Accessed: 1 August 2024).

Usunier, J. C. (2006). Relevance in business research: The case of country-of-origin research in marketing. *European Management Review, 3*(1), 60–73.

Verlegh, P. W., & Steenkamp, J. B. E. (1999). A review and meta-analysis of country-of-origin research. *Journal of Economic Psychology, 20*(5), 521–546.

Visbal, O., Herrera-Mendoza, K., Orozco-Acosta, E., & Herzberg, M. (2017). The influence of country of origin cues on product evaluation: Evidence from Swiss and German consumers. *Journal of Technology Management & Innovation, 12*(2), 18–25.

Voci, D., & Karmasin, M. (2023). Sustainability communication: How to communicate an inconvenient truth in the era of scientific mistrust. *Journal of Communication Management, 28*, 15.

Wall, M., Liefeld, J., & Heslop, L. A. (1991). Impact of country-of-origin cues on consumer judgments in multi-cue situations: A covariance analysis. *Journal of the Academy of Marketing Science, 19*, 105–113.

Wang, H., Gibson, C., & Zander, U. (2020). Editors' comments: Is research on corporate social responsibility undertheorized? *Academy of Management Review, 45*(1), 1–6.

Watson, D., Clark, L. A., & Tellegen, A. (1988). Development and validation of brief measures of positive and negative affect: The PANAS scales. *Journal of Personality and Social Psychology, 54*(6), 1063–1070.

Weder, F. (2023). The evolution of the sustainability story: Strategic sustainability communication as niche construction. *International Journal of Strategic Communication, 17*(3), 228–244.

Weder, F., Karmasin, M., Krainer, L., & Voci, D. (2021). Sustainability communication as critical perspective in media and communication studies—An introduction. In *The sustainability communication reader: A reflective compendium* (pp. 1–12).

Weder, F., Lemke, S., & Tungarat, A. (2019). (Re) storying sustainability: The use of story cubes in narrative inquiries to understand individual perceptions of sustainability. *Sustainability, 11*(19), 5264.

Zain, O. M., & Yasin, N. M. (1997). The importance of country-of-origin information and perceived product quality in Uzbekistan. *International Journal of Retail & Distribution Management, 25*(4), 138–145.

Ziemann, A. (2011). Communication theory and sustainability discourse. In I. T. Godemann & G. Michelsen (Eds.), *Sustainability communication* (pp. 89–96). Springer.

Further Reading

Carvill, M., Butler, G., & Evans, G. (2021). *Sustainable marketing: How to drive profits with purpose*. Bloomsbury Publishing.

Schooler, R. D. (1965). Product bias in the central American common market. *Journal of Marketing Research, 2*(4), 394–397.

Florentina Höhs is a Praedoc teaching and research assistant at the Vienna University of Economics and Business (WU) at the Institute for Strategic Organizational Communication at the Department of Business Communication. There, she conducts research on sustainability communication together with Prof. Franzisca Weder. Florentina holds a Bachelor's degree in Journalism and Communication Science from the University of Vienna and a Bachelor's degree in Economics and Social Sciences (major in Business Administration), as well as a Master's degree in Export and Internationalization Management both from WU Vienna. Her Master's thesis was about the country-of-origin effect on consumers from a marketing perspective. In her research, she deals with sustainability communication with a focus on energy communication and renewable energy sources. In her dissertation, Florentina investigates how regional and local proximity can be established in energy/sustainability communication. Also, she addresses social representations of renewable energy (e.g. low-carbon energy technology) in a local and regional sustainability communication setting. Both theoretical concepts and empirical observations are examined in the dissertation to develop a new perspective on localization and regionalization tendencies in strategic sustainability communication and thus to further develop the research field of strategic sustainability communication.

Community Identity and Evaluation of Oppositional Framing of Renewable Energy Projects: Learnings for Strategic Sustainability Communication

Kumchong Lee, Ned Watt, Jasmine Burdon, and Franzisca Weder

Abstract The use of renewable energy sources is rapidly expanding on a global scale. The climate crisis and recent political conflicts have accelerated the development and introduction of innovative technologies. Hydrogen is one such innovation. Debated internationally in political and industry networks, especially the production of green hydrogen is seen as the way to transform renewable energy such as wind and solar into a form of energy that is not only easier to transport but also easier to store. However, on the ground, the realization of new plants and hydrogen production facilities often causes conflicts with, for example, environmental or community groups. With the case study at hand, we explore community opposition to the development plan of the Hydrogen Park "Gladstone" in Australia. The qualitative content analysis of a text corpus drawn from Facebook postings and comments between December 2021 and April 2022 shows that the community went through a process of identity building by creating an opposition narrative of protecting place identity. Frames that have been activated were distrust and doubt in business activities and political decision-making and a lack of public safety. The chapter complements the existing framing studies on renewable energy with a new set of *oppositional frames* used in renewable energy communication. Despite the limitations of a case study and therefore a small sample of communication content, the insights from the study and the conceptual framework developed complement existing research in the fields of climate change and environment but also the growing area of strategic sustainability communication and broaden the understanding of frames in relation to a broader narrative of transition or transformation.

K. Lee (✉) · J. Burdon
TAFE Centre of Excellence Clean Energy Batteries, TAFE Queensland, Brisbane, Australia
e-mail: k.lee1@uq.edu.au; jasmine.burdon@uq.net.au

N. Watt
Digital Media Research Center, QUT, Brisbane, Australia

F. Weder
Department of Business Communication, WU Vienna University of Business and Economics, Vienna, Austria
e-mail: franzisca.weder@wu.ac.at

F. Weder (ed.), *Strategic Sustainability Communication*, CSR, Sustainability, Ethics & Governance, https://doi.org/10.1007/978-3-031-89486-2_17

1 Introduction

Fossil fuels were once celebrated as the primary drivers of industrialization and economic development. However, their connection to the current environmental and climate crisis has placed them under scrutiny. Renewable energy production has emerged as a key solution to global challenges, attracting attention from industry, global, national, and regional governments, and political players. While there may be disagreements on methods, technological challenges as well as timeframes and deadlines (2030? 2050?), there is a global consensus that we must transition away from fossil fuels and switch to renewable energy sources. The impetus for this transition has been further accelerated since the pandemic, the war between Russia and Ukraine, and the rise of energy security issues in Central Europe and across the globe (Yergin, 2022; Zakeri et al., 2022).

According to the International Renewable Energy Agency's statistics (2022), the share of renewable energy in global electricity generation has been expanding continuously, reaching 28% in 2020. The increasing share can be attributed to scientific advancement and technological innovations, for instance, the production of *green hydrogen* through electrolysis driven by renewable energy sources such as wind or solar. However, to achieve a meaningful transition to a mix of renewable energy and, thus, a reduction of CO2 emissions, such innovation needs to be endorsed by various stakeholders in the markets, regulatory bodies, and wider societies (Geels & Verhees, 2011). While the perceived benefit of renewable energy can be a key factor in predicting its acceptance in the markets and regulatory bodies, its acceptance in wider societies is more difficult to analyze and predict. Despite general support for renewable energy, it is not always welcomed by the community where it is to be rolled out as demonstrated in previous research on the installation of solar and wind parks (i.e., Devine-Wright, 2008, 2011; Batel, 2020; Segreto et al., 2020; Gareiou et al., 2021). The case of the hydrogen facility discussed in this chapter will complement these studies from a sustainability communication perspective.

The contradiction between public support for renewable energy production and related infrastructure and its difficult rollout in the community has led to a more granular conceptualization of acceptance but also opposition and resonance. Batel (2020), for example, described the development of this field in line with the vicissitude of the popular concept of "not in my backyard" (NIMBY), something also researched in the areas of public engagement, marketing, and campaigning (i.e., Devine-Wright, 2011; Benjamins, 2014; Mcclymont & O'hare, 2008; Burningham, 2000). Early studies, which were based on the binary assumption of good (acceptance) and bad (opposition), defined the community's resistance as selfish and irresponsible. Its simplistic and reductionist assumptions were subjected to criticism, leading to new research that introduced alternative views on community opposition (Devine-Wright, 2009). The criticism was directed not only toward the concept but also the methodologies that often rely on self-report-focused surveys to capture the community beliefs and attitudes and their perception of their "backyard" (Haggett & Futak-Campbell, 2011).

However, as the concept of NIMBY also suggests the prioritization of self-interests over the greater good of the wider society, NIMBY opposition is explained as the counterpoint of the idea of participation and engagement (Burningham et al., 2006). However, from a rather constructivist and, for example, development communication or communication for social change perspective, this approach seems too limited. Exploring the resonance of new technologies on the ground and in communities, all conversations and practices in a specific locality can be seen as capacity to promote social change (Servaes, 2012; 2016, Thomas & van de Fliert, 2014; Dutta, 2015). Development and change are described in two ideal-typical communication models as top-down diffusion (see the marketing and campaigning approaches mentioned above) or bottom-up participation (see i.e., Wilkins et al., 2014). Accordingly, next to interpersonal communication, mass, social media, and other public communication processes and structures are studied to understand sustainable development, transition processes, and social change. Also from a broader media and communication studies and discourse perspective, we can state that mediated discourses (in traditional but also social media) play a key role in social life in communities and drive (or hinder) change (Fairclough, 1992). Acceptance and unacceptance, denial or opposition are therefore linked to participation and the process of politicization (Curran, 2017). Thus, community opposition to renewable energy can be better understood as a form of contestation over the meaning of its rollout in the given community (Weder, 2021).

This chapter seeks to better understand the process of communicative participation and politicization of the debate via an analysis of the arguments that the community leads in the context of the installation of a new hydrogen plant. To analyze the arguments, we use framing concepts to particularly examine the community's objection to the building of a hydrogen facility in a rural community in Australia. We focus on how the community group frames the issue of hydrogen and how it is contextualized within existing renewable energy narratives of transition or transformation.

The analysis of Facebook postings (n = 63) and comments (n = 308) within a specific timeframe (from December 2021 to April 2022) shows that the community mobilized by using predominantly *protection* and *safety* frames, which triggered a broader narrative of distrust in business and political processes. Such narratives politicized the issue and promoted the community's participation in challenging the institutionalized consensus on the development of hydrogen facilities in the region. The chapter complements the current status quo of framing research in the areas of environmental, climate change, and sustainability communication. It sheds light on new oppositional patterns in the renewable energy discourse which can be applied to other projects (hydrogen, wind, solar parks, etc.) and community engagement, acceptance, and discourse studies in the future.

2 Framing and Narratives of Renewable Energy: A Literature Review

Communication is essential in understanding social acceptance of renewable energy—from a rather pragmatic, campaigning perspective but also from a social change perspective. Most research studying acceptance of renewable energy and related technology looks at one-way communication processes from and driven by the industry to inform and persuade the public (Johnston-Billings et al., 2019). Hydrogen, a relatively new player in renewable energy sector, has not been a subject of heightened controversy, and thus the research on its social acceptance is still evolving, with the majority of studies looking at public understanding and attitudes (Achterberg, 2014, 2017; Martin et al., 2021; Schmidt & Donsbach, 2016), as well as willingness to accommodate hydrogen facilities and to switch to hydrogen energy (Dumbrell et al., 2022; Han et al., 2022).

Next to a functionalist approach to communication and research that describes the dissemination of information to a specific target audience as the key to acceptance of, i.e., a new technology like wind power or hydrogen production, a social-constructivist perspective understands communication as sense- and meaning-making (Weder & Eriksen, 2023). A focus on participatory communication processes does not merely include individual perceptions and support (acceptance) but also all social processes and practices involving various stakeholders and their interactions and conversations, even the opposing ones (Huijts et al., 2007). From this perspective, individuals use various strategies to combine personal experiences, common sense, and mediated information, as emphasized, for example, in symbolic interactionism, thereby challenging the predetermined meanings and assigning their own for specific issues (Gamson, 1992). As highlighted in the introduction, this is also the perspective on communication in development communication or social change communication, where participatory approaches dominate the field (Servaes, 2016; Thomas & van de Fliert, 2014).

For the project at hand, we are particularly interested in the arguments that determine the discourses on renewable energy, rather than one-way information processes that are strategically set up with the objective of creating awareness and getting acceptance. We are taking a bottom-up perspective (Weder & Eriksen, 2023) and seek to trace back the arguments which define the discourse around a renewable energy project to develop a new perspective on the resonance of new energy projects in communities. Arguments can be either captured by following the discourse from a broader perspective (discourse analysis, see, i.e., Fairclough, 1992) or by identifying the linguistic devices, the organizing principles in a conversation, the so-called "frames" and therefore the interpretative perspectives on an issue represented in this conversation. We follow this second approach and differentiate between overarching master frames or narratives (Weder, 2021) and issue-specific frames (de Vreese, 2005; Brüggemann & D'Angelo, 2018). A master frame or claim is conceptualized as a narrative that emerges along a specific issue life cycle or even cycle of protest (Benford & Snow, 2000). Issue-specific frames are the "engine" that drives the emergence of a narrative (Weder, 2021). A master frame is also described as a generic frame or "collective action frame" (i.e., environmental justice and exploited worker frame) having a wide scope, resonance, and influence but also flexibility and inclusiveness to be adopted in various forms of communication (Snow & Benford, 1992). As well, issue-specific frames have a certain "linguistic elasticity" which enables the researcher to capture the meaning-making process (framing) and the conditions and consequences of the practices (frame) (Snow et al., 2014). While individuals have their own ways of processing information about renewable energy, they may not always be attentive to the issue, nor they are actively constructing its meaning. Unless the issues are of direct relevance to individuals, they often unconsciously process the information through a system of frames (Lakoff, 2010). Given their usefulness, the concepts of frame and framing have been employed in various studies across disciplines and epistemologies (D'Angelo, 2002; Entman et al., 2009; Tewksbury & Scheufele, 2008). For Goffman (1986), for example, frames are an important part of the culture, as they allow individuals to unconsciously make sense of the world by locating, perceiving, identifying, and labelling various occurrences and issues. In other words, frames function as inter-textual references, helping people to place unfamiliar occurrences into familiar cultural patterns or schemas and providing a framework for interpreting and understanding the world around them. For instance, it activates common sense and familiar cultural narratives to associate with the unfamiliar (Byhring & Knain, 2016). Therefore, when framing renewable energy is concerned, it is not just concerned about the immediate information and facts about the energy, technology, and its related projects but also about larger narratives and sets of beliefs within which people situate and relate these issues. This explains in part the discrepancy between general support for renewable energy and objections to its rollout at the local level.

Nevertheless, most of the literature on renewable energy, which referred to frame or framing, used the concepts in a generic way to denote how immediate issues are presented, especially in the media (Delshad, 2013; Kim et al., 2014; Lyytimäki, 2018; Smith et al., 2016; Zukas, 2017). These media studies investigating the framing of renewable energy can be viewed as being based on normative assumptions about energy transition and on the theoretical assumptions about framing that strongly shape and structure public discourses (Batel, 2020). As a result, the media's positive and negative framing of renewable energy has been a common line of inquiry (Haigh, 2010; Kim et al., 2014; Qu et al., 2009), as if they were signals of social acceptance or rejection.

Within the framing research in general, the tendency of directly associating media framing with the audience has already been criticized for some time (Gamson & Modigliani, 1989; van Gorp, 2007). However, this criticism has not translated into the research on audience framing (i.e., community framing) in the field of social acceptance and renewable energy (cf. Aklin et al., 2018). Despite the conceptual affinity between framing and the diverse views about the opposition, the research on community's framing of its opposition still remains scarce and is not based on the real-world case (cf. Curran, 2017) but on the experiment and focus group designed for the sake of framing analysis (Aklin et al., 2018; Benighaus &

Bleicher, 2018). Given the interplay between individual framing and the context in which the framing occurs, the best way to capture social (un)acceptance through framing analysis is with case studies which help to explore spoken or written text where people spontaneously express(ed) their thoughts.

Thus, the aim of the study at hand was to better understand the dynamics of a local controversy related to a hydrogen project and the renewable energy narrative(s) that emerge from certain issue-specific frames used in the discourse. Therefore, we ask the following overarching research question: how is the opposition to a renewable energy project expressed and what are the arguments in the discourse? The explorative character of this question led to a case study, which needs to be further complemented by further quantitative research in the future. In the following, the "Gladstone Case" (hydrogen park in Australia) will be introduced as well as the methodological approach to study the related community discourse and emerging renewable energy frames. After that, the findings will be presented and discussed before we offer an outlook for future research.

3 Hydrogen Park Gladstone: The Case

The Hydrogen Park Gladstone was proposed by the Australian Gas Infrastructure Group (AGIG) to be built as a demonstration facility aimed at contributing to the decarbonization of the domestic gas supply in Gladstone. The facility was intended to produce green hydrogen using renewable electricity and water (Australian Gas Infrastructure Group, 2022). The resulting hydrogen would be blended with natural gas to create a renewable gas blend that makes up 10% of the total gas supply in the region. This blended gas was planned to be distributed through the company's existing natural gas distribution network to homes and businesses. The project dates to 2020 when the Queensland Government announced grant funding of up to $1.78 million in support of the project. The AGIG lodged a development application (DA 25/2021) for building the hydrogen production and blending facilities in Derby Street, South Gladstone, and submissions from the community were invited by the council.

Situated in the vicinity of the planned hydrogen production and blending facilities on Derby Street, South Gladstone found itself at the center of discussions about the potential impact of the project on the community. This affected community is a suburb of Gladstone, a coastal city located 500 km northwest of Brisbane. With a population of over 65,000 people, Gladstone is well-acquainted with industrial projects, as the manufacturing industry contributes to nearly 50% of the region's economic output and accounts for 15% of the total employment in the area (REMPLAN, 2023). Being one of the "coal belt" regions in Central Queensland, with the state's largest multicommodity port, Gladstone enjoyed a mining boom during the 2000s until decarbonization became a more prominent issue—at least on the public agenda (Connolly & Orsmond, 2011). In response to socio-political pressure for net-zero emissions, major gas companies in the region have begun to increase their renewable energy portfolio, and Gladstone has been keen on seizing the opportunity kept its eye on the new opportunity to break through the economic recession in recent years (Connolly & Orsmond, 2011). The Hydrogen Park Gladstone was one of the projects proposed within this broad context of climate change action and the region's economic transformation.

Unlike other projects located in the two State Development Areas in the north-west of the city, the Hydrogen Park was proposed in the eastern part of the region close to residential and community space, involving the source of feud from the onset. Following multiple submissions against the development application, concerned community members created a Facebook Group in December 2021 to continue mobilizing opposition to the development plan. They used the online space to share information and voice their concerns. The group was created in December 2021 and had been active until the mid-April 2022 when the local councilors unanimously voted to refuse the application. During the 5-month period, the group produced 63 postings with 308 comments, and the postings were shared 222 times. On average, each posting was read by 103 members (see Table 1).

A case study is characterized by a rich and in-depth description of a specific case, treated as a bounded system in its own right rather than as a part or sample of a broader investigation (Merriam & Tisdell, 2016). In the context of this study, a case study approach is well-suited for understanding the community's opposition to the hydrogen facilities within a specific socio-cultural context. The case under examination is unique due to the relatively recent emergence of green hydrogen technology and the generally positive public attitude toward it. Furthermore, the local history of the mining industry and the presence of other renewable energy projects in the region also create a distinct boundary for the case, making it a valuable subject for a case study.

A common criticism against a case study often revolves around issues of validity, typically concerning external validity and statistical generalizability. However, in contrast, analytical generalizability in a case study can be achieved through triangulation of data, investigators, theories, and methodologies (Tellis, 1997). In this study, given its exploratory nature, the primary focus is not on establishing causal inferences. Nonetheless, the relationship between oppositional framing and collective action is established based on existing theory from the onset. To enhance the validity of findings and insights generated from the analysis, this chapter establishes a methodological framework that guides and documents the analytical process. Additionally, data triangulation is employed, albeit within its limitations, by including industry and government-produced documents alongside all postings and comments from the Facebook group.

Table 1 Overview of the Facebook group

Members in April 2022	Active members (at least 1 posting or comment)	No. of postings	No. of comments	Postings being shared	Avg. number of members read postings
185	44	63	308	222	103

Following the constructionist approach to framing (Gamson & Modigliani, 1989; van Gorp, 2007), the content of all postings has been analyzed with a qualitative approach of two-way categorization (Mayring, 2021). The main focus was to identify frames as reasoning devices and drivers of a certain narrative and therefore as a unit of analysis, examining how they were woven together to develop frames and connected to a familiar narrative and common senses. While we didn't analyze language devices, such as metaphors, catchphrases, and depictions, or visual cues of frames used by the members to shape rhetorical attributes of their postings and comments, we aimed at analyzing "reasoning devices" which can be differentiated in diagnosis and prognosis or mobilizing frames (Snow & Benford, 1992), strengthening the interpretive approach to and mechanism of framing. After initial identification of framing and reasoning devices, they were grouped together according to central organizing ideas and activated to more familiar and common-sensical narratives. Before manually analyzing the collected data, a word cloud was generated to visually present the frequency of words used in the postings and comments. It prevents the researcher from over-interpreting particular framing and reasoning devices that were used less often, thereby helping to reduce the subjectivity inherent in qualitative and manual framing analysis (Van Gorp, 2010). In the following, the findings are presented and discussed.

4 Results and Discussion

The community objection to the development of hydrogen facilities as they were proposed by AGIG was constructed by the combination of frames that defined and redefined the issue of hydrogen (overview in Fig. 1); with a new set of arguments and frames related to a source of renewable energy and the way it is produced the group legitimized its opposition and provoked the local councilors to respond.

Fig. 1 Frequency of words used in the postings and comments

4.1 Right Project Wrong Location

The dominant frame was *right project, wrong location*, which emphasized that the objection was not to the project or hydrogen production but to the proposed location of the facilities. While quantification of each unit of analysis went beyond the scope of this research, the catchphrase "right project, wrong location" or "right project, (to) right location" appeared 23 times in the postings and comments. It was also reflected in the word cloud, where each four word was frequently found. By emphasizing the "right project" dimension, the community group distanced themselves from a blanket objection to the project. As previous research noted, it is also crucial for the affected community of the renewable energy rollout to employ diagnostic framing to emphasize that they oppose safeguarding the innate value of the site and that they are not simply against it being in their backyard (Haggett & Futak-Campbell, 2011).

Although the community's emphasis on the proximity to a residential area could have degraded their opposition as stake management, the place identity of Gladstone region, which the group repeatedly affirmed, helped define their opposition as a place-protective action (Devine-Wright, 2009). Gladstone Regional Council promoted a narrative of regional identity where it envisages itself becoming a "Centre of Hydrogen Excellence" (The Next Economy, 2022). The group expressed their pride in the region's industrial heritage and support for the region's vision for a hydrogen hub and specific hydrogen projects proposed in the nearby suburbs of Aldoga and Yarwun. By resonating with the dominant narrative of locality promoted by the council, the group's diagnostic framing of "wrong location" could be considered as a genuine effort to maintain the *place identity* that distinguishes the industrial North and the residential South. Hence, it prevented the trigger for NIMBY criticism.

This frame was not simply defending their genuineness of opposition but also used to denounce AGIG's decision as undermining the community's pride in place identity. Several reasoning devices were found, which activated a common narrative of distrust in big business. The comments below are exemplary, along with many framing devices such as "glossy response," "done deal," and "true intention of this project."

> This position chosen because it's easy and most cost effective for AGIG.

> That's the first phase & then it expands to fill more of the rest of the parcel of land.

Although commonly known to discourage political participation, distrust also has the potential to mobilize collective action, challenging the process or the motivations, credibility, or competence of the distrusted (Laurian, 2009). Therefore, the narrative of distrust is often found in protests where opponents are relatively less resourceful and use the perceived distrust as a means to delegitimate the others (Price et al., 2012). Given the distrust of the project proposer shared among the group, AGIG's official response in January to the community opposition could at best reinforce the distrust. The letter extenuated the opposition as "concerns raised by a small number of the community" while emphasizing that AGIG "have proactively engaged with the community and have experienced broad community support" (Australian Gas Infrastructure Group, 2022, p. 1). As previous studies noted, providing quality information or ticking many checkboxes are simply not enough for social acceptance (Flint & Blyth, 2021). It was echoed in the comments, for instance:

> Just because a DA is advertised, IS NOT CONSULTATION.

> Ask for objections AFTER a decision was made. No real consultation.

The apparent divergence between the group and AGIG regarding the meaning of "consultation" echoed the crisis of participation in the context of local development (Monno & Khakee, 2012). However, institutionalized and tokenistic forms of participation could also induce politicization of the issue and further foster the community's involvement in opposition efforts (Legacy, 2017). As it will be demonstrated (Sect. 4.3), the protection frame led to another frame that served as a trigger to engage local politicians, where the narrative of distrust was constantly activated and leveraged to garner more support.

4.2 *Public Safety*

Another commonly used frame by the group was *public safety.* As shown in the word cloud, the keywords "risk" and "safety" were frequently mentioned in the postings and comments. This frame functioned as a rhetorical device to relate the meaning of safety with the proximity of the proposed facility to residential areas. Previous research on framing renewable energy has indicated that the lens of safety is commonly utilized in the media and stakeholder discourses (Bauer & Bogner, 2020; Schmidt & Donsbach, 2016; Stauffacher et al., 2015). This lens is often directed toward technological flaws such as geothermal energy triggering earthquakes (Stauffacher et al., 2015) and synthetic biology risking biosafety (Bauer & Bogner, 2020). However, when it comes to framing hydrogen, the *risk* frame was less prominent compared to the *economic* frame (Schmidt & Donsbach, 2016). The issue of hydrogen has been largely driven by industry and policy sectors, concerning economic viability, and thus safety has not received much emphasis and has been communicated ambiguously (Derempouka et al., 2022).

Despite the relative lack of framing of hydrogen safety in the mainstream discourse, there is one historical exemplar that readily activates a broader narrative of technological flaws and catastrophes. The Hindenburg disaster, broadcasted and photographed in the age of so-called powerful media, has been ingrained in the minds of the general public about hydrogen (Cadwallader & Herring, 1999). The common perception was mirrored in the public opinion survey where the respondents indicated that safety is the most important consideration for them to accept or reject hydrogen energy at home (Martin et al., 2021). Many framing devices such as "explosive," "flammable," "ignite," and "projectile" appeared in the group's postings and comments, which suggests the frame that shapes the meaning of the hydrogen facility in relation to the risk of explosion. The framing devices were bound up with the reasoning devices to insist that risk must be eliminated, if not, relocated along with the facility. The safety frame has a strong politicization potential as it highlights uncertainties surrounding the science and technology of hydrogen, thereby instilling doubt among the public (Bolsen & Druckman, 2015).

As mentioned earlier, the narrative of distrust can help mobilize opposition within the community. However, if the politicization of hydrogen facility development extends to the politicization of the science and technology behind hydrogen, it could undermine the social change potential of the community opposition. Although the dramatic image of the Hindenburg has long loomed over the cultural meaning of hydrogen, safety concerns did not cause outright opposition to hydrogen facilities elsewhere (Ricci et al., 2007; Saxe et al., 2007). Public sensitivity to hydrogen safety varies depending on the context, for instance, whether hydrogen is produced for domestic use or export (Martin et al., 2021). As mentioned previously, the region's industrial legacy and the vision of establishing a hydrogen hub were important contextual factors that aligned with the group's protection frame, which supported the project but not the location. While expressing concerns about safety, the group focused their framing of safety in relation to the location, thereby avoiding the full politicization of the issue beyond the development decision. By doing so, the community steered clear of potential accusations, such as misinformed scientific skeptics and opposition for opposition's sake.

4.3 *Lack of Due Process*

The *lack of due process* frame diagnosed the problem through the lens of the democratic process. It stressed that the selection of the site and associated process were not transparent and that the community was left out. The frame broadened the definition of the problem from the project location (chosen by AGIG) to the process by which the location was approved. In so doing, it infused the issue with a political flavor and made it relevant to the governments and councilors as demonstrated partly by the frequent usage of the word "council." The frame consisted of various framing devices depicting politicians and the political decision-making process as untrustworthy, for instance:

> Backhanders at all levels of government doing under the table handshakes.

> When it walks like a duck and quacks, it's probably a duck.

Although the notion of public participation in the context of renewable energy decisions has been popular, it is often used in an idealized manner without much detail on its scope, purpose, degree, types, and process (Bidwell, 2016). The current case also demonstrated that the different stakeholders had different understandings of participation, and the lack of consensus on the meaning of participation was central to this framing. The local councilors repeatedly stressed that they have no right to request that AGIG relocate the project to a different site. AGIG chose the project location based on its needs and planning law, which was then approved or rejected by the council. Therefore, there was little scope for community participation for the council and AGIG before the lodgment of the development application. On the other hand, the community viewed participation as consultation before the project location was chosen.

While the community's understanding of participation was dismissed under the current planning law, the *lack of due process* frame transposed the issue into the broader narrative of democracy and popular sovereignty. Consequently, the prognosis for the problem was derived from the narrative where the group is not just passive invitees of top-down consultation but active electorates, for instance:

> The Councilors will all be publicly judged at the next election.

> They were voted in by us. They will listen to us & do what's best for suburbia and greater Gladstone.

This resonates with Curran's (2017) observation of the local community's reframing of the notion of social license popularized by the industry and using it as a contestation tool against the industry. In the current case, it was the Facebook group's appropriation of the institutionalized narrative of participation through which the opposition could be legitimized. The group's self-representation as disenfranchised members of the community and the representation of local politicians as untrustworthy and colluding with big business were, therefore, potential subversions, to which the local councilors had to respond seriously.

5 Conclusion

The concept of framing is a useful tool to be used to understand the social acceptance of renewable energy as an interactive, participatory process. For the study at hand, we differentiated between issue-specific frames and master frames or broader narratives. This enabled us to analyze a concrete discourse on one hydrogen plant to be built and also to generate learnings for renewable energy-related conversations and community communication processes in general. We therefore complement not only the existing research on public acceptance of a new technology or renewable energy installation (windmill etc.; Devine-Wright, 2022; Evans et al., 2011; Lu et al., 2023; Burningham et al., 2006; Armstrong, 2021) targeted at with communication campaigns by industry or political institutions with a new, constructivist and therefore "bottom-up" perspective on how interpretations and narratives emerge (Benford & Snow, 2000; Weder, 2021; Weder & Eriksen, 2023). We also complement existing framing research that seeks new ideas to study the conditions under which a particular collective action frame evolves into a master frame and thus an issue cycle of protest (Benford & Snow, 2000).

The Facebook group opposed the AGIG's plan to build hydrogen facilities in close proximity to a residential area in Gladstone by framing the project as necessary but inadequately located. The group's framing aligned with the narrative of locality constructed by the local government. By endorsing the region's vision of a hydrogen hub while framing the project location as conflicting with the existing industrial and residential locality, the group's opposition could avoid NIMBY accusations. In the meantime, the group framed AGIG's decision within the common narrative of distrust in big business, thereby discrediting the developer's proposal and mobilizing the opposition. The framing of public safety in relation to the site location further justified the opposition and maintained consistency in its definitions of the project. The disagreement on the meaning of participation between the AGIG and the community group led to the *lack of due process* frame that redefined the issue as the absence of a democratic process and marginalization of the community. As the frame echoed a common narrative of community inclusion embedded in democratic institutions, the opposition could appropriate the institutionalized narrative to legitimize their dissent and mobilize support.

Despite the emphasis made by the different levels of government on community involvement in the decision-making process for issues affecting the community, the scope and degree of participation are limited when it comes to the decision of renewable energy rollout in local communities. Community concerns about the potential impact of renewable energy rollout are known to be amplified or minimized by the level of trust and confidence in the institutions involved rather than by the knowledge and information about the energy (O'Garra et al., 2008). Therefore, the limited opportunity for genuine participation that promotes the narrative of distrust could hamper social acceptance of renewable energy rollout in the community elsewhere.

Therefore, the contribution of this study can be twofold. First, it applies framing as conceptual and analytical tools to the social acceptance of hydrogen rollout in the community. By doing so, this chapter reveals the construction of community opposition with a due emphasis on how they frame the issue in such a way as to place it in a familiar narrative and common sense. Second, the case of community objection to hydrogen is rare, and therefore this study contributes to filling the gap in the literature on the social unacceptance of hydrogen. The findings are also expected to inform industry and government of the effective communication with local communities, which is important as the community's experience of framing renewable energy could turn into a frame to be used by other oppositions elsewhere in the future.

While the study provides new insights into framing strategies in a renewable energy scenario, this study has limitations due to its reliance on a single case study for exploratory purposes, which has left questions unanswered that require to be picked up in larger, potentially rather descriptive studies to generate results that allow better predictions for future renewable energy communication and stakeholder engagement. The study solely focused on the opposition group, which may not fully capture the diversity of perspectives within the affected community. It is possible that counter frames to the opposition group exist within the community. Also, the impact or effectiveness of oppositional framing and its counter-use of institutionalized narratives were only established at the theoretical level. For instance, this study assumed that the politicization of the issue through oppositional framing triggered local councilors to refuse the development application, which needs further evidence. In the future, an explorative analysis of the councilors' decision-making processes, including the examination of meeting minutes and conducting expert interviews, may reveal how the community opposition was understood by the decision makers and, vice versa, if and how they were influenced. As the hydrogen rollout in Australia continues, local opposition may also increase, making comparative case studies with projects in other countries in the future desirable.

References

Achterberg, P. (2014). The changing face of public support for hydrogen technology explaining declining support among the Dutch (2008–2013). *International Journal of Hydrogen Energy, 39*(33), 18711–18717. https://doi.org/10.1016/j.ijhydene.2014.08.053

Achterberg, P. (2017). Knowing hydrogen and loving it too? Information provision, cultural predispositions, and support for hydrogen technology among the Dutch. *Public Understanding of Science, 23*(4), 445–453. https://doi.org/10.1177/0963662512453117

Aklin, M., Cheng, C.-Y., & Urpelainen, J. (2018). Social acceptance of new energy technology in developing countries: A framing experiment in rural India. *Energy Policy, 113*, 466–477. https://doi.org/10.1016/j.enpol.2017.10.059

Armstrong, J. (2021). People and power: Expanding the role and scale of public engagement in energy transitions. *Energy Research & Social Science, 78*, 102136. https://doi.org/10.1016/j.erss.2021.102136

Australian Gas Infrastructure Group. (2022). *Hydrogen Park Glasstone: Safety summary for Gladstone Regional Council*. https://www.gladstone.qld.gov.au/downloads/file/4352/item-3-1-attach-06-additional-material-submitted-by-the-applicant.

Batel, S. (2020). Research on the social acceptance of renewable energy technologies: Past present and future. *Energy Research & Social Science 24*(3), 311–325. https://doi.org/10.1016/j.erss.2020.101544

Bauer, A., & Bogner, A. (2020). Let's (not) talk about synthetic biology: Framing an emerging technology in public and stakeholder dialogues. *Public Understanding of Science, 29*(5), 492–507. https://doi.org/10.1177/0963662520907255

Benighaus, C., & Bleicher, A. (2018). Neither risky technology nor renewable electricity: Contested frames in the development of geothermal energy in Germany. *Energy Research & Social Science, 47*, 45–55. https://doi.org/10.1016/j.erss.2018.08.022

Benjamins, M. (2014). International actors in NIMBY controversies: Obstacle or opportunity for environmental campaigns? *China Information, 28*(3), 338–361. https://doi.org/10.1177/0920203X14546732

Benford, R., & Snow, D. (2000). Framing processes and social movements: An overview and assessment. *Annual Review of Sociology, 26*(1), 611–639. https://doi.org/10.1146/annurev.soc.26.1.611

Bidwell, D. (2016). Thinking through participation in renewable energy decisions. *Nature Energy, 1*, 1–4. https://doi.org/10.1038/nenergy.2016.51

Bolsen, T., & Druckman, J. N. (2015). Counteracting the politicization of science. *Journal of Communication, 65*(5), 745–769. https://doi.org/10.1111/jcom.12171

Brüggemann, M., & D'Angelo, P. (2018). Defragmenting news framing research: Reconciling generic and issue-specific frames. In P. D'Angelo (Eds.)., *Doing news framing analysis II: Empirical and theoretical perspectives* (pp. 9–111). Rouledge.

Burningham, K. (2000). Using the Language of NIMBY: A topic for research, not an activity for researchers. *Local Environment, 5*(1), 55–67. https://doi.org/10.1080/135498300113264

Burningham, K., Barnett, J., & Thrush, D. (2006). *The limitations of the NIMBY concept for understanding public engagement with renewable energy technologies: A literature review.* School of Environment and Development, University of Manchester.

Byhring, A. K., & Knain, E. (2016). Intertextuality for handling complex environmental issues. *Research in Science Education, 46*, 1–19. https://doi.org/10.1007/s11165-014-9454-6

Cadwallader, L. C., & Herring, J. S. (1999). *Safety issues with hydrogen as a vehicle fuel.* Idaho National Lab.

Connolly, E., & Orsmond, D. (2011). *The mining industry: From bust to boom.* RBA Annual Conference, https://www.rba.gov.au/publications/confs/2011/connolly-orsmond.html

Curran, G. (2017). Social licence, corporate social responsibility and coal seam gas: Framing the new political dynamics of contestation. *Energy Policy, 101*, 427. https://doi.org/10.1016/j.enpol.2016.10.042

D'Angelo, P. (2002). News framing as a multiparadigmatic research program: A response to Entman. *Journal of Communication, 52*(4), 870–888. https://doi.org/10.1111/j.1460-2466.2002.tb02578.x

Delshad, A. B. (2013). Telling stories: Exogenous influences on media, presidential, and congressional framing of biofuels. *Politics & Policy, 41*(2), 213–240. https://doi.org/10.1111/polp.12009

Derempouka, E., Skjold, T., Njå, O., & Haarstad, H. (2022). The role of safety in the framing of the hydrogen economy by selected groups of stakeholders. *Chemical Engineering Transactions, 90*, 757–762.

Devine-Wright, P. (2008). Reconsidering public acceptance of renewable energy technologies: a critical review. In M. Grubb, T. Jamasb, & M. G. Pollitt (Eds.), *Delivering a low carbon electricity system: Technologies, economics and policy* (pp. 443–461). Cambridge University Press.

Devine-Wright, P. (2009). Rethinking NIMBYism: The role of place attachment and place identity in explaining place-protective action. *Journal of Community & Applied Social Psychology, 19*(6), 426–441. https://doi.org/10.1002/casp.1004

Devine-Wright, P. (2011). Place attachment and public acceptance of renewable energy: A tidal energy case study. *Journal of Environmental Psychology,* 31(4), 336–343. https://doi.org/10.1016/j.jenvp.2011.07.001

Devine-Wright, P. (2022). Decarbonisation of industrial clusters: A place-based research agenda. *Energy Research & Social Science, 91*, 10275. https://doi.org/10.1016/j.erss.2022.102725

de Vreese, C. (2005). News framing: Theor and typology. *Information Design Journal, 13*(1), 51–62. https://doi.org/10.1075/idjdd.13.1.06vre

Dumbrell, N. P., Wheeler, S. A., Zuo, A., & Adamson, D. (2022). Public willingness to make trade-offs in the development of a hydrogen industry in Australia. *Energy Policy, 165*, 112987. https://doi.org/10.1016/j.enpol.2022.112987

Dutta, M. (2015). Decolonizing communication for social change: A culture-centered approach. *Communication Theory,* 25(2), 123–143. https://doi.org/10.1111/comt.12067

Entman, R. M., Matthes, J., & Pellicano, L. (2009). Nature, sources, and effects of news framing. In K. Wahl-Jorgensen & T. Hanitzsch (Eds.), *The handbook of journalism studies* (pp. 175–190). Routledge.

Evans, B., Parks, J., & Theobald, K. (2011). Urban wind power and the private sector: Community benefits social acceptance and public engagement. *Journal of Environmental Planning and Management,* 54(2), 227–244. https://doi.org/10.1080/09640568.2010.505829

Fairclough, N. (1992). Discourse and text: Linguistic and intertextual analysis within discourse analysis. *Discourse & Society, 3*(2), 193–217.

Flint, A., & Blyth, S. (2021). Facilitating genuine community participation: Can development learn from design? *Development Studies Research, 8*(1), 63–72. https://doi.org/10.1080/21665095.2021.1884112

Gamson, W. A. (1992). *Talking politics*. Cambridge University Press.

Gamson, W. A., & Modigliani, A. (1989). Media discourse and public opinion on nuclear power: A constructionist approach. *American Journal of Sociology, 95*(1), 1–37. https://doi.org/10.1086/229213

Gareiou, Z., Drimili, E., & Zervas, E. (2021). Public acceptance of renewable energy sources. In K. Grigorios (Eds.), *Low carbon energy technologies in sustainable energy systems* (pp. 309–327). Elsevier. https://doi.org/10.1016/B978-0-12-822897-5.00016-X

Geels, F. W., & Verhees, B. (2011). Cultural legitimacy and framing struggles in innovation journeys: A cultural-performative perspective and a case study of Dutch nuclear energy (1945–1986). *Technological Forecasting and Social Change, 78*, 910–930. https://doi.org/10.1016/j.techfore.2010.12.004

Goffman, E. (1986). *Frame analysis: An essay on the organization of experience; with a forward by Bennett M. Berger*. Northeastern University Press (Original work published in 1974. Harper & Row).

Haggett, C., & Futak-Campbell, B. (2011). Tilting at windmills? Using discourse analysis to understand the attitude-behaviour gap in renewable energy conflicts. *Mekhanizm Rehuluvannya Economiky, 1*(51), 207–220.

Haigh, M. M. (2010). Newpapers use three frames to cover alternative energy. *Newspaper Research Journal, 31*(2), 47–62.

Han, S.-M., Kim, J.-H., & Yoo, S.-H. (2022). The public's acceptance toward building a hydrogen fueling station near their residences: The case of South Korea. *International Journal of Hydrogen Energy, 47*, 4284–4293. https://doi.org/10.1016/j.ijhydene.2021.11.106

Huijts, N. M., Midden, C. J., & Meijnders, A. L. (2007). Social acceptance of carbon dioxide storage. *Energy Policy, 35*(5), 2780–2789.

International Renewable Energy Agency. (2022). *Renewable energy statistics 2022*. IRENA.

Johnston-Billings, A., Pogmore, L., & Kaiser, M. (2019). Social licence has changed. It is time to build trust with society using the tools of today. *The APPEA Journal, 59*(2), 615–618.

Kim, S.-H., Besley, J. C., Oh, S.-H., & Kim, S. Y. (2014). Talking about bio-fuel in the news: Newspaper framing of ethanol stories in the United States. *Journalism Studies, 15*(2), 218–234. https://doi.org/10.1080/1461670X.2013.809193

Lakoff, G. (2010). Why it matters how we frame the environment. *Environmental Communication, 4*(10), 70–81. https://doi.org/10.1080/17524030903529749

Laurian, L. (2009). Trust in planning: Theoretical and practical considerations for participatory and deliberative planning. *Planning Theory & Practice, 10*(3), 369–391. https://doi.org/10.1080/14649350903229810

Legacy, C. (2017). Is there a crisis of participatory planning? *Planning Theory, 16*(4), 425–442. https://doi.org/10.1177/1473095216667433

Lu, J., Rong, D., Lev, B., Liang, M., Zhang, C., & Gao, Y. (2023). Constraints affecting the promotion of waste incineration power generation project in China: A perspective of improved technology acceptance model. *Technological Forecasting and Social Change, 186*, 122165. https://doi.org/10.1016/j.techfore.2022.122165

Lyytimäki, J. (2018). Renewable energy in the news: Environmental, economic, policy and technology discussion of biogas. *Sustainable Production and Consumption, 15*, 65–73. https://doi.org/10.1016/j.spc.2018.04.004

Martin, V., Ashworth, P., Petrova, S., Wade, B., & Witt, K. (2021). Public perceptions of hydrogen: 2021 national survey results.

Mayring, P. (2021). *Qualitative content analysis: A step-by-step guide*. Sage.

Mcclymont. K., & O'hare, P. (2008). "We're not NIMBYs!" Contrasting local protest groups with idealised conceptions of sustainable communities. *Local Environment, 13*(4), 321–335. https://doi.org/10.1080/13549830701803273

Merriam, S. B., & Tisdell, E. J. (2016). *Qualitative research: A guide to design and implementation* (4th ed.). EBSCO Publishing.

Monno, V., & Khakee, A. (2012). Tokenism or political activism? Some reflections on participatory planning. *International Planning Studies, 17*(1), 85–101. https://doi.org/10.1080/13563475.2011.638181

O'Garra, T., Mourato, S., & Pearson, P. (2008). Investigating attitudes to hydrogen refuelling facilities and the social cost to local residents. *Energy Policy, 36*, 2074–2085. https://doi.org/10.1016/j.enpol.2008.02.026

Price, J., Fielding, K., & Leviston, Z. (2012). Supporters and opponents of potable recycled water: Culture and cognition in the Toowoomba referendum. *Society & Natural Resources, 25*(10), 980–995. https://doi.org/10.1080/08941920.2012.656185

Qu, M., Tahvanainen, L., Ahponen, P., & Pelkonen, P. (2009). Bio-energy in China: Content alysis of news articles on Chinese professional internet platforms. *Energy Policy, 37*, 2300–2309. https://doi.org/10.1016/j.enpol.2009.02.024

REMPLAN. (2023). *Economy, jobs and business insights*. Gladstone Regional Council. https://app.remplan.com.au/gladstone/economy/summary?state=genQC3NLEUe4KKJUvD1qaBtBI1IpRP

Ricci, M., Bellaby, P., & Flynn, R. (2007). Stakeholders' and publics' perceptions of hydrogen energy technologies. In R. Flynn & P. Bellaby (Eds.), *Risk and the public acceptance of new technologies*. Palgrave Macmillan.

Saxe, M., Folkesson, A., & Alvfors, P. (2007). A follow-up and conclusive report on the attitude towards hydrogen fuel cell buses in the CUTE project—From passengers in Stockholm to bus operators in Europe. *International Journal of Hydrogen Energy, 32*(17), 4295–4305. https://doi.org/10.1016/j.ijhydene.2007.07.050

Schmidt, A., & Donsbach, W. (2016). Acceptance factors of hydrogen and their use by relevant stakeholders and the media. *International Journal of Hydrogen Energy, 41*(8), 4509–4520. https://doi.org/10.1016/j.ijhydene.2016.01.058

Segreto, M., Principe, L., Desormeaux, A., Torre, M., Tomassetti, L., Tratzi, P., Paolini, V., & Petracchini, F. (2020). Trends in social acceptance of renewable energy across Europe - a literature review. *International Journal of Environmental Research and Public Health, 17*(24), 9161. http://dx.doi.org/10.3390/ijerph17249161

Servaes, J. (2012). Introduction to 'Green ICT'. *Telematics and Informatics, 29*(4), 335–336. https://doi.org/10.1016/j.tele.2012.05.001

Servaes, J. (2016). How 'sustainable' is development communication research? *International Communication Gazette, 78*(7), 701–710. https://doi.org/10.1177/1748048516655732

Smith, H. M., Smith, J. W., Silka, L., Lindenfeld, L., & Gilbert, C. (2016). Media and policy in a complex adaptive system: Insights from wind energy legislation in the United States. *Energy Research & Social Science, 19*, 53–60. https://doi.org/10.1016/j.erss.2016.05.016

Snow, D., & Benford, R. (1992). Master frames and cycles of protest. In A. D. Morris & C. M. Mueller (Eds.), *Frontiers in social movement theory* (pp. 133–155). Yale University Press.

Snow, D. A., Benford, R. D., McCammon, H., Hewitt, L., & Fitzgerald, S. (2014). The emergence, development, and future of the framing perspective: 25+ years since "frame alignment". *Mobilization: An International Quarterly, 19*(1), 23–46. https://doi.org/10.17813/maiq.19.1.x74278226830m691

Stauffacher, M., Muggli, N., Scolobig, A., & Moser, C. (2015). Framing deep geothermal energy in mass media: The case of Switzerland. *Technological Forecasting and Social Change, 98*, 60–70. https://doi.org/10.1016/j.techfore.2015.05.018

Tellis, W. (1997). Introduction to case study. *The Qualitative Report, 3*(2), 1–14. https://doi.org/10.46743/2160-3715/1997.2024

Tewksbury, D., & Scheufele, D. A. (2008). News framing theory and research. In J. Bryant & M. B. Oliver (Eds.), *Media effects: Advances in theory and research* (pp. 17–33). Routledge.

The Next Economy. (2022). *Gladstone Region economic transition readmap*. https://www.gladstone.qld.gov.au/economic-information

Thomas, P., & Van de Fliert, E. (2014). *Interrogating the theory and practice of communication for social change: The basis for a renewal.* Springer.

van Gorp, B. (2007). The constructionist approach to framing: Bringing culture back in. *Journal of Communication, 57*(1), 60–78. https://doi.org/10.1111/j.0021-9916.2007.00329.x

Van Gorp, B. (2010). Strategies to take subjectivity out of framing analysis. In *Doing news framing analysis* (pp. 100–125). Routledge.

Weder, F. (2021). Sustainability as master frame of the future? Potency and limits of sustainability as normative framework in corporate, political and NGO communication. In F. Weder, L. Krainer, & M. Karmasin (Eds.), *The sustainability communication reader* (pp. 103–118). Springer.

Weder, F., & Eriksen, M. (2023). *CSR communication and cultures of sustainability*. The University of Queensland.

Wilkins, K., Tufte T., & Obregon R. (2014). *The handbook of development communication and social change*. Wiley. https://doi.org/10.1002/9781118505328

Yergin, D. (2022). Bumps in the energy transition: Despite a growing global consensus, obstacles to reducing net carbon emissions to zero are stark. *Finance and Development, 59*, 8–13.

Zakeri, B., Paulavets, K., Barreto-Gomez, L., Echeverri, L. G., Pachauri, S., Boza-Kiss, B., Zimm, C., Rogelj, J., Creutzig, F., Ürge-Vorsatz, D., Victor, D. G., Bazilian, M. D., Fritz, S., Gielen, D., McCollum, D. L., Srivastava, L., & Pouya, J. D. H. (2022). Pandemic, war, and global energy transitions. *Energies, 15*(17), 6114. https://doi.org/10.3390/en15176114

Zukas, K. J. (2017). Framing wind energy: Strategic communication influences on journalistic coverage. *Mass Communication and Society, 20*(3), 427–449. https://doi.org/10.1080/15205436.2016.1266660

Kumchong Lee received his PhD in communication from the University of Queensland, Australia. His research interests lie at the intersection of communication, social, and cultural change. He has been involved in both academic and community development sectors, coordinating the Speak My Language project in Queensland (Department of Social Services) and supporting Future Fuel CRC Project in 2022, and Australian Research Council Linkage Project on integrating community and family aged care for diverse Australians (2022–2025). He is currently Research Associate at the TAFE Centre of Excellence Clean Energy Batteries, overseeing the Centre's applied research.

Ned Watt, PhD candidate at Queensland University of Technology, Brisbane, Australia. We work for the project "FactCheck–GPT: An exploration of how publicly funded fact-checkers perceive implications of large language models (LLMs) on their labour and operational effectiveness."

Jasmine Burdon is a former Master student at the University of Queensland, School of Communication and Arts, Brisbane, Australia, and project manager of the Future Fuel CRC Project in 2022. She is now working for the Australian Green party.

Franzisca Weder, Chair for International Organizational Communication at the Vienna University of Economics and Business, Department of Business Communication, Institute for Strategic Organizational Communication, is a leading scholar in the field of sustainability communication. She is an experienced qualitative researcher with over 20 years of experience in exploring, studying, and evaluating communication strategies and campaigns of all kinds of organizations and, in particular, mediated sustainability discourses, corporate social responsibility (CSR), and sustainability-related policies and their impact on business, NGOs, and political institutions.

Strategic Corporate Sustainability Communications of MNCs in Local Contexts: The Case Study of Stakeholders' Co-Creation of Dove's Award-Winning #LetHerGrow Campaign in Thailand

Chanapa Itdhiamornkulchai and Parichart Sthapitanonda

Abstract In the Sustainable Development Goals (SDGs) era, stakeholders place increased values on sustainability communication and expect measurable actions from organizations toward sustainability. Hence, multinational corporations (MNCs) must adapt their sustainability initiatives, including corporate social responsibility, to the local contexts of host countries where they operate and communicate effectively at local, national, and global levels. To enhance the understanding of sustainability communication of MNCs in local contexts, this chapter is organized into four parts. *Part 1* provides an overview of existing sustainability-related terminologies, followed by proposing conceptualizations of three terms coined by the authors: (1) *strategic corporate sustainability communications*, (2) *"glocal" sustainability communication* ("think global, act local"), and (3) *"lobal" sustainability communication* ("think local, act global"). *Part 2* identifies *"glocal"* and *"lobal" strategic corporate sustainability communications* that Dove and Unilever used in Dove's local award-winning #LetHerGrow campaign in Thailand as a part of Unilever's global "Purpose in action" on "Equity, Diversity, and Inclusion" that aims to contribute to SDG#5 and SDG#10. To explore the benefits of dialogic communication, *Part 3* analyzes stakeholders' co-creation of Dove's #LetHerGrow campaign by applying stakeholder information, stakeholder response, and stakeholder involvement strategies, as well as communication *of* sustainability (CoS: one-way communication), communication *about* sustainability (CaS: two-way communication), and communication *for* sustainability (CfS: CaS and/or CoS with measurable sustainable actions). In *Part 4*, the authors propose *The Glocal and Lobal Infinite Interaction Model* for strategic corporate sustainability communications of MNCs. This model offers theoretical implications for future research and practical implications for organizations to develop communication strategies.

C. Itdhiamornkulchai (✉) · P. Sthapitanonda
Faculty of Communication Arts, Chulalongkorn University, Bangkok, Thailand
e-mail: Chanapa.i@chula.ac.th; Parichart.s@chula.ac.th

F. Weder (ed.), *Strategic Sustainability Communication*, CSR, Sustainability, Ethics & Governance, https://doi.org/10.1007/978-3-031-89486-2_18

1 Conceptualization of Strategic Corporate Sustainability Communications

People have placed considerably higher values of importance on sustainability communication in the past decade, especially after the expansion of the 17 Sustainable Development Goals (SDGs) by the United Nations in 2015 under the Agenda 2030 framework (Sahadev et al., 2022). Organizations have used strategic communication to gain legitimacy among stakeholders regarding sustainability (Allen, 2016), leading to emerging conceptualizations of strategic sustainability communication (Weder et al., 2021). This chapter begins with an overview of the development of relevant terminologies, specifically (1) sustainability, (2) sustainability communication, (3) corporate sustainability communications, and (4) strategic communication, followed by proposing three terms conceptualized by the authors to contribute to a better understanding of sustainability communication: (1) *strategic corporate sustainability communications*, (2) *"glocal" sustainability communication* ("think global, act local"), and (3) *"lobal" sustainability communication* ("think local, act global"). These three terms, along with the typologies of Morsing and Schultz (2006) (stakeholder information, stakeholder response, and stakeholder involvement strategies) and Newig et al. (2013) (communication *of* sustainability "CoS," communication *about* sustainability "CaS," and communication *of* sustainability "CfS"), are then used to analyze the award-winning Dove #LetHerGrow campaign. Ultimately, the authors propose a preliminary model for strategic corporate sustainability communications of multinational corporations (MNCs) on digital media.

1.1 *Sustainability*

The sustainability paradigm is based on the worldview of efficient long-term allocation and preservation of environmental, social, and economic aspects. Originally, the World Commission on Environmental Development defined sustainable development as "development that meets the needs of the present without compromising the ability of future generations to meet their own needs" (1987, p. 41). This definition has been quoted in terms of *sustainability* and *sustainable development* by many scholars (e.g., Allen, 2016; Godemann & Michelsen, 2011; Itdhiamornkulchai & Taiphapoon, 2022; Signitzer & Prexl, 2008).

Up to the present time, the conceptualization and operationalization of sustainability have been based on the triple bottom line (Siano et al., 2016; Sivarajah et al., 2020). Originally coined by Elkington (1994), the *triple bottom line* comprises environmental, social, and economic dimensions (Elkington, 2004). Morioka et al. (2016) defined *corporate sustainability* as the "capacity of firms to contribute to global sustainable development and all the challenges regarding economic, social and environmental interconnections together with short, medium and long term aligned and conflicting demands" (p. 659).

In 2015, the United Nations adopted the 17 Sustainable Development Goals (SDGs) that called for cooperation from all countries to achieve all SDGs by 2030 with the ultimate goal of reaching a balance of social, economic, and environmental sustainability (The United Nations Development Programme, 2024).

1.2 Sustainability Communication and Corporate Sustainability Communications

Conceptualizations of sustainability communication emphasize dialogues about sustainable behaviors and development (Newig et al., 2013; Siano et al., 2016). *Sustainability communication* is defined as "an understanding of the world, that is of the relationship between humans and their environment, into social discourse, developing a critical awareness of the problems about this relationship and then relating them to social values and norms" (Godemann & Michelsen, 2011, p. 6), and this definition has been cited in several studies (e.g., Fischer et al., 2016; Stumpf et al., 2019). *Strategic communication*, which refers to the use of goal-oriented communication by organizations (Hallahan et al., 2007) developed from research findings on the environment, situation, and significant public opinions (Botan, 2021), has been used by organizations to gain *legitimacy* among *stakeholders* regarding sustainability (Allen, 2016). Digital corporate communications have been used to enhance the co-participation, co-creation, and co-production of the SDGs (Canel, 2023). Organizations should establish message credibility through transparency and comprehensiveness in their strategic communication by providing specific details regarding their sustainability initiatives to avoid being perceived as greenwash among stakeholders (Allen, 2016). *Greenwashing* is defined as the exaggeration of the environmental friendliness of a brand. Marketers face the challenge of consumers' expectations regarding sustainability (Mahoney & Tang, 2017).

From a corporate perspective, Siano et al. (2016) defined sustainability communication as "all types of corporate and marketing communications about sustainability issues" (p. 2). *Corporate sustainability communications* is defined as "an evolving concept that refers to corporate communications about sustainability issues" (Signitzer & Prexl, 2008, p. 2). These include sustainability reports, stakeholder dialogues, public debates on sustainability issues, and information campaigns to gain public attention that could be integrative elements of business strategy with balanced environmental, economic, and social values and objectives (Signitzer & Prexl, 2008). Hence, "corporate sustainability becomes a narrative construct in which the discourse serves to frame and perform specified sustainable actions (Herrick & Pratt, 2013; Schoeneborn & Trittin, 2013)" (Siano et al., 2016, p. 2). Recently, Itdhiamornkulchai and Taiphapoon (2022) conceptualized "the role of corporate sustainability communication as the blood vessel that facilitates all strategic processes of corporate sustainability initiatives and implementations by enabling the flow and exchange of information among various internal and external stakeholders" (p. 184). Therefore, it can be concluded that, while the content and processes from organizations are emphasized, stakeholder dialogues are essential for corporate sustainability communications.

A recognized conceptualization of sustainability communication is *the typology of sustainability communication modes* by Newig et al. (2013), comprising (1) communication *of* sustainability (CoS), (2) communication *about* sustainability (CaS), and (3) communication *for* sustainability (CfS). International scholars have used this typology to examine sustainability communication up to the present time (e.g., Golob et al., 2023; Sahadev et al., 2022; Slutskiy & Itdhiamornkulchai, 2024; Weder & Eriksen, 2023). Communication *of* sustainability (CoS) is limited to one-way communication from experts, such as companies, scientists, journalists, and educators, to intended audiences. Examples of CoS include sustainability reports, mass media marketing, and documentaries (Newig et al., 2013). Several studies on the development of sustainability communication have identified the paradigm shift from one-way communication to passive receivers to dialogic and interactive communication among stakeholders (e.g., Siano et al., 2016; Sivarajah et al., 2020), which is consistent with the typology of Newig et al. (2013) that sustainability communication is shifting from CoS toward CaS in various industries.

Communication *about* sustainability (CaS) refers to horizontal communication of exchange and discussions of information, interpretations, and opinions concerning sustainability, for instance, discourses, workshops, conferences, and online conversations. Framings of messages, attention, and effectiveness of communication across different systems (such as from science to the political system) are considered when evaluating the quality of CaS, while CoS is assessed based on reach, comprehension, and changes in values and behavior of the audience (Newig et al., 2013). Consequently, Slutskiy and Itdhiamornkulchai (2024) identified that, in corporate contexts, "when we evaluate sustainability communications, it is essential to acknowledge that while all sustainability communications have sustainability-related content, not all are intended for achieving sustainability-related impacts" (p. 1559).

Finally, due to the goal of driving social transformation, communication *for* sustainability (CfS) can be considered the highest level of communication in terms of impact on sustainable development by emphasizing normative aspects toward sustainable development. While CfS tends to use the flows of communication of both CaS and CoS for providing knowledge and creating collaborations, it is evaluated at a deeper level in terms of "measurable action towards sustainable development" (Newig et al., 2013, p. 2980). An example of *CfS by using CoS* is on-site texts and videos to encourage e-waste recycling among university students without talking to them, while an example of *CfS by using CaS* is the use of social media to persuade consumers to use sustainable techniques of washing and mending (Slutskiy & Itdhiamornkulchai, 2024).

1.3 Strategic Corporate Sustainability Communications in "Glocal" and "Lobal" Contexts

Applying the preceding conceptualizations of existing sustainability-related terminologies, the authors propose the term *"strategic corporate sustainability communications"* as organizations' communication regarding the appropriate balance of their goals toward sustainable development, the sustainable impact of their implementations, and the tangible sustainable benefits of their products and/or processes that are integrated across all channels of communications as well as facilitating interactive dialogues concerning sustainability with stakeholders.

Concerning the localization of the SDGs, collaborations between international organizations and national governments concerning sustainability initiatives tend to be effective in engaging local stakeholders and communities (Ansell et al., 2022). In line with the renowned "think globally, act locally" slogan in UN's Agenda 21 (Ansell et al., 2022), the adaptation of global strategies to local contexts to reach sustainable outcomes and actively enhance stakeholder engagement has been conceptualized by communication scholars as *glocal CSR communication* (Jain & De Moya, 2013; Sthapitanonda & Itdhiamornkulchai, 2022). *Corporate social responsibility (CSR) communication* is defined as, "a communicative practice, which corporations undertake to integrate social, environmental, ethical, human rights, and consumer concerns into their business operations and core strategy in close collaboration with their stakeholders" (Nielsen & Thomsen, 2018, p. 493), where "CSR is a term whereby businesses actively incorporate social and environmental issues into their business operations and communicate with their stakeholders" (Abdullah & Abdulrahaman, 2020 p. 600). Ad hoc CSR activities are insufficient nowadays, and businesses should systematically integrate CSR initiatives and genuinely incorporate sustainability proactively into their business models to succeed in the long term (Einarsdottir & Christiansen, 2014). Since Siano et al. (2016) identified that "Sustainability communication thus includes CSR communication that exclusively concerns issues of environmental protection and social responsibility in relation to economic success" (p. 2) and "findings from previous CSR communication studies can be usefully "transferred" to sustainability communication" (p. 2), this chapter extends previous conceptualizations of glocal CSR communication to define *"glocal sustainability communication"* as the adaptation of global sustainability-related goals and strategies to implement and communicate sustainability initiatives in each local context appropriately and favorably with measurable impact toward sustainable development.

In contrast to glocal strategies, the lobal marketing strategy corresponds to the concept of "think local, act global," which is defined from a marketing perspective by Kiliç (2011) as the production of products and services in local markets that are expanded to global markets for consumption. This chapter applies the lobal strategy to conceptualize *"lobal sustainability communication"* as the process of communicating sustainability-related initiatives implemented in local contexts to worldwide audiences at the global level by using media channels and communication strategies that are globally widespread and enable dialogues across continents. The next section explores *"glocal" and "lobal" strategic corporate sustainability communications* of Unilever as a corporate brand, as well as Dove, one of Unilever's product brands.

2 Part 2: Unilever's "Glocal" and "Lobal" Strategic Corporate Sustainability Communications for SDGs

Unilever's global sustainable business strategy, entitled the Unilever Compass, consists of Unilever's 10 "Purposes in action" (Unilever, 2024b), each of which contributes to specific SDGs, aiming for a total of 15 out of 17 SDGs. Specifically, one of Unilever's "Purpose in action," entitled "Equity, Diversity and Inclusion" aims to contribute to SDG#5: "Gender equality" and SDG#10: "Reduced inequalities" (Unilever, 2024a). Recently, Dove, one of Unilever's product brands, implemented several glocal sustainability campaigns in local contexts that are in line with Unilever's global purpose of "Equity, Diversity and Inclusion," including petitions "to end hair-race based discrimination" in the US and the UK and Dove's #LetHerGrow campaign in Thailand to end forced haircut regulations at Thai schools (Unilever, 2024c). By communicating these glocal campaigns not only on local websites and social media accounts of Dove in each country but also on Unilever's global website and official social media accounts, both *glocal and lobal strategic corporate sustainability communications* are evident as displayed in Tables 1 and 2, respectively.

Among these three glocal sustainability-related campaigns, this chapter focuses on corporate sustainability communications and stakeholders' digital co-creation of Dove's award-winning #LetHerGrow campaign in Thailand, which received two Bronze awards at the 2023 Cannes Lions: in the Health and Wellness category (Bangkok Post, 2023) in the "Brand-led Education & Awareness > PR" sub-category and the "Glass: The Lion for Change" category (Edelman, 2023). Since 1972, public schools in Thailand have adhered to nationwide policies of strict short hairstyles for Thai primary and secondary school students: "a crew cut for boys and pudding-bowl bob for girls" despite that these regulations were revoked in 1975 (Khaosod English, 2020), including the requirement of shorter hair than the earlobe for female students (Unilever, 2022). After many protests (Bangkok Post, 2020), the Ministry of Education in Thailand officially eased hairstyle regulations in 2020 by allowing "neat" short/long hair (The Standard, 2022) and forbidding physical punishments that involve cutting students' hair (Bangkok Post, 2022a; The Nation, 2022). Still, schools have the right to set haircut rules, which leads to diverse interpretations of the scope of "neat" hairstyles (Bangkok Post, 2022a; The Standard, 2022). Furthermore, physical haircut punishments persist against the Ministry of Education's 2020 policy (The Standard, 2022). Hence, students continually protest against haircut regulations (Khaosod English, 2020), including over 150,000 tweets,

Table 1 Examples of "glocal" sustainability communication of Dove and Unilever via digital media for Dove's "glocal" sustainability-related campaigns in the USA, the UK, and Thailand

"Glocal" channels	Dove US	Dove UK	Dove Thailand
Dove's website for each country: the webpage with the call to action	"Race-based hair discrimination starts early" "Sign the petition now"	"Untangling black hair discrimination" "Signing a petition"	"Let Her Grow" "Click to end haircut rules"
Dove's website for each country: the webpage that communicates each campaign's impact	"RACIAL EQUITY: The CROWN Act" "516,929 of 600,000 signatures"	"Uprooting hair discrimination with the Halo Code"	https://www.dovelethergrow.com/ "709 Comments, 2.79 K Supports" (A QR code could not be generated due to website maintenance) (Screenshots of the website are available upon request)
Dove's YouTube channel for each country: the main video clip of each campaign	"Dove | As Early As Five | End Race-Based Hair Discrimination" (January 26, 2022)	"Dove | Hair Power: Me and My Afro | A Black History Month documentary" (October 23, 2020)	#LetHerGrow (May 3, 2022)
Dove's official Facebook account for each country: selected post from each campaign	Facebook Post (February 24, 2022)	Facebook Post (May 11, 2021)	Facebook Post (May 3, 2022)

Table 2 Examples of "lobal" sustainability communication of Dove and Unilever via digital media for Dove's "glocal" sustainability-related campaigns in the US, the UK, and Thailand

"Lobal" channels	Dove US	Dove UK	Dove Thailand
Unilever's global website: "Sustainability" "Take Action" Page	"Create a respectful and open world for natural hair" (February 4, 2021)	"Legal Freedom for UK women to wear hair their way" (February 4, 2021)	"Forced haircuts in Thailand—let's liberate not dictate" (September 23, 2022)
Unilever's global Facebook page	#AsEarlyAsFive (February 22, 2022)	(Not found)	#LetHerGrow (September 26, 2022)
Unilever's global X (formerly Twitter)	#AsEarlyAsFive (February 22, 2022)	(Not found)	#LetHerGrow (September 26, 2022)
Dove's global X (formerly Twitter)	#AsEarlyAsFive (January 26, 2022)	(Not found)	#LetHerGrow (September 27, 2022)

where a comment received over 12,000 likes (Thai PBS World, 2022). Dove's research conducted by YouGov (Thailand) Co., Ltd. revealed that 74 percent of respondents answered that forced haircut rules still exist at Thai schools (The Nation, 2022; The Standard, 2022). Then, Dove launched the #LetHerGrow campaign to join the social movement in Thailand to end forced haircut policies at Thai schools.

Thailand is a notable Southeast Asian country to explore sustainability communication in the local context. Multinational corporations (MNCs) in Southeast Asian countries are noteworthy to examine due to their unique local cultures and the MNCs' increased interest in these emerging markets (Mak et al., 2015). Sustainable development in Thailand has been inspired by King Rama IX through his Sufficiency Economy philosophy, making Thai stakeholders involved in sustainability issues and community development (Sison & Sarabia-Panol, 2018). Moreover, Thailand has developed a 20-year national development strategy (2018–2037) that integrates the SDGs (National Strategy Secretariat Office, 2019).

3 Part 3: Analysis of Dove #LetHerGrow in Thailand and Stakeholders' Co-Creation

Dove's #LetHerGrow campaign to advocate for the elimination of forced haircut regulations at Thai schools consists of both *glocal and lobal strategic corporate sustainability communications* that correspond to the communication *of* sustainability (CoS), communication about sustainability (CaS), and communication *for* sustainability (CfS) in the typology of sustainability communication modes of Newig et al. (2013) explained earlier in this chapter, as well as the three CSR communication strategies developed by Morsing and Schultz (2006), comprising the *stakeholder information strategy* (*one-way communication* from organizations to stakeholders), the *stakeholder response strategy* (*two-way asymmetric communication* by receiving stakeholders' responses), and the *stakeholder involvement strategy* (*two-way symmetric communication* with stakeholders' participatory co-creation of corporate actions). In line with the insufficiency of one-way communication (Morsing & Schultz, 2006) and the paradigm shift from CoS (one-way communication) to CaS (two-way communication) (Newig et al., 2013), the unique communication strategy of Dove and Unilever is the integration of direct links to CaS channels in every CoS channel.

Initially, using CaS via digital media to enable stakeholder responses, Dove launched the #LetHerGrow campaign as a 1.59-min video on the official YouTube channel entitled @Dovethailand with over 118,000 subscribers, as follows:

1. The Thai subtitle version on May 3, 2022 (18,904,192 Views, 1.2 K Likes, 147 Comments as of July 2023), posted on Dove's official Facebook account for Thailand (8.1 M Views, 2.6 K Reactions, 183 Comments, 1.8 K Shares as of July 2023) on the same day (glocal sustainability communication)
2. The English subtitle version on May 5, 2022 (3615 Views, 79 Likes, 8 Comments as of July 2023), subsequently posted on Unilever's global Facebook and Twitter accounts on September 26, 2022 (lobal sustainability communication)
3. A 15-s clipping of the 1.59-min Thai subtitle clip on May 9, 2022 (4,686,976 Views, 89 Likes, 29 Comments as of July 2023) (glocal sustainability communication)

All three YouTube videos contain the link to Dove's #LetHerGrow website in Thai (glocal sustainability communication). This website at https://www.dovelethergrow.com/ contains the embedded 1.59-min YouTube clip (Thai subtitle version) to encourage external stakeholders to click on "Support" and write "Comments" about their experiences regarding Thai school haircut regulations and punishments (stakeholder response strategy). Dove's #LetHerGrow website received over 700 comments and over 2700 supports as of September 23, 2022. This achievement was featured on Unilever's global website as an article dated September 23, 2022, in the "Equity, diversity & inclusivity news" section in the "Take Action" webpage in the "Sustainability" tab with the link to Dove's English subtitle version clip on YouTube (Unilever, 2022) (lobal sustainability communication).

Dove also used traditional mass media (CoS) to reach local and global stakeholders by creating full-page advertisements on the front pages of two national newspapers in Thailand: Thairath (glocal sustainability communication in Thai) and Bangkok Post (lobal sustainability communication in English) (stakeholder information strategy), both advertisements contain the QR code that links to #LetHerGrow webpage on Dove's website for Thailand at https://www.dove.com/th/LetHerGrow.html that contains the call-to-action entitled "Click to end haircut rules" with Dove's #LetHerGrow website at https://www.dovelethergrow.com/ as the landing page for stakeholders to share personal experiences publicly (CaS; stakeholder response strategy).

Ultimately, to achieve CfS by directly collaborating with Thai stakeholders to end forced haircuts, Dove integrated on-site and online channels of CaS to co-create this social change for sustainable development. On May 24, 2022, Dove organized an on-site panel discussion entitled "#LetHerGrow – Creating a Future Where Students Can Grow into the Greatest Version of Themselves" at the Town Hall of Unilever House in Thailand. The same QR code was featured on this event's backdrops to encourage attendees to write public comments on Dove's #LetHerGrow website. Dove invited key internal and external Thai stakeholders to this panel discussion to share solutions to end haircut regulations at Thai schools (stakeholder involvement strategy), including a representative from the Ministry of Education, the head of the training program of the Girl Guides Association of Thailand, an adolescent medicine specialist, and several influencers with personal experiences on forced haircuts (Bangkok Post, 2022b; ThaiPR.NET, 2022). The latter two stakeholders are Thai influencers who used the hashtags: #DovePartner and #LetHerGrow to post about this panel discussion on social media and other posts that initiated conversations with external stakeholders concerning the pain points of forced haircuts (stakeholder response strategy).

Two of the influencers who used #DovePartner also extended stakeholder dialogues (CaS) by using #LetHerGrow for other posts beyond forced haircuts at schools a few months after posting about the panel discussion. One influencer engaged in conversations in Thai with the audience about the potential of hair loss after giving birth to children, where many comments shared personal experiences. Another #DovePartner shared her decision-making processes and discussions with her daughter regarding potential physical health risks before permitting her daughter to dye her hair, leading to lots of dialogues in Thai among stakeholders in the comments regarding the safety of different hair dyeing ingredients and reasons for and against permitting children to dye their hair. These conversations reflect that, although Dove's glocal #LetHerGrow campaign emphasizes communication of SDG#5: "Gender Equality" (empowering girls) and SDG#10: "Reduced Inequalities" (equal rights for everyone to choose hairstyles), which are the two main SDGs that Unilever's "Purpose in action," entitled "Equity, Diversity and Inclusion" aims to contribute to, Dove's communication that is related to SDG#3: "Good Health and Well-Being" with the focus on mental health (confidence and self-esteem) has been extended to discussions among stakeholders on physical health.

Regarding SDG#3: "Good Health and Well-Being," while some stakeholders in the comments of some #LetHerGrow posts still agree with the traditional Thai belief that short hair is beneficial for children's physical health, for instance, "Countryside kids have a lot of lice, so they are instructed to keep their hair short. When they grow up at the end of high school to love beauty and maintain hygiene, they can have long hair. Haha" (Thai comment on Dove's YouTube clip translated into English), many stakeholders' comments shared different perspectives that, rather than controlling hair length, children should be educated about personal hygiene, especially in the comments on the #LetHerGrow Facebook Live (9.8 K Views, 445 Reactions, 209 Comments, and 34 Shares) of a Thai influencer who did not use the #DovePartner hashtag and invited one of the #DovePartner influencers to share experiences in the Live video. These comments in Thai include, "Instead of telling children to cut their hair, is it better to tell students to maintain personal hygiene instead?" and "I have waist-length hair that is fragrant, soft, smooth, and very clean."

Regarding SDG#5: "Gender Equality" and SDG#10: "Reduced Inequalities," while Dove's #LetHerGrow campaign focuses on girls and women, many stakeholders discussed that the #LetHerGrow movement should not be limited to females. Before the preceding Facebook Live, the influencer wrote in Thai in his previous post (1.3 M Views, 52 K Reactions, 984 Comments, 5.2 K Shares) that, "...I'd like to add that boys or children of any gender may have experienced something [forced haircuts] like this, so I'd like to add #LetHimGrow and #LetThemGrow and let's continue talking on [Facebook] Live soon." Then in his Facebook Live, he added in Thai, "We're not just talking about girls having their hair cut to the size of their earlobes. We're talking about the boys whose heads were almost shaved...unacceptably...at Thai schools" (both SDG#5 and SDG#10). Furthermore, a Thai comment with 101 Likes on an Instagram post by a news/media publisher stated, "And it is not only girls. Every LGBTQIAN+ also feels like their arms and legs are cut. Press this issue" (both SDG#5 and SDG#3).

On X (formerly Twitter), several Tweets in English from Westerners with #LetHerGrow contained visuals of the campaign, including Dove's video clips and newspaper advertisements, along with compliments in the captions, while comments/replies from Westerners in English also shared direct experiences concerning SDG#10, including "The nursery cut my daughter's hair without asking for pernision [sic] to do it. They couldn't even understand why I was in disbelief for what they did" and opinions related to SDG#5, such as "It's both boy and girl," supported by the reply, "Absolutely agree, and they should do one tomorrow about not shaving boys' heads."

4 Part 4: Proposed Model for Strategic Corporate Sustainability Communications of MNCs

After analyzing case studies of Dove's glocal sustainability-related campaigns, especially the award-winning Dove #LetHerGrow campaign in Thailand, this chapter proposes *"The Glocal and Lobal Infinite Interaction Model"* in Fig. 1, by integrating three main concepts: (1) the typology of sustainability communication modes (Newig et al., 2013), (2) the three CSR communication strategies (Morsing & Schultz, 2006), and (3) glocal and lobal sustainability communication (developed by the authors earlier in this chapter).

As shown in Fig. 1, *The Glocal and Lobal Infinite Interaction Model* developed by the authors in this chapter portrays the infinite dialogic loop of strategic corporate sustainability communications between MNCs and local and global stakeholders in six stages. In *Stage 1: "Listen to the local stakeholders' concerns in each local context,"* Dove conducted surveys among Thai students, women, parents, and teachers regarding current perceptions and experiences of forced haircuts at Thai schools. Although the stakeholder response strategy is used to obtain insights, this stage focuses on top-down managerial decisions since the insights are applied to initiate strategic communications from organizations to stakeholders.

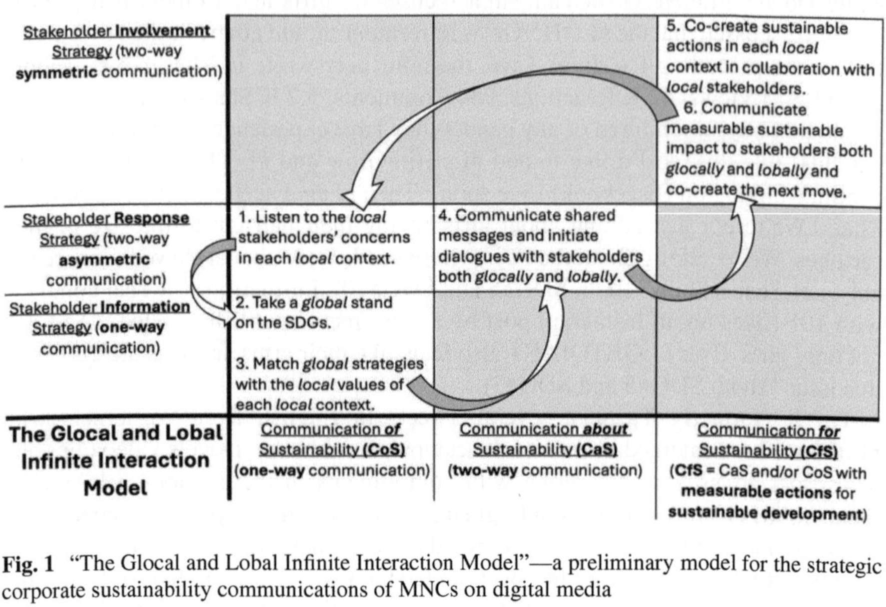

Fig. 1 "The Glocal and Lobal Infinite Interaction Model"—a preliminary model for the strategic corporate sustainability communications of MNCs on digital media

In *Stage 2: "Take a global stand on the SDGs,"* Dove's #LetHerGrow campaign contributes to Unilever's global purpose on "Equity, Diversity and Inclusion" (SDG#5 and SDG#10). In *Stage 3: "Match global strategies with the local values of each local context,"* Dove localized these SDGs to end forced haircuts at Thai schools to enhance girls' confidence. One-way communication *of* sustainability (CoS) is used in both stages to provide information to stakeholders prior to initiating two-way communication in subsequent stages.

In *Stage 4: "Communicate shared messages and initiate dialogues with stakeholders both glocally and lobally,"* Dove launched the #LetHerGrow campaign on websites and social media to encourage stakeholders to share opinions and experiences (stakeholder response strategy; CaS). In *Stage 5: "Co-create sustainable actions in each local context in collaboration with local stakeholders,"* Dove organized a panel discussion by inviting Thai stakeholders to discuss solutions to end haircut regulations, and #DovePartner influencers communicated this via social media (stakeholder involvement strategy; CfS). Finally, in *Stage 6*: *"Communicate measurable sustainable impact to stakeholders both glocally and lobally and co-create the next move,"* the impact of Dove's #LetHerGrow campaign was communicated on Dove's #LetHerGrow Thai website and Unilever's global website. The authors suggest using this opportunity to encourage stakeholders to share initiatives for future co-creations.

In conclusion, the successful case study of Dove's #LetHerGrow campaign in Thailand demonstrates that strategic sustainability communication should empower local values by providing virtual spaces for stakeholders to co-create social change agendas. By applying *The Glocal and Lobal Infinite Interaction Model*, organizations could make managerial decisions regarding the processes and synthesis of global and local channels for participatory co-creation of strategic corporate sustainability communications. Future research could apply this model to analyze other case studies or develop extended models to conceptualize and operationalize glocal and lobal sustainability communication.

References

Abdullah, Z., & Abdulrahaman, A. A. (2020). Antecedents and consequences of CSR communication management (CSRCM): An agenda for future research. *Journal of Critical Reviews, 7*(09), 600. https://doi.org/10.31838/jcr.07.09.119

Allen, M. (2016). *Strategic communication for sustainable organizations: Theory and practice.* Springer. https://doi.org/10.1007/978-3-319-18005-2

Ansell, C., Sørensen, E., & Torfing, J. (2022). *Co-creation for sustainability* (1st ed.). Emerald Publishing Limited. https://doi.org/10.1108/978-1-80043-798-220220001

Bangkok Post. (2020, November 8). *Education Ministry shakes up school dress, hairstyle rules.* https://www.bangkokpost.com/thailand/general/2015827/education-ministry-shakes-up-school-dress-hairstyle-rules

Bangkok Post. (2022a, May 4). *Forced haircut in ad sparks discussions on freedom.* https://www.bangkokpost.com/learning/advanced/2304562/forced-haircut-in-ad-sparks-discussions-on-freedom

Bangkok Post. (2022b, May 26). *Forced haircuts 'belittle kids' self-esteem'—Seminar pushes social change on hairstyle regulations in schools.* https://www.bangkokpost.com/thailand/general/2316138/forced-haircuts-belittle-kids-self-esteem

Bangkok Post. (2023, June 21). *'Let Her Grow' campaign wins global award: Multimedia focus on school hairstyles wins Bronze at 'Oscars of advertising'.* https://www.bangkokpost.com/life/social-and-lifestyle/2596556/let-her-grow-campaign-wins-global-award

Botan, C. H. (2021). Strategic communication: Field, concepts, and the cocreational model. In C. H. Botan (Ed.), *The handbook of strategic communication* (pp. 6–13). Wiley-Blackwell.

Canel, M. J. (2023). A call for the "public relations-imperative for sustainability": Roles for public relations in advancing the 2030 agenda. *Public Relations Review, 49*(4), 1–7. https://doi.org/10.1016/j.pubrev.2023.102368

Edelman. (2023, June 22). *2023 Cannes lions festival: Edelman wins gold in brand experience & activation along with four silver and four bronze lions.* https://www.edelman.com/news-awards/2023-cannes-lions-edelman-wins

Einarsdottir, S., & Christiansen, A. M. (2014). All to win. All to lose. *Sustainability Trends: Five Trends Shaping the Future.* https://www2.deloitte.com/content/dam/Deloitte/is/Documents/risk/Deloitte%20Sustainability.pdf?fbclid=IwAR2akVCXimWydnh3ljdfyl2FhPDAQFF9vfAXsmNc1l6LbU8W2ABq5vYbgqI

Elkington, J. (1994). Towards the Sustainable Corporation: Win-Win-Win Business Strategies for Sustainable Development. *California Management Review, 36*(2), 90–100. https://doi.org/10.2307/41165746

Elkington, J. (2004). Enter the triple bottom line. In A. Henriques & J. Richardson (Eds.), *The triple bottom line: Does it all add up?* (pp. 1–16). Earthscan.

Fischer, D., Lüdecke, G., Godemann, J., Michelsen, G., Newig, J., Rieckmann, M., & Schulz, D. (2016). Sustainability communication. In H. Heinrichs, P. Martens, G. Michelsen, & A. Wiek (Eds.), *Sustainability science* (pp. 139–148). Springer.

Godemann, J., & Michelsen, G. (2011). Sustainability communication—An introduction. In J. Godemann & G. Michelsen (Eds.), *Sustainability communication: Interdisciplinary perspectives and theoretical foundations* (pp. 3–11). Springer. https://doi.org/10.1007/978-94-007-1697-1_1

Golob, U., Podnar, K., & Zabkar, V. (2023). Sustainability communication. *International Journal of Advertising, 42*(1), 42–51. https://doi.org/10.1080/02650487.2022.2144035

Hallahan, K., Holtzhausen, D., van Ruler, B., Verčič, D., & Sriramesh, K. (2007). Defining strategic communication. *International Journal of Strategic Communication, 1*(1), 3–35. https://doi.org/10.1080/15531180701285244

Herrick, C. N., & Pratt, J. L. (2013). Communication and the Narrative Basis of Sustainability: Observations from the Municipal Water Sector. *Sustainability, 5*(10), 4428–4443. https://doi.org/10.3390/su5104428

Itdhiamornkulchai, C., & Taiphapoon, T. (2022). A review of frameworks for evaluating online corporate sustainability communications. In *Proceedings of the 2022 KU social sciences international conference: Roles of social sciences in sustainability transformation in the global south* (pp. 172–189) https://ssic.soc.ku.ac.th/ebook.pdf

Jain, R., & De Moya, M. (2013). Global, local, or glocal: Investigating CSR strategies of best corporate citizens in India. *International Journal of Strategic Communication, 7*(3), 207–226. https://doi.org/10.1080/1553118X.2013.782548

Khaosod English. (2020, July 3). *Schools still ignore gov't revised haircut rule.* https://www.khaosodenglish.com/politics/2020/07/03/schools-still-ignore-govt-revised-haircut-rule/

Kiliç, S. (2011). Relationship between lobal marketing and brand equity: The case of Baktat. *International Journal of Humanities and Social Science, 1*(13), 216–229.

Mahoney, L. M., & Tang, T. (2017). *Strategic social media: From marketing to social change.* Wiley-Blackwell.

Mak, A. K. Y., Chaidaroon, S., & Pang, A. (2015). MNCs and CSR engagement in Asia: A dialectical model. *Asia Pacific Public Relations Journal, 16*(1), 37–60.

Morioka, S. N., Evans, S., & de Carvalho, M. M. (2016). Sustainable business model innovation: Exploring evidences in sustainability reporting. *Procedia CIRP, 40*(2016), 659–667. https://doi.org/10.1016/j.procir.2016.01.151

Morsing, M., & Schultz, M. (2006). Corporate social responsibility communication: Stakeholder information, response and involvement strategies. *Business Ethics, the Environment & Responsibility, 14*(6), 323–338. https://doi.org/10.1111/j.1467-8608.2006.00460.x

National Strategy Secretariat Office. (2019). *National Strategy 2018–2037.* http://nscr.nesdc.go.th/wp-content/uploads/2019/10/National-Strategy-Eng-Final-25-OCT-2019.pdf

Newig, J., Schulz, D., Fischer, D., Hetze, K., Laws, N., Lüdecke, G., & Rieckmann, M. (2013). Communication regarding sustainability: Conceptual perspectives and exploration of societal subsystems. *Sustainability, 5*(7), 2976–2990. https://doi.org/10.3390/su5072976

Nielsen, A. E., & Thomsen, C. (2018). Reviewing corporate social responsibility communication: A legitimacy perspective. *Corporate Communications: An International Journal, 23*(4), 492–511. https://doi.org/10.1108/ccij-04-2018-0042

Sahadev, S., Muralidharan, S., & Singh, P. (2022). Introduction to the special issue on marketing communications and sustainability. *Journal of Marketing Communications, 28*(3), 227–231. https://doi.org/10.1080/13527266.2021.1942145

Schoeneborn, D., & Trittin, H. (2013). Transcending transmission. *Corporate Communications: An International Journal, 18*(2), 193–211. https://doi.org/10.1108/13563281311319481

Siano, A., Conte, F., Amabile, S., Vollero, A., & Piciocchi, P. (2016). Communicating sustainability: An operational model for evaluating corporate websites. *Sustainability, 8*(9), 1–16. https://doi.org/10.3390/su8090950

Signitzer, B., & Prexl, A. (2008). Corporate sustainability communications: Aspects of theory and professionalization. *Journal of Public Relations Research, 20*(1), 1–19. https://doi.org/10.1080/10627260701726996

Sison, M. D., & Sarabia-Panol, Z. (2018). Sustainable community development and risk mitigation in Thailand. In M. D. Sison & Z. Sarabia-Panol (Eds.), *Corporate social responsibility, public relations and community engagement: Emerging perspectives from Southeast Asia* (1st ed., pp. 117–136). Routledge. https://doi.org/10.4324/9781315734668

Sivarajah, U., Irani, Z., Gupta, S., & Mahroof, K. (2020). Role of big data and social media analytics for business to business sustainability: A participatory web context. *Industrial Marketing Management, 86*(2020), 163–179. https://doi.org/10.1016/j.indmarman.2019.04.005

Slutskiy, P., & Itdhiamornkulchai, C. (2024). Towards the evaluation of sustainability communication based on the intended goals. *Changjiang Liuyu Ziyuan Yu Huanjing/Resources and Environment in the Yangtze Valley, 40*(04), 1559–1570.

Sthapitanonda, P., & Itdhiamornkulchai, C. (2022). Communicating CSR across cultures: From global and local to glocal challenges. In A. O'Connor (Ed.), *The Routledge handbook of corporate social responsibility communication* (1st ed., pp. 46–59). Routledge. https://doi.org/10.4324/9781003184911-6

Stumpf, M., Brandstätter, M., & Borghoff, J. (2019). Sustainable project management and the role of integrated communications (IC). *European Project Management Journal, 9*(1), 10–20. https://doi.org/10.18485/epmj.2019.9.1.2

Thai PBS World. (2022, May 12). *Thai student hairstyles heat up social media.* https://www.thaipbsworld.com/thai-student-hairstyles-heat-up-social-media/

ThaiPR.NET. (2022, May 31). *Dove #LetHerGrow panel discussion unites leading voices in "creating a future where students can grow into the greatest version of themselves".* https://www.thaipr.net/en/education_en/3196563

The Nation. (2022, May 6). *Dove calls for solidarity in ending forced haircuts in schools.* https://www.nationthailand.com/pr-news/business/40015309

The Standard. (2022, May 3). *Mandatory haircut rules undermine self-confidence and the identity that was decreased with haircuts.* https://thestandard.co/dove-lethergrow/

Unilever. (2022, September 23). *Forced haircuts in Thailand—let's liberate not dictate.* https://www.unilever.com/planet-and-society/take-action/initiative/forced-haircuts-inthailand-lets-liberate-not-dictate/

Unilever. (2024a). *Equity, diversity & inclusivity.* https://www.unilever.com/sustainability/equity-diversity-and-inclusion/

Unilever. (2024b). *Sustainability.* https://www.unilever.com/sustainability/

Unilever. (2024c). *Take action.* https://www.unilever.com/sustainability/take-action/

United Nations Development Programme. (2024). *The SDGs in action.* https://www.undp.org/sustainable-development-goals

Weder, F., & Eriksen, M. (2023). *CSR communication and cultures of sustainability.* The University of Queensland. https://doi.org/10.14264/428d81b

Weder, F., Karmasin, M., Krainer, L., & Voci, D. (2021). Sustainability communication as critical perspective in media and communication studies—An introduction. In F. Weder, M. Karmasin, & L. Krainer (Eds.), *The sustainability communication reader* (pp. 1–12). Springer.

World Commission on Environmental Development. (1987). *Our common future.* http://www.un-documents.net/our-common-future.pdf

Chanapa Itdhiamornkulchai is a part-time lecturer at the General Education Center Chulalongkorn University, a part-time lecturer at the Language Institute Thammasat University (LITU), and a full-time PhD candidate at the Faculty of Communication Arts, Chulalongkorn University (Thailand). During her PhD studies, she has worked in several teaching and research careers at Chulalongkorn University, including being a part-time lecturer at the Bachelor of Arts in Communication Management International Program (BCM), a part-time lecturer at the Chulalongkorn University Institute (CULI), a guest lecturer of the Sustainable Environment Research Institute (SERI), a research assistant at the Research Unit on Communication Innovation for Development of Quality of Life and Sustainability, and a freelance researcher at the Faculty of Communication Arts. Within three years, she has delivered 15 international conference presentations, published two book reviews, and published six full papers, specifically a journal article on green marketing communications in the Journal of Communication Arts, a SCOPUS-indexed journal article on sustainability communication, a book chapter in the Routledge Handbook of Corporate Social Responsibility Communication, and three full international conference proceedings. Her research interests include sustainability communication, CSR communication, strategic communication, green marketing communications, and digital marketing.

Parichart Sthapitanonda, PhD is a Professor and Vice President of Chulalongkorn University in Bangkok (Thailand). Her research focuses on communication and campaigns on various topics, ranging from commercial to noncommercial and political issues.

Food Security and Farmers' Political Participation in Rural Communication, a Strategy of Sustainability Communication: Analysis of Sub-District Level Farmers' Political Behavior in Bangladesh

Mohummad Kamrul Hassan

Abstract This chapter proposes that rural communication, which delves into micro-level analysis such as farming communities in rural areas, can effectively facilitate sustainability communication in rural regions, thereby fostering a sustainable agricultural sector. Understanding farmers' political behavior and attitudes is essential for strategizing sustainability communication in rural contexts. Hence, this chapter offers an overview of food security, sustainability, political participation, rural communication, and sustainability communication concepts. Furthermore, the study investigates the effects of communication on farmers' political interests, political efficacy, and political participation. The study's results support the notion that political interest positively influences communication variables. Rural communication exhibits a negative association with political efficacy and challenges the notion that increased consumption of political information leads to higher political efficacy. Political participation is significantly correlated with rural communication, with communication frequency exerting a greater impact on political engagement than communication source reliability and use. Face-to-face interactions emerge as the most influential communication component in enhancing farmers' political participation.

1 Introduction

Food security, a prominent sustainable development goal (SDG), is closely tied to the political landscape, with democratic governance and free flow of information acting as a safeguard against famine (Sen, 1999), the most extreme form of food

M. K. Hassan (✉)
Green Fund Initiative, New York, NY, USA
e-mail: hassan@greenfundi.org

F. Weder (ed.), *Strategic Sustainability Communication*, CSR, Sustainability, Ethics & Governance, https://doi.org/10.1007/978-3-031-89486-2_19

insecurity. This reflects the view that political and civil rights contribute to the protection of economic and social rights, including the right to food (Zidouemba, 2017). The free dissemination of information plays a major role in prevention policies adverse to food security (Sen, 2000) and shapes the behavior of the citizens including the farmers.

Farmers, who play a central role in the food security system, are regarded as the nucleus of rural political, social, and economic activities but are often labeled as voiceless in the democratic process. Their political participation is elaborated as one of the key elements of functioning democracy in the global south. The notion of political participation refers to activities undertaken by the public to influence political decisions, either directly or by affecting the selection of persons who make policies (Conge et al., 1988; Deth, 2001). Farmers' political participation is not defined distinctly in the literature except there is contextual variance between farmers and "other citizens" who live in urban areas. Given that farming shapes the essence of rural life, the political behavior of farmers significantly impacts sustainable agricultural and rural development, including food security. Additionally, farmers' political efficacy, considered a prerequisite for political participation, involves the belief that individual political actions can influence the political process (Craig & Maggiotto, 1982). Thus, farmers' minimal level of political efficacy and knowledge is crucial for their active involvement in political affairs.

Social interactions, technology, and the environment all rely on communication to gain relevance and meaning in society (Ziemann, 2011). Similarly, the discourse of food security, political participation, and sustainability are not independent of communication. Rural communication, which goes hand in hand with sustainability communication, emerges as a key factor in shaping farmers' attitudes and behaviors, influencing their political participation. Communication for sustainability or sustainability communication is an emerging concept that can be operationalized in any context related to sustainable development. This chapter suggests that rural communication, focusing on micro-level analysis like farming communities in rural areas, can serve as a conduit for sustainability communication in rural regions, promoting a sustainable agricultural sector. Rural communication involves facilitating participatory communication processes to meet the knowledge and information needs of rural populations (Hassan, 2020). By integrating sustainability communication with rural communication, positive impacts on food security can be achieved, as rural communication strategies are tailored to farmers and agriculture.

This study focuses on rural communication and its influence on the political attitudes and behaviors of farmers to gain a comprehensive understanding of their overall political participation process. Understanding farmers' political attitudes and behaviors is significant in strategizing sustainability communication for rural areas.

The study investigates various components of rural communication, a potential strategy of sustainability communication, such as farmers' sources of political information, communication frequency, and the reliability of these information sources. Additionally, farmers' interest in politics is considered, as political interests drive activities such as seeking information, engaging in interpersonal discussions for satisfaction (Bandura, 1997; Salmon, 1986), and sharing information with others. Political participation, in this study, is assessed by farmers' involvement in political discussions both online and offline, sharing political information with others (such as fellow farmers, friends, family, and neighbors), attending political events, contributing financially to political causes, frequency of political engagement, and participation in voting during elections. Political efficacy is measured by farmers' confidence in their ability to engage in politics and their confidence in the political system, which allows them to influence government decisions. The study explores the correlations between communication factors and the processes of political participation and political efficacy to gain a deeper understanding of the impact of rural communication.

2 Rural Communication as a Sustainability Communication Strategy

Understanding "sustainable development" is as varied as its strategies to achieve it in practice (Ziemann, 2011), and it necessarily involves communication (Newig et al., 2013). The notion of sustainability communication is a blend of two disciplines: sustainability and communication, where the goal is to achieve sustainability in a particular context and communication is the process. Similar to sustainability, which spans multiple disciplines, sustainability communication also operates at the crossroads of various fields and disciplines (Godemann, 2011), including political affairs as sustainability serves as a guiding principle within political discourses (Weder et al., 2023). It is a global social process (Ziemann, 2011) of application of communication strategies and principles for better socio-economic and ecological life. This concept is applied at different levels of the public space and between different social systems (Golob et al., 2023) including the rural context. Strategic sustainability communication in rural contexts is the planned dissemination of information aimed at fostering sustainable agricultural practices, enhancing environmental stewardship, and supporting the socio-economic well-being of farming communities. It involves the use of culturally relevant communication channels and methods to engage rural populations, influence their attitudes and behaviors, and encourage active participation in sustainability initiatives. By aligning communication strategies with the unique needs and challenges of rural areas, this approach ensures that sustainability efforts are effective, inclusive, and sustainable over the long term. As an emerging field with somewhat unspecific boundaries (Weder et al., 2021), there is a pressing need to delve deeper into strategic sustainability communication and its potential integration into various domains such as rural communication, food security, and the political participation of rural populations. It could be integrated with any theme related to sustainable development such as food security and political participation supported by theories of rural communication, as Ziemann

(2011) prioritizes sustainability as a central interdisciplinary research topic for economic and political realms.

Communication is embedded in the dynamic process of influencing political participation. Much research focuses on examining the impact of communication on political participation and how communication impacts political behaviors (e.g., McLeod et al., 1994; Price, 1989; Shah & Scheufele, 2006). The concept of rural communication emphasizes micro-level analysis. Rural communication is composed of mass media, interpersonal communication sources, rural traditional media, community media, and new media. The composition of rural media ecology is different because of the existence of rural traditional media, community media, and the issue of low access to ICT. These media have the potential to affect different aspects of democratic participation in different parts of the population in different ways based on their socio-demographic characteristics (Carpini, 2004). The communication structure in rural areas is shaped by agricultural stakeholders whose primary concern is to exchange agricultural and rural development information. However, we cannot deny the existence of implicit and explicit political properties in rural media and communication sources. Moreover, political factors are naturally embedded in every aspect of rural life.

Given the distinct communication structure in rural areas compared to urban settings, sustainability communication strategies must be customized to meet the needs of rural populations. Thus, integrating rural communication can significantly contribute to achieving sustainable development goals, including food security.

3 Implications of Food Security on Political Participation

Food security, a multifaceted issue influenced by various social, political, economic, and technical factors, has garnered attention primarily for its socio-economic dimensions, with political factors often receiving less focus in scholarly studies. While scholars have debated the relationship between political factors such as democracy and food security, recent research sheds light on the nuanced interplay between politics and food security.

Harris (2014) delved into the impact of political aspects, such as democracy and freedom, on food security across Sub-Saharan African nations since 1990. Contrary to conventional wisdom, procedural and institutional aspects of democracy were found to have no direct correlation with hunger levels in Africa. However, the protection of civil liberties emerged as a significant factor associated with improved food security.

Expanding on this, Schmitt-Sands (2017) provided a comprehensive analysis of democracy's dimensions affecting social welfare. Her study revealed that both political rights and civil liberties are positive predictors of food security. Notably, an increase in civil liberties showed a more consistent and robust association with food security compared to political rights.

Additionally, insights from Smith (1998) underscored the role of responsive political systems, viable opposition forces, and a free press in positively influencing food security. Amartya Sen, a Nobel laureate in Economic Science, articulated a seminal theory linking famine, democracy, and a free press (Sen, 1999). Sen's work highlighted the absence of substantial famines in democratic countries with relatively free media, suggesting a protective role of political and civil rights, including the right to food. However, scholarly discourse on Sen's thesis has been varied. While scholars like Rubin (2009) have expressed reservations about the direct role of democracy and a free press in famine prevention, others like Banik (2007) have provided empirical support for Sen's arguments, especially in the context of India's democracy.

Furthermore, reports from organizations like the Famine Early Warning Systems Network (FEWSNET 2013) underscored the political dimensions of famines, attributing them to catastrophic political failures rather than natural phenomena. Sen's theory elucidates democracy's instrumental value in addressing citizens' claims, particularly in famine prevention. Democratic mechanisms, including free elections and media, incentivize authorities to implement preventive policies and facilitate information dissemination crucial for early warning systems.

Moreover, governments wield significant influence in molding the physical, social, and economic aspects of a nation's food security. Thus, endeavors to improve agriculture and food security must recognize the pivotal role of governance which is strongly intertwined with effective political participation as Nekola (2006) identified a strong correlation between the level of political participation and selected indicators of governance performance.

4 Nexus of Food Security, Sustainability, and Political Participation

Food security is a global concern that is significantly correlated with a country's level of development (Pérez-Escamilla, 2017). The concept is impacted by numerous factors including climate change and political stability (Wischnath & Buhaug, 2014), connected to effective political participation. In a broader perspective, food security is interrelated with sustainable food and agricultural sectors, where the needs of present and future generations are met in commitment to environmental, social, and economic dimensions (FAO, 2014). It is one of the prominent sustainable development goals under SDG two. However, it is generally accepted that sustainability needs to be more appropriately addressed in the pillars of food security, maybe as a new dimension (Berry et al., 2015), or integrating other variables such as political factors, where the food crises and climate change (Singh et al., 2021) have its importance. Food security and sustainability are interrelated. However, the associated contexts are complex and uncertain (Eakin et al., 2017). Therefore, these interrelationships deserve unbroken attention from multidisciplinary approaches

(Green et al., 2020). Specifically, the political dimension is crucial, as viable policies and their implementation are also crucial for sustainable food security (Eyduran et al., 2020).

5 Political Participation in Achieving Sustainable Development Goals

Over recent decades, there has been a notable emergence of sustainability and ecological interest groups influencing citizen engagement, political participation, and democratic innovations, particularly at the local level (Kersting, 2021). The Local Agenda 21 strategies introduced at the Rio Conference in the 1990s fostered a transition toward more deliberative forms of political engagement focused on sustainability goals. The 2030 Agenda for sustainable development presents a clear mandate for public participation, outlining a shared vision to "ensure responsive, inclusive, participatory, and representative decision-making at all levels" (United Nations, 2015).

Achieving sustainability relies on political commitment, influenced by citizens' involvement in politics. The interconnection between political engagement and sustainability is profound (Kersting, 2021), as political choices significantly affect environmental, social, and economic sustainability. Therefore, farmers' political participation such as voting, advocacy, and interaction with elected representatives shape the creation and execution of sustainability policies. Furthermore, political participation can sway the passage of sustainability-related legislation. The transition toward a sustainable society involves more than just technical aspects; it requires navigating intricate normative and political decisions (Meadowcroft, 2009). Without citizen involvement and inclusive decision-making, there is a risk of these decisions being perceived as illegitimate (Hendriks, 2009).

6 Methodology of the Study

A survey was conducted among farmers from two sub-districts in northern Bangladesh who were members of a crop intensification project. About 60 respondents were randomly selected from the project's list, and data were collected through face-to-face interviews using a questionnaire during August–September 2022. The collected data were analyzed using IBM's SPSS statistics software (version 29.0.0.0 (241)), Microsoft Excel, and Python Packages. Spearman's correlation analysis was employed to examine the relationship between variables, particularly dealing with ordinal data such as Likert Scale data. The study investigated the relationships between political interest and rural communication, rural communication and political efficacy, and rural communication and political participation.

7 Assessment of Rural Communication's Influence on Political Dynamics

The analysis of the study offers an array of significant insights into how communication variables—rural communication sources, frequency of communication of farmers with those sources, and reliability of those sources—have an impact on the political efficacy and political participation of the farmers in the context of rural communication ecosystem.

7.1 Rural Communication Sources, Frequency, and Reliability

The analysis of rural communication features provides an in-depth overview of farmers' use of communication sources, their frequency of communication sources, and the reliability of those sources. The dominance of political actors as information sources in the study could be explained by the fact that (according to Baldersheim et al., 2001) the rural people are more easily reached by the political actors than big city actors. Mass media (TV) is also a popular information source for farmers as they spend their free time in spaces such as rural tea stalls, a common gathering place for farmers, where they watch TV, chat with others, and buy tea. This is also supported by the study conducted by Ramachandran et al. (2006). Political information received from a social media user has come up as an important information source in the study. The users of social media in rural areas exchange political information with non-social media user farmers. In popular gathering spaces such as tea stalls in Bazaars, it is common for social media users to exchange information with non-users. This exchange often includes experiences from social media, which can spur political participation and decision-making among non-users. This phenomenon resembles the two-step flow of communication theory by Lazarsfeld, where information flows from social media to media users and then to a wider population, such as farmers. The result seems inconsistent in that the farmers communicate more with interpersonal communication sources (i.e., family, friends, other farmers, etc.) but get more political information from political actors. This could be because the communication with the political actors is not reciprocal (i.e., political actors send a message to farmers and it is received, but the farmers do not provide any feedback).

Reliability in this study refers to how trustworthy a political information source is perceived to be. Regarding the reliability of information sources, the farmers appear silent as most of them are neutral on the statements of trustworthiness. The farmers have more trust in interpersonal sources such as family, friends, neighbors, and other farmers; mass media; and community media than political actors, social media, and information from social media users. The reliability of interpersonal sources can be explained by Blau's (1964) social exchange theory, stating that trust grows as people interact, interpret each other's behavior, and reciprocate. Trust is also positively affected by the length of a relationship (Lewicki & Bunker, 1996). As two people interact and learn about each other, their relationship matures and the trust among them grows (Hassan et al., 2016). Communication in rural areas is a complex phenomenon as many other factors are embedded in the process. Political information is often biased and intended to achieve a political goal. Moreover, the trust in political information is also filtered through the farmer's political beliefs and ideology. Therefore, future research should focus on this area of trustworthiness to develop a more precise methodology for understanding this phenomenon.

7.2 *Influence of Political Interest on Rural Communication*

This study found a positive relationship between the political interest of the farmers and the use of political communication sources (r_s = 0.28, p < 0.032), their frequency of communication ($r_s = 0.40, p < 0.001$), and sharing of political information with others ($r_s = 0.30, p < 0.017$). Therefore, this study supports the motivating basis of political interest. Previous studies (Kenski & Stroud, 2006; Shah et al., 2005) produced similar results where political interests significantly encouraged the respondents to pay attention to political information (Zhou, 2009). Consequently, increasing political efficacy and shaping political participation behaviors (Zhou, 2009). More recent data shows that political interest has become a more important determinant of political news consumption (Stromback et al., 2013). Although the analysis has produced similar results in prior research. Testing farmers' political interests is the contribution of this study as no previous research has focused on this in the context of rural communication and the farming community.

7.3 *Influence of Rural Communication on Farmer's Political Efficacy*

Previous research (Park, 2019; Pinkleton et al., 1998) suggests that people who seek more political information and consume news in traditional and social media have higher political efficacy. Zhou and Pinkleton (2012) found that using social media and public affairs websites positively predicts political efficacy. However, this study shows a negative association between communication variables such as communication sources and farmers' political efficacy ($r_s = -0.041$, $p < 0.0757$); between farmers' communication frequency and political efficacy ($r_s = -0.205, p < 0.116$); and farmers' reliability on the communication sources and political efficacy ($r_s = -0.106$, $p < 0.418$). The result is not surprising as other research has also shown similar results with using of media type and political efficacy. For example, online political interaction does not predict political information efficacy and instead shows a negative influence on external efficacy (Zhou, 2009). On the other hand, Tewksbury et al. (2008) demonstrated that newspaper readers exhibited higher levels of political efficacy compared to television and Internet users, whereas Kushin and Yamamoto (2010) found that radio listeners appeared to be more politically efficacious than users of social media platforms. Therefore, it is evident that not all the studies produce similar results. The farmers' political efficacy might be associated with other factors than direct communication intervention such as the prior research showcases, for example, cynicism and apathy are negatively associated with political efficacy (Zhou, 2009). Farmers' rural communication may act as a mediator to produce those feelings of cynicism and apathy, leading to a negative association since Carpini (2004) supports the notion that media use can also foster cynicism, apathy, ignorance, and disengagement. Apart from this theoretical reasoning, future research can conduct in-depth analyses of farmers' political efficacy and rural communication.

7.4 *Influence of Rural Communication on Political Participation*

This study shows that communication variables have a positive effect on farmers' political participation. The communication sources and political participation ($r_s = 0.37$, $p < 0.003$) as well as the reliability of the sources and political participation ($r_s = 0.31$, $p < 0.015$) have indicated a moderate relationship. The frequency of communication with sources has shown comparatively stronger associations with political participation ($r_s = 0.51$, $p < 0.001$). The communication sources in this study are not only mass media and online media but also interpersonal communication sources such as family, friends, neighbors, other farmers, and political actors. Therefore, this study cannot conclude that overall media has an impact on political participation rather political participation of farmers is predominantly influenced by face-to-face interactions. Similar interpretations are also applicable in the case of frequency of communication as farmers communicate more frequently with their reliability on interpersonal sources. The outcome was in line with earlier empirical findings, stating that political mobilization is likely dependent on comprehensive deliberation, or political dialogues among citizens, rather than solitary learning or inward contemplation (Zhou, 2009). Interpersonal communication contains more political information for repeated exposure, helping to create a comprehensive understanding of overcoming the ambiguities of what they read or watch in news media, promoting social responsibilities, or providing grounds for debate. Therefore, it could be concluded that interpersonal communication exerts a much stronger influence on political participation.

8 Conclusion

Tailoring sustainability communication strategies to rural populations is crucial due to the unique communication structures in rural areas. Integrating rural communication into these strategies can contribute to achieving sustainable development goals, such as food security. The connection between communication, political participation, sustainability, and food security is indisputable. This study provides significant insight into farmer's political participation and their political efficacy in the context of rural communication. The dynamic process of influencing political behavior involves communication as an impact factor. The study establishes the theoretical notion and empirical findings that political interest is a stirring factor that positively associates with communication variables. Rural communication shows a negative association with political efficacy and rejects the notion that consuming more political information can cause higher political efficacy. The result of this study supports the concept that media use can enhance cynicism, apathy, ignorance, and disengagement in political environments that may be influencing political efficacy negatively. Rural communication shows a significant positive relationship with political participation. Specifically, the frequency of communication influences political participation more strongly than reliability and the use of communication sources. Face-to-face communication creates the most vital communication elements that enhance farmers' political participation. The small number of respondents posed a limitation, hindering the generalizability of this study's findings. However, the study will still contribute to the body of knowledge on strategic sustainability communication and rural communication, as it focuses on examining a less explored rural context and places greater emphasis on the theoretical relationship between rural communication variables and political dynamics rather than solely estimating population parameters. Furthermore, the study's findings contribute to the development of sustainability communication strategies tailored to rural areas, particularly when accounting for the political aspects of food security sustainability.

References

Baldersheim, H., Jamil, I., & Aminuzzaman, S. (2001). Electoral participation in Bangladesh: Explaining regional variations. *Journal of Commonwealth and Comparative Politics, 39*(2), 51–72. https://doi.org/10.1080/713999547

Bandura, A. (1997). *Self-efficacy: The exercise of control*. Freeman.

Banik, D. (2007). *Starvation and India's democracy*. Routledge. ISBN 978-0-415-40729-8.

Berry, E. M., Dernini, S., Burlingame, B., Meybeck, A., & Conforti, P. (2015). Food security and sustainability: Can one exist without the other? *Public Health Nutrition, 2015*(18), 2293–2302. https://doi.org/10.1017/S136898001500021X

Blau, P. M. (1964). Justice in social exchange. *Sociological Inquiry, 34*, 193–206. https://doi.org/10.1111/j.1475-682X.1964.tb00583.x

Carpini, M. D. (2004). Mediating democratic engagement: The impact of communications on citizens' involvement in political and civic life. In L. L. Kaid (Ed.), *Handbook of political communication research* (pp. 395–434). Lawrence Erlbaum Associates.

Conge, P., et al. (1988). The concept of political participation: Toward a definition. *Comparative Politics, 20/2*. https://doi.org/10.2307/421669

Craig, S., & Maggiotto, M. (1982). Measuring political efficacy. *Political Methodology, 8*(3), 85–109. http://www.jstor.org/stable/25791157

Deth, J. W. v. (2001). Studying political participation: Towards a theory of everything? In *Joint sessions of workshops of the European consortium for political research workshop "electronic democracy: Mobilisation, organisation and participation via new ICTs"*. Grenoble.

Eakin, H., Connors, J. P., Wharton, C., Bertmann, F., Xiong, A., & Stoltzfus, J. (2017). Identifying attributes of food system sustainability: Emerging themes and consensus. *Agriculture and Human Values, 2017*(34), 757–773. https://doi.org/10.1007/s10460-016-9754-8

Eyduran, S. P., Akin, M., Eyduran, E., Celik, S., Erturk, Y. E., & Ercisli, S. (2020). Forecasting Banana harvest area and production in Turkey using time series analysis. *Erwerbs-Obstbau, 2020*(62), 281–291. https://doi.org/10.1007/s10341-020-00490-1

FAO. (2014). *Building a common vision for sustainable food and agriculture: Principles and approaches*. FAO. FEWSNET report: 260,000 people died in the Somalia famine | Oxfam International. Oxfam.org. 2013-05-01. Retrieved 2023-05-20

Godemann, J. (2011). Sustainable communication as an inter-and transdisciplinary discipline. In I. T. Godemann & G. Michelsen (Eds.), *Sustainability communication* (pp. 39–51). Springer.

Golob, U., Podnar, K., & Zabkar, V. (2023). Sustainability communication. *International Journal of Advertising, 42*(1), 42–51. https://doi.org/10.1080/02650487.2022.2144035

Green, A., Nemecek, T., Chaudhary, A., & Mathys, A. (2020). Assessing nutritional, health, and environmental sustainability dimensions of agri-food production. *Global Food Security - Agriculture Policy, 26*, 100406. https://doi.org/10.1016/j.gfs.2020.100406

Harris, K. (2014). Bread and freedom: Linking democracy and food security in sub-Saharan Africa. *African Studies Quarterly., 15*, 13–35.

Hassan, M. K. (2020). *Rural communication services, an approach to development communication in Bangladesh context: A conceptual exploration. UODA studies part II*. University of Development Alternatives. ISSN 1817 4825.

Hassan, M. K., Torres, C. S., Jamias, S. B., Velasco, M. T. H., & Ocampo, M. C. (2016). Rural communication services in the delivery and adoption of saline-tolerant rice variety among farmers in Amtali, Barguna District, Bangladesh. *The Philippine Journal of Development Communication, 8*, 53–79.

Hendriks, C. M. (2009). Policy design without democracy? Making democratic sense of transition management. *Policy Sciences, 2009*(42), 341–368.

Kenski, K., & Stroud, N. (2006). Connections between Internet use and political efficacy, knowledge, and participation. *Journal of Broadcasting & Electronic Media, 50*(2), 173–192.

Kersting, N. (2021). Participatory democracy and sustainability. Deliberative democratic innovation and its acceptance by citizens and german local councilors. *Sustainability, 2021*(13), 7214. https://doi.org/10.3390/su13137214

Kushin, M. J., & Yamamoto, M. (2010). Did social media really matter? College students' use of online media and political decision making in the 2008 election. *Mass Communication & Society, 13*(5), 608–630.

Lewicki, R., & Bunker, B. (1996). *Developing and maintaining trust in working relations*. https://doi.org/10.4135/9781452243610.n7.

McLeod, J. M., Kosicki, G. M., & McLeod, D. M. (1994). The expanding boundaries of political communication effects. In J. Bryant & D. Zillman (Eds.), *Media effects: Advances in theory and research* (pp. 123–162). Erlbaum.

Meadowcroft, J. (2009). What about the politics? Sustainable development, transition management, and long-term energy transitions. *Policy Sciences, 2009*(42), 323–340.

Nekola, M. (2006). *Political participation and governance effectiveness: Does participation matter?* NISPAcee. ISBN: 80-89013-25-2.

Newig, J., Schulz, D., Fischer, D., Hetze, K., Laws, N., Lüdecke, G., & Rieckmann, M. (2013). Communication regarding sustainability: Conceptual perspectives and exploration of societal subsystems. *Sustainability, 5*(7), 2976–2990. https://doi.org/10.3390/su5072976

Park, C. S. (2019). The mediating role of political talk and political efficacy in the effects of news use on expressive and collective participation. *Communication and the Public, 4*(1), 35–52. https://doi.org/10.1177/2057047319829580

Pérez-Escamilla, R. (2017). Food security and the 2015–2030 sustainable development goals: From human to planetary health: Perspectives and opinions. *Current Developments in Nutrition, 2017*(1), e000513. https://doi.org/10.3945/cdn.117.000513

Pinkleton, B. E., Austin, E. W., & Fortman, K. K. (1998). Relationships of media use and political disaffection to political efficacy and voting behavior. *Journal of Broadcasting & Electronic Media, 42*, 34–49.

Price, V. (1989). *Public opinion and communication: Theoretical dependencies among levels*. Presented to the International Communication Association, San Francisco, California.

Ramachandran, R., Jaggarajamma, K., Muniyandi, M., & Balasubramanian, R. (2006). Identifying effective communication channels in a rural community: A field report from South India. *Indian Journal of Tuberculosis, 53*(4), 206–211.

Rubin, O. (2009, December). The merits of democracy in famine protection—Fact or fallacy? *The European Journal of Development Research., 21*(5), 699–717. https://doi.org/10.1057/ejdr.2009.37

Salmon, C. T. (1986). Perspectives on involvement in consumer and communication research. In B. Dervin & M. J. Voigt (Eds.), *Progress in communication sciences* (Vol. VII, pp. 243–268). Ablex.

Schmitt-Sands, C. E. (2017). The political determinants of food security: democracy, decentralization, and federalism. *Wayne State University Dissertations*. 1872. https://digitalcommons.wayne.edu/oa_dissertations/1872

Sen, A. (1999). *Development as freedom*. Oxford University Press.

Sen, A. (2000). Wars and famines: On divisions and incentives. *Peace Economics, Peace Science and Public Policy, 6*(2), 10–26.

Shah, D. V., Cho, J., Eveland, W. P., Jr., & Kwak, N. (2005). Information and expression in a digital age: Modeling Internet effects on civic participation. *Communication Research, 32*, 531–565.

Shah, D. V., & Scheufele, D. A. (2006). Explicating opinion leadership: Non-political dispositions, information consumption, and civic participation. *Political Communication, 23*(1), 1–22.

Singh, N. P., Anand, B., Singh, S., Srivastava, S. K., Rao, C. S., Rao, K. V., & Bal, S. K. (2021). Synergies and trade-offs for climate-resilient agriculture in India: An agro-climatic zone assessment. *Climatic Change, 164*, 11. https://doi.org/10.1007/s10584-021-02969-6

Smith, J. P., (1998, September 11). *Food security and political stability in the Asia-Pacific region*. Asia-Pacific Center for Security Studies, Honolulu. Retrieved from: https://dkiapcss.edu/Publications/Report_Food_Security_98.html.

Stromback, J., Djerf-Pierre, M., & Shehata, A. (2013). The dynamics of political interest and news media consumption: A longitudinal perspective. *International Journal of Public Opinion Research, 25*(4). https://doi.org/10.1093/ijpor/eds018

Tewksbury, D., Hals, M. L., & Bibart, A. (2008). The efficacy of news browsing: The relationship of news consumption style to social and political efficacy. *Journalism & Mass Communication Quarterly, 85*(2), 257–272.

United Nations. (2015). *Transforming our world: The 2030 agenda for sustainable development*. United Nations.

Weder, F., Karmasin, M., Krainer, L., & Voci, D. (2021). Sustainability communication as critical perspective in media and communication studies—An introduction. In F. Weder, M. Karmasin, L. Krainer, & D. Voci (Eds.), *The sustainability communication reader* (pp. 1–12). Springer VS.

Weder, F., Rademacher, L., & Schmidpeter, R. (2023). Sustainable communication? Media and communication responsibility in global transformation processes. In F. Weder, L. Rademacher, & R. Schmidpeter (Eds.), *CSR communication in the media. CSR, sustainability, ethics & governance*. Springer. https://doi.org/10.1007/978-3-031-18976-0_1

Wischnath, G., & Buhaug, H. (2014). On climate variability and civil war in Asia. *Climatic Change, 122*, 709–721. https://doi.org/10.1007/s10584-013-1004-0

Zhou, Y. (2009). *The role of communication in political participation: Exploring the social normative and cognitive processes related to political behaviors*. [Unpublished thesis], Washington State University. Retrieved September 22, 2022, from https://www.semanticscholar.org/paper/THE-ROLE-OF-COMMUNICATION-IN-POLITICAL-EXPLORING-TO-Zhou/be265a0eaf39425df11a2143521d779e466d9f96

Zhou, Y., & Pinkleton, B. E. (2012). Modeling the effects of political information source use and online expression on young adults' political efficacy. *Mass Communication & Society, 15*(6), 813–830.

Zidouemba, P. R. (2017). Climate variability and food crises in Burkina Faso: A computable general equilibrium analysis. *International Journal of Food and Agricultural Economics, 5*(1), 79–95.

Ziemann, A. (2011). Chapter 8; Communication theory and sustainability discourse. In J. Godemann & G. Michelsen (Eds.), *Sustainability communication: Interdisciplinary perspectives and theoretical foundations* (pp. 89–96). Springer.

Mohummad Kamrul Hassan Born and raised in Bangladesh, Mohummad Kamrul Hassan earned his PhD in development communication from the University of the Philippines Los Baños. With over 15 years of experience in the nonprofit sector, he has served as a researcher, consultant, and development practitioner. Hassan works as a consultant and is the Co-founder of Green Fund Initiative, a USA-based nonprofit organization. His research interests include rural communication services (RCS), communication for development, political communication, social network analysis, social media studies, and the migration crisis.

(Using) Technology for Sustainability

Embedding Sustainability in Film and Television Industries: Framing Practices and Narratives for Strategic Sustainability Communication

Marta Rocchi

Abstract This chapter aims to discuss how film and television industries can support and enhance sustainability, specifically highlighting the functions of strategic sustainability communication (SSC). Many scholars and experts have examined the role of screen media narratives and communication in representing and raising awareness of environmental challenges. Some of this research looked at the potential of film and television industries to stimulate critical inquiries and behavioural change through sustainability narratives. Following this line of inquiry, this chapter firstly provides a brief overview of the intersection between environmental concerns and audiovisual narratives through the lens of SSC. The chapter shows that one of the main issues for SSC within the film and television industries involves nurturing a public sustainability mindset throughout the sector, particularly as it pertains to corporate social responsibility (CSR) communications. Subsequently, the chapter delves into the representation of sustainability, exploring the embedding of energy transition issues within two European TV series: the Norwegian *Okkupert* and the Italian *Imma Tataranni - Sostituto procuratore*, focusing on negotiations and interlinkages of such narratives with traditional institutions, including public service broadcasters.

1 Introduction

Effectively communicating narratives, frames, and stories related to climate change and sustainability is an ongoing challenge in our society. Despite their presence in public discourse (Cox, 2013; Cox & Depoe, 2022) and growing audience interest (Act Green, 2023), these narratives and representations, along with broader environmental communication, still linger on the periphery of media production. To enhance public awareness and engagement with environmental challenges, it is imperative to

M. Rocchi (✉)
Department of the Arts, Alma Mater Studiorum – University of Bologna, Bologna, Italy
e-mail: marta.rocchi5@unibo.it

F. Weder (ed.), *Strategic Sustainability Communication*, CSR, Sustainability, Ethics & Governance, https://doi.org/10.1007/978-3-031-89486-2_20

grant them the prominence they deserve and bring them to the forefront of public discourse. Some scholars suggest that the issue is not a lack of sustainability stories, but rather a need for different types of stories (Joosse et al., 2023: 1910). According to several researchers, narratives may have a profound impact on the moral and political structure of society (i.e. Frank, 2017) since they "can construct and reflect knowledge representative of the place-, time- and event-dependent dimensions of life" (ivi: 311). Therefore, it is through (different types of) narratives that new visions of sustainable living might emerge, driving social change from the stories we tell, and the narratives individuals discuss, question, and adopt in their lives. Following this line of research, this chapter aims to discuss how film and television industries can support and enhance sustainability, highlighting the role of strategic sustainability communication (SSC).

The concept of sustainability is an issue of ongoing debate, increasingly permeating and intersecting various fields of knowledge (Godemann & Michelsen, 2011), including media and communication studies (Weder et al., 2021), on which this contribution is based and contributes to. Kuhlman and Farrington (2010) have emphasized how, since the Brundtland Report of 1987, the concept of sustainability has been subjected to an unfortunate shift in meaning (i.e. social, economic and environmental "pillars" of sustainability as the "triple bottom line") that "obscures the real contradiction which exists between long-term sustainability and short-term welfare" (ivi: 3437). They argue that distinguishing between the requirements of the present and those of future generations is a prudent method for analysing scenarios and "propose to call the former well-being and the latter sustainability. This prevents sustainability from becoming an empty phrase indistinguishable from 'goodness'. Well-being mostly corresponds with the social and economic dimensions of the 'triple bottom line', sustainability with the environmental dimension" (ivi: 3344). More recently, Golob et al. (2023) have provided a picture of how communication is conceptualized in terms of sustainability (including sustainability communication, sustainability reporting, sustainability/green advertising, greenwashing) and have underlined the importance of adopting a more comprehensive approach to grasp the role of diverse contexts and institutions in fostering sustainable practices. In alignment with these views, this chapter defines strategic sustainability communication (SSC) specifically within the audiovisual sector as encompassing all communicative practices that pursue a sustainability mission as a guiding principle. From this standpoint, a primary function of SSC is to raise awareness and inspire social change by examining how the human-environment relationship and its challenges are shaped within social discourse. Consequently, the objective of SSC within the film and television industries is also to support sustainability as a "cultural practice" on two levels: (I) corporate social responsibility (CSR) strategies; and (II) on-screen narratives, which provide more space and scope for communication *about* sustainability (Weder, 2023a: 31).

Several scholars have argued that in climate change and sustainability narratives, we need to shift the focus from merely providing information to engaging in storytelling (e.g. Veland et al., 2018). In this context, fictional narratives can also play a key role: while they "cannot or should not fill the 'information deficit', [...] they might fill a 'narrative deficit' that could push the reader to become more curious about real risks and opportunities" (Veland et al., 2018: 44). Given the surge in popularity of fictional narratives surrounding sustainability over the past decade (Bilandzic & Kalch, 2021), and the role of "the sustainability story as the *strategic story* that is today" (Weder, 2023b: 228), this chapter aims to provide a brief overview of the main research strands within the audiovisual sector through the lens of SSC. It then delves specifically into the representation of sustainability stories, examining how energy transition issues might be embedded within two European TV series, and the underlying logic behind these representations. This work is based on an overarching question: how can film and television industries shape SSC? Specifically, how are sustainability narratives, such as energy transition themes, integrated into contemporary European fictional narratives? Within this specific sub-strand, the analysis will be articulated along two axes: firstly, efforts will be made to elucidate how, from a narrative point of view, two European fictional television industries can embed storylines centred around sustainable energy; secondly, there will be an attempt to understand the logic behind considering interlinkages between traditional television institutions—both public service and private—and the specific sustainability challenges they must confront. In particular, the chapter focuses on two case studies: the Norwegian TV series *Okkupert* (TV2, 2015–2020) and the Italian TV series *Imma Tataranni—Sostituto procuratore* (Rai 1, 2019–). The selected case studies were chosen as they centre on constructions and mediations of sustainable energy. In particular, climate content is a recurring theme in Norwegian culture, and some instances of Norwegian climate fiction have achieved notable success internationally (Furuseth et al., 2020). Norwegian climate change-related works often embody four recurring features: "notions of Nordicity, an (imagined) intimate connection between Norwegianness and nature, references to Norwegian petroculture, and an atmosphere of Nordic gloom and melancholia" (Furuseth et al., 2020: 9). In this perspective, the analysis focuses on *Okkupert*, which stands out as a notable audiovisual production addressing both petroculture and sustainability. Conversely, climate change is not a commonly explored theme in Italian media products. In 2022, the Observatory of Pavia started a collaboration with Greenpeace Italy to systematically monitor media coverage of the climate crisis in Italian news outlets. The results revealed a frequently superficial attention to the topic, often mentioned but not deeply explored in many cases.[1] A year of monitoring also exposed how the media often failed to establish causes and consequences in their reporting on the crisis, neglecting to assign responsibility (Azzalini &

[1] The lack of attention to on-screen representation—excluding recent research related to branded entertainment and social washing, considering, for instance, Enel's involvement in the third season of *Doc—Nelle tue mani* (Rai 1, 2020-) (Brembilla & Guerini Rocco, 2024)—is counterbalanced by policies and productive initiatives implemented by local governments. One example is the Green Film protocol developed by the Trentino Film Commission (Cucco & D'Urso, 2024). In addition, Carelli et al. (2022) have shown a discernible trend that has emerged in recent years within Italian media productions, placing emphasis on the portrayal of the role of green areas and natural resources in audiovisual representation, particularly in the unscripted macro-genre. However, this trend leans more towards promoting tourism aiming at "disseminating a culture of sustainability in its various forms" (ivi: 201).

Marchese, 2023). In this perspective, *Imma Tataranni—Sostituto procuratore* stands out as an exception within Italian fictional production. Both TV series serve as distinct poles to analyse the interplay between the narrative embedding of sustainability themes and their connections with productive and distributive contexts.

2 Exploring the Intersection of Environmental Concerns and the Audiovisual Industries

The centrality of environmental and sustainability issues has led to the emergence of various areas of study within the audiovisual sector, exploring, for example, the convergence of cultural and creative industries with the environmental crisis (e.g. Miller, 2017; Maxwell & Miller, 2017; Oakley & Banks, 2020). Furthermore, the audiovisual sector serves as a focal point in relation to environmental issues in media policy studies (Kääpä, 2018; Sørensen & Noonan, 2022; Shanahan et al., 2022). Numerous contributions have also examined how various media institutions, including public service broadcasters, influence societal discourse, perceptions, and actions regarding climate change and sustainability (e.g. Spellerberg et al., 2006; Debrett, 2017; Smith, 2017). Screen media narratives indeed wield significant potential to influence audience awareness, knowledge, beliefs, and, critically, behaviours—from initial content production through consumption and engagement (e.g. Hansen, 2011; McGreavy & Lindenfeld, 2014; Smith, 2017; Graves & Morris, 2023; Dales et al., 2024). Nevertheless, comprehending the factors that influence public perceptions of nature and, consequently, fostering environmental awareness poses significant challenges (e.g. Berenguer, 2007; Fernández-Bellon & Kane, 2019; Dunn et al., 2020), as the links between information provided and changes in behaviour are intricate and uncertain (Kollmuss & Agyeman, 2002).

To understand how the film and television industries can shape SSC practices, we need to consider two primary facets at the intersection of environmental concerns and the audiovisual industries. The first pertains to the industry's own role in exacerbating the environmental crisis, a matter increasingly underscored in contemporary research (Kääpä & Vaughan, 2022, 2024). Despite varying definitions of sustainability and estimates and habits related to the production, distribution, and preservation of audiovisual content (e.g. Gassmann & Gouttefarde, 2021; Albert, 2022; Marks & Przedpełski, 2022; Antoniazzi, 2024), it is now acknowledged that the film and television industries exhibit polluting and wasteful practices. Lupu et al. (2023: 13) have underlined how, despite various green initiatives, the lack of clarity regarding the carbon footprint of audiovisual production allows stakeholders to perceive carbon emissions as someone else's problem.[2] While we await a

[2] This observation extends to documentary filmmakers who, in the process of "planning and shooting a scene, […] still do not ask themselves habitually if the approach that they have chosen for portraying that certain aspect of the real is the most eco-sustainable possible" (Formenti, 2022: 193).

consistent and standardized evaluation of global emissions in the audiovisual industry (Vaughan, 2022), several scholars have highlighted available solutions to enhance the sustainability of film and television production. While Gassmann and Gouttefarde (2021) have pointed out that these solutions hinge on three key components, such as awareness, knowledge, and action, Lupu et al. (2023) have suggested an expanded framework for examining green innovation within the audiovisual sector, distinguishing among solution-, content-, and mindset-driven innovation. In this perspective, it becomes feasible to enact and implement SSC practices and strategies that effectively engage various stakeholders in the audiovisual industry, encompassing both productive and distributive aspects. A key asset of SSC involves CSR strategies and the communication of sustainability practices within the industry. However, it is crucial to be aware of two possible poles as Weder and Marte (2023) have underlined. CSR communications

> are sometimes driven by intrinsic ethical motives or the perception that sustainable development is a deeper process of ethical reflection […], but much more often it is stimulated by the expectation of possible business returns that companies may reap from CSR initiatives, namely building trust and reputation, customer loyalty, brand equity as well as generating positive effects in employee recruitment (ivi: 64).

Therefore, to avoid falling into greenwashing pitfalls (Golob et al., 2023), even before considering business returns, it is essential to prioritize sustainable principles and primarily foster the development of a "public sustainability mindset" (Lupu et al., 2023). This mindset can facilitate a systemic shift within the sector and promote SSC in the framework of CSR communication within the film and television industries, creating space and opportunities for sustainable action and fostering a culture of sustainability (Weder, 2023a, 2023b: 31). This shift in mindset encompasses solution-driven innovations that are necessary for reducing environmental impact across various stages of the audiovisual process.

The second dimension of the intersection between environmental concerns and the audiovisual industries focuses on the on-screen representation of climate and sustainability content. The rise of streaming services and the audience's regular consumption of environmental media content hold the potential to foster public engagement and, under certain circumstances, may heighten environmental awareness and encourage behavioural changes (e.g. Hansen, 2011; Cox, 2013; Silk et al., 2021). It is noteworthy that audiovisual narratives, particularly serial narratives, have often demonstrated commendable receptivity to socially significant and relevant subjects (e.g. Sowards & Pineda, 2013; Sisson & Kimport, 2014; Nærland, 2019). While numerous studies have investigated the prevalence and impact of various health and societal issues in scripted entertainment (e.g. immigration, criminal justice), until recently, little attention was given to climate and sustainability themes (Giaccardi et al., 2022). A 2022 report has shown the "*glaring absence*" of these topics within US scripted entertainment (Giaccardi et al., 2022). The study highlighted a significant lack of climate change discussions in US scripted media, revealing that only 2.8% of all scripts included any climate-related terminology, and a mere 0.6% of scripted TV and films used the specific term "climate change". One of the primary challenges in addressing sustainability issues lies in narrating such a complex topic within the framework of scientific and socio-cultural discourse while avoiding the reinforcement of environmentally problematic ideas and behaviours (Jones et al., 2019; Vaughan, 2019; Geal, 2021; Rocchi, 2022). Notably, Joosse et al. (2023) recently performed a frame analysis of how storytellers describe their storytelling for sustainability, while Dales et al. (2024) introduced a valuable tool called Critical Sustainability Stories (CriSS), designed to assist professional storytellers in decision-making in order to improve their narratives. Through the examination of the case studies in the subsequent section, the chapter delves into this second dimension, focusing on on-screen representation and its interconnections with production/distribution contexts.

3 Fictional Narratives: The Embedding of Energy Transition Issues

Considering the growing body of work and demand for environmentally conscious and climate change-themed content (Graves et al., 2023), this section specifically analyses the embedding of energy transition narratives within two European fictional productions: the Norwegian TV series *Okkupert* (TV2, 2015–2020) and the Italian TV series *Imma Tataranni—Sostituto procuratore* (Rai 1, 2019–). From a methodological perspective, the analysis will be conducted considering both narrative and productive standpoints. Regarding the narrative dimension, to comprehend how the TV series articulates sustainability from the perspective of SSC, we examine two distinct layers: sustainability representation and sustainability principles. The first layer (sustainability representation) is closely tied to the media text and its portrayal of sustainability, wherein we can differentiate between various categories such as visual, audio, and audiovisual representations (i.e. plot embedding). The second layer (sustainability principles) identifies how the media text engages with sustainability principles. We consider those principles outlined by Godemann (2021: 25). Specifically, the analysis explores and identifies how the fictional narrative: (I) encourages proactive and future-oriented action, (II) facilitates dialogue and reflection on values and attitudes, (III) promotes decision-making in determining how to act, (IV) connects with social practices and daily life, and (V) reveals positions or contradictory interpretations (Godemann, 2021: 25).

The Norwegian TV series *Okkupert* delves into sustainability and energy transition themes, depicting the emergence of the Green Party following a devastating hurricane in Scandinavia. This event prompts the new Prime Minister, Jesper Berg, to adopt a pro-environment agenda, which includes discontinuing oil and gas extraction in favour of sustainable energy sources. The emergence of a fictional Russian invasion is a consequence of this shift. Leyda (2018) notes that the series portrays the political Green Party as not yet ready to lead and considering "an era of extreme weather and anthropogenic climate change, the series posits petropolitics as a potential catalyst for national and regional political instabilities" (ivi: 84–85). While the TV series primarily focuses on the political and economic consequences of the Russian occupation, it also carries a strong environmental message, depicting Norway as a leader in the transition to sustainable energy. However, climate change and sustainability are rarely explicitly mentioned beyond the initial premise. The lack of explicit visual representation of the energy transition issue in *Okkupert* (except for the thorium power station, which the Prime Minister is seen flying to in his helicopter to inaugurate it) is compensated for by its strong embedding into the narrative, shaping the script to convey its urgency and relevance to the audience. As Graves and Morris (2023) have recently asserted, modelling pro-environmental attitudes and behaviours can be achieved through emotionally resonant characters, settings, and storylines in both factual and fictional content. Their study has demonstrated that these elements not only augment environmental knowledge but also enable viewers to empathize and connect with climate sustainability content. In *Okkupert,* such storytelling strategies extend to characterizations and personality traits. For instance, through the character of Bente, the owner of a struggling restaurant, *Okkupert* presents the risks associated with her pursuit of short-term gratification without fully vilifying her self-interest. By embodying Bente's short-sighted self-centeredness, the series exposes some of the factors that enable Norway to evade accountability for its involvement in the fossil fuel industry and denounces such evasion as unpatriotic (Leyda, 2018: 93). All these elements highlight how in *Okkupert* there is a narrative that supports four out of the five previously mentioned sustainability guiding principles identified by Godemann (2021). Indeed, the TV series, through its storytelling, encourages proactive and future-oriented action, also revealing risks (e.g. Russian occupation) and positions or contradictory interpretations (e.g. the unpreparedness of the Green Party). It facilitates dialogue and reflection on values and attitudes through the in-depth exploration of the political protagonists of the story (e.g. the Prime Minister Jesper Berg), as well as ordinary people (e.g. Bente), thus connecting with social practices and daily life.

From a production standpoint, Norway boasts a well-established tradition of public service broadcasting, grounded in the belief that both factual and fictional content play a vital role in fostering public awareness (Dodds & Hochscherf, 2020). As Sauders (2020) underscores, "going beyond the artistic merits, the choices that showrunners make about how to represent space and place serve specific geopolitical agendas. Unlike the vast majority of American and British productions, series from Denmark, Sweden, Norway and Finland, tend to receive significant levels of public funding, which comes with the informal stipulation that the content serves the nation in some way" (ivi: 66). Consequently, the Norwegian Broadcasting Corporation is mandated "to promote public debate and play its part in ensuring that the entire population receives sufficient information to enable it to actively participate in democratic processes" (Norwegian Broadcasting Corporation, 2008: 1). Given the intentional nature of Norwegian fictional television production, it deserves serious consideration and the subjects depicted in television dramas are not merely used for entertainment: they are crafted to spark discussions and provide reasons for reflection (Dodds & Hochscherf, 2020). TV2, the channel that aired *Okkupert*, is the second national broadcaster with some public service obligations (Enli & Syvertsen, 2016). Its narratives reveal several direct and indirect elements that align with the previously mentioned four features frequently present in Norwegian climate change-related content (Furuseth et al., 2020: 9), fostering discourse on energy transition and sustainability while encouraging audience participation in public debate.

When examining Italian audiovisual serial production, the exploratory analysis underlying this chapter highlights that the theme of sustainable energy is often either inadequately explored or only briefly depicted in fictional narratives. A recent exception is found in *Imma Tataranni—Sostituto procuratore*, an Italian TV series chronicling the experiences of a young and ambitious public prosecutor dedicated to combating organized crime while challenging prevalent stereotypes associated with women's ageing (Casoli, 2021). In exploring content related to sustainability and energy transition, my focus turns towards a specific episode from the second season titled "La donna che perse la testa", originally aired on 6 October 2022. This episode delves into the investigative pursuits of the protagonist, compelling her to probe into the narrative of a character. Here, the storyline introduces a debated and controversial topic concerning the noise generated by wind turbine blades and their potential impact on wildlife. The character argues that while the practice of populating the countryside with wind turbines may have initially stemmed from ecological intentions, ultimately it has evolved into a business that impoverishes the territory, disrupts ecosystems, and adversely affects the nervous systems of nearby residents.[3] The episode delves into debates concerning environmental and social issues linked to wind energy, touching upon conflicts between local communities, developers, and environmental activists. These few minutes provide some elements that highlight how in *Imma Tataranni—Sostituto procuratore,* there is a narrative that aligns with two out of the five sustainability guiding principles identified by Godemann (2021). The TV series, in a short timeframe, connects with social practices and daily life, revealing different perspectives, but it does not encourage proactive and future-oriented action, nor does it facilitate dialogue and reflection on values and attitudes, as it does not present contradictions or promote decision-making in determining how to act.

To analyse the productive logic for this case study, it is necessary to broaden the scope and observe the interplay between reception and production. Indeed, some environmental associations (e.g. VV. AA., 2022a) have expressed their gratitude to the authors of the TV series, commending their courage in finally providing a voice to those who have long criticized the use of wind power and its impact on Italian landscapes, natural environments, and wildlife. Conversely, ANEV (the National Wind Energy Association) urgently called for a response from Rai (the Italian public broadcaster that aired the TV series) regarding the episode, contending that it presents an inaccurate and biased portrayal of the wind power sector, which, in reality, brings environmental and economic benefits (VV. AA., 2022b). They state that

---

[3] The link to the video is available at https://youtu.be/FssiFYBnmTc?feature=shared (last accessed 29/03/24).

"it is surprising that the Rai public service, instead of providing accurate information on such important topics as renewables, clean and inexhaustible wind energy that does not damage the environment, disseminates such a misleading and scientifically unfounded message, and we hope that measures will be taken to rectify what was broadcast last night" (VV. AA., 2022b, author's translation). In response to ANEV, Maria Pia Ammirati, director of Rai Fiction, contended that *Imma Tataranni—Sostituto procuratore* belongs to the genre of pure fantasy works and the narration of events therefore disregards the requirement of truth (ANSA, 2022). It is interesting to note that just a few weeks before this discussion there was an event organized by Rai for Sustainability-ESG, in collaboration with Rai Fiction, in which Antonio Parlati, head of the Rai TV and Radio Production Centre in Naples, highlighted the importance of social issues being addressed in fictional stories because they have more time to better assimilate certain concepts (VV. AA., 2022c). According to Roberto Natale, director of Rai for Sustainability-ESG, sustainability is the main theme that public service should address. Natale emphasized that the new Rai direction for sustainability aims to push forward these themes, which, however, must permeate the entire public service (VV. AA., 2022c). Francesco Nardella, deputy vice director of Rai Fiction, spoke about how sustainable topics are effective narrative prompts to create a story. He highlighted how serial storytelling requires "energetic material" to create plots and a collective experience, and that collective experience is also represented by sustainability. But Nardella added that RAI does not do product placement for sustainability and that each theme is presented in a non-didactic manner, inherently intertwined with the narrative, so that it can be accepted by the audience.

4 Discussion and Conclusions

Considering the overarching question underpinning this contribution, the film and television industries might play a dual role in shaping SSC: on one level, they can enhance their CSR strategies and communication about industries' sustainability practises; on another level, through their narratives they can provide more and diverse space and scope for communication *about* sustainability (Weder, 2023a: 31).

Regarding the first aspect, CSR communication presents an intriguing challenge and a valuable opportunity for the film and television sector, positioning sustainability as a guiding principle for decision-making, practices, and programmes. This approach requires a public sustainability mindset that allows both the creation of "a *strategic story* that businesses select on purpose to create their own socio-ecological niches" (Weder, 2023b: 229) and enables businesses to seamlessly "integrate social and environmental concerns into their operational framework and communicate effectively with stakeholders" (Weder & Marte, 2023: 21), thereby *also* enhancing their social reputation (see more on communicative responsibility as impacted oriented-approach in Weder & Marte, 2023: 64).

Regarding the second issue, the contribution has specifically highlighted two distinct place-based approaches to embedding and mediating on-screen sustainability representations within European fictional TV series. Indeed, as Boykoff (2011) emphasizes, "putting the complexities of climate science and governance into context is critically important for successfully communicating climate change and for achieving greater connectivity between communities of science, policy, media and the public citizenry" (ivi: 98). However, science and scientific reasoning represent only "part of the story" (Frank, 2017), stressing the need for diverse narratives to foster sustainability as "cultural practice". In this perspective, each case study demonstrates varying levels of adherence to the sustainability guiding principles identified by Godemann (2021: 25) and offers distinct views on how narratives can address life's context-specific dimensions, as discussed by Frank (2017). Within this framework, it is crucial to recognize the significance of how we discuss the environment (Lakoff, 2010) since media representation has the potential to drive change by normalizing sustainable practices, thanks to the facilitative role that both public service and private broadcasters can play in SSC. By continuously incorporating sustainability issues into their programming, they maintain an ongoing discourse and reaffirm their roles in the cultural arena (Enli & Syvertsen, 2016; Smith, 2017). The chapter underscores how Norwegian broadcasters are acutely aware of this aspect, utilizing audiovisual products to foster discourse and encourage audience participation in public debates. While *Okkupert* continues the Norwegian tradition of employing popular culture as a tool for national identity-building, even in the sustainable energy arena (Saunders, 2020), the Italian public broadcaster appears inclined to follow a similar path for fictional products, despite some statements and the case exemplified by the episode of *Imma Tataranni—Sostituto procuratore*. Indeed, the controversy surrounding climate and sustainability content poses a dilemma for many public service broadcasters, challenging the established remit of informing, educating, and entertaining while upholding their principle of "due impartiality" (Debrett, 2017: 462). Moreover, providing factually accurate information to audiences, particularly when fiction intersects with reality, demands a higher degree of scrutiny. Programme makers must distinguish between well-established facts and opinions in science coverage, ensuring clarity for the audience (Jones in Debrett, 2017: 463), particularly when addressing "controversial issues". Numerous authors emphasize the importance of involving conservationists and experts in the production of audiovisual narratives addressing climate and sustainability issues. This collaboration not only ensures the dissemination of accurate ecological content but also aims to generate measurable, positive conservation impacts (e.g. Reinermann et al., 2014; Jones et al., 2019; Dunn et al., 2020; Yeo et al., 2015).

As recently highlighted by Podkalicka et al. (2024), "environmental media also includes products that may not overtly label themselves as 'environmental' or 'green' but are driven by environmental aims, values, or contain environmental messages. They too can be influential" (ivi: 12). Rather than generating standalone narratives, themes related to sustainability and eco-friendliness are often seamlessly integrated into existing storylines, adhering to specific genre and format conventions. Therefore, considering SSC within audiovisual industry both the CSR communication and the on-screen embedding lines analysed in the chapter, along with others that will need to be worked on more extensively in future studies, may help to enhance the diversity of sustainability narrative forms (i.e. from scientific reports and scenario planning, to fictional writing; Veland et al., 2018: 45), contributing to "extend polyvocality on the topic of climate change in the media" (Smith et al., 2018: 603).

A thorough examination of sustainability issues within the film and television industries requires a rigorous assessment of production practices, textual content, and an in-depth analysis of the processes of assimilation and reinterpretation. Clearly, this requirement surpasses the scope of this chapter, which nonetheless strives to better position the film and television industries within the realm of SSC, reconnecting the primary focuses of communication research on media and environmental issues, such as the study of production, content, and the social/political implications of environmental communication (Hansen, 2011: 21). This contribution aims to foster interdisciplinary research, extending its implications to fields such as CSR Communication, Public Relations, and Science Communication. Future research will seek to deepen our comprehension of how SSC operates within the specific local and global contexts of the film and television industries where sustainability is communicated, debated and negotiated. These objectives encompass investigating the impact of the sector in terms of emissions and unsustainable practices, alongside exploring the role and influence of storytelling. The audiovisual sector, in its broadest scope, can play a pivotal and enduring role in advancing effective communication pertaining to sustainability. This has implications for all forms of strategic communication and narratives *of*, *about* and *for* sustainability since "today sustainability is a principle of action that *comes into action with & by communication*—that is not only *communicatively constructed* but also *communicatively cultivated* in our society" (Weder & Marte, 2023: 2). As a guiding principle for action, sustainability holds the potential to shape the film and television industries, urging them to strengthen their dimension of responsibility towards the environment, organizations, societies, and individuals.

References

Act Green. (2023). *Understanding audience attitudes towards the role of cultural organisations in tackling the climate emergency*. Benchmark Report by Indigo. Accessed Feb 27, 2024, from https://s3-eu-west-1.amazonaws.com/supercool-indigo/Act-Green-2023-Benchmark-report-c-Indigo-Ltd.pdf

Albert. (2022). *A step-change on sustainability*. BAFTA albert annual review 2022. Accessed Feb 27, 2024, from https://wearealbert.org/wp-content/uploads/2023/10/06_ALBERT-ANNUAL-REPORT-v8.pdf

ANSA. (2022). *Rai a imprese eolico, Imma Tataranni è pura fantasia*. Accessed Feb 27, 2024, from https://www.ansa.it/basilicata/notizie/2022/10/20/rai-a-imprese-eolico-imma-tataranni-e-pura-fantasia_0f88a15f-4fbc-40ba-9b6b-72fe4cde071f.html

Antoniazzi, L. (2024). Film heritage and sustainability revised. In S. Belotti, L. Cesaro, & C. Formenti (Eds.), *The (un)bearable lightness of media: Critical approaches to sustainability in film and audiovisual production, circulation and preservation* (pp. 143–147). Mimesis.

Azzalini, M. & Marchese, M.. (2023). L'informazione sulla crisi climatica in Italia 2022. Rapporto di ricerca. Accessed Feb 27, 2024, from https://www.greenpeace.org/static/planet4-italy-stateless/2023/04/3eac43b3-report-2022-linformazione-sulla-crisi-climatica-in-italia-odp-gp-def.pdf

Berenguer, J. (2007). The effect of empathy in pro-environmental attitudes and behaviors. *Environment and Behavior, 39*(2), 269–283. https://doi.org/10.1177/0013916506292937

Bilandzic, H., & Kalch, A. (2021). Fictional narratives for environmental sustainability communication. In F. Weder, L. Krainer, & M. Karmasin (Eds.), *The sustainability communication reader: A reflective compendium* (pp. 123–142). Springer.

Boykoff, M. T. (2011). *Who speaks for the climate?: Making sense of media reporting on climate change*. Cambridge University Press.

Brembilla, P., & Guerini Rocco, S. (2024). *La fiction Rai tra brand integration e social washing: i casi di Doc 3 e Mare fuori 4*. Paper presented at the Screen Culture Reloaded symposium, University of Urbino Carlo Bo, Urbino, 10–11 July 2024.

Carelli, P., Scaglioni, M., & Sfardini, A. (2022). Green is better: Representation of territories and landscapes in italian audiovisual products. In M. Cucco, M. Scaglioni, A. Sfardini, & G. Willems (Eds.), *Green is better: Representation of territories and landscapes in italian audiovisual products* (pp. 194–202). Vita e Pensiero.

Casoli, S. (2021). Imma Tataranni e l'amore transgenerazionale. La rappresentazione della sessualità femminile in età matura nella serialità televisiva italiana. In P. De Rosa, V. Re, & E. Mandelli (Eds.), *Aging girls. Identità femminile, sessualità e invecchiamento nella cultura mediale italiana* (pp. 229–249). Meltemi.

Cox, R. (2013). *Environmental communication and the public sphere* (6th ed.). Sage Publications.

Cox, R., & Depoe, S. (2022). Emergence and growth of the field of environmental communication. In A. Hansen & R. Cox (Eds.), *The Routledge handbook of environment and communication* (pp. 13–32). Routledge.

Cucco, M., & D'Urso, F. (2024). Prove generali di politica pubblica fra ambiente e audiovisivo. In P. Carelli & M. P. Pasini (Eds.), *Green Italy. Esperienze, media e culture per un turismo sostenibile* (pp. 13–24). Vita e Pensiero.

Debrett, M. (2017). Representing climate change on public service television: A case study. *Public Understanding of Science, 26*(4), 452–466. https://doi.org/10.1177/0963662515597187

Dales, A., Padfield, R., & Bridge, G. (2024). Critical sustainability stories: How communicators can translate the complexity of the climate emergency and shape public narratives. *Geoforum, 150*, 103963. https://doi.org/10.1016/j.geoforum.2024.103963

Dodds, K., & Hochscherf, T. (2020). The geopolitics of Nordic noir: Representations of current threats and vigilantes in contemporary Danish and Norwegian serial drama. *Nordicom review, 41*(1), 43–61. https://doi.org/10.2478/nor-2020-0015

Dunn, M. E., Mills, M., & Veríssimo, D. (2020). Evaluating the impact of the documentary series blue planet II on viewers' plastic consumption behaviors. *Conservation Science and Practice, 2*(10), e280. https://doi.org/10.1111/csp2.280

Enli, G., & Syvertsen, T. (2016). The end of television—Again! How TV is still influenced by cultural factors in the age of digital intermediaries. *Media and Communication, 4*(3), 142–153. https://doi.org/10.17645/mac.v4i3.547

Fernández-Bellon, D., & Kane, A. (2019). Natural history films raise species awareness—A big data approach. *Conservation Letters, 13*(1), e12678. https://doi.org/10.1111/conl.12678

Formenti, C. (2022). Beyond greenwashed protocols: The reconversion of existing modes of representation as a new possible approach to the cutback of the documentary film's environmental footprint. In M. Cucco, M. Scaglioni, A. Sfardini, & G. Willems (Eds.), *Green is better: Representation of territories and landscapes in italian audiovisual products* (pp. 180–193). Vita e Pensiero.

Frank, A. K. (2017). What is the story with sustainability? A narrative analysis of diverse and contested understandings. *Journal of Environmental Studies and Sciences, 7*, 310–323. https://doi.org/10.1007/s13412-016-0388-3

Furuseth, S., Gjelsvik, A., Gürata, A., Hennig, R., Leyda, J., & Ritson, K. (2020). Climate change in literature, television and film from Norway. *Ecozona, 11*(2), 8–16. https://doi.org/10.37536/ECOZONA.2020.11.2.3468

Gassmann, P., & Gouttefarde, M.. (2021). *Greening the European audiovisual industry. The best strategies and their costs.* Report for the European Commission. Accessed Feb 27, 2024, from https://digital-strategy.ec.europa.eu/en/library/greening-european-audiovisual-industry

Geal, R. (2021). *Ecological film theory and psychoanalysis. Surviving the environmental apocalypse in cinema.* Routledge.

Giaccardi, S., Rogers, A., & Rosenthal, E. L. (2022). *A glaring absence. The climate crisis is virtually nonexistent in scripted entertainment.* USC Norman Lear Center Media Impact Project. Accessed Feb 27, 2024, from https://learcenter.s3.us-west-1.amazonaws.com/GlaringAbsence_NormanLearCenter.pdf

Godemann, J., & Michelsen, G. (2011). Sustainability communication–an introduction. In J. Godemann & G. Michelsen (Eds.), *Sustainability communication* (pp. 3–11). Springer.

Godemann, J. (2021). Communicating sustainability. Some thoughts and recommendations for enhancing sustainability communication. In F. Weder, L. Krainer, & M. Karmasin (Eds.), *The sustainability communication reader: A reflective compendium* (pp. 15–29). Springer.

Golob, U., Podnar, K., & Zabkar, V. (2023). Sustainability communication. *International Journal of Advertising, 42*(1), 42–51. https://doi.org/10.1080/02650487.2022.2144035

Graves, C., & Morris, L. (2023). *What do we know? The impact of climate content on audiences' pro-environmental behaviour.* albert. Accessed Feb 27, 2024, from https://wearealbert.org/wp-content/uploads/2023/05/Albert-Pres-Doc-Impact-Report-v3-MSW20.pdf

Graves, C., Swords, J., & Morris, L. (2023). *The impact of climate content.* albert. Accessed Feb 27, 2024, from https://wearealbert.org/wp-content/uploads/2023/12/Albert-Climate-Report-v1-MSW-EL-V4.pdf

Hansen, A. (2011). Communication, media and environment: Towards reconnecting research on the production, content and social implications of environmental communication. *International Communication Gazette, 73*(1–2), 7–25. https://doi.org/10.1177/1748048510386739

Jones, J. P. G., Thomas-Walters, L., Rust, N. A., & Veríssimo, D. (2019). Nature documentaries and saving nature: Reflections on the new Netflix series our planet. *People and Nature, 1*(4), 420–425. https://doi.org/10.1002/pan3.10052

Joosse, S., Westin, M., Möckel, F., Keasey, H., & Lorenzen, S. (2023). Storytelling to save the planet: Who gets to say what is sustainable, who tells the stories, and who should listen and change? *Journal of Environmental Planning and Management, 67*(9), 1–19. https://doi.org/10.1080/09640568.2023.2258276

Kääpä, P. (2018). *Environmental Management of the Media: Policy, industry, practice.* Routledge.

Kääpä, P., & Vaughan, H. (2022). *Film and television production in the age of climate crisis: Towards a greener screen.* Springer International Publishing.

Kääpä, P., & Vaughan, H. (2024). Environmental media management: Overcoming the responsibility deficit. In A. López et al. (Eds.), *The Routledge handbook of Ecomedia studies* (pp. 179–186). Routledge.

Kollmuss, A., & Agyeman, J. (2002). Mind the gap: Why do people act environmentally and what are the barriers to pro-environmental behavior? *Environmental Education Research, 8*(3), 239–260. https://doi.org/10.1080/13504620220145401

Kuhlman, T., & Farrington, J. (2010). What is sustainability? *Sustainability, 2*(11), 3436–3448. https://doi.org/10.3390/su2113436

Lakoff, G. (2010). Why it matters how we frame the environment. *Environmental Communication, 4*(1), 70–81. https://doi.org/10.1080/17524030903529749

Leyda, J. (2018). Petropolitics, cli-fi and occupied. *Journal of Scandinavian Cinema, 8*(2), 83–101. https://doi.org/10.1386/jsca.8.2.83_1

Lupu, R., Komorowski, M., Lewis, J., Mothersdale, G., & Pepper, S. (2023). Greening the audiovisual sector: Towards a new understanding through innovation practices in Wales and beyond. *Sustainability, 15*(4), 2975. https://doi.org/10.3390/su15042975
Marks, L. U., & Przedpełski, R. (2022). The carbon footprint of streaming media: Problems, calculations, solutions. In P. Kääpä & H. Vaughan (Eds.), *Film and television production in the age of climate crisis: Towards a greener screen* (pp. 207–234). Springer International Publishing.
Maxwell, R., & Miller, T. (2017). Greening cultural policy. *International Journal of Cultural Policy, 23*(2), 174–185. https://doi.org/10.1080/10286632.2017.1280786
McGreavy, B., & Lindenfeld, L. (2014). Entertaining our way to engagement? Climate change films and sustainable development values. *International Journal of Sustainable Development, 17*(2), 123–136. https://doi.org/10.1504/IJSD.2014.061766
Miller, T. (2017). *Greenwashing culture*. Routledge.
Nærland, T. U. (2019). Fictional entertainment and public connection: Audiences and the everyday use of TV-series. *Television & New Media, 20*(7), 651–669. https://doi.org/10.1177/1527476418796484
Oakley, K., & Banks, M. (2020). Cultural industries and the environmental crisis. In *New approaches for policy*. Springer.
Podkalicka, A., Nilsson, D., & Troon, S. (2024). Popular environmental Media in Australia: Reflections on audience engagement and impact. *Media International Australia.* https://doi.org/10.1177/1329878X241232
Reinermann, J.-L., Lubjuhn, S., Bouman, M., & Singhal, A. (2014). Entertainment-education: Storytelling for the greater, greener good. *International Journal of Sustainable Development, 17*(2), 176–191. https://doi.org/10.1504/IJSD.2014.061781
Rocchi, M. (2022). Environmental misinformation and audiovisual serial narratives. An automatic analysis of the twitter social discursiveness on *Seaspiracy*. *Elephant & Castle, 28*, 124–134.
Saunders, R. A. (2020). Landscape, geopolitics, and National Identity in the Norwegian thrillers *occupied* and *Nobel*. *Nordicom Review, 41*(1), 63–83. https://doi.org/10.2478/nor-2020-0006
Shanahan, J., McComas, K., & Deline, M. B. (2022). Representations of the environment on television, and their effects. In A. Hansen & R. Cox (Eds.), *The Routledge handbook of environment and communication* (2nd ed., pp. 306–328). Routledge.
Silk, M., Correia, R., Veríssimo, D., Verma, A., & Crowley, S. L. (2021). The implications of digital visual media for human-nature relationships. *People and Nature, 3*(6), 1130–1137. https://doi.org/10.1002/pan3.10284
Sisson, G., & Kimport, K. (2014). Telling stories about abortion: Abortion-related plots in American film and television, 1916–2013. *Contraception, 89*(5), 413–418. https://doi.org/10.1016/j.contraception.2013.12.015
Smith, J. (2017). Demanding stories: Television coverage of sustainability, climate change and material demand. *Philosophical Transactions of the Royal Society A: Mathematical, Physical and Engineering Sciences, 375*(2095), 20160375. https://doi.org/10.1098/rsta.2016.0375
Smith, J., Revill, G., & Hammond, K. (2018). Voicing climate change? Television, public engagement and the politics of voice. *Transactions of the Institute of British Geographers, 43*(4), 601–614. https://doi.org/10.1111/tran.12250
Sowards, S. K., & Pineda, R. D. (2013). Immigrant narratives and popular culture in the United States: Border spectacle, unmotivated sympathies, and individualized responsibilities. *Western Journal of Communication, 77*(1), 72–91. https://doi.org/10.1080/10570314.2012.693648
Sørensen, I. E., & Noonan, C. (2022). Production, policy and power: The screen industry's response to the environmental crisis. *Media, Culture & Society, 44*(1), 172–184. https://doi.org/10.1177/01634437211065697
Spellerberg, I., Buchan, G., & Early, N. (2006). Television and environmental sustainability: Arguing a case for a code of standards in NZ. *Pacific Journalism Review, 12*(2), 137–147.
Vaughan, H. (2019). *Hollywood's dirtiest secret: The hidden environmental costs of the movies*. Columbia University Press.

Vaughan, H. (2022). Policy approaches to Green film practices: Local solutions for a planetary problem. In P. Kääpä & H. Vaughan (Eds.), *Film and television production in the age of climate crisis: Towards a greener screen* (pp. 43–68). Springer International Publishing.

Veland, S., Scoville-Simonds, M., Gram-Hanssen, I., Schorre, A. K., El Khoury, A., Nordbø, M. J., Lynch, A. H., Hochachka, G., & Bjørkan, M. (2018). Narrative matters for sustainability: The transformative role of storytelling in realizing 1.5 C futures. *Current Opinion in Environmental Sustainability, 31*, 41–47. https://doi.org/10.1016/j.cosust.2017.12.005

VV. AA. (2022a). *Comunicato Stampa in margine alla Fiction Rai1*. "IMMA TATARANNI, SOSTITUTO PROCURATORE". Accessed Feb 27, 2024, from https://www.assotuscania.it/?p=3561

VV. AA. (2022b). *Rinnovabili, la disinformazione sull'eolico arriva nelle fiction Rai*. Accessed Feb 27, 2024, from https://greenreport.it/news/economia-ecologica/rinnovabili-la-disinformazione-sulleolico-arriva-nelle-fiction-rai/

VV. AA. (2022c). *Festival dello sviluppo sostenibile*. La sostenibilità in Rai diventa pop - Il caso "Un posto al sole". Accessed Feb 27, 2024, from https://www.raiplay.it/video/2023/05/La-sostenibilita-in-Rai-diventa-pop-Il-caso-Un-posto-al-sole-899a45cf-9f24-4df5-8784-0502f15ec03a.html

Weder, F., Krainer, L., & Karmasin, M. (2021). *The sustainability communication reader: A reflective compendium*. Springer.

Weder, F., & Marte, E. (2023). *CSR communication and cultures of sustainability*. The University of Queensland. https://doi.org/10.14264/428d81b

Weder, F. (2023a). Sustainability as cultural practice and media as institutions of change. In F. Weder, L. Rademacher, & R. Schmidpete (Eds.), *CSR communication in the media: Media management on sustainability at a global level* (pp. 27–33). Springer International Publishing.

Weder, F. (2023b). The evolution of the sustainability story: Strategic sustainability communication as niche construction. *International Journal of Strategic Communication, 17*(3), 228–244. https://doi.org/10.1080/1553118X.2023.2229304

Yeo, S. K., Xenos, M. A., Brossard, D., & Scheufele, D. A. (2015). Selecting our own science: How communication contexts and individual traits shape information seeking. *The Annals of the American Academy of Political and Social Science, 658*(1), 172–191. https://doi.org/10.1177/0002716214557782

Marta Rocchi is an assistant professor in tenure track in Cinema, Photography, Radio, Television and Digital Media at the University of Bologna. Leveraging her scientific background and research experience in Media Studies as a data scientist, she has developed expertise in executing projects characterized by interdisciplinary and systemic approaches. Her research interests primarily focus on quali-quantitative methods for analysing contemporary audiovisual seriality, investigating gender inequalities within the audiovisual industry, and exploring the audiovisual communication of science. Among her most recent publications are: Audiovisual Data: Data-Driven Perspectives for Media Studies (2024 edited with G. Avezzù); Investigating Medical Drama Tv Series: Approaches and Perspectives (2024 edited with S. Antonioni); "Environmental Misinformation and Audiovisual Serial Narratives: An Automatic Analysis of the Twitter Social Discursiveness on Seaspiracy" (2022); and "Women's Labour in TV Series Production: A Comparative Analysis of Italian Generalist TV and Pay Platforms (2016-2019)" (2023). https://www.unibo.it/sitoweb/marta.rocchi5/en

Musicians as Sustainability Communication Agents: Insights from a Structured Survey and Content Analysis

Francisco Tigre Moura and Brigitte Huber

Abstract This book chapter presents two studies which aim to investigate the role of musicians as sustainability communicators and their impact on promoting sustainable behaviors among their fan bases. Study one consisted of an online survey with 1008 active concert goers in Germany, revealing that respondents hold positive attitudes toward musicians communicating sustainability, believed they can influence the behavior of fans, and have a responsibility to communicate sustainability issues. Study two analyzed 9600 Instagram posts by musicians on the Top 100 singles chart in Germany and identified that only 0.5% of all posts addressed environmental topics. Accordingly, our findings indicate a gap between expectations and the actual behavior of musicians in promoting sustainability. Implications for sustainability communication and directions for future studies on the relationship between music and sustainability are discussed.

1 Introduction

Sustainability communication is a complex and challenging task that requires strategic implementation. Despite growing awareness and concerns involving sustainability issues, current communication practices often fall short in effectively engaging and inspiring individuals to embrace sustainable behaviors (Villarino & Font, 2015). The magnitude and urgency of the matter demand a strategic shift, as traditional methods have proven inadequate in conveying the gravity of the situation and motivating meaningful action. Exploring new ways of fostering sustainable consumption (Sesini et al., 2020) but also sustainable behavior in general is needed. To address this gap, there is a growing urge for innovative and impactful

F. Tigre Moura · B. Huber (✉)
Department of Marketing and Communication, IU University of Applied Sciences, Bad Honnef, Germany
e-mail: francisco.tigre-moura@iu.org; brigitte.huber1@iu.org

F. Weder (ed.), *Strategic Sustainability Communication*, CSR, Sustainability, Ethics & Governance, https://doi.org/10.1007/978-3-031-89486-2_21

communication strategies that can effectively promote sustainable practices, empower individuals, and mobilize collective efforts.

While scholars have already started investigating new ways of sustainability communication in environmental and climate change communication, research also focuses increasingly on new communicator roles and the resonance they create as well as the impact they potentially have on individual—here: sustainable—behavior. A specific research focus lays on *greenfluencers* and *eco-influencers* (Bentley et al., 2021; Breves & Liebers, 2022; Huber et al., 2022; Kapoor et al., 2023; Knupfer et al., 2023; Schmuck, 2021; Schorn et al., 2022; Yalcin et al., 2020). However, less is known about the role of attractiveness source models and communicators such as artists or musicians in promoting sustainable behavior (Brennan et al., 2019; Jacoby & Ji, 2010). This is particularly astonishing considering the influence of music on human behavior and personality development (Hallam, 2010; Cheremisova, 2017; Cuadrado, 2019) and the abundance of ongoing initiatives within the music industry at present. Examples include:

- Climate Music:
- Music Climate Pact:
- Music Climate Revolution:
- Music Declares Emergency:
- Sound Future:

Bentz (2020) argues that artistic and creative practices can help expand our imaginaries of the future by opening our minds to new scenarios of change. Similarly, art-based interventions such as eco-jams were explored as an innovative way of engaging the public with sustainability issues (Weder, 2022). Reid and Petocz (2021) highlight that art in general and music in particular "provides us with a powerful way of modifying and counteracting the non-sustainable ways of living and being that have increasingly proliferated during the last hundred years" (p. 3). Accordingly, Østergaard (2019) calls for empirical studies that investigate the relation between music and sustainability. Building on some initial studies on the role of musicians in communicating climate change (Prior, 2022; Wodak, 2018) or environmental topics in general (Publicover et al., 2018), our study specifically focuses on the role of musicians in communicating sustainability-related topics and issues. Building on Weder (2023), strategic sustainability communication of musicians can be conceptualized as "a communicative niche construction" which can be located at the intersection of musicians and their environment; musicians might adapt to the constantly changing environment but also impact their environment. In view of this scenario, this research aims to understand the role of musicians in helping promote sustainable behaviors and the current state of such practices. To do so, in the next sections we will provide an overview of relevant literature on music and sustainability and present own empirical data from (1) a quantitative survey of music fans in Germany on their attitudes toward sustainability and the communication of musicians' sustainability communication, and (2) a quantitative content analysis that shows the prevalence of sustainability messages in Instagram posts by artists listed in the Top 100 singles chart in Germany.

2 Musicians and Sustainability Communication

In the following, musicians are conceptualized as attractiveness source models and as communicators. Based on this theoretical framework and some initial empirical research in the field, research questions for the empirical part are presented.

2.1 *Sustainability and Attractiveness Models*

In view of the urgent environmental challenges, the need to develop and implement innovative communication strategies for effectively conveying sustainability messages to both broad audiences and specific niches is more pressing than ever. Importantly, for a faster behavioral change, it is relevant that such messages are communicated in a positive and appealing way. One strategy which is often applied in marketing to create positive associations with brands and products is by adopting individuals who are somehow admired for their physical appearance, charisma, or humor to serve as communication agents (Frank & Mitsumoto, 2023). These figures are referred to as "attractiveness source models" (Seiler & Kucza, 2017). The attractiveness source model posits that its effectiveness in communication is based on how likable or physically attractive an individual is to an audience (Caballero et al., 1989). Its investigation in the sustainability realm has shown that this form of communication strategy is very effective for green products, although product type plays an important role (Eren-Erdogmus et al., 2016). As examples of source attractiveness models (Cuomo et al., 2019; Kumar et al., 2022), musicians represent a suitable group to communicate sustainability to their followers and influence behavioral change. Indeed, music psychology research shows strong evidence for the power of music to change people's emotions, moods, and beliefs (Prior, 2022). Nevertheless, there is a concern that fans might not be willing to be confronted with topics perceived as negative or concerning during hedonistic moments, given that escapism is an important psychological function of listening to music (Schäfer et al.

2013). Scholars so far were mainly interested in investigating sustainability in the context of music events (e.g., Bär et al., 2022; Brennan et al., 2019; Brooks et al., 2009; Cummings, 2016; Gohoungodji & Amara, 2024; O'Rourke et al., 2011) and tried to identify ways of making music festivals "greener" by, for example, evaluating strategies on how to reduce waste. Mair and Smith (2022) argue that making music events more sustainable is not sufficient and call for research that investigates the broader implications of such events on changing attitudes and fostering sustainable behavior. One way of doing so might be to foster musicians as sustainability agents.

2.2 Musicians as Sustainability Communicators

Musicians can also be conceptualized as one example of a key communicator in constructing and communicating ideas about sustainability. Recent work points to the key role of peers and opinion leaders for communication about and for sustainability (Golob et al., 2023; Weder & Erikson, 2023). "Communicators have now new ways to participate in the sustainability discourse. The widening of channels opens new opportunities to listen to people's concerns and understanding of certain issues. On the other hand, it allows more sustainability communicators to reach a greater audience" (Godemann, 2021, p. 20). There is some initial empirical research exploring the role of musicians as sustainability communicators. Publicover et al. (2018) interviewed 14 environmentally aware musicians who were part of the "Playlist for the Planet," a compilation album released by a Canadian environmental non-profit organization in 2011. Findings show that the interviewed musicians followed different strategies on how to convey their messages: While many chose simple, catchy tunes and phrases, others used more complex arrangements and tried to create musical tension to express the tension that they feel regarding environmental topics. Some preferred angry statements and others focused on hope and empowerment in their songs. Musicians agreed on the importance of making their fans "feel" something when they try to use their music for environmental education. However, which kinds of feelings they aim to evoke also depends on the specific audience. One musician stressed that when playing in front of kids, it should be rather "light" and not too negative. In some settings, humor was used to convey environmental messages. The results also show that musicians are aware of a certain risk that some audience might turn away when musicians speak out about serious topics such environmental issues. The findings highlight that musicians see potential in their music as a way of addressing sustainability messages. Wodak (2018) differentiates between two ways of addressing environmental topics in music—*communicating* versus *conveying* topics. He argues that while *communicating* climate change in music is about literal and data-based representation and thus more likely to "accurately represent the phenomenon," *conveying* climate change topics means using a more poetic and evocative style which applies to popular music about climate change. "In so doing this approach is more amenable to wider societal engagement as conveyance may engage audiences more emotionally than intellectually" (p. 7). In our research, both ways of addressing sustainability-related topics are included when we talk about the sustainability messages of musicians. Building on the literature discussed in this section, we formulated the following open research questions:

RQ1: *What are the attitudes of music fans toward musicians addressing sustainability-related issues?*
RQ2: *To what extent do musicians use their social media platforms to address sustainability-related issues?*

3 Methodology

The research presented in this book chapter comprises two independent studies to address both research questions.

3.1 Study One: Quantitative Survey

The first study aimed at addressing RQ1. It focused on attitudes of music fans toward musicians communicating sustainability-related issues and consisted of a cross-sectional, structured online survey with music fans in Germany. The questions regarding sustainability communication were part of a larger project, titled "Live FM: Fan Monitor" (Tigre Moura & Leschik, 2023). It aimed at investigating respondents' (1) Attitudes toward the role of musicians in communicating sustainability, (2) to assess to what extent individuals see risks for musicians that communicate sustainability, and (3) to explore individuals' characteristics that foster positive attitudes toward musicians' sustainability communication. All factors were measured through 5-point Likert-type scales. The survey was directly and indirectly shared online via university mailing lists, shared across social media platforms associated with both the institutions and the researchers, as well as with industry stakeholders, and on LiveInnovation.org. Additionally, data was collected through the Talk Online Panel, utilizing a voluntary response sampling method. A total of 1107 respondents answered the survey. Following a data screening phase to remove invalid responses, a final valid sample of 1008 respondents was included in the analysis. The final sample consisted of 52.2% of males, 47.5% of females, and 0.2% of diverse respondents. 38.4% ranged between 18 and 34 years old, and 38.8% between 25 and 52 years of age. Furthermore, 18.6% held a bachelor's degree, 21.7% a postgraduate degree, and 33.4% did not hold an academic degree. For more details on the survey and sample, see https://liveinnovation.org/fan-monitor

3.2 Study Two: Content Analysis

The second study addressed RQ2, which investigated the extent to which musicians use their social media platforms to communicate sustainability-related issues. It consisted of a quantitative content analysis (Kassarjian, 1977) of social media posts on Instagram of prominent musicians. The sample of artists for the analysis was derived from the official "Top 100 singles" chart in Germany (www.offiziellecharts.de/charts). The artists were selected over a range of months in 2021. An initial screening was conducted to select only artists with active Instagram profiles. Following this process, a final sample of 408 artists was chosen, from whom a maximum of 25 latest posts were analyzed. The unit of analysis consisted of posts in the artists' feed, and thus, content shared through Instagram stories (a popular feature of the application) was not included. The final total sample of posts analyzed was of $N = 9600$, as some artists contained profiles with less than 25 posts. The categories of analysis were part of a larger content analysis study that investigated Instagram profiles of musicians. For the scope of this chapter, results that will be shared next will address categories such as, "type of cause-related content" (e.g., political, social, environmental, and educational) and "type of environmental content" (e.g., awareness toward climate change, sustainable actions and initiatives, environmental protection). Prior to the data collection, a pre-test was conducted by two separate coders to ensure intercoder reliability, an essential step in content analysis (Lombard et al., 2002). Once a satisfactory reliability result was achieved, the final data collection took place.

4 Results

4.1 Music Fans' Characteristics and Attitudes Toward Sustainability Communication

In total, 1008 respondents were surveyed. 73% of respondents had attended 1 to 3 concerts in the 12 months prior to the survey, while 6.8% had attended 7 or more concerts in the same period. Regarding favorite music genres, "Pop/dance" (59.6%), "Rock/Hard Rock" (46.6%), and "Rap/Hip-Hop" (33.2%) were selected as the most preferred ones. Overall, respondents reported a moderately high interest in the topic of sustainability ($M = 3.54$; $SD = 1.11$), as 57.3% referred to being very or extremely interested in the topic. Moreover, there was a strong perception that climate change represents a serious threat to humankind ($M = 3.97$; $SD = 0.45$), as only 13.5% disagreed or fully disagreed that it represents such a threat. We also investigated respondents' self-evaluation of changes made to their personal consumption behaviors to be more sustainable. Results indicated a rather low evaluation ($M = 2.36$;

SD = 1.00). 13.8% reported they do not wish to make changes to their consumption behaviors to be more sustainable, 50.3% reported they had made minor changes, and only 16.4% said they had made major changes. Further analysis also revealed that within respondents who had made major changes to their behaviors, 63.5% were males and 54.4% held a bachelor's or master's degree. Results also indicated a moderate perception among respondents as to whether their individual consumption behaviors have little impact on the environment (M = 3.00; SD = 1.24), as 37.7% agree or fully agree their behaviors cause little impact on the environment.

Regarding sustainability communication, our findings showed that, overall, respondents believe musicians should raise awareness toward climate and sustainability issues (M = 3.27; SD = 0.29). Importantly, 50.2% believe that it is the role of artists to raise awareness toward climate and sustainability issues. Moreover, respondents believe that musicians can influence fans to have more sustainable consumption behaviors (M = 3.78; SD = 1.02), as shown in Fig. 1.

The study also indicated that 42.5% of respondents agree or fully agree that their favorite artists raise awareness toward climate and sustainability issues (M = 3.23; SD = 1.13). Results also indicated a positive affective response to when musicians share sustainability-related messages (M = 3.42; SD = 1.21). However, only 9.1% fully agree that their favorite artist frequently raises awareness toward climate and sustainability issues. Further analysis also revealed that, among respondents who do not enjoy when their favorite musicians share sustainability-related messages, 53.8% held up to the equivalent of a high school degree, highlighting the influence of educational level on the acceptance of sustainability messages in the music sector.

Finally, 21.5% of respondents found it extremely or very risky for the image of artists to be associated with climate change and sustainability issues (M = 2.60; SD = 1.19). Moreover, further analyses revealed that gender played a role in predicting the acceptance of musicians communicating sustainability-related issues, as 84.2% of respondents who found it risky for artists to be associated with climate change and sustainability issues were male.

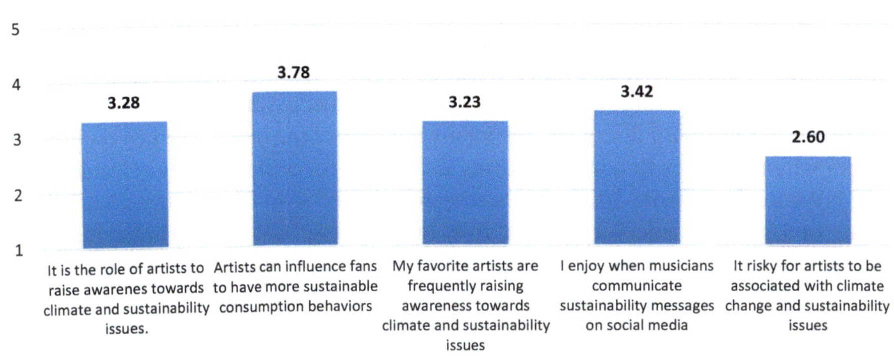

Fig. 1 Overview of music fans' perceptions toward sustainability communication and musicians

4.2 *Musicians' Sustainability Communication on Social Media*

In total, 9600 Instagram posts were analyzed by means of a quantitative content analysis (study two). The study analyzed the extent to which musicians shared Instagram posts to communicate sustainability and climate change-related issues. Results showed that all "cause-related" posts combined (e.g., political, social, environmental, and educational) accounted for only 2.9% of all Instagram posts analyzed. Of these, only 0.5% ($N = 45$) were somehow related to sustainability and climate change topics. 55.2% of posts were career-related content, and 28.0% portrayed their private life with content not related to their musical career.

5 Discussion and Future Research Agenda

Social media offers musicians an extraordinary opportunity to reach out to and mobilize their community (Haynes & Marshall, 2018) in various ways. Findings from study one reinforced this factor and revealed that the entertainment sector (music in specific) has the potential to promote a new narrative of sustainability as a "revolution, cultural change, of innovation and of doing things differently" (Weder et al., 2021, p. 4). The positive attitude of respondents toward accepting sustainability messages emphasized the suitability of the music sector and musicians as communication agents for sustainability and climate change-related topics. Importantly, the low perception of risk from respondents toward sustainability communication by artists should encourage musicians to actively engage in such discussions without hindrance.

However, the extremely low number of sustainability-related social media posts shown in study two should be viewed with concern. It also reveals that the conflict between attitude and behavior in green consumption (Nguyen et al., 2019) is also extended to the case of communication by attractiveness source models, such as musicians. Future research must address this issue and focus on understanding the underlying reasons for such low engagement and factors which may prevent musicians from doing so. For example, investigate ways in which society can establish engaging and motivating online environments for discussing sustainability within the realm of music. By doing so, musicians might become one of the key communicators when it comes to communicating about and for sustainability (Golob et al., 2023; Weder & Erikson, 2023). More research is needed to further investigate these types of opinion leaders and their role in fostering sustainable behavior (e.g., Dalrymple et al., 2013; Jansson et al., 2017).

Although findings from study two may suggest an alarming sign of disconnect between musicians and the environmental challenge society faces, one must interpret them with care. There are many other modes of communication which musicians may adopt to reach their audiences, and that often involve sustainability-related messages. It may include, for example, written messages in clothes worn during public appearances, concerts, and merchandise sold, verbal messages during concerts and interviews, and the use of other social media channels (including Instagram stories and lives, which were not part of the analysis). Therefore, future research should also extend the investigation to account for other possible communication modes and channels. Furthermore, study two focused solely on musicians who reached the Top 100 Singles chart in Germany. As they mostly represented established artists (who often need to represent or adhere to external commercial interests from record labels, for example), future studies should focus on developing comparisons with independent artists, not tied to record labels. This may reveal important conclusions involving popularity and commercial ties in the music sector and the propensity to engage in sustainability communication. Finally, it would be advisable for future studies to develop cross-genre comparisons (e.g., hip-hop, pop, indie rock, electronic). This would allow the understanding of how music fans from different genres, and the dynamics of various subcultures in music, accept and perceive sustainability communication.

Moreover, the study presented in this chapter contains limitations that need to be addressed. First, results from study one should be interpreted considering social desirability bias, which is a common factor in sustainability and sustainable consumption behavior studies (Cerri et al., 2019). In short, it refers to the tendency of individuals to respond in a manner that is socially acceptable or favorable rather than providing honest or accurate responses. In the context of the study presented here, respondents may have felt pressured to conform to socially desirable attitudes or behaviors, such as adopting sustainable behaviors or having a positive attitude toward their favorite musicians communicating sustainability messages on social media.

Also, regarding data collection, study one made use of various sampling techniques, including convenience sampling and voluntary response sampling through a paid online panel, which may represent a limitation (Lefever et al., 2007; Jawale, 2012). Moreover, study two presents two relevant limitations: first, the data obsolescence as it was gathered in 2021. Second, as the data collection was conducted by two separate reviewers instead of a triangulation of coders, which would have allowed for a more rigorous investigation.

Despite these limitations, our findings make an important contribution to the understanding of the relationship between music and sustainability and provide a solid base for future studies interested in the role of musicians in sustainability communication.

References

Bär, S., Korrmann, L., & Kurscheidt, M. (2022). How nudging inspires sustainable behavior among event attendees: A qualitative analysis of selected music festivals. *Sustainability, 14*(10), 6321. https://doi.org/10.3390/su14106321

Bentley, K., Chu, C., Nistor, C., Pehlivan, E., & Yalcin, T. (2021). Social media engagement for global influencers. *Journal of Global Marketing, 34*(3), 205–219. https://doi.org/10.1080/08911762.2021.1895403

Bentz, J. (2020). Learning about climate change in, with and through art. *Climatic Change, 162*(3), 1595–1612.

Brennan, M., Scott, J. C., Connelly, A., & Lawrence, G. (2019). Do music festival communities address environmental sustainability and how? *A Scottish case study. Popular Music, 38*(2), 252–275. https://doi.org/10.1017/S0261143019000035

Breves, P., & Liebers, N. (2022). #Greenfluencing. The impact of parasocial relationships with social media influencers on advertising effectiveness and followers' pro-environmental intentions. Environmental. *Communication, 16*(6), 773–787. https://doi.org/10.1080/17524032.2022.2109708

Brooks, S., Magnin, A., & O'Halloran, D. (2009). Rock On!: Bringing strategic sustainable development to music festivals. *Progress in Industrial Ecology –an International Journal, 6*(3), 285–306.

Caballero, M. J., Lumpkin, J. R., & Madden, C. S. (1989). Using physical attractiveness as an advertising tool: An empirical test of the attraction phenomenon. *Journal of Advertising Research, 29*, 16.

Cerri, J., Thøgersen, J., & Testa, F. (2019). Social desirability and sustainable food research: A systematic literature review. *Food Quality and Preference, 71*, 136–140. https://doi.org/10.1016/j.foodqual.2018.06.013

Cheremisova, I. (2017). Music and development of creative personality within higher education environment. *Aesthetic and semiotic approach. European Journal of Multidisciplinary Studies, 2*(1), 78–87.

Cuadrado, F. (2019). Music and Talent: An experimental project for personal development and well-being through music. *International Journal of Music Education, 37*(1), 156–174.

Cummings, J. (2016). The greening of the music festival scene: an exploration of sustainable practices and their influence on youth culture. In A. Bennett, J. Taylor, & I. Woodward (Eds.), *The festivalization of culture* (pp. 169–186). Routledge.

Cuomo, M. T., Foroudi, P., Tortora, D., Hussain, S., & Melewar, T. C. (2019). Celebrity endorsement and the attitude towards luxury brands for sustainable consumption. *Sustainability, 11*(23), 6791. https://doi.org/10.3390/su11236791

Dalrymple, K. E., Shaw, B. R., & Brossard, D. (2013). Following the leader: Using opinion leaders in environmental strategic communication. *Society & Natural Resources, 26*(12), 1438–1453. https://doi.org/10.1080/08941920.2013.820812

Eren-Erdogmus, İ., Lak, H. S., & Çiçek, M. (2016). Attractive or credible celebrities: who endorses green products better? *Procedia-Social and Behavioral Sciences, 235*, 587–594.

Frank, B., & Mitsumoto, S. (2023). An extended source attractiveness model: the advertising effectiveness of distinct athlete endorser attractiveness types and its contextual variation. *European Sport Management Quarterly, 23*(4), 1091–1114.

Godemann, J. (2021). Communicating sustainability. Some thoughts and recommendations for enhancing sustainability communication. In F. Weder, L. Krainer, & M. Karmasin (Eds.), *The sustainability communication reader* (pp. 15–30). Springer VS. https://doi.org/10.1007/978-3-658-31883-3_2

Gohoungodji, P., & Amara, N. (2024). Greening live shows: What factors matter?: A systematic review of factors affecting sustainability practices in music festivals. *International Journal of Arts Management, 26*(2), 69–90.

Golob, U., Podnar, K., & Zabkar, V. (2023). Sustainability communication. *International Journal of Advertising, 42*(1), 42–51. https://doi.org/10.1080/02650487.2022.2144035

Hallam, S. (2010). The power of music: Its impact on the intellectual, social and personal development of children and young people. *International Journal of Music Education, 28*(3), 269–289. https://doi.org/10.1177/0255761410370658

Haynes, J., & Marshall, L. (2018). Beats and tweets: Social media in the careers of independent musicians. *New Media & Society, 20*(5), 1973–1993. https://doi.org/10.1177/1461444817711404

Huber, B., Lepenies, R., Quesada Baena, L., & Allgaier, J. (2022). Beyond individualized responsibility attributions? How eco influencers communicate sustainability on TikTok. *Environmental Communication, 16*(6), 713–722. https://doi.org/10.1080/17524032.2022.2131868

Jacoby, J. B., & Ji, X. (2010). Artists as transformative leaders for sustainability. In B. W. Redekop (Ed.), *Leadership for environmental sustainability* (pp. 147–158). Routledge.

Jansson, J., Nordlund, A., & Westin, K. (2017). Examining drivers of sustainable consumption: The influence of norms and opinion leadership on electric vehicle adoption in Sweden. *Journal of Cleaner Production, 154*, 176–187. https://doi.org/10.1016/j.jclepro.2017.03.186

Jawale, K. V. (2012). Methods of sampling design in the legal research: Advantages and disadvantages. *Online International Interdisciplinary Research Journal, 2*(6), 183–190.

Kapoor, P. S., Balaji, M. S., & Jiang, Y. (2023). Greenfluencers as agents of social change: the effectiveness of sponsored messages in driving sustainable consumption. *European Journal of Marketing, 57*(2), 533–561. https://doi.org/10.1108/EJM-10-2021-0776

Kassarjian, H. H. (1977). Content analysis in consumer research. *Journal of Consumer Research, 4*(1), 8–18. https://www.jstor.org/stable/2488631

Knupfer, H., Neureiter, A., & Matthes, J. (2023). From social media diet to public riot? Engagement with "greenfluencers" and young social media users' environmental activism. *Computers in Human Behavior, 139*, 107527. https://doi.org/10.1016/j.chb.2022.107527

Kumar, R., Kaushal, S. K., & Kumar, K. (2022). Does source credibility matter in promoting sustainable consumption? Developing an integrated model. *Social Responsibility Journal, 19*(7), 1320–1347. https://doi.org/10.1108/SRJ-07-2021-0257

Lefever, S., Dal, M., & Matthíasdóttir, Á. (2007). Online data collection in academic research: advantages and limitations. *British Journal of Educational Technology, 38*(4), 574–582. https://doi.org/10.1111/j.1467-8535.2006.00638.x

Lombard, M., Snyder-Duch, J., & Bracken, C. C. (2002). Content analysis in mass communication: Assessment and reporting of intercoder reliability. *Human Communication Research, 28*(4), 587–604. https://doi.org/10.1111/j.1468-2958.2002.tb00826.x

Mair, J., & Smith, A. (2022). Events and sustainability: why making events more sustainable is not enough. In J. Mair & A. Smith (Eds.), *Events and sustainability: can events make places more inclusive, resilient and sustainable?* (pp. 1–17). Routledge. https://doi.org/10.4324/9781003314295

Nguyen, H. V., Nguyen, C. H., & Hoang, T. T. B. (2019). Green consumption: Closing the intention-behavior gap. *Sustainable Development, 27*(1), 118–129. https://doi.org/10.1002/sd.1875

O'Rourke, S., Irwin, D., & Straker, J. (2011). Dancing to sustainable tunes: an exploration of music festivals and sustainable practices in Aotearoa. *Annals of Leisure Research, 14*(4), 341–354. https://doi.org/10.1080/11745398.2011.639383

Østergaard, E. (2019). Music and sustainability education–a contradiction? *Acta Didactica Norge, 13*(2), 2–20. https://doi.org/10.5617/adno.6452

Prior, H. M. (2022). How can music help us to address the Climate Crisis? *Music & Science, 5*. https://doi.org/10.1177/20592043221075725

Publicover, J. L., Wright, T. S., Baur, S., & Duinker, P. N. (2018). Music as a tool for environmental education and advocacy: artistic perspectives from musicians of the Playlist for the Planet. *Environmental Education Research, 24*(7), 925–936. https://doi.org/10.1080/13504622.2017.1365356

Reid, A., & Petocz, P. (2021). *Educating musicians for sustainability*. Routledge.

Schäfer, T., Sedlmeier, P., Städtler, C., & Huron, D. (2013). The psychological functions of music listening. *Frontiers in Psychology, 4*, 511. https://doi.org/10.3389/fpsyg.2013.00511

Schmuck, D. (2021). Social media influencers and environmental communication. In B. Takahashi, J. Metag, J. Thaker, & S. E. Comfort (Eds.), *The handbook of international trends in environmental communication* (pp. 373–387). Routledge.

Schorn, A., Vinzenz, F., & Wirth, W. (2022). Promoting sustainability on Instagram: How sponsorship disclosures and benefit appeals affect the credibility of sinnfluencers. *Young Consumers, 23*(3), 345–361. https://doi.org/10.1108/YC-07-2021-1355

Seiler, R., & Kucza, G. (2017). Source credibility model, source attractiveness model and match-up-hypothesis: An integrated model. *Journal Of International Scientific Publications: Economy & Business*, 11.

Sesini, G., Castiglioni, C., & Lozza, E. (2020). New trends and patterns in sustainable consumption: A systematic review and research agenda. *Sustainability, 12*(15), 5935. https://doi.org/10.3390/su12155935

Tigre Moura F, Leschik D (2023) Live FM: Fan Monitor (2023 Edition). *An analysis of attitudes and behaviors of music fans in Germany towards technology innovation and sustainable consumption*. Retrieved from Live FM: Fan Monitor project, Germany: https://liveinnovation.org/fan-monitor

Villarino, J., & Font, X. (2015). Sustainability marketing myopia: The lack of persuasiveness in sustainability communication. *Journal of Vacation Marketing, 21*(4), 326–335. https://doi.org/10.1177/1356766715589428

Weder, F. (2022). #finaltrashtination. An art-based intervention to collaboratively generate conversations about climate change. *Journal of Science Communication, 21*(2), N05. https://jcom.sissa.it/article/pubid/JCOM_2102_2022_N05/

Weder, F. (2023). The evolution of the sustainability story: Strategic sustainability communication as niche construction. *International Journal of Strategic Communication, 17*(3), 228–244. https://doi.org/10.1080/1553118X.2023.2229304

Weder, F., Karmasin, M., Larissa, K., & Voci, D. (2021). Sustainability communication as critical perspective in media and communication studies—an introduction. In F. Weder, L. Krainer, & M. Karmasin (Eds.), *The sustainability communication reader* (pp. 1–14). Springer Fachmedien.

Weder F., & Erikson M. B. (2023) *CSR communication and cultures of sustainability*. https://doi.org/10.14264/428d81b

Wodak, J. (2018). Shifting baselines: Conveying climate change in popular music. *Environmental Communication, 12*(1), 58–70. https://doi.org/10.1080/17524032.2017.1371051

Yalcin, T., Nistor, C., & Pehlivan, E. (2020). Sustainability influencers: Between marketers and educators. *Business Forum, 28*(1).

Francisco Tigre Moura holds a PhD in Marketing from the University of Otago (New Zealand) and has worked since 2013 as Professor of Marketing at IU International University of Applied Sciences, Germany. He is the founder of LiveInnovation.org, a research and education website where, among other projects, leads a monitor project named "Live FM: Fan Monitor," which investigates behavioral trends of music fans in Germany, including sustainable consumption in the live music sector. Research interests also include artificial creativity and technology experiences.

Brigitte Huber (Ph.D., University of Vienna) is Professor of Marketing and Head of Studies Marketing B.A. at IU International University of Applied Sciences, Germany, and external lecturer at the University of Vienna, Austria. Before, she was a postdoctoral researcher at the Department of Communication at the University of Vienna and at the Department of Communication Science and Media Research at the Ludwig Maximilian University of Munich. Her research interests include social media, influencer communication, science communication, journalism studies, and political communication.

Envisioning a Canadian Durability Index for Electronic Devices Through Communication Design: The Case of the Smartphone

Marie Reumont, Geneviève Raîche-Savoie, and Claudia Déméné

1 Introduction

Over the last few years, there has been a growing interest among the population to keep their electrical and electronic equipment (EEE) for a longer period of time. This focus on durability was heightened, in part, by the unethical practices of some manufacturers who intentionally shorten the lifespan of products or prematurely release new versions or software updates, among other tactics, to increase replacement rates. This phenomenon, known as planned obsolescence, is one of many challenges that limit or compromise the ability of future generations to meet their own needs (Brundtland 1987). One of the numerous communication strategies for sustainability that were developed to encourage consumers to purchase more durable products involved the introduction of a set of voluntary indicators known as repairability and durability indexes. This, in turn, aimed to motivate manufacturers to produce, and retailers to distribute, more long-lasting goods.

The repairability index, initially introduced by French public authorities, is a mandatory rating that must be displayed at the time of purchase.[1] Unfortunately, the durability index was not adopted by the European Union, as it conflicted with their

[1] This measure was implemented on 1 January 2021 as part of the anti-waste law enacted on 10 February 2020 and covers several typologies of appliances (smartphone, laptop, washing machine, TV, lawnmower, dishwasher, vacuum cleaner, pressure washer, etc.)

M. Reumont (✉)
École des Technologies supérieures, Montreal, Canada
e-mail: marie.reumont@etsmtl.ca

G. Raîche-Savoie · C. Déméné
École de Design, FAAAD, Université Laval, Québec, Canada
e-mail: Claudia.Demene@design.ulaval.ca

F. Weder (ed.), *Strategic Sustainability Communication*, CSR, Sustainability, Ethics & Governance, https://doi.org/10.1007/978-3-031-89486-2_22

forthcoming revised Energy label, which will include durability criteria such as repairability (HOP, 2023).

In Canada, specifically in the province of Quebec, a groundbreaking law was enacted in 2023 to protect consumers against planned obsolescence. Inspired by French regulations, this law also aims to encourage the durability, repairability, and maintenance of goods among individuals, as well as private and public organizations. However, the translation of this law into a concrete set of actions has yet to be defined and implemented by Quebec's public authorities. This information is typically conveyed to consumers through an environmental labelling system, a set of symbols that communicates the durability performance of products.

To explore the possibilities of implementing such a label system in Canada, we conducted a "research through design" study to determine the conditions that influence the effectiveness of a sign system displaying durability information for an EEE (a smartphone). The two-phased research strategy consisted of 1) co-constructing three durability index potential "design solutions"[2] (proposals displaying numeric, alphabetic, or alphanumeric symbols and 2) testing these proposals through focus groups. The findings reveal, among other relevant results, that a durability index, displaying a recognizable and universal iconography with a numeric symbol in years, would be the most effective sign system for communicating the lifespan performance of a smartphone. Despite the perceived usefulness of this sign system, participants would have preferred a more detailed score breakdown of the criteria composing the final durability evaluation (repairability, maintenance, etc.).

We recommend, among other things, designing a durability index that displays real-time information for the actual or future user of EEE, within an environmental label system that includes a detailed footprint. These recommendations, if considered by public authorities accountable for the implementation of a durability index for EEE, could be a strategic communication opportunity to cultivate a new form of awareness among consumers to buy, use and maintain more durable EEE, countering at the same time different forms of obsolescence.

2 Displaying Durability as a Strategic Communication Opportunity for Sustainability

According to various policymakers, environmental action groups, and scholars (Équiterre, 2018; Heidenstrøm et al., 2021; Jacobs & Hörisch, 2021), displaying information about the durability and repairability of electrical and electronic equipment (EEE) could potentially decelerate their acquisition rates and the electronic

[2] The expression "design solution" refers directly to the creative design process which is defined, according to Cross and Dorst (2001), as the "co-evolution of problem/solution spaces" (p. 425). This process consists of "developing and refining together both the formulation of a problem and ideas for a solution, with constant iteration of analysis, synthesis and evaluation processes between the two notional design 'spaces'—problem space and solution space" (p. 434, authors' emphasis).

waste production (e-waste) (Cooper, 2020). As reported by HOP[3] (2023), when consumers are informed about durability and/or repairability, they are three times more inclined to select an EEE with the longest lifespan and twice as likely to opt for the most repairable option. Durability is "the ability of a product to perform as required, under defined conditions of use, maintenance and repair, until a limiting condition is reached (European standard EN 45 552)" (HOP, 2023, p. 7, our translation).

As a result, there is growing interest in promoting sustainability (Weder, 2021) through visual narratives aimed at consumers, retailers, and manufacturers, highlighting the durability and repairability of electronic and electrical equipment (EEE). This voluntary or mandatory discourse for change could be conveyed by an index integrated into an environmental label system. According to Peirce's theory of semiotics (Atkin, 2008), the index is a sign that should be strong enough to emerge from the context in which it is found and, more importantly, coherent and meaningful to reach the interests and needs of the intended audience. Furthermore, it should prompt the desired reaction from this audience (Frascara, 2004). In other words, the repairability or durability index's agency—its capacity to act, that is, to make a difference (Castor & Cooren, 2006)—should influence the consumption attitudes, behaviours, or perceptions of consumers and the manufacturing process of such EEE.

In this context, this chapter seeks to contribute to the growing field of sustainability communication by demonstrating how the co-construction of a tangible visual narrative, specifically a lifespan durability index for EEE, can play a role in the "evolution of the sustainability story" as a strategy to cultivate responsible consumer habits (Weder, 2023b, p. 228). According to Weder (2023a), sustainability communication can be defined as "communication about sustainability to ensure the legitimacy of sustainable development" (p. 592, emphasis by author). Given the complexity and uncertainty surrounding sustainability and its related challenges, communication requires the implementation of various strategies—such as mass media, risk and science communication, as well as corporate and political communication—to effectively convey sustainability messages and engage the public (Weder, 2023a). For example, one of the key challenges of our research project is to convey environmental information that is clear, transparent, and allows consumers to compare products, with the goal of encouraging the extension of electronic products' lifespans.

As such, the case study described in this chapter presents a potential public authorities' strategic communication approach for sustainability aimed at Canadian citizens. Hallahan et al. (2007, p. 3) define strategic communication "as the purposeful use of communication by an organization to fulfill its mission". Specifically, it relates to "Information/Social Marketing Campaigns" which purpose is to "reduce the incidence of risky behaviors or to promote social causes important to the betterment of the community" (Hallahan et al., 2007, p. 6). In Canada, energy

---

[3] HOP (Halte à l'Obsolescence Programmée) is a French environmental advocacy group which limits to "Stop Planned Obsolescence".

consumption is the only sustainability information that is communicated to consumers through the mandatory *EnerGuide* energy label and the *EnergyStar* environmental certification, which highlights the most energy-efficient products on the market. This focus on energy consumption may be due to the fact that energy use during the use of electronic equipment represents one of the most significant environmental impacts. While technological innovations have improved the energy efficiency of electronic devices, concerns about the decreasing lifespan of these products remain pressing. In recent decades, there has been an acceleration in the acquisition and disposal cycles of electronic goods, resulting in a tremendous amount of electronic waste.

In this regard, Canadian public authorities could implement the durability index as an innovative strategic sustainability communication effort to address consumption behaviours and empower future owners of EEE to make informed decisions that directly influence sustainable development. According to Weder (2023a), this strategic sustainability communication approach differs from other forms of communication. In contrast to "communication *of* sustainability," (p. 228) which focuses on informing and educating people about sustainable products, and "communication *about* sustainability" which involves discussions on more or less sustainable choices, "communication *for* sustainability" aims to provide citizens with a deeper sense of purpose and meaning (Weder, 2023a).

Therefore, the main objective of this chapter is to uncover, through a "research through design" study (Findeli et al., 2008; Prochner & Godin, 2022; Raîche-Savoie & Déméné, 2022), how we investigated the conditions that influence the effectiveness of a lifespan durability index for a smartphone. Such a type of research, which integrates both research and a design project (such as the collaborative design and evaluation of an environmental label), typically generates knowledge that benefits the educational, research, and practitioner communities within the design field (Findeli et al., 2008).

3 From Obsolescence to Durability

Well documented in the scientific literature, the issues surrounding the voluntary lifespan's reduction of products, better known as planned obsolescence, have been the subject of much debate in recent years (Sung. et al. 2021; Taffel, 2023). Planned obsolescence, which impacts the electronics sector, remains highly problematic from an environmental, social, and ethical point of view (Cooper, 2020). Their production puts considerable pressure on natural resources and causes significant pollution and greenhouse gas emissions (HOP, 2022, p. 2).

To tackle various forms of obsolescence[4] and minimize the production of e-waste, France (ADEME et al., 2016; ADEME, 2022), Germany (German Environment Agency, 2017), and Norway (Heidenstrøm et al., 2021) have emphasized the relevance of designing and using voluntary or mandatory environmental label systems that inform consumers about the primary environmental impacts of EEE. Several studies and reports had also highlighted the positive influence on purchasing decisions favouring products with longer lifespans through an appropriate environmental label system (EESC 2016; Jacobs & Hörisch, 2021). These labels encourage EEE consumers to buy more durable products and to better inform them about the maintenance, repair, donation, or sale of EEE (EESC 2016), all strategies supporting a circular economy (Delchet-Cochet, 2021; Sung et al., 2021).

As concluded by the French Environment and Energy Management Agency (ADEME et al., 2020): "Extending the total usage duration of household equipment always generates a benefit, whether at the economic or environmental level" (p. 24, our translation).

4 Environmental Label System as a Communication Strategy to Raise Awareness Towards EEE Durability

Environmental label systems, usually constituted of signs that can easily be decoded (letters, numbers, colours, etc.), seek to inform consumers about the environmental impacts of a product or a service throughout its life cycle. This label system should provide consumers with verified and verifiable information to guide them towards more responsible choices on various communication platforms, such as manufacturers' or distributors' packaging, point of sale displays, and websites. While prevalent in cosmetics, food and household cleaners, such environmental information is less common for electronic devices, except for their energy use, recyclability and the toxic or dangerous aspect of certain materials they contain. Raising public awareness and educating people about better consumption choices through clear and concise labels are usually communication strategies that political authorities support (Bernard, 2014; Eisenriegler, 2020).

This is the case in France, which has been a pioneer in implementing a legal framework[5] and a label system to encourage owners to keep their EEE for a longer

[4] There are many forms of obsolescence: Economic obsolescence linked to the quality/price ratio of the product, the fall in its price and the cost of repairs; Technological obsolescence linked to improvements in the characteristics of the product; Psychological obsolescence linked to the image and changing needs of consumers (Équiterre, 2018, p.15).

[5] The 2020 law on waste reduction and promotion of the Circular Economy introduced a voluntary environmental labelling system that includes both a reparability index and a durability index. The 2015 law on the energy transition for green growth relied on clear, legible, verifiable, and trustworthy information on environmental impacts.

period of time. As part of the strategy, a repairability index (for specific EEE) was introduced in 2021 to inform consumers, at the time of purchase, of an overall score with a value out of ten. This score was based on five criteria: (1) Availability of documentation, (2) Simplicity of disassembly, (3) Availability and accessibility of spare parts, (4) Affordability of spare parts, and (5) A criterion based on the category specificity of the product. However, even though 75% of the French population find this index useful (when purchasing certain products), criticisms have been made towards the manufacturers' self-declaration system which does not guarantee their optimal transparency. Instead, it is suggested to provide more detailed fact sheets, rather than a summary (HOP, 2022).

Moreover, according to Déméné and Marchand (2015), environmental labels displaying a repairability or a durability index would only be effective if a consumer considers the information provided as a selection criterion at the time of purchase, if he/she succeeds in creating an attachment to his/her product by using and maintaining it properly or if he/she has other alternatives to premature end-of-life (or relative obsolescence), such as repairs, donation, or resale. None of these strategies had been concretely implemented to create a shift in the overall consumption of EEE.

In Quebec (Canada), the law n°21[6] protecting consumers against planned obsolescence and promoting the durability, repairability, and maintenance of goods was adopted in October 2023. Additionally, the government of Quebec has announced a revision of the Quebec Policy on Residual Materials Management, with the goal of enhancing the circularity of its economy by 2028 (Gouvernement du Québec, 2024). Quebec becomes the first government in North America to legislate against planned obsolescence, and the second in the world after France. Nevertheless, this is only the first step towards promoting durability through concrete measures, such as an environmental label programme.

5 The Strategic Sustainability Challenge of Communicating Lifespan Duration of EEE

Even though reducing EEE waste has been a growing concern, studies to determine more precisely the effectiveness of an environmental label displaying the lifespan of EEE are still scarce, despite the French authority's intent to transform the repairability index into a durability index. According to a recent publication by HOP (Nebelsztein, 2024), while this innovative index is expected to be launched by the end of 2024, starting with washing machines and televisions, its implementation for smartphones has been abandoned. This decision comes as the EU energy label for smartphones is set to incorporate a repairability index using different criteria.

[6] Originally known in Quebec as Projet de loi 29 (for more information, see Assemblée Nationale du Québec, 2023) https://www.assnat.qc.ca/fr/travaux-parlementaires/projets-loi/projet-loi-29-43-1.html

As of today, most of the measures concern repairability (e.g. affordability and availability of spare parts, etc.), which should support, from our perspective, the overall product's durability. Typically, the durability information is not communicated directly to EEE owners but aggregated with other criteria such as repairs or energy. So far, the *Longtime* environmental label is the first and only independent European ecolabel that promotes products designed to last (Longtime, 2023).

6 Cross-Disciplinary Collaboration as an Effective Strategic Sustainability Communication Strategy

Designing an effective environmental label system to urge consumers to choose a more durable smartphone is a complex task, as it requires the integration of numerous informational criteria into a visually simple (and small) format. Firstly, what are the essential criteria to define the concept of durability regarding electronic devices? Secondly, how to make sure consumers will recognize, at a glimpse of an eye, a visual system that represents an intangible concept such as durability? Thirdly, how to make this visual system adaptable to all types of devices and lastly, more importantly, how could this system support the comparison between devices before, during, and after purchase?

Deciphering this visual ecosystem requires the involvement of multiple stakeholders, encompassing individuals engaged in manufacturing, distribution, sales, repair, recycling, disposal, legislation, and other relevant aspects related to smartphones. In other words, when Communication Designers aiming to visually represent the concept of durability entails a cross-disciplinary collaboration that encourages stakeholders to exchange and deliberate on their individual "professional visions" (Comi et al., 2019) regarding what must be considered when calculating, comparing, representing, and so forth, the lifespan performance of a smartphone. The key elements that will emerge from that "design conversation" will then become the main "signs" of the future label system. This configuration of signs must be strong enough to emerge from the context (e.g. point of sale display, packaging, websites, etc.) in which it is found and, more importantly, coherent to meet the interests or needs of the audience with which it interacts (Frascara, 2004). In this context, communication design "involves making information understandable, usable, interesting, and, if possible, pleasing" (p. 61). In addition, "This [visual] communication should generate the [anticipated] reaction from the public" (p. 61). It also relates to the "imagined user" to whom these visual representations are targeted and the signs they convey (Massanari, 2010, p. 403).

Fig. 1 Typology of signs in the French Repairability Index (Ministère de la Transition écologique et de la Cohésion des territoires, 2024)

7 The Anticipated and Actual Agency of Signs

Signs carry meaning by combining shapes, colours, and content (textual/visual). In the context of a label which must communicate a specific environmental information on a product, such as the France Repairability Index (Fig. 1), meaning is usually carried by a configuration of distinct types of signs. Combined, these signs will speak directly and differently to the intended audience (consumer, manufacturer, distributor, etc.).

Chandler (2017) reminds us that:

> all meaningful phenomena (including words and images) are signs. To interpret something is to treat it as a sign. All experience is mediated by signs, and communication depends on them (p. 2).

Therefore, the design of any sign (alone or in configuration such as ecolabels), as Peirce suggests, corresponds to a typology of signs: the icon, the index, and the symbol. The icon is a sign whose perception, through the senses, reminds or imitates the object, for example, the gear and the tool in Fig. 1. Secondly, the index is a sign that points to its object. In this case, the combination of signs directs attention to the specific object being referenced by the index: the repairability score associated with it. Finally, the symbol is a sign which has an arbitrary link with the object. The relationship between the two entities is conveyed by convention and must therefore be learned. For example, the score (6/10) and the green colour of the label refer to the traffic light conventions.

To design and configure those signs regardless of the typology, one must acquire an in-depth knowledge of the audience and the context of intervention to promote the proper reception of the message and, potentially, induce the desired action. Moreover, Frascara (2006) suggests including the "public" (consumer, manufacturer, etc.) in the design process to anticipate the best reception of the message through signs (icon, index, and symbols), for example, environmental indicators such as a repairability or durability index. Consequently, communication design is a two-way act of transmission that could generate a fruitful dialogue towards a new "sustainability story" (Weder, 2023a, 2023b) between the designers and the target audience, considering that more than 80% of consumers believe that durability is an important criterion for making a choice, sometimes even ahead of price (Vie publique, 2023).

8 The Case of the Smartphone Durability Index Label System

This contextual and academic knowledge led us to develop a "research through design" (Findeli et al., 2008; Prochner & Godin, 2022; Raîche-Savoie & Déméné, 2022) project seeking ultimately to recommend a sustainability communication strategy to public authorities to encourage the implementation of an environmental label displaying the lifespan of electrical and electronic equipment.

9 Research Through Design as a Research Strategy

This type of research combines both research and a design project (e.g., building, product, label system, etc.). Thus, the design project becomes the equivalent of the fieldwork in the humanities and social sciences to collect data and generate results (Findeli, 2004). RTD requires the researcher to also take on its own role as a designer. The designed artefact—in our case an environmental label—is not an end in itself, but rather a means to create knowledge for the disciplinary fields associated with it (Findeli et al., 2008).

10 RTD Objective and Sub-Objectives

This research consists of two main phases. The first, an exploratory phase, was conducted to develop various design solutions (hereafter referred to as "proposals") for a durability index (sub-objective 1). The second phase involved testing and evaluating these proposals (sub-objective 2). Table 1 summarizes the main objective of the research and the two sub-objectives.

To tackle the main research objective—identify the conditions that influence the effectiveness of a sign system displaying a durability index for EEE in Canada—the detailed research strategy consisted of reuniting:

Table 1 Research objective and sub-objectives

	PHASE 1: EXPLORATION (2018–2020)	PHASE 2: TESTING (2021–2022)
Main objective	Identify the conditions that influence the effectiveness of a sign system displaying a durability lifespan information for an electronic device (a smartphone) in Quebec (Canada).	
Sub-objectives	Design durability proposals, using different sign systems, through a cross-disciplinary approach.	Test the durability index proposals through focus groups.

1. A RTD team composed of a researcher and communication designer (Author 1), a product design professor (Author 3), and an industrial ecodesigner, **designed three durability index proposals** in collaboration with an advisory committee composed of ten cross-disciplinary professionals (environment, design, engineering, waste management and repair sector) (sub-objective 1).
2. A RTD team composed of two researchers and communication designers (Authors 1 and 2), **tested the durability index proposals** through focus groups to analyse the effectiveness of its agency (sub-objective 2).

11 The Case of the Smartphone

The smartphone was chosen as the most accurate device to conduct this RTD since this *Information and Communication Technology* (ICT) frequently falls victim to various forms of obsolescence and is rarely used for more than 2 years. In addition, the smartphone is the most purchased EEE by Canadians (Équiterre, 2018). Its manufacturing requires up to 600 times its weight in natural resources (Schmidt-Bleek, 2015), and the equivalent of 85 bathtubs of water to extract the minerals necessary to produce the phone's electronic components (Équiterre, 2022). Owners typically replace it every 2 years, seduced by new technical (technological obsolescence) and aesthetic (psychological obsolescence) performances. Motivated by service providers offering new plans every 2 years with newer devices, there are few incentives to keep one's smartphone. Regarding its repairability, the smartphone is the first device to present defects after 2 years (Équiterre, 2022). More often a victim of premature end-of-life rather than technical obsolescence (absolute obsolescence), it regularly ends its life in a consumer's drawer or in a landfill (even if it is forbidden in Canada), with no hope of reuse or upcycling.

12 Data Collection

The research spanned over 4 years (2018–2022) in the province of Quebec in Canada. It was divided into two principal phases, as described below (see also Table 1).

13 Phase 1 (2018–2020): Designing Durability Index Proposals

The advisory committee took part in six collaborative design workshops (3 h each) which led to the design of three durability index proposals. These potential solutions were iteratively designed by the RTD team in the first 2 years of the research

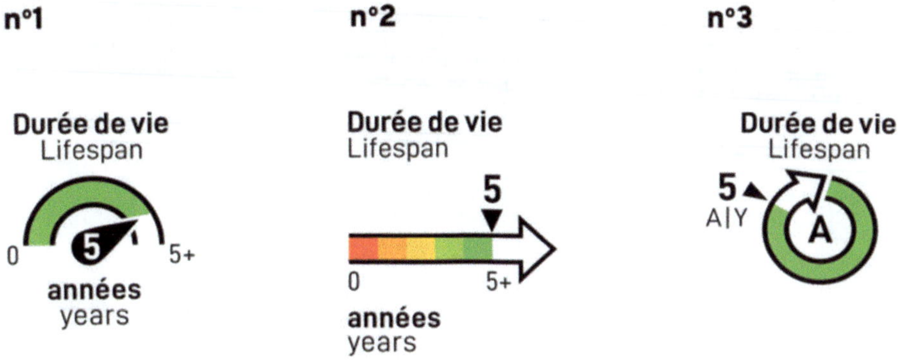

Fig. 2 Canadian durability index proposals

(Fig. 2). The results of this 20-month exploration phase have been published separately (Reumont et al., 2024; Reumont, 2021).

14 Phase 2 (2022): Evaluating Durability Index Proposals

This phase consisted of testing the three durability index proposals (Fig. 2) designed by the RTD team (in collaboration with the advisory committee) among members of the academic community.

The first two proposals evaluated by participants (No. 1 and No. 2) were of a smartphone lifespan shown in years by shape a semicircle (No. 1) or straight arrow (No. 2). The third proposal, the duration in years combined with a circular arrow, displayed the information as an alphanumeric value, "A" being the highest duration and "E" the lowest (No. 3).

These proposals were tested through three online focus group meetings with participants from the academic community (teachers, administrative staff as well as master's and doctoral students). Twenty-three individuals were selected that they use a smartphone.

For 90 min, participants were prompted with open-ended questions to share their feedback on various aspects. These aspects include the accuracy of each proposal (three different configurations of signs) in conveying the durability of electronic devices through icons, indices, and symbols (shapes, typography, colours, etc.), as well as the perceived significance of each sign's configuration. Participants were also invited to comment on the product itself, its packaging, and the corresponding website (Figs. 3 and 4).

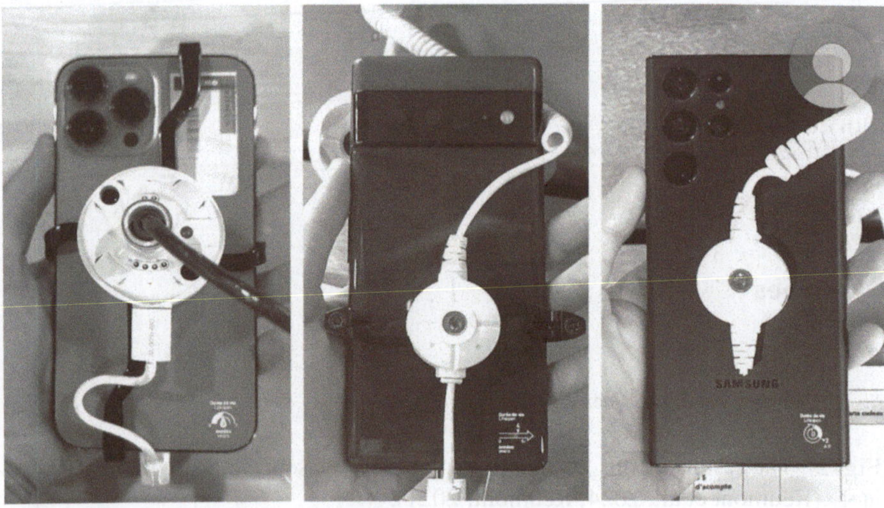

Fig. 3 Canadian durability index proposals on the smartphone itself

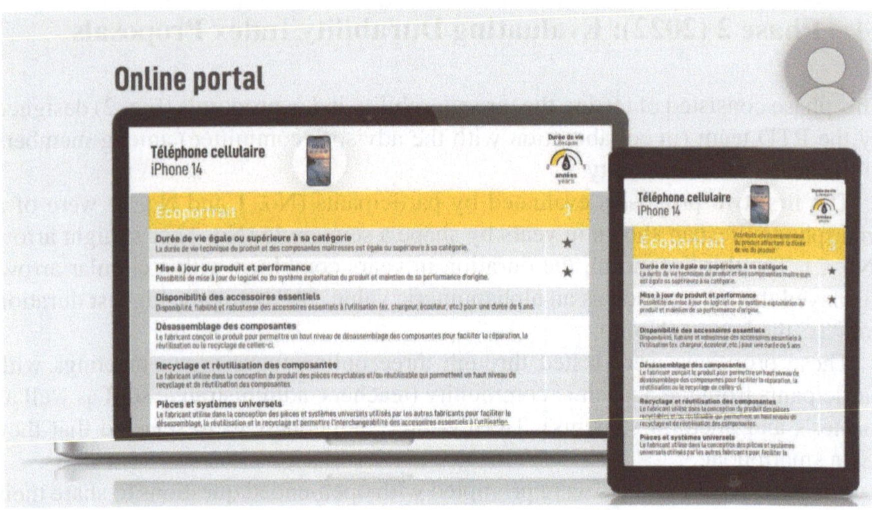

Fig. 4 Durability index proposal No. 1 in a fictional corporate website

15 Findings

To recommend a sign system to political authorities to encourage the implementation of an environmental label displaying the lifespan of EEE—a *durability index*—we conducted focus group meetings to collect comments about three proposals of a durability index that were designed through a cross-disciplinary collaborative design approach.

16 Proposal No. 1: Semicircle (the "Speedometer")

The first proposal was the most effective option (clarity, conciseness, interpretability), according to most of the participants. The *"speedometer"* recognizable shape conveyed, according to them, a universal sign that could be easily associated with the idea of durability. Many participants could easily identify that the smartphone's lifespan had a beginning and an end. They could also effortlessly pinpoint the device's score, which was indicated by the sign system.

For some participants, the shape of the semicircle had a more technological or environmentally friendly aspect (paradoxically) than proposed designs No. 2 and No. 3. However, other participants also commented on the analogy of the icon with a speedometer, a forest fire gauge or the intensity of use which might lead to a negative connotation that could potentially undermine the strength of the message. For example, a participant explained that *"the shape, the speedometer, still gives me the impression that it is speed rather than time that is being expressed"*.

The colour code system appeared useful and clear for several participants, particularly since the chosen colours evoked the universal connotation of traffic light (green: go, yellow: warning, red: attention). This proposal was also the most legible one, even at a smaller size, especially when the index is located on a smartphone.

17 Proposal No. 2: Arrow (the *"Timeline"*)

This option appealed for its conciseness and legibility, despite its common appearance and, for some, its lack of originality. The arrow icon recalled a timeline, allowing participants to easily associate it with the concept of durability. Furthermore, participants also noted that the smartphone's durability increment was easy to understand. It was also suggesting that the smartphone's "actual" durability could surpass the number of years declared by the manufacturer. In short, the arrow, which connotes a timeline, appeared to be noticeably clear to all participants.

Regarding the colour-coding system of this proposal, the incremental red to green (poor to excellent durability) system reminded participants of icons representing battery charge or the colour scale rating system of the Canadian *EnerGuide* label (Government of Canada s.d.). This additional colour system gave participants an alternative option to interpret the index's meaning.

When the "timeline" shape index was located on a smartphone, in the context of an eventual purchase, it became difficult for some participants to read it. Furthermore, the icon could confuse the point of sales' consumers since the arrow appears to point to something else situated to the right of the device.

18 Proposal No. 3: Circle (the "Cycle")

Among the three potential design solutions, proposal No. 3 was the most dubious option for participants since it raised many questions and comments. According to most of the participants, the icon was hard to decode when compared with the two other proposed designs. The principal reason for that confusion appears to be related to the diversity of symbols displayed altogether to communicate durability: a number, a letter, a circle, an arrow, a colour code, etc. Some participants would require more information and more time to understand. More precisely, the use of a letter, in the centre of the symbol (the cycle), appeared to be a confusing element that overloaded the configuration of the overall durability index system. Some participants also asserted that a legend should accompany the index to understand the meaning of the alphabetical system. Therefore, since familiarization with another system of evaluation (e.g. alphabetical system) can take some time for consumers, it could also lead to a loss of interest from them, instead of onboarding them.

When it came to the suggested cycle icon, represented by the circular arrow, many participants associated the shape (representing the idea of durability) with that of recycling. Nevertheless, despite the potentiality of this symbol, the cycle shape appeared to create too much ambiguity with other symbols (such as, of course, the immensely popular recycle symbol). It was also mentioned that, if that proposal was indexed to a smartphone "which will be discarded, with some small possibility of being recycled, [it] would be, somehow, lying to consumers".

Many other elements raised questions among the participants, especially the absence of a "+" after the "5" (5 years and more), which could make it difficult to determine the precise technical lifespan (as planned by the manufacturer) of a particular device. In addition, the absence of the "+" symbol could also indicate the maximum usage and, therefore, it could be a sign (the absence of signs) that the EEE has reached the end of its life, even though it is still working. However, it was also said that the overall symbols' diversity (cycle, alphanumeric indicators) could become its strength since it displays a lot of information all at once. Although the focus groups' discussion revolved around the durability of smartphones, this proposal was also discussed in the context of displaying the durability of other devices. According to one participant:

> The idea of the lifecycle, the circle, the letter added to the number, being able to have more information, which complicates the message, but which is [...] more suitable [for EEE].

In other words, the durability performance represented in years by a numeric and an alphabetical symbol, as well as a colour code, provides too much information. For example, one participant questioned whether "3 years" was good or bad. For example, the combination of the symbol "3" (which represents an average durability score), the green colour and the letter "A" (which both represents a good durability score) could also be used to demonstrate that, even though a device has a durability score of "3 years", it could still be a good "3 years" performance.

Going back to the first phase of the research, the anticipated sign's ambiguity was also a concern for the members of the consultative group, which took part in the cross-disciplinary collaborative design phase (see Reumont, 2021 and Reumont et al., 2024 for more information). It raised questions of the effectiveness of the signs themselves and their configuration relating to the notion of durability. Will this simple sign system be able to convince consumers to keep their smartphones for a longer period of time? Displaying EEE durability in years can be stated as the most efficient "design solution".

Overall, for many participants, representing EEE durability performance in a yearly format was the most clear, concise, and interpretable option, according to Bernard (2014) design criteria. Moreover, this durability indicator should display a number (compared to a letter) because it conveyed more general, concrete, and legible information to participants already familiar with this type of evaluation. Nevertheless, the choice of units or system (letter or number) to display durability could change depending on the category of EEE. For example, in the case of a washing machine, the durability could be indicated in "cycles" rather than "years". However, this simple performance information remains ambiguous, as one participant explained:

> The duration of a smartphone or a computer, which are devices that usually stay on during the day and are turned off at night, a cycle doesn't mean much to me.

Therefore, for some products categories, it would be interesting to obtain more information than just a duration number in years, as it was also discussed during the design phase of the environmental label proposed design (to learn about this research first phase, see Reumont, Cooren and Déméné 2024; Reumont, 2021). One participant specifically raised this issue about a washing machine:

> [...] I am a heavy user. I do not know if this three-year period, the average three-year lifespan, applies to me. I have a large family; I do a lot of laundry. Is the three-year lifespan of a washing machine the same [for me]?

Other types of indicator units (e.g. cycles), relating to the type of usage of various categories of EEE were also discussed, for example, the number of cycles for washing machines as we saw before. In the view of participants, it seems much easier, for a manufacturer, to calculate and display the duration in cycles since *"everyone uses a washing machine differently"*. Participants welcomed the "honesty" of this type of information but suggested that performing a mathematical calculation on usage frequency could be tedious or confusing for many consumers.

19 The Agency of the "+" Symbol

Besides the discussions about the durability indicator (number, letter, colours, etc.) and the shape of the icon, the "+" symbol displayed in proposed designs No. 1 and No. 2 raised contradictory comments. The "5+" in the first proposal (the *speedometer*) gave the participants the impression that a device displaying that sign configuration could last beyond 5 years, whereas for the second proposal (the *timeline*), the product seems to have reached its five-year usage limit. This comment shifted the discussion towards an even more important topic: Who is responsible for deciding the duration of a device (5 years, in our proposals)? Is "5 years" an ideal figure for all EEE of the same category, or is it the actual lifespan of a specific exemplar in particular? Is there a possibility that it could last longer than it was originally intended to last?

The "+" symbol, in this case, had been associated with the possibility of maintaining or using a product for a longer period. This simple symbol was seen as an incentive to keep the device longer, in the case of a device that has an average score (e.g. "3"). As this participant explained, "If it lasts three years on average in the industry, but if I take care of it, why couldn't I make it to four?". This minimal symbol could either encourage or discourage a consumer/user depending on the other signs the "+" is combined with.

20 Comparing the Durability Performance of Different EEE Categories

The last topic that caught participants' attention related to the comparison of the durability performance of different devices for different users' needs. In other words, how would the durability performance between EEE of the same category be displayed and differentiated? The durability index could display specific information for each device so that any user can choose which type of EEE best suits them. However, this more complex sign system could also detract from the information's concise nature requirement.

This discussion paved the way to a completely different type of reflection about the nature of the information that should be displayed by a durability index and on the sign system requirement to remain clear, concise, and interpretable. Is a standardized "mono" symbol indicator and more importantly "static" communication strategy (as the case of the French repairability index) sufficient to highlight the purchase decision process of each potential consumer profile?

The need to display, in real time, to each consumer and for each EEE, the different criteria influencing its durability could become a relevant incentive for consumers to understand why and how they could keep their smartphone. Furthermore, displaying this information could also become an incentive for manufacturers to design and produce more durable EEE.

21 Transforming Durability Index Potential "Design Solutions" into Future Actions

As we saw through focus groups, proposal No. 1 had the most powerful agency for its legibility, followed by proposal No. 2, whose timeline was easily associated with the lifespan idea. Solution No. 3 was the least intelligible one due to its alphanumeric sign system. On the other hand, this last proposal raised many useful questions relating to the nature of information to be displayed. For example, what could be added to display more information about each EEE's environmental footprint? Moreover, the discussions gave us great insights regarding the future iterations to be designed, in a subsequent phase. It also enabled us to collect information on both the configuration of a future durability index, its content, and the media by which it could be communicated. Our strategy consisted in synthesizing all results in the following recommendations. Some recommendations to public authorities as the first step of a strategic communication for sustainability.

The close analysis of focus group data resulted in three principal sets of recommendations. Used as a strategic sustainability communication effort, it could motivate the implementation of an environmental label displaying the lifespan of electrical and electronic equipment. This environmental label—the durability index—should facilitate the comparison between products in the same category.

1. Communicate EEE durability information within a sign system featuring:
   (a) the performance of the EEE in *years* using a recognizable iconography that suggests the notion of "better or worse" performance and the notion of time (for example, a speedometer icon).
   (b) a secondary code to reinforce the legibility of the durability performance of the EEE (for example, the green light colour code system).
   (c) an indicator unit for the various categories of devices to help users to concretely visualize the device's durability (for example, cycles for washing machines or dishwashers).
2. Implement an awareness and incentive programme through an online portal and a strong branding to:
   (a) familiarize the different audiences (distinct categories of users, manufacturers, distributors, interest groups, etc.) to the meaning of the different signs in use within the durability index.
   (b) provide consumers with information about reusing, repairing (or self-repairing) and, as a last resort, recycling.
3. Design a "connected and augmented" durability index to:
   (a) prompt consumers to choose a more sustainable device with real-time durability information for each unit.
   (b) prevent consumers from perceiving this sign as an expiration date, due to the static nature of the information on the product and its packaging.

(c) provide information on how each EEE model device should be used, maintained and repaired for a longer lifespan.
(d) supply the product's environmental footprint, e.g., quantified information on the main environmental impacts calculated over its entire lifespan.

22 Conclusion

The implementation of a durability index for electrical and electronic equipment (EEE) in Quebec (Canada) represents a strategic sustainability communication effort that could significantly influence consumer behaviour and contribute to *cultivate* more responsible consumer and organizations (private and public) habits. By providing clear, transparent, and detailed information about the lifespan of products such as EEE, a durability index could empower consumers to make more informed decisions by helping them to choose products with a longer lifespan potential, thereby encouraging them to repair, maintain and discard their products in a more sustainable way. Ultimately, this could lead to the reduction of e-waste which severely contribute to the depravation of the environment. The case study outlined in this chapter demonstrates the potential impacts of a Strategic Sustainability *"Visual"* Communication based on a sign system that was collaboratively designed with consumers and stakeholders. The sets of recommendations to co-construct an environmental label displaying a durability index (the lifespan of EEE) we are proposing to public authorities could contribute to the "sustainability story" by strategically addressing the challenges of obsolescence and e-waste, transforming them into "new opportunities and potential for actions" (Weder, 2a, p. 229).

Acknowledgements We would like to thank Dr. Robert Butler for proofreading this chapter. Artificial intelligence (AI) was also used to assist in the writing and translation of this chapter.

Further Readings

Reumont, M., Cooren, F., and Déméné, C. (2024). Envisioning sustainability through (un)shared professional visions of the "visual" materials of a design situation: A CCO approach. *Journal of Communication Management, 28*(1), 110–133. https://doi.org/10.1108/JCOM-07-2022-0084

References

ADEME, Chauvin, C., & Fangeat, E. (2016). *Allongement de la durée de vie des produits*. https://www.ademe.fr/allongement-duree-vie-produits

ADEME, Michel, F., Dulbecco, J. R., & Lhotellier, J., RDC Environment. (2020, April). *Évaluation environnementale et économique de l'allongement de la durée d'usage de produits de consom-*

mation ou biens d'équipements à l'échelle d'un foyer français. ADEME. https://www.actu-environnement.com/media/pdf/news-35637-eval-eco-allongement-duree-foyer.pdf

ADEME. (2022). *Communiquer sur la performance environnementale de vos produits, procédés et services écoconçus*. ADEME. https://www.ademe.fr/expertises/consommer-autrement/passer-a-laction/reconnaitre-produit-plus-respectueux-lenvironnement/dossier/laffichage-environnemental/affichage-environnemental-contexte-objectifs

Assemblée Nationale du Québec. (2023), *Projet de loi n° 29, Loi protégeant les consommateurs contre l'obsolescence programmée et favorisant la durabilité, la réparabilité et l'entretien des biens*. https://www.assnat.qc.ca/fr/travaux-parlementaires/projets-loi/projet-loi-29-43-1.html.

Atkin, A. (2008). *Peirce's theory of signs*. Stanford Encyclopedia of Philosophy.

Bernard, Y. (2014, March 27). *Les conditions de l'efficacité des dispositifs d'étiquetage environnemental des produits de consommation: une synthèse de la littérature*. Conférence invitée IREGE, Annecy.

Castor, T., & Cooren, F. (2006). Organizations as hybrid forms of life: The implications of the selection of agency in problem formulation. *Management Communication Quarterly, 19*(4), 570–600. https://doi.org/10.1177/0893318905284764

Chandler, D. (2017). *Semiotics: The basics* (3rd ed.). Taylor & Francis.

Comi, A., Jaradat, S., & Whyte, J. (2019). Constructing shared professional vision in design work: The role of visual objects and their material mediation. *Design Studies, 64*, 90–123. https://doi.org/10.1016/j.destud.2019.06.003

Cooper, T. (2020). Slower cycles: An essential characteristic of the circular economy. In S. Eisenriegler (Ed.), *The circular economy in the European Union: An interim review*. Springer International Publishing. https://doi.org/10.1007/978-3-030-50239-3_9

Delchet-Cochet, K. (2021). *Économie circulaire: De la réduction des déchets à la création de valeur* (Vol. 3). ISTE Group.

Déméné, C., & Marchand, A. (2015). L'obsolescence des produits électroniques: Des responsabilités partagées. *Les ateliers de l'éthique/The Ethics Forum, 10*(1), 4–32. https://doi.org/10.7202/1032726ar

Denoncourt, J., Deshaies, J., Déméné, C., & Raîche-Savoie, G. (2024). Vers une bonification de la réparation professionnelle et de l'autoréparation: l'expérience de deux personas propriétaires d'un réfrigérateur et d'un téléphone intelligent. *Sciences du Design, 19*, 180–205. https://doi.org/10.3917/sdd.019.0180

Eisenriegler, S. (2020). *The circular economy in the European Union: An interim review*. Springer International Publishing.

Équiterre. (2022). Pour des appareils électroménagers et électroniques réparables au Canada. Diagnostic, enjeux et solutions. *Équiterre*. https://www.equiterre.org/fr/ressources/rapport-pour-des-appareils-electromenagers-et-electroniques-reparables-au-canada

Équiterre. (2018). Obsolescence of home appliances and electronics. *Équiterre*. https://www.equiterre.org/en/resources/obsolescence-des-appareils-electromenagers-et-electroniques

Findeli, A. (2004). *La recherche-projet: Une méthode pour la recherche en design*. http://www.blog.designfiction.ch/wp-content/uploaded/Symp_Findeli.rtf.

Findeli, A., Brouillet, D., Martin, S., Moineau, C., & Tarrago, R. (2008). Research through design and transdisciplinarity: A tentative contribution to the methodology of design research. Dans *Swiss design network symposium*, https://swissdesignnetwork.ch/src/publication/focused-current-design-research-projects-and-methods-2008/SDN-Publication-2008_Focused.pdf

Frascara, J. (2004). *Communication design: Principles, methods, and practice*. Allworth Press.

Frascara, J. (2006). *Designing effective communications: Creating contexts for clarity and meaning*. Allworth Press.

German Environment Agency. (2017). *Strategies against obsolescence*. Ensuring a minimum product lifetime and improving product service life as well as consumer information. https://www.umweltbundesamt.de/publikationen

Gouvernement du Québec. (2024, May 17). *Feuille de route gouvernementale en économie circulaire* 2024–2028. https://www.quebec.ca/nouvelles/actualites/details/feuille-de-route-gouvernementale-en-economie-circulaire-2024-2028-55940

Hallahan, K., Holtzhausen, D., Ruler, B., Verčič, D., & Sriramesh, K. (2007). Defining strategic communication. *International Journal of Strategic Communication., 1*, 3–35. https://doi.org/10.1080/15531180701285244

Heidenstrøm, N., Strandbakken, P., Haugrønning, V., & Laitala, K. (2021). Product lifetime in European and Norwegian policies (201822; p. 139). *OsloMet*. www.oslomet.no/en/about/sifo

HOP. (2022). The French repairability index: A first assessment—One year after its implementation. *HOP*. https://www.halteobsolescence.org/wp-content/uploads/2022/02/Rapport-indice-de-reparabilite.pdf

HOP. (2023). *Livre Blanc*. Élaborer un indice de durabilité fiable et ambitieux. HOP. https://www.halteobsolescence.org/wp-content/uploads/2023/07/Livre-blanc-Elaborer-un-indice-de-durabilite-fiable-et-ambitieux.pdf

Jacobs, K., & Hörisch, J. (2021). The importance of product lifetime labelling for purchase decisions: Strategic implications for corporate sustainability based on a conjoint analysis in Germany. *Business Strategy and the Environment, 31*(4), 1275. https://doi.org/10.1002/bse.2954

Longtime. (2023). What is the LONGTIME® label? *Longtime*. https://longtimelabel.com/en/what-is-the-longtime-label/

Massanari, A. L. (2010). Designing for imaginary friends: Information architecture, personas and the politics of user-centered design. *New Media & Society, 12*(3), 401–416. https://doi.org/10.1177/1461444809346722

Ministère de la Transition écologique et de la Cohésion des territoires. (2024). *Indice de réparabilité. Consulté le 12 aout*: https://www.ecologie.gouv.fr/politiques-publiques/indice-reparabilite

Nebelsztein, M. (2024, February 16). *L'indice de durabilité smartphone ne verra pas le jour*. HOP. https://www.halteobsolescence.org/lindice-durabilite-smartphone/

Prochner, I., & Godin, D. (2022). Quality in research through design projects: Recommendations for evaluation and enhancement. *Design Studies, 78*(C), 1–26. https://doi.org/10.1016/j.destud.2021.101061

Raîche-Savoie, G., & Déméné, C. (2022). La pluralité de la recherche en design: tentative de clarification et de modélisation de la recherche-action, de la recherche-création et de la recherche-projet. *Sciences du Design, 16*, 10–29. https://doi.org/10.3917/sdd.016.0010

Reumont, M., (2021). *La conception collaborative et multidisciplinaire de l'agentivité anticipée d'une représentation visuelle. Le cas de l'affichage environnemental de la durée de vie d'un appareil électronique*. Université de Montréal. https://hdl.handle.net/1866/26258

Reumont, M., Cooren, F., & Déméné, C. (2024). Envisioning sustainability through (un)shared professional visions of the "visual" materials of a design situation: A CCO approach. *Journal of Communication Management, 28*(1), 110–133. https://doi.org/10.1108/JCOM-07-2022-0084

Schmidt-Bleek, F. (2015). *Interview: Friedrich Schmidt-Bleek–Le lourd fardeau que représente l'utilisation irrationnelle des ressources/Interviewer: Commission européenne*. https://ec.europa.eu/environment/ecoap/about-eco-innovation/experts-interviews/friedrich-schmidtbleek_fr.

Sung, K., Singh, J., Bridgens, B., & Cooper, T. (2021). Introduction: State-of-the-art upcycling research and practice. In K. Sung, J. Singh, & B. Bridgens (Eds.), *State-of-the-art upcycling research and practice* (Lecture notes in production engineering). Springer. https://doi.org/10.1007/978-3-030-72640-9_1

Taffel, S. (2023). AirPods and the earth: Digital technologies, planned obsolescence and the Capitalocene. *Environment and Planning E: Nature and Space, 6*(1), 433–454. https://doi.org/10.1177/25148486221076136

Vie publique. (2023). *Projet de textes règlementaires relatifs à l'indice de durabilité, pris en application de l'article 16-II de la loi n° 2020–105 du 10 février 2020 relative à la lutte contre le gaspillage et à l'économie circulaire*. Vie publique. https://www.vie-publique.fr/consultations/290777-projet-de-textes-reglementaires-sur-indice-de-durabilite-art-16-loi-agec

Weder, F. (2023a). Cultivation of sustainability in a discourse of change: Perspectives on communication for sustainability as new "norm" and principle of action in socio-ecological transformation processes. *Journal of Language and Politics, 22*(5), 577–600. https://doi.org/10.1075/jlp.22122.wed

Weder, F. (2023b). The evolution of the sustainability story: Strategic sustainability communication as niche construction. *International Journal of Strategic Communication, 17*(3), 228–244. https://doi.org/10.1080/1553118X.2023.2229304

Marie Reumont holds a PhD in organizational communication and design from Université de Montréal. With a career spanning over 35 years, she is a recognized practitioner of communication design with the creation, among other things, of a not-for-profit organization promoting sustainability among her community of practice. Today, her research focuses on the cross-disciplinary and collaborative design process of complex projects such as the digital transformation in the context of large public construction project or environmental communication, etc. Mobilizing "research through design" as her principal research approach, she seeks to understand how the cross-disciplinary collaborative process of designing "things" such as 2D or 3D visuals through analogic or digital technologies has the capacity to positively transform the environment in which we live.

Geneviève Raîche-Savoie , currently pursuing her doctoral studies at Université Laval, brings over a decade and a half of experience teaching design at both college and university levels across various Canadian institutions. Geneviève is also involved in an independent communication design practice focused on promoting environmental awareness and fostering positive social change. This blend of teaching and practice has led her to embrace the role of a "practitioner-researcher", bridging the realms of Education, Design, and Entrepreneurship. Her research is dedicated to nurturing an entrepreneurial mindset among graphic designers, aiming to foster a more reflective, socially, and environmentally conscious approach to their profession.

Claudia Déméné is a professor of sustainable design at the School of Design, Université Laval, Canada. Her expertise focuses on the social and environmental issues surrounding the product lifecycle phases. Her main research aims to develop an environmental label of the lifespan of electronic devices to defy different obsolescence forms. She also explores different alternatives that support responsible consumption and production patterns such as Do-It-Yourself practices and local design. Parallel to her research activities, she is a collaborative member of the Interdisciplinary Research Center for Operationalization of Sustainable Development (CIRODD), and she collaborates in a major research project on sustainable mobility aimed at understanding psychosocial factors that contribute to solo driving attitudes.

Boundary Crossing Approaches Through Education, Sustainability and Communication

Mark Pope, Elizabeth Hauke, Milán Páczai, Berin Pinar, and Eliott Stoclet

Abstract This chapter identifies communication strategies that may be used to enable sustainable thinking and practice within our Higher Education Institution. Recognising the contingent nature of knowledge surrounding sustainability, we adopt a critical and integrated approach. This leads to our proposition that strategic sustainability communication can be improved in Higher Education (and beyond) through open and dialogical approaches that fully engage with complexity, uncertainty and a range of perspectives, rather than avoiding or feigning mastery of them. We call for more transdisciplinary connections, across sectors and regions, drawing on personal experiences, and that give especially heightened voice and agency to students, or other disempowered parties. These proposals require new modes of communication that cross established boundaries. To enable these modes, the interlocutors need the spaces and crucially the competencies to engage, including the potential to be reflexive. Finally, we discuss a short case study surrounding an App, designed to realise these strategies.

1 Introduction

This chapter looks to provide a better understanding of sustainability communication by presenting insight from an educational perspective. We identify specific communication strategies that enable sustainable thinking and practice within our setting—in a Higher Education Institution in the United Kingdom.

Strategic sustainability communication aims to facilitate sustainable thinking and practice in the longer term. In turn, education is a forward-looking endeavour, helping people to develop, look to and prepare for their future—sustainability and education are thereby inextricably linked. By exploring the challenges and

M. Pope (✉) · E. Hauke · M. Páczai · B. Pinar · E. Stoclet
Centre for Languages, Culture and Communication, Imperial College London, London, UK
e-mail: mark.pope@imperial.ac.uk; e.hauke@imperial.ac.uk; milan.paczai20@imperial.ac.uk; berin.pinar21@imperial.ac.uk; eliott.stoclet19@imperial.ac.uk

F. Weder (ed.), *Strategic Sustainability Communication*, CSR, Sustainability, Ethics & Governance, https://doi.org/10.1007/978-3-031-89486-2_23

opportunities that we encounter in our roles as students, educators and leaders within higher education, we leverage examples and insights of use to anyone interested in strategic sustainability communications more broadly.

Fundamentally, we adopt an integrated, holistic and critical approach to sustainability, and this requires the enabling of new communications that cross established boundaries surrounding the structures and fields of academia and education. Moreover, in our educational context, sustainability and its applications are multifaceted and dynamic. Therefore, our approaches continuously evolve and adapt to meet new challenges and serve our community.

Weder et al. (2021) highlight how, in contrast, the context provided by public sustainability communication can present crude perspectives on sustainability, lacking in nuance or transformative potential. We therefore see sustainability education as providing an opportunity to foster the sensibilities, perspectives and criticality to go beyond this.

In the pages below, we start by providing context on sustainability education, before expanding on our use of dialogical pedagogies, transdisciplinarity and critical, transformative practice. We then provide a short case study surrounding a platform for educational communication that underlines how our approach provides students with a particularly heightened level of individual and collective agency and reflexivity. We believe this is essential if they are to learn how to create the change required for sustainability now and in the future.

2 Sustainability Education

Sustainability education sets out to develop knowledge, skills, values and competences appropriate for the complex socio-ecological issues of the twenty-first century (Corres et al., 2020). It is often linked with sustainable development, further prompting questions about what level and type of change and transformation is required (Ibid. 2020). Indeed, the fourth United Nations Sustainable Development Goal is to: 'Ensure inclusive and equitable quality education and promote lifelong learning opportunities for all' and it includes a target that 'all learners acquire the knowledge and skills needed to promote sustainable development' (United Nations, 2015). Accordingly, sustainability education literature provides insight into how progress could be made in these areas, but also highlights the significant challenges faced (see UNESCO, 2015, Tilbury, 2011, 2016; Sterling, 2006; Orr, 2003; or the edited collection from Jones et al., 2010a).

From an international perspective, UNESCO (2015) advocates for more focus on sustainable development in education as a whole, and highlights the need for more humanistic and participatory education that draws on multiple perspectives and looks beyond economic goals. However, historically, changes to address the interconnected ecological, social and economic demands of education for sustainability have been slow. Wals and Blewitt (2010) outline how, initially, in the 1990s, universities began introducing quite discrete or technical fields such as environmental studies and environmental engineering to the curriculum, and then focus shifted to improving Higher Education Institutions (HEIs) own ecological footprints.

Subsequently, Lozano et al.'s (2015) literature review of articles on sustainability practices in HEIs did find that education received the most attention. Indeed, the dominant, didactical pedagogies in Higher Education, such as lecturing, have been increasingly criticised (Tilbury, 2011; Orr, 2003). There have been calls for the adoption of more constructivist, student-centred pedagogies—and ones that also foster changes within the students' personal approaches to sustainability (Cotton & Winter, 2010).

This is in line with writing on sustainability education that calls for more transformational approaches to pedagogy (Wals & Blewitt, 2010). Seatter and Ceulemans (2017: 62) summarise how transformative learning 'refers to a change in the frame of reference, moving from a personal framework of individual ideas into one that references multiple views', suggesting this is essential for such a complex and changing phenomenon like sustainability. Sustainability education therefore requires ambition.

However, while Arruda Filho et al.'s (2019) survey-based research offered some valuable cause for enthusiasm, it also prompted further questions. They found that a training module in sustainability and globally responsible leadership did improve university students' knowledge of elements of a sustainability mindset (including systems thinking, learning community and ecological footprint). Nonetheless, their study was focused on output, and they recognise that more qualitative research could reveal more about how pedagogical processes could improve sustainability education. Accordingly, we present examples below that aim to do this through our dialogic, transformative practice.

More broadly, Lozano et al. (2015) observed a lack of integration in HEIs. They found that key domains—such as those related to education and pedagogy, assessment and reporting, and research—continued to be compartmentalised. A striking example of this is disciplinary silos (Orr, 2003; Tilbury, 2011). To counter this, Jones et al. (2010b) stress the importance of interdisciplinarity and transdisciplinarity. Their value is exacerbated when grappling with issues surrounding sustainability, with its complex ever-changing socio-economic-ecological real-world context. Leal Filho et al. (2017) thus called for HEIs to integrate sustainable development into education and research, and, indeed, across all their systems holistically.

2.1 *Sustainability as a Guiding Principle*

Following the literature above, we aspire to provide an integrated, pluralistic and dynamic approach to sustainability education, and we are keen to avoid a restrictive definition of 'sustainability' so that we leave room for a plurality of approaches. UNESCO (2015: 20) offers a starting point with their statement that 'sustainability is understood as the responsible action of individuals and societies towards a better future for all, locally and globally—one in which social justice and environmental stewardship guide socioeconomic development'. We appreciate their recognition of the global and local, the individual and the collective, social justice and the environment.

However, we place particular stress on the interrelated, contingent and constructed qualities, and our critical awareness of the modernising implications of 'development' and the potential depoliticising nature of universalising language. Morcover, we draw from 'sustainability science' as a discipline (Komiyama & Takeuchi, 2006) where sustainability requires consideration of the interactions between planetary, human and social systems, without immediately inferring the superior agency or importance of any. We therefore offer a different definition. It remains broad, to avoid prescribing how it should be applied, whilst drawing on the above points, and also our view that sustainability can be both an evaluative tool and a principle for action:

> Sustainability is a guiding principle for actions or evaluations that aim to preserve, affirm or enrich the integrated qualities and diversities of human and non-human life and their environments. These actions enable social and ecological justice, for now and in the future, from the local to the global scale.

We propose that this definition provides scope for communication and critical consideration of diverse perspectives that can enable positive action for sustainability.

3 Our Context

At Imperial College, students study degree programmes in science, technology, engineering, maths, medicine and business (STEMMB). In our Change Makers teaching team, we provide optional modules for students across the institution. We offer students opportunities to combine exploring their own experiences, interests and position in the world with topics related to sustainability and global challenges.

We link academic study with authentic case studies and aim to develop students' abilities to create meaningful change. We have created a pedagogical approach called Live, Love, Learn (Hauke, 2017) which draws together various educational perspectives under these three aspects (as outlined in Hauke, 2019):

- Live: prior learning, individual and collective experiences are valued as a foundation for new knowledge and understanding of the lived world.
- Love: empathic engagement with others critically anchors learning to the real, messy, intricate needs of individuals and communities around the world.
- Learn: pursuing challenging, active learning is vital to create independent, critical thinkers who approach complexity with confidence and self-awareness.

Explore our Live, Love, Learn Website here

We aim to provide an educational and communicative space where students can develop these three aspects. We want them to engage empathically and take on challenging tasks that allow them to understand how they can be critical in a way that coheres with their understanding of themselves—while recognising that learning is a process without end, where more open dialogue and communications can facilitate its continuation.

4 Participatory, Dialogical Pedagogical Processes

Of key importance, therefore, is the process that is followed: *how* sustainability, learning and teaching are approached. Communication is key to this. If students can participate in a dialogue where they can negotiate and construct meanings with multiple interlocutors, including their peers or educators, they will be better placed to engage with the complexity of sustainability issues. However, traditional approaches to university teaching, where information is transmitted through one-way communication in the form of a lecture or textbook, provide passive experiences for the learner.

We move beyond traditional approaches that view a curriculum as content to be transmitted or a product to be achieved (Tyler, 1949), and although we focus on the process (Stenhouse, 1975), we aim to ensure that learning is translatable for future experiences. Therefore, instead, we see curriculum as *praxis* (Grundy, 1987), and view curriculum as necessarily providing students with experiences that equip them to transform themselves and the world, going forward beyond their official studies. This involves a more critical approach, giving space for dialogue that encourages a plurality of perspectives and alternative thinking (Wals & Blewitt, 2010).

We recognise how 'knowing' about sustainability can be a process that is developed through a social and cultural context. Hauke (2019: 1) describes this recognition of emergent qualities of knowing as a 'complex, integrative, reciprocal process that unites the knower with the to-be-known'. This is something that we bring into our broader approach to sustainability, communication and education. A recognition of the contingent and transient nature of knowing can help us see the necessity of crossing many boundaries—such as those boundaries that can be erected between human and non-human systems, the present and the future, and very broadly between a person, people and the world around them.

Our approach does not resolve questions surrounding these indefinite and contingent aspects of sustainability. Instead, it highlights the types of questions that need to be asked, sometimes just to recognise the un-resolvability of these issues in practice, where attempting to resolve them can lead to a reductive loss of nuance and sensitivity to complexities. Crucially, we propose that these questions should be addressed in communications between a range of people from multiple perspectives.

We are, therefore, keen to avoid what Seatter and Ceulemans (2017) call 'steering practices', where teachers impose a particular agenda on students. This does challenge the approach of educators such as Morris (2002), whose aim is for students to adopt an eco-centric world view. Instead, our aim is to enable students to develop their own positions and agency, cognizant of complexities of the changing world. We therefore value Seatter and Ceulemans' (2017) proposal for a 'trialogue-style' of teaching where three-way dialogue between 'teacher', 'student' and 'domain of knowledge' facilitates communication to co-construct understandings.

We go further by encouraging students' involvement in co-creating key structures and aspects of the learning experience. In our teaching team, we have introduced dialogical feedback, which has a positive impact on student participation and inclusion (for example, see Hauke, 2022) and also co-created assessment and assessment criteria (for example, see Pope, 2020), again with positive effects crucial to improving students' agency and opportunities to contribute into the dynamic processes involved in their own education.

Read our blog on inclusive, dialogical feedback here

Read our blog on co-creating assessment criteria with students here

The mode of learning and research, and their communication are integral to what is learned, not just in terms of the content, knowledge or skills. Other understandings and practices are developed or acquired—this includes meta-learning (students' learning of how learning takes place), acceptance of power relations, ideas surrounding knowledge construction and the culture surrounding the practices experienced at university. David Orr (2003: 13) suggests that through the example set to them with traditional teaching, students are inadvertently taught that passivity and dominance are expected norms.

We therefore advocate a deconstruction of expertise—seeing the educator as expert with established knowledge also enforces a static reductive view of sustainability. The educator must show, share and transparently experience uncertainty, wrestle with complexity and hold space for imperfect understandings in order to model, validate and engage with this form of knowledge and practice. The key is a transparency and an openness that allows for communication of all of these aspects. This can allow for a more authentic meta-learning, better suited for the complexities of sustainability and the contingent knowledge that our approach is particularly aware of.

5 Transdisciplinary Approaches

One of the most significant areas where communication could be more open is between academic disciplines within education, and more broadly between academia and the real lived world. However, there has long been a tendency in Higher Education for the focus to remain inward-looking within disciplines or institutions. Orr criticises the silos that the disciplinary structure of the education system has created. He states (2003: 11):

> We have fragmented the world into bits and pieces called disciplines and subdisciplines, hermetically sealed from other such disciplines. As a result, after 12 or 16 or 20 years of education, most students graduate without any sense of the unity of things. The consequences for personhood and planet are large.

Our approach to pedagogy and sustainable development is shaped by a recognition of how rigid disciplinary approaches juxtapose with rapid and social, technological and environmental change in the twenty-first century, and the corresponding rise of complexity and 'Wicked Problems' (Rittel & Webber, 1973). Wicked problems are problems that involve a myriad of factors and can be difficult to conceive, let alone resolve. Consequently, the need for agile and innovative thinking beyond traditional academic and educational boundaries has increased.

We therefore promote multidisciplinary, interdisciplinary and transdisciplinary approaches that can, respectively, combine, integrate and traverse disciplinary boundaries. Moreover, we believe there is a special role for transdisciplinarity in providing holistic approaches. Jones et al. (2010b) note how it is particularly valuable for facilitating the testing of theories and concepts across disciplines, like those related to sustainability. And, not only can transdisciplinary approaches combine and integrate academic disciplines, but they can encourage engagement with stakeholders' perspectives from outside the university (Thompson-Klein, 2015: 59).

Transdisciplinarity can allow for real-world situations to become central to academic studies, and for insight from disciplines to be drawn on, as and when needed. Starting with the real-world case study or lived experiences first, rather than the discipline, can be liberating of imposed boundaries and facilitate creative discussion of the complex issues. This does place responsibility on those applying transdisciplinary methods to decide when and how they draw on their disciplinary or extra-disciplinary insights—and we believe this is a valuable skill for students. This aligns with Jahn et al. (2012), who highlight how transdisciplinarity can benefit from reflexivity, as those applying it reflect on their key role in its practice.

Our modules are open to students from across the university, and studying different degree disciplines, albeit within the STEMMB fields. We are therefore well-placed to draw on our students' experiences from their degrees, but also their whole lives beyond their degrees, and beyond any one country, or region. In broaching these new areas, it is more important to foster environments where students feel safe to explore ideas and innovate with each other in an open communication environment. We give them heightened levels of responsibility and agency in this regard and encourage them to reflect on their practice. This can be difficult, and we scaffold our approaches carefully to support our students in learning how to do this, as individuals and as a collective, using dialogical and often experimental approaches.

6 Critical Awareness for Change and Transformation

Many students, and particularly those studying in STEMMB disciplines, are accustomed to working in a positivist paradigm. This can mean that they adopt a removed disposition and attempt to pursue an objective approach to the topics they study. They may remove a consideration of emotions, or critiques that require a level of interpretation of social, political, cultural or even economic and technological issues. This can limit the engagement that they have themselves with their studies, and the work that they produce, (Sterling, 2001, 2006), and also their reflexivity surrounding their own position.

By bringing in dialogical approaches to generating change, we discuss and explore these aspects. This is not to say that dialogicality is without issues. For instance, we are aware of some of the problems that dialogicality can bring, particularly in terms of consensus or norms being established that exclude or fail to take into account certain perspectives. Wegerif et al. (2023) highlight how it could lead to norms surrounding communication that privilege some cultures or social groups over others. Therefore, highlighting power relations and using exercises such as those reflecting on empathy and listening skills become more important (see below).

Vogt and Weber (2020) also call for a broadening of what is considered to be rational by scientists—and the inclusion of wisdom, where practical considerations, values, ethics, norms and competing interests are also considered, and where communication between disciplines and different perspective takers can flourish. We aim to develop wisdom, and for a form of 'phronesis' to develop. Phronesis can be understood as a craft where wisdom is derived from practical analysis, often iteratively and dialogically, informed by personal values and intentions (Frank, 2012). We therefore strive to support students so that they have the competencies to develop such wisdom on sustainability issues (see below).

We recognise that asking for the development of wisdom on complex issues is not a straightforward task, and it requires consultation and negotiation. We therefore ask for students to engage in discussion and contemplation on these aspects. We scaffold activities so that they are led and initiated by students, giving them control and agency, but ensuring all participants can be heard and valued—thereby maximising their potential to develop their own competencies in these areas.

6.1 *Competencies*

David Orr (2003: 12) argues that 'the goal of education is not mastery of subject matter but mastery of one's person'. In our pedagogy, we not only aim to enable students to better contribute to sustainable development projects, but also to develop themselves personally. With our dialogical approach, of immediate importance are communication skills, and we work on these, as a core skill.

We endeavour to promote other appropriate competencies related to sustainability education too. Corres et al. (2020) conducted a systemic review of educational academic papers on sustainability competencies focused on change and transformation. They found that competencies focused on 'doing' were most popular—so Critical Thinking, Participation in Community, and Connections were widely recognised as important. However, competencies focused on 'being', like Emotions Management, Futures and Achieving Transformation, could benefit from more attention.

We suggest that from an educational perspective, meta-learning and reflexivity could further enable the achievement of all of these—to ensure that students have the ability to learn how they learn, and how they work going forward into a variety of situations in the future. Again, we have enacted these in our practice. For example, we have promoted metacognition and more holistic and emotional reflection through multi-modal reflective practices (Pope et al., 2024) where students created and exchanged postcards; and, we ask students to assess their own competences against a Change Makers skill set (see below).

7 Putting Our Approaches into Practice

In the final sections here, we will discuss the App that we created for students to facilitate communication and the sharing of resources for sustainable development projects. Our 'Change Makers App' was designed to promote our principles surrounding sustainability education—particularly surrounding participation and dialogue, transdisciplinarity and change making.

By employing an App that users could access on their phones, we aimed to disrupt the traditional mode of communication in Higher Education. As discussed above, we wanted to move away from more didactic approaches to learning and teaching, and curriculum design focused on content or product. By giving users the choice of when and how to apply the features on the App, we aimed to give them the potential to design their own learning paths, and to take charge of their own projects.

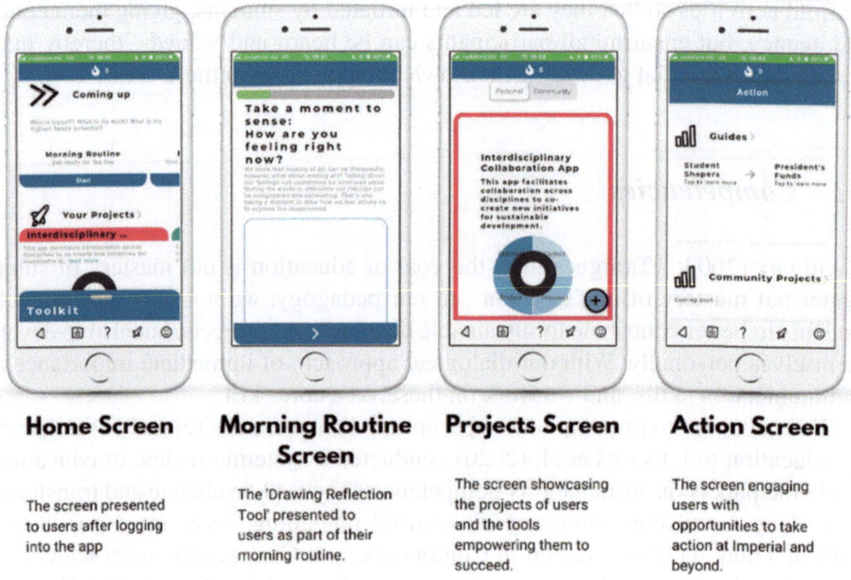

Fig. 1 Visualisation of app screens (Paczai et al., 2023)

The App itself features a range of tools and explanations related to the research, design and implementation of sustainable development projects. Some of these were developed by teaching staff and students and some were accessible guides to established methods or methodologies such as research methods, or soft systems methodology (Checkland & Poulter, 2006).

In order to design and create the App, our team of students and academic staff worked together drawing extensively on each other's perspectives and expertise. We pursued the project through our university's 'StudentShapers' programme (that pairs students and staff to work together on educational research and development projects)—in a manner that bridges student–staff and educational–research divides. The students were proficient in coding and developed a user-friendly Home Screen and variety of other accessible features (see Fig. 1).

7.1 Routines and Reflections

The routines features provide a variety of options for daily self-reflection. Users are prompted to reflect on how they are feeling, encouraged to engage in diaphragmatic breathing, before engaging in more holistic reflection, including gratitude journaling. One of the features for Morning and Evening Routines (pictured above in Fig. 1 for the 'Morning Routine Screen') prompts users to reflect on how they are feeling visually by creating a drawing—thereby engaging with multi-modal means of reflection and reflexivity (as we have advocated before, see Pope et al., 2024). These

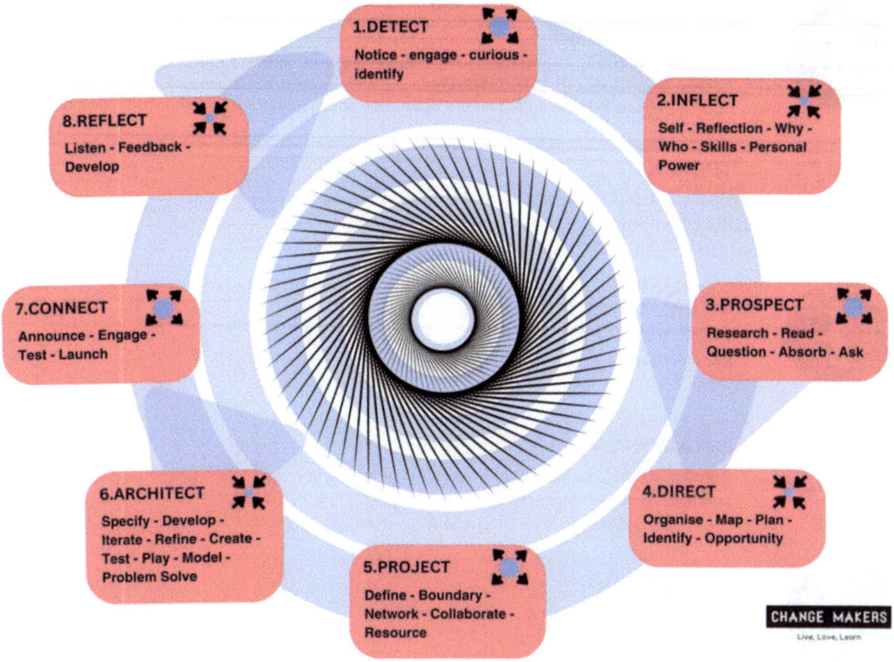

Fig. 2 The Change Makers process

optional features help users develop a better awareness of their own physical and emotional dispositions on a daily basis.

In addition, inspired by techniques advocated by the Presencing Institute at MIT (Scharmer, 2016), we developed a feature where users can reflect on how well they have applied listening skills on a daily basis. Users are prompted to evaluate whether they learned new information on a factual basis, engaged empathically with others, sensed emotional connections, or even engaged in generative listening, where they listen to others to generate new ideas and identify emergent opportunities (Scharmer, 2016). Users' own ratings of their performance over time can be recorded, thereby allowing them to monitor and even *transform* their longer-term approach to these listening skills. This emphasis on a key element of dialogue—listening—can better cultivate users' ability to consider multiple perspectives and therefore their criticality and transformative potential.

7.2 *Projects*

The App also featured an indicative visual representation of a potential process that could be followed by students when they engage in a project to further sustainable development. This was inspired by our own pedagogical principles of critical

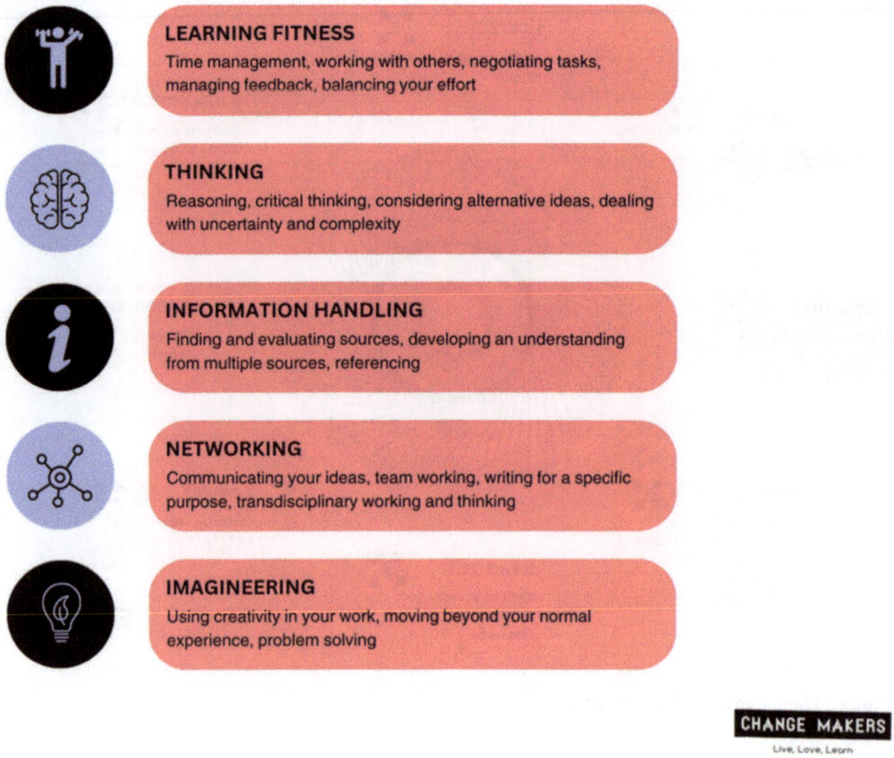

Fig. 3 The Change Makers skillset

change making, combined with design processes that stressed the roles of both divergent and convergent thinking (Design Council, 2017), and also the practice of 'sensing' with an open heart and mind (Scharmer, 2016). The 'process' is depicted as a circle to underscore how the steps do not need to be followed sequentially, and the pairs of inflection and detection indicate how students can recursively look within themselves and the world beyond, aligned with our particular approach to integration and also trandisciplinarity where disciplinary knowledge is not necessarily the starting point (Fig. 2).

7.3 *Change Makers Skillset*

To further enhance the capacity of users to reflect, we refer to five skillsets and competencies that we also use in our regular Change Makers teaching (see Fig. 3). As a set, the skills bring in a student-friendly style and integrated approach, relevant to our take on sustainability education. Following on from the discussion above surrounding competencies, ostensibly these competencies are focused on 'doing', but they are set in the broader context of reflection, transformation and 'being' a change maker: 'Learning Fitness' brings a holistic slant to study skills, considering aspects such as work–life balance; 'Thinking' is encouraged that engages with the complex, contingent contemporary world; 'Information Handling' relates to a range of research and digital literacy skills; 'Networking' as a concept encompasses many of the communication and relation building skills required in the twenty-first century; and finally, 'Imagineering', promotes heightened levels and modes of creativity. Each tool on the App includes an indication of which skills are likely to be practised when that tool is employed—thereby providing the students with another opportunity to monitor their use of competencies, and their capacity to make change themselves.

We believe that the App is a good space to locate an archive of methods and tools that students can access easily to support them in sustainability projects and education. The tools themselves encourage collaborative or dialogical approaches with other people and oneself in a holistic manner, with the agency placed metaphorically, and literally, in the hands of the users of the App.

Principally, the App has been an effective medium for sharing our principles surrounding sustainability education for those who have engaged with it. Its creators and users have gone on to be involved in various sustainable development projects at our university and engaged in action outside of the digital space too, all following the principles of connection, collaboration, co-creation and recognising the contingent and exciting opportunities that sustainable education and action can present.

View the Imperial+ linktr.ee here

An example of these sustainable development projects is Imperial+. Convened by the authors, it is a unique student-led initiative uniting sustainability leaders, student societies and venturers and their initiatives into a thriving ecosystem from across the institution and beyond. Challenging business-as-usual by shifting culture from silos and competition to conversations and collaboration, it aims to accelerate the sustainability transition through systems transformation.

8 Conclusion

In this chapter, we argued for more participatory approaches to pedagogy where students are given more agency. We proposed that the highly contingent nature of knowledge surrounding sustainability demands that we look beyond established boundaries and structures of current communication practices in HEIs. This can be realised through more dialogical approaches, and we call for openness and transparency in communication between educators and students.

We consider it to be an imperative to cross academic silos with more interdisciplinary and transdisciplinary approaches. The discussion above outlined how personal positions and experiences from outside the academy must be strongly drawn on too. This can heighten the complexity of communication, and traversing this range of perspectives may not be easy. We therefore underline how a supportive environment is essential—where student agency and experimentation is encouraged, and difference, nuance and uncertainty are validated and engaged with.

Finally, we advocate for education that is grounded in a desire to improve the world, and that moves away from solely objectivist approaches. It can also be better informed by the values and ethics, and motivations and emotions that people hold. We argue that developing students' competencies holistically—particularly meta competencies surrounding their capacity to learn how they learn and be reflexive—can better prepare them for the opportunities and challenges of sustainability and its communication going forwards.

We recognise that the cultural context, resources and institutional set up is key to establishing sustainability education within a university (Leal Filho et al., 2017). Our own position is reasonably conducive to dialogical, transdisciplinary learning. Like many other universities in the United Kingdom, we have an international range of students and staff, bringing international perspectives. However, our cohorts are still unrepresentative of the world. Therefore, we advocate drawing from external sources to complement this. We are provided with the space to deliver optional modules to undergraduate students from across the university. We would encourage other institutions, to create such spaces across faculties and departments—and to publicise their practice. Weiss and Barth (2019) also stress how more sharing of experiences, particularly from those in the Global South would valuably add to the literature, in this regard.

From this chapter, we suggest that strategic sustainability communication more broadly can draw the following conclusions. Primarily, there is value in media and communication that bridges silos and divides between stakeholders. Moreover, where voice and agency can be provided to those who have less, the value is even greater. Overcoming boundaries and barriers to communication is key to this—whether they are structural or based on the competencies or attitudes of people themselves. Development of competencies needs to include the potential to be reflexive, to be actively receptive and open to other perspectives, and to embrace uncertain or contingent positions. The analysis above highlighted how these are not simple tasks but educators, and all those involved in strategic sustainability communication, can play a role in developing a culture where they are encouraged, so that broader sustainability communication is less reductive and more critically nuanced to facilitate positive change.

Recommendations for Further Reading

1. For further insight into key themes and applications of sustainability education, see this landmark edited collection: Jones, P., Selby, D., & Sterling, S. (Eds) (2010) *Sustainability Education: Perspectives and Practice Across Higher Education*, Earthscan: Oxon.
2. To explore our educational approach and its applications in more detail, please see our website: http://livelovelearn.education/.

References

Arruda Filho, N., Cassitas Hino, M., Przybylowicz Beuter, B., (2019) Including SDGs in the education of globally responsible leaders. *International Journal of Sustainability in Higher Education*, *20*(5), 856–870. https://doi.org/10.1108/IJSHE-01-2019-0032

Checkland, P. B., & Poulter, J. (2006). *Learning for action: A short definitive account of soft systems methodology and its use for practitioners, teachers and students*. Wiley.

Corres, A., Rieckmann, M., Espasa, A., & Ruiz-Mallén, I. (2020). Educator competences in sustainability education: A systematic review of frameworks. *Sustainability, 12*(23), 9858.

Cotton, D., & Winter, J. (2010). 'It's Not Just Bits of Paper and Light Bulbs': A review of sustainability pedagogies and their potential for use in higher education. In P. Jones, D. Selby, & S. Sterling (Eds.), *Sustainability education: Perspectives and practice across higher education*. Earthscan.

Design Council. (2017). *The double diamond*. Accessed February 10, 2024, from https://www.designcouncil.org.uk/our-resources/the-double-diamond/

Frank, A. W. (2012). Practicing dialogical narrative analysis. In J. A. Holstein & J. F. Gubrium (Eds.), *Varieties of narrative analysis* (pp. 41–44). Sage.

Grundy, S. (1987). *Curriculum: Product or praxis*. Routledge.

Hauke, E. (2017). Our approach. *Live, Love, Learn*. Accessed February 23, 2024, from http://livelovelearn.education/

Hauke, E. (2019). Understanding the world today: The roles of knowledge and knowing in higher education. *Teaching in Higher Education, 24*(3), 378–393.

Hauke, E. (2022). *Practical steps to create a positive and inclusive feedback culture in the classroom*. Accessed March 20, 2024, from http://livelovelearn.education/2022/08/05/practical-steps-to-create-a-positive-and-inclusive-feedback-culture-in-the-classroom/

Jahn, T., Bergmann, M., & Keil, F. (2012). Transdisciplinarity: Between mainstreaming and marginalization. *Ecological Economics, 79*, 1–10.

Jones, P., Selby, D., & Sterling, S. (Eds.), (2010a) *Sustainability education: Perspectives and practice across higher education*. Earthscan.

Jones, P., Selby, D., & Sterling, S. (2010b). More than the Sum of their Parts? Interdisciplinarity and Sustainability. In P. Jones, D. Selby, & S. Sterling (Eds.), *Sustainability education: Perspectives and practice across higher education*. Earthscan.

Komiyama, H., & Takeuchi, K. (2006). Sustainability science: Building a new discipline. *Sustainability Science, 1*, 1–6.

Leal Filho, W., Jim, W., & Y., Brandli, L., Veiga Ávila, L., Miranda Azeiteiro, U., Caeiro, S., & Rejane da Rosa Gama Madruga, L. (2017). Identifying and overcoming obstacles to the implementation of sustainable development at universities. *Journal of Integrative Environmental Sciences, 14*(1), 93–108.

Lozano, R., Ceulemans, K., Alonso-Almeida, M., Huisingh, D., Lozano, F., Waas, T., Lambrechts, W., Lukman, R., & Hugé, J. (2015). A review of commitment and implementation of sustainable development in higher education: Results from a worldwide survey. *Journal of Cleaner Production, 108*(Part A), 1–18.

Morris, M. (2002). Ecological consciousness and curriculum. *Journal of Curriculum Studies, 34*(5), 571–578.

Orr, D. (2003). *Earth in mind: On education, environment, and the human prospect*. First Island Press.

Paczai, M., Pinar, B., Mang, K., & Pope, M. (2023). *Change Makers App (Version 1.1)* [Mobile App]. Accessed June 1, 2023.

Pope, M. (2020). *How do you make assessment criteria more student-centric? Co-create it with students*. Accessed March 2, 2024, from http://livelovelearn.education/2020/01/21/how-do-you-make-assessment-criteria-more-student-centred-co-create-it-with-the-students/

Pope, M., Hauke, E., Davis, N., Kale, R., Kolesnikova, A., & Lee, T. (2024). Pushing the boundaries of reflection. *Exchanges: The Interdisciplinary Journal, 11*(2), 1–28.

Rittel, H., & Webber, M. (1973). Dilemmas in a general theory of planning. *Policy Sciences, 4*(2), 155–169.

Scharmer, O. (2016). *Theory, U. Leading from the future as it emerges: The social technology of presencing* (2nd ed.). Berrett-Koehler.

Seatter, C., & Ceulemans, K. (2017). Teaching sustainability in higher education: Pedagogical styles that make a difference. *Canadian Journal of Higher Education, 47*(2), 47–70.

Stenhouse, L. (1975). *An introduction to curriculum research and development*. Heinemann.

Sterling, S. (2006). *Sustainable education*. Accessed March 2, 2024, from https://www.sustainableeducation.co.uk/wp-content/uploads/2019/05/Contrasting_paradigms_in_teachingandlearning.pdf

Sterling, S. (2001). *Sustainable education – re-visioning learning and change*. Green Books.

Thompson Klein, J. (2015). *Interdisciplining digital humanities: Boundary work in an emerging field*. University of Michigan Press.

Tilbury, D. (2011). Higher education for sustainability: A global overview of commitment and progress. In GUNI (Ed.), *Higher education in the World 4. Higher education's commitment to sustainability: From understanding to action* (pp. 18–28). Palgrave.

Tilbury, D. (2016). Student engagement and leadership in higher education for sustainability. In M. Barth, G. Michelsen, M. Rieckmann, & I. Thomas (Eds.), *Routledge handbook of higher education for sustainable development* (pp. 241–260). Routledge.

Tyler, R. W. (1949). *Basic principles of curriculum and instruction*. University of Chicago Press.

UNESCO. (2015). *Rethinking education: towards a global common good?* Accessed March 2, 2024, from https://unevoc.unesco.org/e-forum/RethinkingEducation.pdf

United Nations. (2015). *Goal 4*. Accessed August 4, 2024, from https://sdgs.un.org/goals/goal4#targets_and_indicators

Vogt, M., & Weber, C. (2020). The role of universities in a sustainable society. Why value-free research is neither possible nor desirable. *Sustainability, 12*(7), 2811.

Wals, A., & Blewitt, J. (2010). Third wave sustainability in higher education: Some (inter)national trends and developments. In P. Jones, D. Selby, & S. Sterling (Eds.), S*ustainability education: Perspectives and practice across higher education*. Earthscan.

Weder, F., Karmasin, M., Krainer, L., & Voci, D. (2021). Sustainability communication as critical perspective in media and communication studies – An introduction. In F. Weder, L. Krainer, & M. Karmasin (Eds.), *The sustainability communication reader: A reflective compendium* (1st ed., pp. 1–12).

Weiss, M., & Barth, M. (2019). Global research landscape of sustainability curricula implementation in higher education. *International Journal of Sustainability in Higher Education, 20*(4).

Wegerif, R., Shi, S., Rubio-Jimenez, A., Long, Y., Liu, Q., & Chang, C. (2023). Dialogic education: Tensions and dilemmas. In *International encyclopedia of education* (4th ed.). Elsevier.

Mark Pope teaches in the Change Makers team at Imperial College London. He employs critical and transdisciplinary methods in his research on communications, sustainability and pedagogy.

Elizabeth Hauke is a practitioner-researcher in Higher Education working to promote student-centred active learning. She founded and leads Change Makers at Imperial College London, and developed the Live, Love, Learn approach to sustainability education. Her research interests focus on authenticity in education.

Milán Páczai recently graduated with an Electrical Engineering degree. He initiated the Change Makers App and founded multiple student-led projects including Imperial+.

Berin Pinar recently graduated with a Biotechnology degree. She founded the Change Makers society and designed the Change Makers App.

Eliott Stoclet recently graduated with a degree in Bioengineering. He founded multiple student-led projects including Imperial+.

Visual Storytelling for Sustainability Communication. Practice-Based Design Techniques

Arlene Birt

Abstract Our ability to emotionally engage a wide variety of audiences with climate science is critical to our continued survival on this planet. Visual communication can facilitate this engagement—particularly when we aim to show humanity within the scientific data.

Visuals are increasingly used to translate complex scientific and technical content for non-expert audiences. There is also a growing focus on storytelling as a method of communication to inspire behavior change. This chapter outlines the benefits of communicating environmental content through visual storytelling techniques. "Visual storytelling" is a method of communication that emphasizes narrative while using graphics to convey information and context. Using practice-based examples of visual storytelling as case studies, this article catalogs the benefits of visual storytelling and identifies key techniques of visual communication.

The more communicators are able to creatively engage non-specialists in complex topics like climate, the sooner society can mitigate emissions and adapt to our changing climate.

1 Visual Stories as Strategic Communication

While words and numbers excel at providing details, pictures are optimal for showing context and relationships (Tufte, 1990). Modern life is inundated with information, and simply sharing information is not enough to motivate change. Making content accessible and engaging using visuals is a key part of strategic communication (Goransson & Fagerholm, 2018). When aiming to motivate behavior change on social and environmental topics, clarifying complex information is even more

A. Birt (✉)
Minneapolis College of Art and Design, Minneapolis, MN, USA

Background Stories (Design Consultancy), Minneapolis, MN, USA
e-mail: arlene@backgroundstories.com; https://www.backgroundstories.com; https://www.arlenebirt.com

F. Weder (ed.), *Strategic Sustainability Communication*, CSR, Sustainability, Ethics & Governance, https://doi.org/10.1007/978-3-031-89486-2_24

critical. So, how can we deliver information to reach audiences in a way that has an impact? Stories have been critical tools for sharing knowledge throughout human history (Eisner, 2008). Whether it is about the best hunting grounds in prehistoric times, or navigating a health insurance plan today, stories frame and convey meaning through connecting a series of events. Narrative strategies, such as stories, are effectively used to communicate science to policymakers (D. Jones & Anderson Crow, 2017). Climate change research and policy is increasingly embracing storytelling (Moezzi et al., 2017) and presenting climate information in a narrative way also increases emotional response among audiences (Wong-Parodi & Feygina, 2021).

Paired together, "visual stories" are a powerful tool to help audiences "see" how social and environmental topics relate to themselves, their families, communities, and the world. Visual narratives increase the perception of the importance of an issue among policymaking audiences (Boscarino, 2022) and aid collective memory surrounding events (Zelizer, 2004).

2 Defining Visual Stories

Stories structure content into narratives across both written and spoken language (Eisner, 2008). Articles, stories, presentations, and speeches each have a beginning, a middle, and an end, and this structure can be used with visuals, as well (Eisner, 2008). In visuals, the "beginning" is the part of the visual that the designer wants the viewer to see first: typically a headline or another main idea that can be conveyed through text or graphics. (Designers refer to this as hierarchy.) The ending typically includes a call to action for the audience.

Visuals can extend the imagination in a way that shows a stylized or conceptual diagram of a situation—to enable viewers to see the big picture—rather than focus on every realistic detail (Ware, 2004).

Visual stories can include comics, films, animations, infographics, illustration, even data visualization—when they are intentionally structured as a story. Motion-based stories make it easy to present information in a way that bridges time and space. With careful consideration and use of design principles, static visuals can also capture the passage of time and space. When developing motion-based deliverables, static visuals are generally developed as a first step, so the same basic principles apply. While a variety of types of visuals convey narrative climate messages (O'Neill & Smith, 2014), this chapter focuses on illustration and graphic design approaches.

In contrast to non-story-focused visuals, visual stories are intended to structure information through narrative. For example: a non-story visual may be a poster that contains factoids or a flier using a bullet-pointed style to convey content. If the facts are shown in a way that indicates connections between them, or, if they are put into the context of a bigger picture, they can *then* be considered a story. Both non-story visuals and visual stories can be well-designed, beautiful, and eye-catching—but if they are not structured as a narrative, they do not qualify as a visual story. This article outlines approaches to structure content into visual stories, by identifying whether the content is process-based, scenario-based, or a combination of both.

3 Benefits of Visual Storytelling

Visual stories can *put data into context* and *establish emotional connection* (Segel & Heer, 2011) by showing how specific actions relate to the bigger picture of environmental change. Information in context can help audiences "see"' how social and environmental topics connect to their own lives. This relationship is important to connect with audience member's values and help them understand how they fit within a seemingly abstract set of data, or to "see themselves within the data" (Kearney, 1994). Also, showing climatic impacts on the human condition through visuals provides a strong base to connect with individual experiences in order to motivate action (Birt, 2024). For example, the Intergovernmental Panel on Climate Change (IPCC) project referenced below shows how abstract information can be shown relevant to audiences by placing generations of human lifespans alongside the data.

Visuals can also *decrease psychological distance* among individuals. Psychological distance refers to the feeling that climate change is happening "*here*, *now*, to *people like me*, and with high *probability*" (Trope & Liberman, 2010; Spence et al., 2012). This article argues that graphics have the power to bridge psychological distance in ways that more realism-based visual media (such as photos) cannot: For example, illustration can connect distant places within the same frame, showing how an event in one place can have impact around the world.

Visual narrative can be more *easily approachable* than complex data visualizations, for non-specialist audiences (Hullman & Diakopoulos, 2011). While topics like climate science are inherently complex, visuals can serve as a middle ground between showing all the complicated details and oversimplifying. Visuals can also convey information (like data) that is impractical to convey in a narrative, word-based format. Stories are a method of conveying linear or process-based information in a format that makes content accessible to a broad range of audiences because people are familiar with the basic sequences of stories.

Graphics also provide at-a-glance reference to values and culture in ways that words cannot: for example, by visually depicting styles of clothing, transportation, housing, and other identifiable characteristics (Tufte, 1990).

IPCC Example

A key figure in the Intergovernmental Panel on Climate Change (IPCC) 2023 Synthesis Report (Lee et al., 2023) *Figure AR6 SYR SPM1 panel c*, captured global media attention and inspired spin-offs on social media. The figure breaks new ground for IPCC as it ties historical and projected temperatures to human lifespans in a way that makes climate science accessible to non-expert audiences. This figure is a visual story because it aligns visuals of human life span along with data on annual temperature averages. This parallel set of information triggers the realization that generations are differently impacted by the effects of climate change. For example, a grandmother lives her life in a very different world than her granddaughter will experience throughout her lifespan (Birt, 2024).

4 How to Structure Visual Stories

To create a visual story, a designer combines techniques of graphic design with techniques of storytelling. The following process can help structure complex content in a visual format.

4.1 Understand the Content and the Audience

The first steps of creating visual stories parallel with how one would create a word-based story. It is important to build a visual narrative around the key message that you want to convey to this audience and the audiences' values. To do this, it is helpful to be very specific about which audience you are trying to reach, their current level of understanding of a topic, and what you want them to do next. Once you determine the specific audience and message, the content of an infographic or illustration can be structured in a story format— putting the information into context. For example, this could be an outline of each piece of content in a step-by-step process, or how various points relate to a "bigger picture."

With a clear outline of content for a specific audience, one can begin putting content into a visual narrative—or story—structure. While much of Western narrative tend to focus on the journey of an individual "hero" following Joseph Campbell's well-known work (Campbell, 2008), the author's experience is that visual stories do not necessarily have to feature a human main character. Due to the nature of mass communication, audiences are frequently larger than a single individual. Likewise, main characters—while they ought to relate to individual members of an audience—can go beyond a single person to represent collective effort. When communicating about social and environmental topics, the featured character of a visual can be, for example: a watershed, the health of a community, the community itself, the electric grid, or a changing climate.

4.2 Put Information into Context

Once you have clarity on the specific audience and key messages for this audience, the next step is to structure the content. When structuring content into a visual narrative format, it helps to understand that visual stories are uniquely positioned to bridge time or space—or both—related to psychological distance (Birt, 2021). For example, visuals can show a full sequence of events that occur at different times in one visual frame—this is bridging time. Likewise, an illustration that shows an event happening many places at once is bridging space. The ability of visuals to

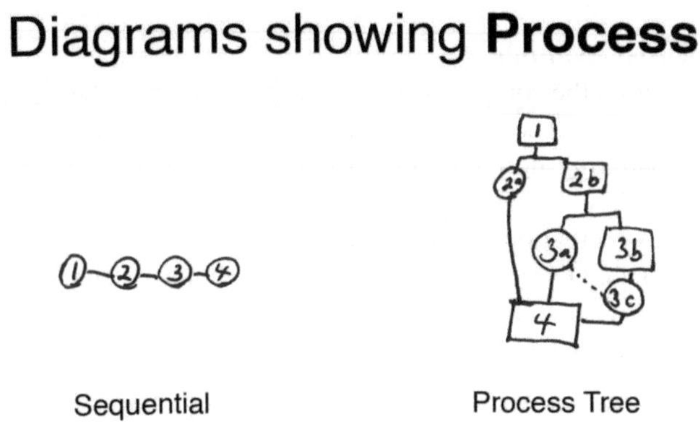

Fig. 1 Sequential and process model

span across time or space is important because these are key aspects of bridging psychological distance, which helps audiences connect with climate information, leading to behavior change (Maiella et al., 2020).

- To transcend "time"': a visual story can show a sequence of moments in time. This could be in the structure of a beginning, middle, and end. It is often useful for these visuals to follow a *process*-focused visual approach.
- To transcend "space": a visual story shows how individual pieces of content relate to one another, or a larger concept. It is often useful for these visuals to follow a *scenario*-focused visual approach.

Some concepts can fall under either approach and for these you will need to follow the approach that best fits the key messages for your audience. For example, a "before"' vs. "after" state can be presented as a process (showing the process of retrofitting a building), or as different grouped parts of the same visual scenario (a view of a retrofit building with cut-aways showing "before", or vice versa: the old building with cut-aways showing potential future states).

Here is some guidance to help determine when to select each approach:

Processes visuals (see Fig. 1) include timelines and step-by-step flows of information. These can be visualized through any of the following sketches. A process has an entry/exit point (or points) and an order to the steps (at least some of them) (Birt, 2021). To determine if a Process visual makes sense for your story, examine your content and ask:

- Are we guiding the viewer through a series of steps?
- Is the order of the elements the most important aspect of the message we want to communicate?

If either of these are true, you might have a visual process on your hands.

Process-Focused Example

This signage shows the complex system of water flows built beneath a school's favorite sledding hill. The infographic helps visitors see what happens underground by illustrating how the system collects polluted rainwater from the neighborhood, filters it, and then delivers cleaner water into a local lake that is known for its recreation. This infographic features process-focused narrative visuals showing the progress of water through the system. It also bridges time and space by showing a process that's typically geographically distributed within a single frame (Fig. 2).

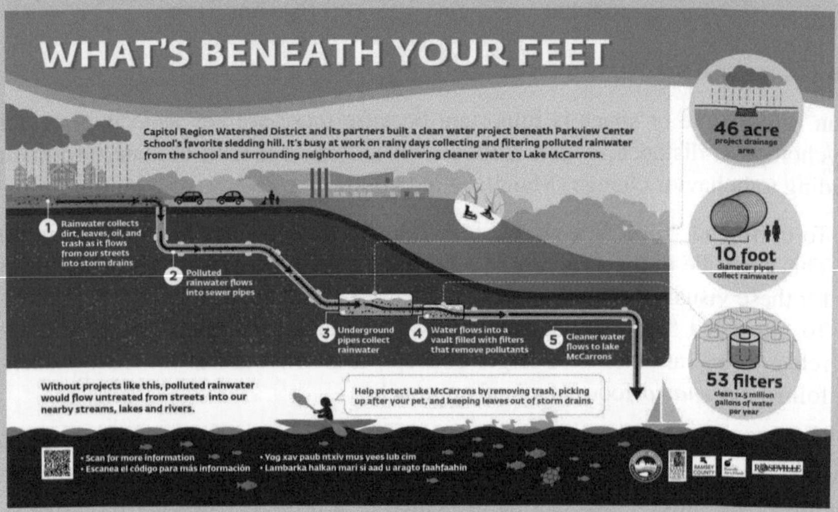

Fig. 2 This infographic sign shows a behind-the-scenes process and is an example of a process-focused visual story—that condenses time and space into a single visual. Designed by Background Stories for the Capital Region Watershed District

Scenario-based visuals (see Fig. 3) include maps (both geographic and conceptual), systems, networks, and lay-of-the-lands. These show how pieces of info relate to other pieces and whole contexts (Birt, 2021). Scenarios are often annotated to help viewers understand the components of the visual, and how they relate to one another. Examine your content and ask yourself if any of these are true:

- Is it most important that viewers understand the big picture of this message?
- Do pieces of content relate to one another in ways that are critical to the message?
- Is zooming into a detail or zooming out of the information to show context the most important aspect of the message?

If any of these are true, a visual that shows the scenario could be useful.

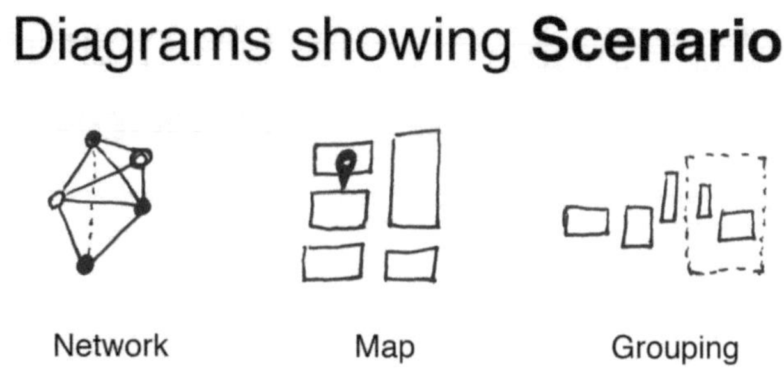

Fig. 3 Scenario-based visuals

Scenario-Focused Example

The visual narrative across this set of infographics helps the Minnesota Department of Health (MDH) explain how its Statewide Health Improvement Program improves the health of Minnesotans. A set of infographics highlight four pillars of the MDH's program and the research-based methods that support this work. This example is a scenario-based visual story because it connects seemingly separate elements of the program through the context of a community. Seeing these elements situated within the community, viewers can better understand—and relate to—how the program impacts life in Minnesota, see Fig. 4).

Fig. 4 This set of infographics are examples of scenario-focused visual stories. These visuals connect seemingly separate elements of content within a big-picture view of the scenario. Designed by Background Stories for the Minnesota Department of Health Statewide Health Improvement Program

Process-based and scenario-based approaches can also be combined, see Fig. 5. This leads to the most engaging, layered, and educational visuals (Birt, 2021). In these cases, typically one approach becomes the most prominent, the other approach being nested inside frames or callout windows to show more detail on a specific part of the larger visual.

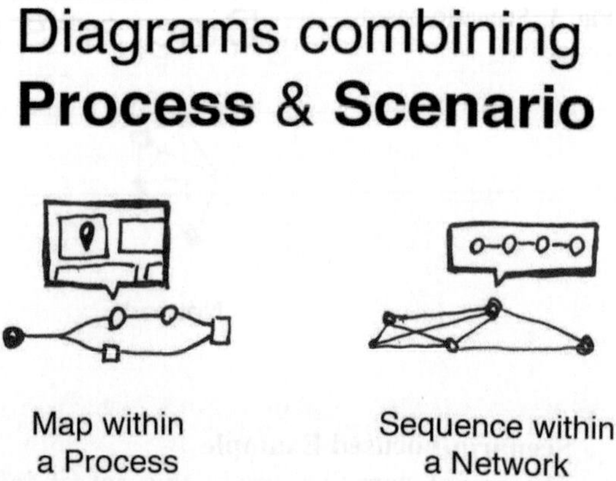

Fig. 5 Maps and sequences, combinations

When creating visuals that combine process and scenario approaches, the more complex the visual becomes, the more careful a designer needs to be about using design principles to add emphasis and intentionally guide the eye through the complexity. Design principles including scale, balance, hierarchy, contrast, and tools like typography, alignment, and white space can all help guide viewers.

Combination Process and Scenario-Focused Example

This infographic shows a process (24 h of variable energy demand) embedded within a scenario-focused illustration (a network that connects a battery, energy sources, cities, and industry across a landscape). While consistent colors are used throughout the infographic, the inset box that contains the process makes it clear that this information is connected to the battery and is a different type of information from the landscape surrounding it (Fig. 6).

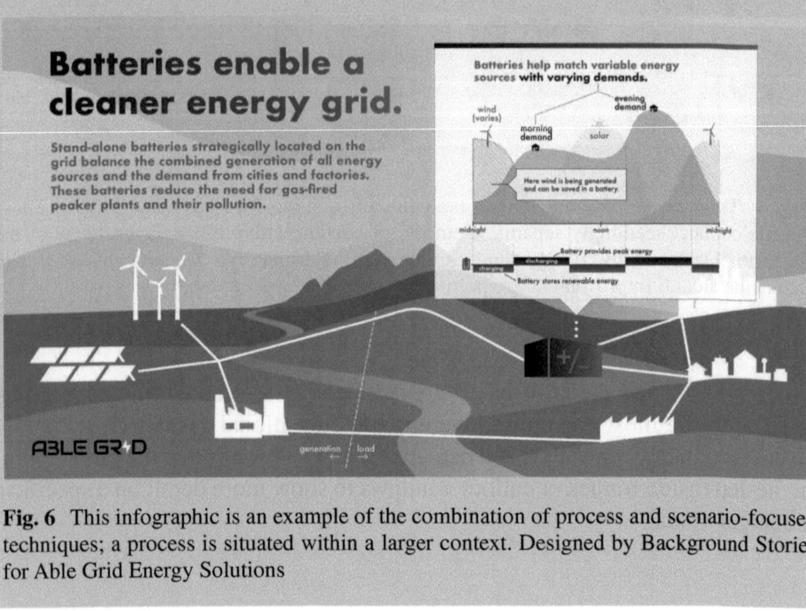

Fig. 6 This infographic is an example of the combination of process and scenario-focused techniques; a process is situated within a larger context. Designed by Background Stories for Able Grid Energy Solutions

4.3 *Use Consistent Visual Language*

Another important aspect of visual stories is the consistency of visuals—particularly when referring to the same concept or category multiple times. This is the visual equivalent of "wording" in writing. When the same aspect of a visual is referenced multiple times, assign a visual language to orient viewers—particularly when that aspect is part of the key message. Visual language can include colors, styles, icons, and textures: anything that visually repeats together with the repetition of specific content. By employing these and other visual elements consistently, communicators can guide audiences through complex information with clarity and coherence.

Consistent visual language also reduces cognitive load. Just as using the same words when referring to a specific topic, repeating visual elements make content more clear to non-expert audiences, and more easy to follow.

Visual Language Example
Figures throughout the IPCC 2023 Synthesis Report were designed to incorporate consistent visual language. This included colors, styles, icons, textures, as well as consistent treatment of positive/negative data and common topics throughout the visuals (Birt, 2024). Small icons on graph axes orient viewers at-a-glance to key variables that reoccur throughout figures. Also, handwritten-style annotations are used throughout the report's figures to highlight and explain points that may be obvious to scientists/experts. Both of these techniques aim to reduce cognitive load for viewers.

5 Visual Storytelling for Strategic Sustainability Communication[1]

Integrating techniques of visual storytelling into sustainability communication holds immense promise for engaging audiences in understanding complex topics like climate change. Visual stories not only convey scientific data but also provide context and establish emotional connections, ultimately fostering a deeper understanding of the interplay between environmental issues and individual lives.

By understanding the content being communicated and the audience it targets, communicators can craft compelling visual narratives to orient viewers within complex information. Whether through process-focused visuals that illustrate sequences

[1] Artificial Intelligence (AI) support was used in summarizing key points from the article, in order to formulate the conclusion. All input provided by AI was reviewed and substantially edited by the author.

of events or scenario-based narratives that contextualize complex systems, visual storytelling offers a versatile toolkit for communicating sustainability messages.

By harnessing the power of visuals to capture attention and inspire action, we can help audiences cultivate an understanding of the complex and interconnected narratives of environmental issues. These benefits make visual storytelling an important strategy to motivate behavioral change on the urgent challenges posed by climate change.

References[2]

Birt, A. (2021, January 9). Process vs. scenario visuals. *Background Stories*. https://backgroundstories.com/process_vs_scenario/

Birt, A. (2024, February 8). Create your Own climate generation visual. *Background Stories*. https://backgroundstories.com/create-your-own-climate-generation-visual/

Boscarino, J. E. (2022). Is seeing believing? Comparing the influence of textual and visual policy narratives on public opinion. *Politics & Policy, 50*(4), 720–751. https://doi.org/10.1111/polp.12471

Campbell, J. (2008). *The hero with a thousand faces*. New World Library.

Eisner, W. (2008). *Graphic storytelling and visual narrative*. W. W. Norton.

Goransson, K., & Fagerholm, A.-S. (2018). Towards visual strategic communications: An innovative interdisciplinary perspective on visual dimensions within the strategic communications field. *Journal of Communication Management, 22*(1), 46–66. https://doi.org/10.1108/JCOM-12-2016-0098

Hullman, J., & Diakopoulos, N. (2011). Visualization rhetoric: Framing effects in narrative visualization. *IEEE Transactions on Visualization and Computer Graphics, 17*, 2231–2240. https://doi.org/10.1109/TVCG.2011.255

Jones, M. D., & Anderson Crow, D. (2017). How can we use the 'science of stories' to produce persuasive scientific stories? *Palgrave Communications, 3*(1), 1–9. https://doi.org/10.1057/s41599-017-0047-7

Kearney, A. (1994). Understanding global change: A cognitive perspective on communicating through stories. *Climate Change, 27*. https://doi.org/10.1007/BF01096270

Lee, H., Calvin, K., Dasgupta, D., Krinner, G., Mukherji, A., Thorne, P., Trisos, C., Romero, J., Aldunce, P., Barret, K., Blanco, G., Cheung, W. W. L., Connors, S. L., Denton, F., Diongue-Niang, A., Dodman, D., Garschagen, M., Geden, O., Hayward, B., … Park, Y. (2023). In Core Writing Team, H. Lee, & J. Romero (Eds.), *IPCC, 2023: Climate Change 2023: Synthesis Report, Summary for Policymakers. Contribution of Working Groups I, II and III to the Sixth Assessment Report of the Intergovernmental Panel on Climate Change*. IPCC [Monograph]. https://doi.org/10.59327/IPCC/AR6-9789291691647.001

Maiella, R., La Malva, P., Marchetti, D., Pomarico, E., Di Crosta, A., Palumbo, R., Cetara, L., Di Domenico, A., & Verrocchio, M. C. (2020). The psychological distance and climate change: A systematic review on the mitigation and adaptation behaviors. *Frontiers in Psychology, 11*. https://doi.org/10.3389/fpsyg.2020.568899

Moezzi, M., Janda, K. B., & Rotmann, S. (2017). Using stories, narratives, and storytelling in energy and climate change research. *Energy Research & Social Science, 31*, 1–10. https://doi.org/10.1016/j.erss.2017.06.034

---

[2] Artificial Intelligence (AI) supported the identification of resources used in literature review. All input provided by AI was reviewed and substantially edited by the author.

O'Neill, S. J., & Smith, N. (2014). Climate change and visual imagery. *WIREs Climate Change, 5*(1), 73–87. https://doi.org/10.1002/wcc.249

Segel, E., & Heer, J. (2011). Narrative visualization: Telling stories with data. *IEEE Transactions on Visualization and Computer Graphics, 16*, 1139–1148. https://doi.org/10.1109/TVCG.2010.179

Spence, A., Poortinga, W., & Pidgeon, N. (2012). The psychological distance of climate change. *Risk Analysis, 32*(6), 957–972. https://doi.org/10.1111/j.1539-6924.2011.01695.x

Trope, Y., & Liberman, N. (2010). Construal-level theory of psychological distance. *Psychological Review, 117*(2), 440–463. https://doi.org/10.1037/a0018963

Tufte, E. (1990). *Envisioning information*. Graphics Press.

Ware, C. (2004). *Information visualization: Perception for design* (2nd ed.).

Wong-Parodi, G., & Feygina, I. (2021). Engaging people on climate change: The role of emotional responses. *Environmental Communication, 0*, 1–23. https://doi.org/10.1080/17524032.2020.1871051

Zelizer, B. (2004). *The voice of the visual in memory*. https://core.ac.uk/reader/76383410

Arlene Birt is a visual storyteller, speaker, educator, and public artist. An expert in information visualization and narrative visuals, she puts data into context and helps individuals emotionally connect to environmental concepts.

Arlene holds roles that span design practice and creative research: she is founder and creative director at Background Stories, an infodesign consultancy that translates complex ideas, systems, and metrics into clear visuals for non-expert audiences. As graphics officer for the 2023 UN Intergovernmental Panel on Climate Change (IPCC) Synthesis Report, Arlene collaboratively designed visuals to explain climate science together with 60+ climate scientists and 195 governments. As professor in Creative Entrepreneurship, she teaches sustainability and visual communication at the Minneapolis College of Art and Design (MCAD).

Her work has been featured in the Washington Post, Financial Times, Le Monde, Creative Review (UK), U.S. News and World Report, numerous other global media, and at the Barcelona Design Museum.

www.arlenebirt.com; www.backgroundstories.com

Strategic Sustainability Communication for and Challenges in the Future

Sustainability as a Guiding Principle of Strategic Communication. A Heuristic for Future-Proof Strategic Communication

Ulrike Röttger and Lennart Rettler

Abstract Strategic communication of organizations has numerous points of contact with sustainability issues. In addition to its mediating role, which refers to the reporting on how an organization shows commitment to sustainability, strategic communication can also be assigned a promoter role for a sustainable transformation both within and beyond the organization. Another aspect to consider is how strategic communication can itself be sustainable. In this contribution, we introduce the concept of *sustainable strategic communication*, which combines ideas regarding how strategic communication itself is sustainable while promoting sustainable transformation. We first characterized sustainable strategic communication by proposing a comprehensive heuristic that systematizes the various connections of strategic communication to sustainability issues. This heuristic was then discussed with 14 experts in the field in guided interviews and subsequently developed further in a workshop with 21 communication professionals. The results of both steps are outlined in this paper. Through this multilevel characterization of sustainable strategic communication, we put forward a concept that may guide professionals and researchers of strategic communication through the Great Transformation towards Sustainability.

1 Introduction

The proclaimed *Great Transformation* towards sustainable development[1] (Schneidewind & Singer-Brodowski, 2015; Bruckmeier, 2024) implies a comprehensive reorientation of action in all areas of society towards sustainability.

[1] Borrowing the term from Polanyi (2001).

U. Röttger (✉) · L. Rettler
Department of Communication, University of Münster, Münster, Germany
e-mail: ulrike.roettger@uni-muenster.de; lennart.rettler@uni-muenster.de

F. Weder (ed.), *Strategic Sustainability Communication*, CSR, Sustainability, Ethics & Governance, https://doi.org/10.1007/978-3-031-89486-2_25

Consequently, it also affects strategic communication. We heed the call of Schneidewind and Singer-Brodowski (2015) to establish "new forms of literacy and knowledge integration" (p. 189) in research that reflect a transformative approach and in which sustainability is seen as a target of processes of strategic communication. Thus, we introduce the concept of *sustainable strategic communication*, which shall provide an orientation to professionals and researchers in reflecting on how sustainability is treated within strategic communication processes.

We understand strategic communication as referring to all intentionally used forms of communication with which organizations participate in conversations and opinion-forming processes that are critical to their goals (see Nothhaft & Zerfass, 2020). Accordingly, we view strategic communication from a functional perspective as a communication function of organizations. This communication function is orchestrated by the communication department. Therefore, the activities of the communication department and their preconditions and effects are at the center of our consideration.

The concept of sustainable strategic communication will be defined and outlined from the perspective of communication studies by discussing the following research question: *What characterizes sustainable strategic communication?*

We attempt to answer this question using a threefold approach. First, we present a heuristic of sustainable strategic communication building on the existing literature. Then, we report the findings of expert interviews (n = 14) about how experts characterize sustainable strategic communication and how they fill and evaluate the individual dimensions of the heuristic in terms of content. Finally, we lay out the results of an action research workshop in which we discussed the characteristics of sustainable strategic communication mentioned in the expert interviews with communication professionals (n = 21).

2 Modeling a Heuristic of Sustainable Strategic Communication

To characterize sustainable strategic communication as a transformative approach to strategic communication, we first developed a heuristic. We based this heuristic on the existing literature on sustainability in strategic communication. However, most strategic communication research, targeting the topic of sustainability, focuses on strategic communication *about* sustainability (Weder, 2022). Sustainability is merely seen as a topic for communication departments to "talk" about in their communications, often as one element of corporate social responsibility (CSR) communication (Verk et al., 2021; Rasche et al., 2017). However, to align with the demands of the Great Transformation to sustainability, more extensive approaches to sustainability in strategic communication are needed. In the literature, the perspective of communication about sustainability has therefore recently been supplemented by two additional perspectives: the *sustainability of strategic communication* and *strategic communication for sustainability*. We founded our conceptualization of sustainable strategic communication on these two approaches.

2.1 The Sustainability of Strategic Communication

In the Communications Trend Radar 2021 (Zerfass et al., 2021), in which researchers predict trending topics for corporate communications, the authors foresee a change from *sustainability communications* to *sustainable communications*. Businesses need to adjust various practices to sustainability to align with progressing sustainability demands, such as the Sustainable Development Goals (SDGs) or environmental, social, and governance (ESG) standards. Thus, processes of strategic communication also require adjustments. Communication departments often treat sustainability primarily as a "content topic"; instead, a shift is needed towards treating sustainability as a "strategic topic," that is, considering "the sustainability of processes, platforms, media, and products of corporate communications" (Zerfass et al., 2021, p. 19). Along these lines, Prexl (2010) argues for a shift to not only communicating about *sustainability* but also communicating *sustainably*. She argues for explicitly considering how communication products and messages must be designed to be sustainable themselves, which requires accounting for how communication departments are claiming resources. Thus, Prexl (2010) states that communication departments need to reflect on their day-to-day behavior, even though this could possibly change established working processes fundamentally. In practice, two leading German communication agency associations, GPRA (2023) and GWA (2023), started initiatives reflecting on the environmental impact of communications. In this context, Kannengießer (2020) started an academic discussion in communication studies to examine the environmental impact of communication, especially in a digital world where environmental impact often remains invisible.

2.2 Strategic Communication for Sustainability

A slightly different perspective on the relationship between strategic communication and sustainability can be found in approaches that we characterize as *strategic communication for sustainability*. In these, sustainability functions "[...] as [a] mode of evaluation, diagnostic and prognostic attribution" (Weder, 2021, p. 104). The outputs of strategic communication are evaluated regarding the extent to which they contribute to a general, societal sustainable development (Weder, 2022). Sustainability is described as a moral compass (Weder et al., 2019) that guides strategic communication.

2.3 Sustainable Strategic Communication

Merging the aforementioned approaches, we define sustainable strategic communication as the strategic communication of organizations that uses resources in an economically, ecologically, and socially responsible manner (sustainability of strategic communication) and attempts to contribute to a general societal transformation towards sustainability (strategic communication for sustainability). Thus, sustainable strategic communication weighs sustainability as a guiding principle in every action and requires communicating for sustainability in a sustainable manner.

We argue that strategic communication can fully live up to its role in contributing to a Great Transformation towards sustainability only if it is viewed from both viewpoints. Sustainable strategic communication does not need to focus on sustainability as a topic but must instead be seen as an approach to the practice itself. In this sense, our approach is inherently normative.

2.4 The Heuristic of Sustainable Strategic Communication

To better identify the various facets of sustainable strategic communication, we developed a heuristic of sustainable strategic communication based on literature research and evaluation. To model the heuristic, we combined existing models of communication evaluation stages (for an overview, see Macnamara, 2018; see also Macnamara, 2020; Macnamara & Gregory, 2018; Rolke et al., 2020; Rolke & Zerfass, 2010) with the common division of sustainability into the ecological, economic, and social dimensions (e.g., WCED, 1987; Elkington, 1997; Purvis et al., 2019). According to this understanding, a *development*, *action*, or *action result*, can only be assessed as sustainable if the three dimensions are given more or less equal consideration (Elkington, 1997; Pufé, 2017).

Models of communication evaluation systematically map the entire process of value creation of strategic communication and are thus suitable for systematizing the importance of sustainability as a guiding principle throughout this process. Further, these models reflect what professionals aim for in their communication (Macnamara & Gregory, 2018). Although individual models differ partially in their terminology, they essentially rely on the same differentiation of the strategic communication management process into four levels: *input* (deployed resources), *output* (characteristics of communication measures), *outcome* (communication effects on stakeholders), and *outflow* (long-term strategic and/or financial effects). Next, we describe what sustainability means for the four different levels of the strategic communication management process, considering the three dimensions of the concept.

Input On the input level, the focus lies on the use of (financial, ecological, and human) resources throughout the creation of communications products, both by the communications department itself and by contracted service providers. The attention is thus on internal organizational requirements that are necessary to initiate concrete means and measures of (sustainable) strategic communication (e.g., energy consumption at the workplace or emissions from mobility).

Output The output level comprises the pure production performance of strategic communication. The focus at this level is on the concrete characteristics of the services provided by communication management, including characteristics of the product itself as well as the production processes.

For sustainable strategic communication, this means that we ask to what extent the communication products and services created are sustainable themselves. This applies, for example, to the carbon footprint or the resource consumption of individual communication measures.

Outcome The outcome level refers to the intended and achieved effects of communication measures on the knowledge, opinions, attitudes, emotions, and behaviors of target groups, including the use of attention and stakeholder relationships.

Outflow In existing evaluation models, the outflow level (sometimes also referred to as impact) describes the performance contribution of strategic communication to organizational success. Communication departments committing to sustainable strategic communication may foster inner-organizational change towards sustainable development. Furthermore, sustainable strategic communication may support societal change at large. Thus, on the outflow level, the *value* that strategic communication creates for sustainable development can be mapped.

Building on a combination of the triple bottom line and the communication controlling model of planning and evaluation, we proposed a heuristic to categorize different aspects of sustainable strategic communication (Table 1). The dimensions of sustainable strategic communication can be separated along the defined approaches: the sustainability of strategic communication largely targets the input

Table 1 Sustainable strategic communication heuristic

Dimension of sustainable strategic communication	Sustainability of strategic communication		Strategic communication for sustainability	
	Evaluation levels			
Triple bottom line	Input	Output	Outcome	Outflow
Economic	Financial resources required	Economic aspects of communication processes and characteristics	Intended economic effects	Economic value created
Ecological	Ecological resources required	Ecological aspects of communication processes and characteristics	Intended ecological effects	Ecological value created
Social	Human resources required	Social aspects of communication processes and characteristics	Intended social effects	Social value created

and output levels, while strategic communication for sustainability is mostly influential at the outcome and outflow levels.

This heuristic breaks down the multiple aspects through which strategic communication may support sustainable development. It helps professionals identify where and how to improve processes towards sustainability by, for example, reconsidering claimed resources and pursued effects.

3 Methods: Expert Interviews and Action Research Workshop

To further substantiate the sustainable strategic communication heuristic formulated based on a comprehensive literature review and illustrate it with concrete examples from practice, we conducted a two-part study (see Fig. 1).

3.1 Expert Interviews

In a first step, we discussed our heuristic of sustainable strategic communication in expert interviews with 14 participants who play a leading role in questions of sustainability and (strategic) communication. As proposed by Schneidewind and Singer-Brodowski (2015), tackling the Great Transformation towards sustainability in science implies transdisciplinary approaches and cooperation between academics and society. We thus interviewed experts from various academic disciplines as well as professionals. To narrow down the field, we decided to only interview German experts. Sustainability issues have traditionally played a major role in public discourse in Germany, and German companies have been confronted with the sustainability expectations of stakeholders on a large scale for many years. The topic of sustainability in communication is also currently prominent in the communications industry, as evidenced by existing initiatives in the German communications

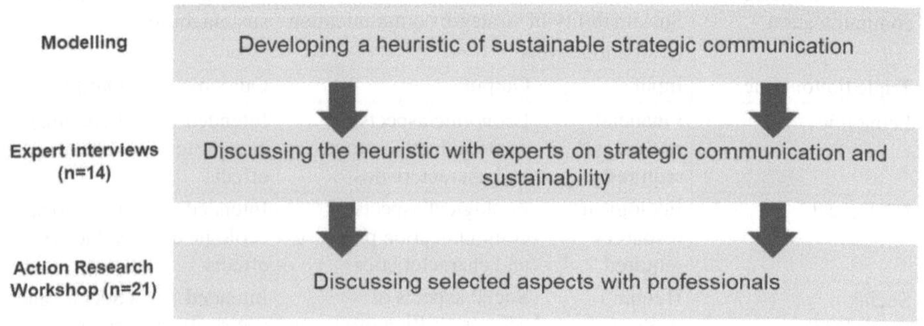

Fig. 1 Overview of the research design

industry (GPRA, 2023; GWA, 2023). We identified experts through purposive sampling (Mergel et al., 2019; Ritchie et al., 2014), following the *knowledge-sociological perspective*: an expert is someone who owns knowledge, problem-solving skills, and/or privileged access to relevant information, which are manifested in education and training, socialization, working experience, responsibilities, and/or certificates (Blöbaum et al., 2014; Meuser & Nagel, 1991).

To find experts, we used two approaches. First, we looked at German academics who published significant work on questions of sustainability and strategic communication. Then, we searched for professionals who teach sustainability in higher education by researching the seminars offered by the German Academy for Public Relations. We also included experts who hold responsibility related to sustainability in popular German professional associations of strategic communication. In the second step, we asked our participants to name additional experts. The sample comprised seven interviewees identifying as women and seven as men. The average age of the participants was approximately 45 years. The sample included five academic professors from the disciplines of strategic communication/communication studies, sustainability management, and organizational development (Interviewees 1–5), four people with leading roles in the communication departments of major German companies (Interviewees 6, 7, 9, and 13), three people with leading roles in large communications agencies (Interviewees 8, 10, and 11), and two people with leading roles in the sustainability departments of major German companies (Interviewees 12 and 14). The interviews lasted between 25 and 61 min, with an average duration of approximately 45 min.

3.2 Action Research Workshop

To deepen the findings from the 14 expert interviews, we subsequently conducted an action research workshop (Eden & Huxham, 1996). An action research workshop involves stakeholders who possess a "genuine concern" (Eden & Huxham, 1996, p. 75) for the topic under investigation in the research process. In our case, professionals in the field of strategic communication and sustainability were identified as such. The main goal of our action research workshop was to evaluate and further develop theoretical ideas together with professionals and introduce these ideas into practice (Avison et al., 1999). The method can be seen as an "iterative process involving researchers and practitioners acting together on a particular cycle of activities, including problem diagnosis, action intervention, and reflective learning" (Avison et al., 1999, p. 94).

We conducted the workshop with 21 German professionals who perform a professional role in the field of communication and sustainability. We invited professionals via the mailing list of an initiative that drives collaboration between strategic communication professionals and academics. Professionals could apply to attend the workshop. We limited the number of possible participants to 24 to keep the group sizes small.[2]

4 Expert Interviews: Discussing the Heuristic

Presenting the results of the expert interviews, we first discuss the general findings. Then, we sort the participants' statements according to our heuristic. The statements were coded following a qualitative content analysis (Mayring, 2020) with MAXQDA 2022 according to the different fields of the sustainable strategic communication heuristic (Table 1).

Overall, the interviewees primarily referred to aspects of the ecological dimension of sustainability, followed by the social dimension. All participants stated the importance of the topic, while many made a distinction between communicating about sustainability and creating communication that is sustainable. Interviewees stressed that sustainability has become a topic that targets every organizational operation, including communication:

> And the other thing that I think is coming into consideration more and more often, either brought to communication from the outside or also internally […] is the ecological sustainability of communication. (Interviewee 11)[3]

4.1 Input Level

At the input level, all interviewees pointed out that it is highly important to constantly examine the day-to-day use of ecological resources (e.g., heating and electricity) in the organizational environment (see also Prexl, 2010).

Further, communication departments should consider the sustainability of their service providers, such as agencies. Other interviewees stated that ecologically sustainable means of transportation should be considered for communication professionals. Such behavior might improve credibility, as stated by Interviewee 9, who described how a journalist told him that it left a lasting impression when he, as the head of corporate communications, arrived at a press conference by bike.

In terms of claimed economic resources, interviewees pointed to the promotion of long-term thinking and development, to which they attributed a massive impact on the use of resources. Meanwhile, short-term thinking was often described as leading to intensive short-term use of resources without creating a long-term impact.

---

[2] Three participants who had applied ended up missing the workshop.

[3] Note that since all interviews were conducted in German, the quotes were translated by the authors.

Interviewees also stated that an ecological use of resources (e.g., saving energy) usually produces economic advantages.

Regarding social resources, strategic communication departments may consider whether working conditions for the professionals are adequate. Our participants discussed the importance of diversity and inclusion issues as well as bettering the work–life balance of employees and enabling part-time work. Interviewee 6 argued that communication professionals often have a strong connection to these topics, making their consideration even more important in communication departments.

Some interviewees also pointed out the fundamental problem that recognizing the negative effects of specific resource use is often difficult, especially when it comes to digital resources (see also Kannengießer, 2020). For instance, Interviewee 10 discussed the difficulty of evaluating resources claimed for strategic communication:

> In PR, you're always asking, 'where can we start?,' because many things we do consist of the spoken and written word, and the resources needed aren't always obvious. (Interviewee 10)

4.2 *Output Level*

Participants emphasized the importance of maintaining a balanced approach when utilizing resources for products and ensuring that resource allocation aligns precisely with the actual need and demand for the product. Thus, interviewees argued that content should be digitalized, the use of paper should be prevented, and, if possible, recycled materials should be used. These tendencies are exemplified in the trend of shutting down employee or customer magazines and relying on digital equivalents instead. One interviewee explained this exemplarily:

> Yes, paper free, […] do I really need a big fat flyer delivered to every household or not? (Interviewee 6)

Moreover, participants felt that it is important to consider when, where, and how communication products are produced:

> To what extent do you absolutely have to photograph in South Africa with a giant film crew? […] You can also save on certain shoots via digitalization. (Interviewee 7)

In particular, the need for evaluating the sustainability of events was highlighted as a major topic by the interviewees. Regarding the profession, a trend emerged towards digitalizing events, as evidenced by *virtual stakeholder dialogues* (Ziegele et al., 2021). However, participants also reported that the outcomes of many digital events do not always meet expectations. This points to a major issue that must be weighed up in the context of sustainable strategic communication: considering sustainability in the production of communicational work leads to cancellation or fundamental changes in existing communicational products, such as employee magazines; however, the seemingly more sustainable product may not always lead to the same results (e.g., ecological sustainability of digital events), thus harming other forms of sustainability (e.g., lack of economic sustainability of digital events). Therefore, it is important to determine whether seemingly more sustainable approaches are *functionally equivalent* to seemingly less sustainable products, as described by Interviewee 1:

> Where do I find the sweet spot between 'I don't give up on communicative impact' and 'I still have a big impact on sustainability'? (Interviewee 1)

In most cases, however, considering sustainability throughout the production process may lead to functionally equivalent or even improved results, as stated by several participants.

4.3 Outcome Level

Many interviewees reported that activating internal and external stakeholders to act sustainably should be the central outcome-goal of sustainable strategic communication. Activation can range from small day-to-day decisions to initiating sustainable projects or making projects more sustainable, as Interviewee 9 exemplified:

> When someone says to us in sponsoring, 'Will you sponsor our music festival?', the first question, then, is not who is playing, but what is your environmental concept? How do you want to organize the arrival and departure, because otherwise, we no longer add our brand to it. (Interviewee 9)

Interviewee 11 stated that due to search engine optimization, communication departments nowadays try to keep users on a website as long as possible. Dwell time on a website is often considered an important success metric even though it may not be sustainable: by encouraging spending time on a website, organizations try to claim as much attention as possible, even if it is not useful to the users. Interviewee 11 considered such metrics to be rather unsustainable:

> But in the sense of ecological sustainability, we should also be user-centered [...] because we want to give users the opportunity to get the information they want as quickly as possible and then disappear again. (Interviewee 11)

According to the interviewees, sustainable strategic communication should make use of its ability to bundle attention and focus it on the topic of sustainability. Communication departments should aim to prevent information fatigue by selecting important topics to communicate about and thus promote the well-being of stakeholders. "It would probably be more sustainable to take that back a bit [...] instead of really feeling like we have to send messages all day, both internally and externally" (Interviewee 13). Interviewee 4 summarized this as follows:

> In my opinion, sustainable communication is efficient communication. (Interviewee 4)

Simultaneously, it should be considered that communication professionals frequently need to adapt to short attention spans and adjust to communicational trends.

Strategic communication claims users' attention heavily. How to do so sustainably must therefore be discussed continually. Interviewees further stated that communication departments could strive to create long-term relationships with stakeholders, equaling sustainability with stability. While such long-term relationships ensure the economic sustainability of an organization, they might also lead to social sustainability when stakeholders' connection to and trust in an organization are high. These answers, which stress the importance of securing the longevity of an organization, reflect an organization-centric view of sustainability that may not align with the public perception of sustainability.

4.4 *Outflow Level*

Interviewees argued that sustainable strategic communication may lead to the creation of and establishment in new markets, as put by Interviewee 5:

> Sustainability is a zeitgeist, and that also opens up a new market. (Interviewee 5)

Communication agencies, for example, can distinguish themselves from competitors, by presenting themselves as explicitly regarding sustainability in their services.

Many interviewees felt that sustainable strategic communication can contribute positively to an organization's monetary performance, thus fostering economic sustainability. In this regard, the participants again adopted an organization-centric perspective.

Going one step further, Interviewee 1 contended that strategic communication should also publicly address the question of how companies should do business:

> I think it is also a bit of the task of sustainability communication to answer such questions as: 'What is business actually supposed to do?' (Interviewee 1)

Similarly, Interviewee 2 thought that strategic communication offers opportunities to open up discussions and question existing assumptions.

In nearly all interviews, participants mentioned that sustainable strategic communication may contribute to intra-organizational change towards sustainability, as formulated by Interviewee 7:

> I'm sure that communication plays an important, if not central, role in this transformation process, externally as well as internally. (Interviewee 7)

At the same time, strategic communication is also heavily bound to and dependent on other actors, such as sustainability, strategy, and executive departments, in driving change.

Moving beyond the confines of creating organizational value, sustainable strategic communication could also contribute to social change at the outflow level. Strategic communication can notably create a space in which sustainability can be negotiated and may also help to contribute to the behavioral change of stakeholders—an argument put forward by Interviewee 8:

I see it as our responsibility to bring about behavioral change or lead that actively (Interviewee 8).

Below, Table 2 provides an overview of thc results of the expert interview study. It illustrates and characterizes our heuristic of sustainable strategic communication.

5 Action Research Workshop: Discussing Selected Aspects with Professionals

Based on the findings from the expert interviews, we conducted an action research workshop to further develop the heuristic. During this workshop, we explored in detail aspects of both dimensions of our proposed model (see "Dimensions" in Fig. 2). We concentrated on discussing findings regarding the output and outflow levels (see "Level" in Fig. 2). The workshop thus contained two different sessions. In each session, 10 or 11 participants worked together with one of the authors of this paper. Each session lasted approximately 1 h. The discussions were recorded and transcribed. Each session followed a central question (see "Question" in Fig. 2). We applied two different formats (see "Format" in Fig. 2), one being the collection of measures and rating of their usefulness by the professionals and the other being the discussion of two theses drawn from the expert interviews.

5.1 *Session One: Discussing the Output Level*

In session one, we focused on the output level of the heuristic, discussing the question "*How can organizations shape communication towards becoming more sustainable without losing communicative impact?*" This question directly aligns with the findings of study one, where interviewees discussed ways to find sustainable solutions when creating communicative output and whether such approaches threaten communicative impact. Participants were asked to find solutions to this question by first writing down personal ideas and then debating them with a partner before detailing their ideas in groups of four or six people. All ideas were pinned to a board. After the solutions were presented, each participant had to rate them based on how important they perceived each solution to be. Each participant had a maximum of eight points to distribute among all the ideas. The participants were allowed to save points (e.g., using only six of their eight points) and to give as many points as they wished to a solution (e.g., allocating all eight points to one solution).[4]

In our subsequent evaluation, thematically identical or similar solutions were grouped. Table 3 provides an overview of the clustered solutions proposed by the

[4] The maximum number of points that were attributed to a single solution was two.

Table 2 Sustainable strategic communication heuristic with input from expert interviews

	Input level	Output level	Outcome level	Outflow level
Meaning	Sustainability in present conditions	Sustainability in communicational products	Sustainability in communicational effects	Sustainability in communicational impacts
Example	Dealing with electricity, mobility, labor	Digitalizing communicational products (magazines, events)	Being more user centered	Aiming for change (in organizations, society)
Further Question	How are communication departments dealing with present resources?	Is there a tradeoff, between taking into account sustainability and reaching communicational goals?	Should communication departments adjust metrics towards sustainability?	How exactly does strategic communication contribute to intra-organizational or societal change towards sustainability?
Appropriate quotation	"*Be it emission reduction targets concerning our mobility. […] But also, our everyday behavior in the office with regard to our ecological footprint.*" (Interviewee 13)	"*Where do I find the sweet spot between 'I don't give up on communicative impact' and 'I still have a big impact on sustainability?'*" (Interviewee 1)	"*But in the sense of ecological sustainability, we should also be user centered, […] because we want to give users the opportunity to get the information they want as quickly as possible and then disappear again.*" (Interviewee 11)	"*I see it as our responsibility to bring about behavioral change or lead that actively.*" (Interviewee 8)

	Session 1	Session 2
Dimension	Sustainability of Strategic Communication	Strategic Communication for Sustainability
Level	Output	Outflow
Question	How can organizations shape communication towards being more sustainable, without losing impact?	How can strategic communication aid organizations internally in their transformation towards sustainability?
Format	Collecting and rating of measures	Discussion of two theses

Fig. 2 Overview of the two sessions of the action research workshop

participants. Solutions that did not receive at least one point do not appear in the table.

Table 3 shows that from the perspective of practitioners, efficient and targeted communication measures with the lowest possible waste (target group-specific communication: 35 mentions) and measures with a good environmental footprint (20 mentions) are particularly important for the development of more sustainable communication.

Regarding target group-specific communication, the importance of targeting ("We need to know exactly who we are talking to, then the communication materials can also be prepared precisely"[5]) and strategic channel selection are highlighted, among other things.

Practitioners believe that digital formats play a particularly important role in ecological measures. However, they also spoke about reducing the amount of travel required when working with customers and service providers. In this context, references were made to the principle of "zero-kilometer communication," whereby care is taken to ensure that the communications department works primarily with service providers that are located nearby.

Overall, the discussion exhibited strong parallels to other thematic sustainability discourses (e.g., in the field of heating and heating technology): The focus was on using resources as efficiently as possible while achieving goals as well as possible. This logic also included suggestions such as using synergies at the measure level (17 mentions), increasing collaboration with other departments to save resources and learn from each other (13 mentions), and recycling communication measures (six mentions). Meanwhile, the suggestion that "less is more" (16 mentions)

[5] Note that all quotations are translated arguments presented by the professionals.

Table 3 Overview of the prioritization of measures for the sustainable design of strategic communication (basis: 146 points awarded by 21 people)

Solutions[a]	Total points
• Target group-specific communication – "individualization (avoid scattering loss)" – "efficient and target group-specific communication" – "targeting: target groups and channels" – "strategic channel selection"	35
• Implementing measures with a good environmental footprint – "zero-kilometer communication" – "reduction of travel in the collaboration" – "green IT" – "putting print to the test" – "using more digital formats"	20
• "Utilizing synergies at the measure level"	17
• Less communication ("less is more")	16
• Collaboration – "sharing economy in communication" – "transparency and collaboration" – "best practice sharing"	13
• "Using clear, simple language"	10
• Considering sustainability in measurements	9
• "Benchmarking"	6
• "Recycling of communication measures"	6
• "Guiding and enabling"	5
• Others	9

[a]Note that all solutions are translations from German phrases given that the workshop was held in German. Terms in quotation marks are translated quotes from the workshop, and all other terms were paraphrased by the researchers

questions the basic mechanisms of strategic communication to a much greater extent. The idea, to generally reduce the amount of communication, put out by communication departments, was also raised in the expert interviews.

Finally, the suggestion to consider sustainability in measurements (nine mentions) highlights that sustainability will only have lasting relevance in strategic communication if it is also reflected in key performance indicators (KPIs). It is therefore important to question existing KPIs and, if necessary, establish new ones that are sensitive to sustainability.

The following suggestions reflect the communications departments' views of external target groups. First, participants pointed out that it is particularly important to use clear and simple language when dealing with often complex sustainability issues (10 mentions). Second, the task of guiding and enabling internal and external stakeholders to act sustainably was brought up (five mentions).

5.2 *Session Two: Discussing the Outflow Level*

In session two, we focused on the question "*How can strategic communication aid organizations internally in their transformation towards sustainability?*" This session focused on the role of strategic communication and professionals in cultivating sustainable practices across an organization (for a further discussion of the roles of strategic communication professionals, see Zerfass & Volk, 2019). We especially wanted to discuss the opportunities and limitations of strategic communication in departments. To do so, we presented two theses to the participants of the workshop, which we derived from the interviews:

- Thesis 1: Communicators have the strongest potential to establish a culture of sustainability within an organization.
- Thesis 2: Most communications departments lack the necessary sustainability know-how.

5.3 *Discussing Thesis 1*

Regarding Thesis 1, the professionals agreed that the role of communication departments is that of an *incubator*. This word was introduced in the discussion by one of the professionals. An incubator—a term mostly related to start-up businesses—can be seen (and was described) as a supportive institution that uses its capabilities to support others "in need" and helps them reach a goal. As incubators, strategic communication professionals connect organizational members and collect ideas. They provide information and education, aid in tackling questions of sustainability, and ultimately enable organizational members to create or sustain a sustainability culture. They also help create a sense of community among employees and essential *momentum* when needed.

Although most participants saw communication professionals as fulfilling an important role, they disagreed with Thesis 1: to them, strategic communication does not hold the strongest potential for establishing a culture of sustainability within an organization. Instead, they considered top-level/management commitment important in striving for a culture of sustainability, as well as outside pressure from competitors, consumers, organizational units in other countries, and national and international governance institutions such as the European Union. Additionally, human resources departments were seen as important in establishing a culture of sustainability because they initially recruit potential workers seeking a culture of sustainability and are in charge of employee retention. Lastly, the participants deemed sustainability departments to be important actors: "Actually, it feels like everyone has to work with the sustainability department to create a culture that fuels this". For establishing a culture of sustainability, professionals described the building of competency groups ("sustainability councils") as an important step. These groups, where strategic communication should occupy a key position, could help in placing the topic of sustainability on the agenda of an organization and leading the way. The discussion of Thesis 1 with the professionals produced slightly different results compared to the expert interviews. Communicators were characterized as incubators, possessing a certain importance yet less influence than other organizational actors in establishing a culture of sustainability within an organization. For their part, the expert interviews stressed the importance of the interplay of different organizational actors.

5.4 *Discussing Thesis 2*

Thesis 2 was chosen to initiate a discussion on competencies existing and lacking in communication departments to support an organizational transformation towards sustainability. The conversations revealed an ambivalence regarding sustainability know-how. As one professional put it, "Too much know-how can also sometimes be a hindrance. If you are the technical expert yourself, then you are also a technical idiot." The professionals saw themselves as translators who adjust sustainability-related information needs to reach all stakeholders and organizational members, such as outside actors. Most of these actors need an understandable framing of sustainability issues aside from specified expert knowledge. Sustainability being a broad term, participants questioned what sustainability know-how really comprises: social topics, financial topics, and topics of bio-diversity all require completely different base knowledge. While the professionals seemed to feel pressure to be "experts on everything" (as one stated), they stressed that general communicative know-how (e.g., collecting information, choosing adequate words and tonality) remains more important: "You have to build up a bit of expertise, but you can't neglect the main function of communication for that reason."

The professionals underlined the productivity associated with working in interdisciplinary teams, in which they can profit from different perspectives, for instance, by also integrating professionals from different backgrounds into communication departments (two professionals, for example, hold a degree in natural science). Here, again, the importance of working together with a sustainability department was highlighted. In discussing the second thesis, the professionals put forward hands-on ideas reflecting their day-to-day work, whereas experts approached a more meta perspective in the interviews.

6 Conclusion

The Great Transformation towards sustainability requires alternative approaches to practices in all areas of society, including strategic communication. Accordingly, academia must reflect on how it can accompany and support this transformation.

The exploration of sustainable strategic communication in this paper underscores the critical role of sustainability as a guiding principle integrated into the fabric of strategic communication processes. By examining sustainability at the input, output, outcome, and outflow levels, the study proposes a comprehensive heuristic that highlights the multifaceted relationships between strategic communication and sustainability.

The insights gained from expert interviews and the action research workshop highlight the importance of sustainable practices within communication departments, emphasizing the need for efficient resource use, targeted communication, and the potential for strategic communication to foster organizational and societal change. The professionals' practical suggestions, such as implementing measures with a favorable environmental footprint and adopting a "less is more" approach, illustrate tangible steps towards integrating sustainability into communication strategies.

However, the journey towards sustainable strategic communication is fraught with challenges, notably the necessity for top-level commitment and intra- and inter-organizational collaboration. The evolving role of sustainability departments and the emerging prominence of sustainability spokespersons underline the dynamic nature of this field. Evaluating the sustainability of communication actions, particularly in the digital realm, remains a persistent challenge.

This paper's normative approach, which merges the sustainability of communication processes and strategic communication for sustainability, offers a starting point for further research and practical application. The heuristic presented here is a valuable tool for professionals aiming to align their communication practices with sustainability goals and for academics seeking to expand the discourse on sustainable strategic communication. Studying how professionals treat the topic of sustainability in their daily actions is one important area of follow-up research to our rather explorative approach. Academic discussions may also reflect the usefulness of the triple bottom line as a way to describe sustainability (for a detailed critique, see Henriques, 2013).

In conclusion, sustainable strategic communication requires a concerted effort by academics, professionals, and society to navigate the complexities of sustainability and communication. Thus, by adopting sustainability as a guiding principle in strategic communication, the profession can effectively contribute to mastering the Great Transformation towards sustainable development.

References

Avison, D. E., Lau, F., Myers, M. D., & Nielsen, P. A. (1999). Action research. *Communications of the ACM, 42*(1), 94–97. https://doi.org/10.1145/291469.291479

Blöbaum, B., Nölleke, D., & Scheu, A. M. (2014). Das Experteninterview in der Kommunikationswissenschaft [The expert interview in communication studies]. In S. Averbeck-Lietz & M. Meyen (Eds.), *Handbuch nicht standardisierte Methoden in der*

Kommunikationswissenschaft [Handbook on not-standardized methods in communication studies] (pp. 1–13). Springer VS. https://doi.org/10.1007/978-3-658-05723-7_11-1

Bruckmeier, K. (2024). *The anthropocene and its future*. Palgrave Macmillan. https://doi.org/10.1007/978-3-031-56649-3_3

Eden, C., & Huxham, C. (1996). Action research for management research. *British Journal of Management, 7*(1), 75–86. https://doi.org/10.1111/j.1467-8551.1996.tb00107.x

Elkington, J. (1997). *Cannibals with forks: The triple bottom line of 21st century business*. Capstone.

GPRA. (2023). *Nachhaltigkeit fängt bei uns an. Agenturen als aktive Gestalterinnen der Transformation [Sustainability starts with us. Agencies as active transformation shapers]*. GPRA. https://www.gpra.de/ueber-die-gpra/dafuer-stehen-wir/nachhaltigkeit/attachment/gpra-positionspapier-nachhaltigkeit-2023/

GWA. (2023). *GWA Green Guide – Vier Steps zur nachhaltigen Agentur [GWA green guide – four steps to a sustainable agency]*. GWA. https://www.gwa.de/termine/gwa-forum-nachhaltigkeit-gwa-green-guide-vier-steps-zur-nachhaltigen-agentur/

Henriques, A. (2013). CSR, sustainability and the triple bottom line. In A. Henriques & J. Richardson (Eds.), *The triple bottom line: Does it all add up?* (pp. 26–33). Routledge. https://doi.org/10.4324/9781849773348-3

Kannengießer, S. (2020). Nachhaltigkeit und das "gute Leben". Zur Verantwortung der Kommunikations- und Medienwissenschaft in digitalen Gesellschaften [Sustainability and the "good life". On the responsibility of communication and media science in digital societies]. *Publizistik, 65*(1), 7–20. https://doi.org/10.1007/s11616-019-00536-9

Macnamara, J. (2018). *Evaluating public communication. Exploring new models, standards, and best practice*. Routledge.

Macnamara, J. (2020). New developments in best practice evaluation. Approaches, frameworks, models, and methods. In V. Luoma-aho & M.-J. Canel (Eds.), *The handbook of public sector communication* (pp. 435–454). Wiley. https://doi.org/10.1002/9781119263203.ch28

Macnamara, J., & Gregory, A. (2018). Expanding evaluation to progress strategic communication: Beyond message tracking to open listening. *International Journal of Strategic Communication, 12*(4), 469–486. https://doi.org/10.1080/1553118X.2018.1450255

Mayring, P. (2020). Qualitative Inhaltsanalyse [Qualitative content analysis]. In G. Mey & K. Mruck (Eds.), *Handbuch Qualitative Forschung in der Psychologie [Handbook on qualitative research in psychology]* (pp. 1–17). Springer. https://doi.org/10.1007/978-3-658-18387-5_52-2

Mergel, I., Edelmann, N., & Haug, N. (2019). Defining digital transformation: Results from expert interviews. *Government Information Quarterly, 36*(4), 101385. https://doi.org/10.1016/j.giq.2019.06.002

Meuser, M., & Nagel, U. (1991). ExpertInneninterviews – vielfach erprobt, wenig bedacht [Expert interviews – often done, seldomly considered]. In D. Garz & K. Kraimer (Eds.), *Qualitativ-empirische Sozialforschung. Konzepte, Methoden, Analysen [Qualitative-empirical social sciences. Concepts, methods, analyses]* (pp. 441–471). VS Verlag für Sozialwissenschaften. https://doi.org/10.1007/978-3-322-97024-4_14

Nothhaft, H., & Zerfaß, A. (2020). Strategische Kommunikation: Begriffliche und konzeptionelle Grundlagen aus internationaler Perspektive [Strategic communication: Terminology and conceptual foundations from an international perspective]. In A. Zerfaß, M. Piwinger, & U. Röttger (Eds.), *Handbuch Unternehmenskommunikation [Handbook corporate communication]* (pp. 1–27). Springer Gabler. https://doi.org/10.1007/978-3-658-03894-6_5-1

Polanyi, K. (2001). *The great transformation: The political and economic origins of our time* (2nd ed.). Beacon Paperback.

Prexl, A. (2010). *Nachhaltigkeit kommunizieren – nachhaltig kommunizieren. Analyse des Potenzials der Public Relations für eine nachhaltige Unternehmens- und Gesellschaftsentwicklung [Communicating sustainability – communicating sustainably. Analysis of the potential of public relations for a sustainable corporate and social development]*. VS Verlag. https://doi.org/10.1007/978-3-531-92471-7

Pufé, I. (2017). *Nachhaltigkeit [Sustainability]*. UTB. https://doi.org/10.36198/9783838587059

Purvis, B., Mao, Y., & Robinson, D. (2019). Three pillars of sustainability: In search of conceptual origins. *Sustainability Science, 14*(3), 681–695. https://doi.org/10.1007/s11625-018-0627-5

Rasche, A., Morsing, M., & Moon, J. (2017). The changing role of business in global society: CSR and beyond. In A. Rasche, M. Morsing, & J. Moon (Eds.), *Corporate social responsibility: Strategy, communication, governance* (pp. 1–28). Cambridge University Press.

Ritchie, J., Lewis, J., McNaughton Nicholls, C., & Ormston, R. (2014). *Qualitative research practise: A guide for social science students and researchers*. SAGE Publications.

Rolke, L., & Zerfaß, A. (2010). Wirkungsdimensionen der Kommunikation: Ressourceneinsatz und Wertschöpfung im DPRG/ICV-Bezugsrahmen [Impact dimensions of communication: Use of resources and added value in the DPRG/ICV reference framework]. In J. Pfannenberg & A. Zerfaß (Eds.), *Wertschöpfung durch Kommunikation. Kommunikations-Controlling in der Unternehmenspraxis [Value creation through communication. Communication controlling in corporate practice]* (pp. 50–60). Frankfurter Allgemeine Buch.

Rolke, L., Buhmann, A., & Zerfaß, A. (2020). Evaluation und Controlling der Unternehmenskommunikation [Evaluation and controlling of corporate communication]. In A. Zerfass, M. Piwinger, & U. Röttger (Eds.), *Handbuch Unternehmenskommunikation [Handbook communication]* (pp. 1–21). Springer Gabler. https://doi.org/10.1007/978-3-658-03894-6_27-1

Schneidewind, U., & Singer-Brodowski, M. (2015). Enabling the great transformation: Transdisciplinarity as individual and institutional challenge. In F. Schmidt & N. Nuttall (Eds.), *Contributions towards a sustainable world: In dialogue with Klaus Töpfer* (pp. 189–200). Oekom Verlag.

Verk, N., Golob, U., & Podnar, K. (2021). A dynamic review of the emergence of corporate social responsibility communication. *Journal of Business Ethics, 168*, 491–515. https://doi.org/10.1007/s10551-019-04232-6

WCED (World Commission on Environment and Development). (1987). *Our common future (Brundtland Report)*. Oxford University Press.

Weder, F. (2021). Sustainability as master frame of the future? Potency and limits of sustainability as normative framework in corporate, political and NGO communication. In F. Weder, L. Krainer, & M. Karmasin (Eds.), *The sustainability communication reader. A reflective compendium* (pp. 103–119). Springer VS. https://doi.org/10.1007/978-3-658-31883-3_7

Weder, F. (2022). Strategic problematization of sustainability reframing dissent in strategic communication for transformation. *Public Relations Inquiry, 11*(3), 337–360. https://doi.org/10.1177/2046147X211026857

Weder, F., Lemke, S., & Tungarat, A. (2019). (Re)storying sustainability: The use of story cubes in narrative inquiries to understand individual perceptions of sustainability. *Sustainability, 11*(19), 52–64. https://doi.org/10.3390/su11195264

Zerfass, A., & Volk, S. C. (2019). *Toolbox Kommunikationsmanagement. Denkwerkzeuge und Methoden für die Steuerung der Unternehmenskommunikation*. Springer Gabler.

Zerfass, A., Stieglitz, S., Clausen, S., Ziegele, D., & Berger, K. (2021). *Communications trend radar 2021. Denialism, virtual corporate communications, sustainable communications, digital nudging & voice interaction*. Academic Society for Management & Communication.

Ziegele, D., Kurtze, H., & Zerfaß, A. (2021). From reality to virtuality. Moving stakeholder dialogues from face-to-face to online formats. *Communication Insights, 12*, 1–22. https://hdl.handle.net/10419/246739

Further Reading

Canel, M. J. (2023). A call for the "Public Relations-imperative for sustainability": Roles for public relations in advancing the 2030 agenda. *Public Relations Review, 49*(4), 102368. https://doi.org/10.1016/j.pubrev.2023.10236

Schneidewind, U., & Singer-Brodowski, M. (2015). Enabling the great transformation: Transdisciplinarity as individual and institutional challenge. In F. Schmidt & N. Nuttall (Eds.), *Contributions towards a sustainable world: In dialogue with Klaus Töpfer* (pp. 189–200). Oekom Verlag.

Ulrike Röttger (Dr., University of Zurich) is a Professor of Public Relations Research at the University of Münster. Her research focuses on trust and strategic communication, sustainability and strategic communication, university communication, and communication consulting.

Lennart Rettler is a PhD candidate at the University of Münster. His research takes a relational approach to strategic communication and public relations. He discusses concepts such as community, reciprocity, cooperation, and social capital. His interests further lie in social media-based strategic communication.

How Civil Society Actors Employ Strategic Communication to Delegitimate Corporations in Public Debates on Sustainability

Sonja Fürst, née Walke and Helena Stehle

Abstract In recent years, the legitimacy of corporations has been increasingly challenged by civil society actors, especially in public debates about sustainability. This chapter sheds light on strategic communication of civil society actors that strives to delegitimate corporations engaging in unsustainable practices or displaying (potential) discrepancies between their corporate activities and communication on sustainability efforts. The chapter discusses the characteristics of such communication, particularly its strategies and processes. Based on a theoretical literature review, the chapter focuses on public relations studies that are complemented by literature from neighboring fields, such as organizational and environmental research. It contributes to expanding knowledge on the delegitimation of others in public debates about sustainability through strategic communication (hereon in "delegitimating strategic sustainability communication")—a topic that is of great importance in current debates on sustainability but has so far been treated with restraint—and offers impulses for further research.

1 Introduction

In recent years, the legitimacy of corporations has been increasingly challenged by civil society actors (Palazzo & Scherer, 2006). Sometimes entire industries—such as the airline or fossil fuel industries—come under the scrutiny of NGOs

---

The book chapter has its origin in the bachelor thesis of the author Sonja Fürst (née Walke) at the University of Münster.

S. Fürst, née Walke
Institute of Journalism and Communication Studies, Department of Social Sciences, University of Hamburg, Hamburg, Germany

H. Stehle (✉)
Department of Communication, University of Münster, Münster, Germany
e-mail: helena.stehle@uni-muenster.de

F. Weder (ed.), *Strategic Sustainability Communication*, CSR, Sustainability, Ethics & Governance, https://doi.org/10.1007/978-3-031-89486-2_26

(non-governmental organizations), citizens' initiatives, activists, and protest movements, among others. This is particularly true in the context of public debates about sustainability. Sustainability is understood as a societal development and fundamental transformation of society as a whole that "meets the needs of the present without compromising the ability of future generations to meet their own needs" (United Nations (UN), 1987, p. 15), and that strives to balance and reconcile environmental, social, and economic aspects (Kuhlman & Farrington, 2010; Thomsen, 2018). In public debates about sustainability, for example, environmental pollution and human rights violations by corporations are denounced (Palazzo & Scherer, 2006).

While corporations strive to communicate their own goals or efforts concerning sustainability in order to legitimate themselves and their own actions (Lock & Schulz-Knappe, 2019; Crumley et al., 2022), civil society actors bring new perspectives to public debates on related issues. They work to expose unsustainable practices and potential discrepancies between corporate activities and communication, and challenge the legitimacy of companies in this regard (Woods, 2019). Thus, at a time when sustainability is more important than ever before, civil society actors are important "agents of change" (Buijs et al., 2024, p. 2864) and actors in public debates.

Similar to other organizations and actors that participate in public debates and communicate of, about, or for sustainability (Weder et al., 2021; Weder, 2023), the sustainability communication of civil society actors can also be systematized according to three modes: First, civil society actors, such as Fridays for Future or Greenpeace usually communicate *for* sustainability. Second, they can communicate *of* their own sustainability goals and efforts. Third, they also have an important role to play when it comes to communicating *about* sustainability, which can be defined as deliberative negotiation processes in public debates about whether and to what extent corporations have responsibility for sustainability (Weder & Erikson, 2023, p. 78). This chapter focuses on the role of civil society actors in the latter mode of communication.

In public debates about sustainability, civil society actors can use strategic communication to delegitimate companies (Sandhu, 2018, 2022) in the sense of "denouncing the other's cause" (Hein & Chaudhri, 2019, p. 173). Yet, this "shadow side of legitimation" (Sandhu, 2014, p. 1180, authors' translation) has remained largely unexplored. Furthermore, previous research on strategic communication concerning sustainability has primarily examined civil society actors as stakeholders who are addressed by a corporation's strategic communication. They have rarely been analyzed as actors who themselves communicate strategically and whose skepticism threatens a corporation's legitimacy (Schwarz & Fritsch, 2014; Ciszek, 2015; Sandhu, 2018). Their communication has been mostly neglected by public relations research, which has long focused on the corporate perspective (Wolf, 2018), despite the "pressing need for a greater understanding of activists, how they use public relations to gain power, and their effects on critical corporate assets such as reputations and CSR" (Coombs & Holladay, 2012, p. 885). Thus, in this chapter, civil society actors are considered as active communicators who use strategic sustainability communication—that is, goal-oriented, strategically planned communication on sustainability issues and in public debates about sustainability (Ellerup Nielsen & Thomsen, 2018; Thomsen, 2018; Stehle, 2022)—to delegitimate corporations that engage in unsustainable practices or display (potential) discrepancies between corporate activities and communication on sustainability efforts (hereon in: "delegitimating strategic sustainability communication").

This chapter addresses the research question of how to characterize civil society actors' delegitimation of corporations through strategic communication in public debates on sustainability. It discusses (1) who uses or can use such communication (e.g., NGOs, citizens' initiatives, activists, and protest movements) and for what motives and (2) which strategies and processes are discussed as being used in civil society actors' delegitimating strategic sustainability communication (e.g., doubting an actor's integrity or ability and competence to take action, or influencing background beliefs).

This conceptual chapter is based on a theoretical literature review (Paré et al., 2015). The review focuses on public relations literature which is complemented by literature from neighboring fields, such as organizational and environmental research (e.g., Joutsenvirta & Vaara, 2015; Davis et al., 2016; Glozer et al., 2019; Hein & Chaudhri, 2019; Ross & Rivers, 2019; Schifeling & Hoffman, 2019), as there are as yet few public relations publications that analyze delegitimation through strategic communication (e.g., Smith & Ferguson, 2013; Duong et al., 2019).

To characterize the delegitimating strategic communication of civil society actors in public debates on sustainability, the chapter is structured as follows: First, delegitimation is introduced and discussed as a goal of civil society actors' strategic sustainability communication, which is at the interface with activist communication. Second, delegitimating strategic sustainability communication by civil society actors is characterized. Third, the findings, open questions, and impulses for further research are discussed.

2 Civil Society Actors and Their Strategic Sustainability Communication: Striving for Delegitimation

Civil society can be defined as "the sphere of institutions, organizations and individuals located between the family, the state and the market in which people associate voluntarily to advance common interest" (Anheier, 2004, p. 22). Based on a sectoral approach, civil society and its actors are often distinguished from the state (public actors) and the economy (private actors) (Frantz & Fuchs, 2014; Then & Mildenberger, 2022). This chapter focuses on such non-state, non-market actors—for example, NGOs, citizens' initiatives, activists, advocacy groups, and protest or social movements—who often develop around and deal with morally and/or politically charged issues (Moernaut et al., 2020). These "voluntarily engage in forms of public participation and action around shared interests, purposes or values" (United Nations (UN), Office of the High Commissioner for Human Rights, 2008, p. vii).

Civil society actors "range from highly institutionalised non-governmental organisations (NGOs) to informal place-based initiatives such as grassroots organisations" (Buijs et al., 2024, p. 2864). They can evolve from citizen initiatives to organizations; even individual activists can be included under this definition, as active citizens and organizations can be closely networked in their activities (Buijs et al., 2024). Civil society actors are often perceived positively by the public, especially in the context of sustainability (Frantz & Fuchs, 2014), and the high moral legitimacy often attributed to them is their advantage and "a source of discursive power" (Frantz & Fuchs, 2014, p. 84).

When it comes to sustainability, civil society and market actors are bound to overlap in many ways, such as in the negotiation of voluntary sustainability standards (Lambin & Thorlakson, 2018). These interactions have become less confrontational and more collaborative (Lambin & Thorlakson, 2018). Nevertheless, there are numerous public debates on sustainability in which civil society actors do not seek cooperation but draw attention to market actors alongside or instead of political actors.

Activism is defined as "the process by which groups of people [or individuals] exert pressure on organisations or other institutions to change polices, practices, or conditions that the activists find problematic" (Smith, 2005, p. 6). As activists, civil society actors attempt to compel other players such as corporations to change what they perceive to be problematic practices or approaches; or at the very least they seek to influence public opinion (Stokes & Atkins-Sayre, 2018). With their activism, they "challenge corporations on topics ranging from detrimental environmental practices to stances on controversial social issues, such as gun control" (Woods, 2019, p. 332). To target those who cause sustainability problems, they use communication in a strategic way.

In the context of activist communication—understood as communication that "challenge[s] hegemonic institutions by challenging their legitimacy" (Heath, 2018, p. 2)—strategic sustainability communication of civil society actors aims to undermine the "license to operate" of certain organizations, questioning their practices and approaches in public debates on sustainability. In this way, civil society actors also jeopardize other corporate objectives, many of which are based on stakeholder perceptions, such as reputation (Coombs & Holladay, 2012). To achieve their objective to delegitimate corporations, civil society actors often use "name and shame" communication (Sandhu, 2012, p. 284) or "name and blame" campaigns (Sandhu, 2022, p. 877), which target high-profile organizations and, for example, attribute responsibility to them for an environmental disaster, thus highlighting them as the worst-case scenario (Sandhu, 2012).

While strategic sustainability communication—often referred to as corporate social responsibility (CSR) communication in the corporate context—is an important means of legitimating organizations (Mast & Stehle, 2009, 2016; Sandhu, 2014), attempts of delegitimation are also communicative in nature. Delegitimation, however, has rarely been addressed in research, muss less in strategic communication and public relations research. There are only few studies that explicitly address delegitimation and attempt to define it.

In political science, legitimacy is a key concept and delegitimation is also considered. Bäckstrand and Söderbaum (2018), for example, develop a framework from the field of global governance that encompasses both legitimation and delegitimation and offers empirical insights. They define (de)legitimation as a "social practice" (p. 104) and identify three types of such practice, these being discursive, institutional, and behavioral in nature.

Originating in organization and business research, Ashforth and Gibbs (1990), for example, characterize organizational legitimacy as double-edged, and criticize previous work for implicitly assuming that the means and strategies used for legitimation actually produce the desired effect. They expand this one-sided perspective by addressing those dynamics that often undermine attempts at legitimation and integrating them into a "typology of ways in which organizations protest too much" (Ashforth & Gibbs, 1990, p. 178). While this typology focuses on the failure of an organization's own legitimation efforts, it is not directly applicable to delegitimation attempts by others. Such attempts, however, are addressed by Devers et al. (2009).

Devers et al. (2009) apply Goffman's (1963) concept of stigma to delegitimation attempts directed at corporations. Rooted in organization and business research, too, they (2009) describe organizational stigma as "a negative social evaluation" (p. 157) which is the result of an "attribution process" (p. 157). This process can be divided into an "Individual Labeling Process" (p. 160), where "a single stakeholder may label and make claims about an organization" (p. 162), and a "Collective Labeling Process" (p. 162), where more than one member of a stakeholder group and "collective perceptions" (p. 162) are involved.

Combining communication, organization and business research, Hein and Chaudhri (2019) describe delegimation similarly and define it as a process by which actors can engage in "denouncing the other's cause" (p. 173). They provide an analysis of Greenpeace's efforts to delegitimate Shell, thus looking specifically at market and civil society actors.

Harbo (2022) also looks at civil society actors, in this case social movements, and defines delegitimation as "the act of creating and conveying a negative image of the other, demonstrating the missing alignment with context-specific social values" (p. 3). Communication is inherent in such denouncing processes, but it is not specifically examined, especially in its strategic form, by Harbo, who approaches her analysis from cultural studies and other disciplines.

In comparison, coming from public relations research, Smith and Ferguson (2013) suggest redefining delegitimation as a process consisting of "evaluating an opponent or opposing idea" (p. 385) and questioning it, and as a strategy that can be applied.

These definitions emphasize different aspects. While Hein and Chaudhri (2019), for example, see delegitimation mainly as a process, Harbo (2022) describes it as an act, and Smith and Ferguson (2013) as a strategy. What their definitions have in common is the notion that delegitimation is aimed at others or opponents and includes acts of evaluation, denunciation, or creating negative images (Smith & Ferguson, 2013; Hein & Chaudhri, 2019; Harbo, 2022). These acts can be seen as communicative in nature. The following subchapter focuses on this communication. To this end, a definition of delegitimation will be provided before addressing delegitimating strategic communication.

Based on the aforementioned literature and Sandhu's (2012) consideration of legitimation as "a dynamic process of forming a judgment about a particular object (here: an organization) on the basis of specific criteria" (p. 19, authors' translation), delegitimation is understood in the following as a specific part of this dynamic process. Delegitimation entails the judging of an organization or industry that aims to question its actions or existence "within some socially constructed system of norms, values, beliefs, and definitions" (Suchman, 1995, p. 574), and to label it as undesirable, inappropriate, or reprehensible. Through this, the regulative, pragmatic, moral, or cognitive legitimacy of an organization ("organizational" or "institutional" legitimacy; Suchman, 1995; Boyd, 2000; Sandhu, 2022) or of specific practices, perspectives, or policies ("actional" legitimacy; Boyd, 2000) can be challenged (Suchman, 1995; Sandhu, 2014).

3 Civil Society Actors' Delegitimating Strategic Communication in Public Debates on Sustainability: Strategies and Process

Attacking legitimacy is an important objective of the strategic sustainability communication of civil society actors. However, legitimacy is foremost an attribution of stakeholders. As long as an organization is in line with social norms, values, or expectations, or at least these are not questioned, stakeholders tend to attribute legitimacy (Ashforth & Gibbs, 1990). Therefore, delegitimation can occur as a result of slowly changing norms, values, or expectations or—in the case of legitimacy conflicts or crises—through an organization's own fault, an accident, or attacks on an organization's legitimacy by other actors (Sandhu, 2012). As legitimacy is subject to a discursive process and "dynamic negotiations" (Sandhu, 2012, p. 17, authors' translation) between different actors, strategic communication can influence its attribution or invalidation. This subchapter examines the characterization of civil society actors' strategic communication used to delegitimate other organizations, in this case corporations, in public debates about sustainability. As the topic is not yet well developed in public relations research, research from adjacent disciplines is used to supplement it.

In public relations research, few contributions have been devoted to delegitimating strategic communication, in particular in the context of sustainability or CSR. Smith and Ferguson (2013), for example, analyze the use of legitimacy strategies by activists and the energy industry in a case study involving gas drilling. As mentioned above, they also address delegitimation and look at how federal regulators are delegitimated through communication (e.g., by pointing to the unnecessary nature of certain actions or to the distance of the regulators from those actually affected).

Duong et al. (2019) examine a campaign by activists against deforestation in Vietnam and an authoritarian political system. In their qualitative case study, they focus on issue management and framing, showing how deforestation was framed as a problematic situation leading to protest against the local government.

Other contributions that offer insights for the characterization of delegitimating strategic sustainability communication of civil society actors can be found in interdisciplinary areas such as research on organizations (e.g., Joutsenvirta & Vaara, 2015; Glozer et al., 2019; Schifeling & Hoffman, 2019) or environmental communication (e.g., Davis et al., 2016; Ross & Rivers, 2019). What can be derived from these and the contributions from public relations research for the characterization of delegitimating strategic sustainability communication?

Despite their origins in different disciplines, many contributions have a similar approach: With few exceptions (e.g., Schifeling & Hoffman, 2019; Brulle & Werthman, 2021; Illia et al., 2022; Sovacool & Dunlap, 2022), most contributions are empirical and qualitative. These are often case studies (e.g., Guérard et al., 2013; Bäckstrand & Kylsäter, 2014; Duong et al., 2019) or discourse analyses (e.g., Glozer et al., 2019; van Vuuren-Verkerk et al., 2021; Harbo, 2022). Some authors also conducted in-depth interviews (Guérard et al., 2013; Duong et al., 2019). When considering and characterizing delegitimating strategic sustainability communication of civil society actors, these contributions offer insights, on the one hand, on strategies, and, on the other, on a specific process of such communication.

When it comes to identifying specific strategies of delegitimating strategic sustainability communication, no systematic models or overviews have been found in the literature to date. In order to approach a systematization of such strategies, the four forms of legitimacy—regulative, pragmatic, moral, and cognitive (Sandhu, 2014, 2022)—can be used. They can be adapted to systematize the various delegitimation strategies described in the literature,[1] allowing the strategies to be sorted according to which of the four forms of legitimacy are under attack:

- First, the *legal compliance* of an actor or its actions can be disputed from a regulatory perspective (e.g., Hein & Chaudhri, 2019). This means that the regulative legitimacy of a corporation is attacked. This can be done, for example, by questioning the legality of activities or by pointing out deviations from legal requirements. Although this strategy can be effective, civil society actors seem to use it less often than, for example, market actors when trying to delegitimate other actors (e.g., Joutsenvirta & Vaara, 2015; Koch & Ulver, 2022).
- Second, the *ability* of corporations *to achieve their goals and perform certain tasks* can be questioned (e.g., Joutsenvirta & Vaara, 2015; Hein & Chaudhri, 2019). Here, the pragmatic legitimacy of a company is attacked. For instance, the

[1] In this, the authors of this chapter are inspired by Sandhu (2014, 2022), who has modeled legitimation strategies in a similar way.

realistic achievability of goals or the competence of the company, its managements and its employees are challenged.

- Third, the *integrity* of corporations can be scrutinized at a moral level, thus attacking their moral legitimacy (e.g., Björkvall & Westberg, 2021). Civil society actors can, for example, portray corporations as unwilling to assume social or environmental responsibility. Such normative arguments are often used in legitimacy conflicts in which the issue of sustainability is at the center.
- Fourth, civil society actors can try to *change the societal background beliefs* that are at the heart of cognitive legitimacy and that have not been challenged previously (e.g., Joutsenvirta & Vaara, 2015; Björkvall & Westberg, 2021; Koch & Ulver, 2022). This is an attack on the cognitive legitimacy ascribed to corporations and can be done, for example, by questioning previous principles of the market economy (e.g., profit orientation as a key goal) or by criticizing supposedly self-evident facts of a society (e.g., highly industrialized countries need a lot of energy).

Further strategies can be revealed on a linguistic level, when discursive legitimation strategies are used to characterize delegitimating strategic communication. Discursive strategies are substrategies that operate on the language-level indicating different language-based means of strategic communication (Sandhu, 2014). Four discursive legitimation strategies are distinguished in literature: authorization, rationalization, moralization, and mythopoesis or narration (van Leeuwen, 2007; Sandhu, 2012, 2014). These discursive strategies are also closely linked to the different types of legitimacy: authorization is associated with regulative legitimacy, rationalization with pragmatic and cognitive legitimacy, and moralization with moral legitimacy; mythopoesis or narration must be assigned on a case-by-case basis (Sandhu, 2012). Although the literature analysis has not yet identified any publications that examine these discursive legitimation strategies in delegitimating strategic communication, it can be assumed that they can help to further describe and systematize such communication.

What the literature does suggest, however, is a specific process or sequence of delegitimating strategic sustainability communication that can be described in four phases (e.g., Sandhu, 2012, 2022; Guérard et al., 2013):

1. *Blame:* This phase contains the identification of a problem and its originator(s). Responsibility for an undesired action (or the omission of desired actions) or a circumstance is attributed to an individual or an organization. With regard to sustainability, it is often the social and ecological responsibility of an actor that is under attack.
2. *Name:* The individual or organization is named as the cause of the problem.
3. *Shame:* The actors named are linked to the negatively evaluated behavior. To this end, the aforementioned strategies are used as arguments that can be based on different aspects of legitimacy (regulative, pragmatic, moral, cognitive). Civil society actors can undermine the legitimacy of corporations by shaping issues and introducing frames into the public debate. Ad-hominem attacks on groups, negative portrayals of the attacked actor (e.g., as immoral), and emotions such as

shame are used in a strategic manner. Communication may also include mocking corporate speech or using memes to attack the legitimacy of others.

4. *Justification:* When an individual or an organization is held accountable, this is done with explicit or implicit reference to "context-specific social values" (Harbo, 2022, p. 3) such as environmental or climate protection, animal welfare, or humanism. These values connect to a certain responsibility that has been attributed in phase 1 ("blame").

This process can be used to further describe and characterize delegitimating strategic sustainability communication by civil society actors. However, it must be noted that some questions remain unanswered and that delegitimation has been much less researched than legitimation.

4 Discussion and Conclusion

This chapter aims to characterize delegitimating strategic sustainability communication of civil society actors. The aim is to stimulate further discussion in public relations research and related fields on a topic that is of great importance in the current debate on sustainability, but has thus far not been given its due (e.g., Coombs & Holladay, 2012; Sandhu, 2018). The chapter seeks to contribute to the body of knowledge on delegitimation in strategic sustainability communication, and to provide suggestions for further research on the topic.

Based on a theoretical literature review, the chapter shows that there are only few publications in public relations research that shed light on delegitimating strategic sustainability communication of civil society actors (e.g., Smith & Ferguson, 2013; Duong et al., 2019). Other contributions can be found in related fields such as environmental communication and organization studies (e.g., Davis et al., 2016; Glozer et al., 2019; Ross & Rivers, 2019; Schifeling & Hoffman, 2019). In the literature, different approaches to delegitimation (e.g., as a process or a strategy) emerge. Furthermore, various strategies can be systematized by the different forms of legitimacy that are attacked (e.g., regulative or moral). These include, for example, attempting to change background beliefs, or questioning the integrity of a corporation, the legality of its actions, or its ability and competence to take necessary actions. Moreover, a process or sequence of delegitimating strategic sustainability communication emerges that can be described in four phases.

In view of the state of research, numerous questions remain unanswered when it comes to delegitimating strategic communication of civil society actors in public debates on sustainability. First, given the dynamic negotiation of legitimacy and its attribution, other actors (e.g., employee activists or public relations practitioners themselves; Holtzhausen & Voto, 2002; McCown, 2007; Girschik et al., 2020) and platforms such as journalistic media and social media, must also be considered and included.

Second, there are also open questions when considering the dyad between civil society actors and corporations. For example, corporations under attack can fight back and respond with crisis communication to protect their legitimacy and other assets such as reputation (Coombs & Holladay, 2012; Coombs, 2013; Valentini & Kruckeberg, 2016; Frandsen & Johansen, 2020). Furthermore, the moral legitimacy of civil society actors themselves, which is often implicitly accepted (Frantz & Fuchs, 2014; Ruser, 2022), can also be challenged. This, in turn, can affect the dyad and its power relations. Moreover, when the actor whose legitimacy is challenged responds to critical questions or allegations, conflicting frames compete (Guérard et al., 2013).

This also raises the question, third, of the effects of delegitimating strategic sustainability communication—in principle, but also, for example, when only one form of legitimacy is attacked but the other forms are strongly attributed; or when stakeholders recognize that an actor delegitimating another's cause may be trying to legitimate their own actions. To date, the impact of delegitimating strategic sustainability communication on an organization's legitimacy requires further examination. This also applies to the impact of such communication on the civil society actors themselves, for example in building or maintaining a social movement. According to Harbo (2022), for example, a group can "sustain and inform their collective identity; their knowledge of self" (p. 6) through the "construction of narrative structures" (p. 6). In other words, by distinguishing themselves from others and discrediting them, a group of people can define and consolidate their own values and norms and thus construct their social identity (Devers et al., 2009). What remains unanswered, however, is the extent to which attempts of delegitimation can boomerang back on the actors and their own legitimacy.

This chapter maps the debate in public relations research and related fields on a topic that is of great importance in the current discourse on sustainability, but has so far been treated with restraint (e.g., Coombs & Holladay, 2012; Sandhu, 2018). It contributes to expanding knowledge on delegitimating strategic sustainability communication of civil society actors and offers stimulus for further research on the topic.

References

Anheier, H. K. (2004). *Civil society: Measurement, evaluation, policy*. Routledge. https://doi.org/10.4324/9781849770736

Ashforth, B. E., & Gibbs, B. W. (1990). The double-edge of organizational legitimation. *Organization Science, 1*(2), 177–194.

Bäckstrand, K., & Kylsäter, M. (2014). Old wine in new bottles? The legitimation and delegitimation of UN public–private partnerships for sustainable development from the Johannesburg summit to the Rio+20 summit. *Globalizations, 11*(3), 331–347. https://doi.org/10.1080/14747731.2014.892398

Bäckstrand, K., & Söderbaum, F. (2018). Legitimation and delegitimation in global governance. Discursive, institutional, and behavioral practices. In J. Tallberg, K. Bäckstrand, & J. A. Scholte

(Eds.), *Legitimacy in global governance: Sources, processes, and consequences* (pp. 101–118). Oxford University Press. https://doi.org/10.1093/oso/9780198826873.003.0006

Björkvall, A., & Westberg, G. (2021). Shame and pride in the delegitimization and relegitimization of air travel. *Critical Approaches to Discourse Analysis Across Disciplines, 13*(2), 63–83.

Boyd, J. (2000). Actional legitimation: No crisis necessary. *Journal of Public Relations Research, 12*(4), 341–353. https://doi.org/10.1207/S1532754XJPRR1204_3

Brulle, R. J., & Werthman, C. (2021). The role of public relations firms in climate change politics. *Climatic Change, 169*(1–2), 8. https://doi.org/10.1007/s10584-021-03244-4

Buijs, A. E., et al. (2024). Civil society for sustainable change: strategies of NGOs and active citizens to contribute to sustainability transitions. *Journal of Environmental Planning and Management, 67*(12), 2863–2884. https://doi.org/10.1080/09640568.2023.2205571

Ciszek, E. L. (2015). Bridging the gap: Mapping the relationship between activism and public relations. *Public Relations Review, 41*(4), 447–455. https://doi.org/10.1016/j.pubrev.2015.05.016

Coombs, T. (2013). Future of crisis communication. In A. Thießen (Ed.), *Handbuch Krisenmanagement* (pp. 395–402). Springer. https://doi.org/10.1007/978-3-531-19367-0_23

Coombs, W. T., & Holladay, S. J. (2012). Fringe public relations: How activism moves critical PR toward the mainstream. *Public Relations Review, 38*(5), 880–887. https://doi.org/10.1016/j.pubrev.2012.02.008

Crumley, E. T., et al. (2022). Media interviews as strategic external communication to maintain legitimacy for sustainability activities. *Corporate Communications: An International Journal, 27*(1), 148–166. https://doi.org/10.1108/CCIJ-05-2021-0053

Davis, C. B., Glantz, M., & Novak, D. R. (2016). "You can't run your SUV on cute. Let's go!": Internet memes as delegitimizing discourse. *Environmental Communication, 10*(1), 62–83. https://doi.org/10.1080/17524032.2014.991411

Devers, C. E., et al. (2009). A general theory of organizational stigma. *Organization Science, 20*(1), 154–171.

Duong, H. T., Vu, H. T., & Nguyen, N. (2019). Activists' strategic communication in an authoritarian setting: Integrating social movement framing into issues management. *International Journal of Strategic Communication, 13*(2), 133–151. https://doi.org/10.1080/1553118X.2019.1590366

Ellerup Nielsen, A., & Thomsen, C. (2018). Reviewing corporate social responsibility communication: A legitimacy perspective. *Corporate Communications: An International Journal, 23*(4), 492–511. https://doi.org/10.1108/CCIJ-04-2018-0042

Frandsen, F., & Johansen, W. (2020). Arenas and voices in organizational crisis communication: How far have we come? In F. Frandsen & W. Johansen (Eds.), *Crisis communication* (pp. 195–212). De Gruyter. https://doi.org/10.1515/9783110554236-009

Frantz, C., & Fuchs, D. (2014). The impact of civil society on sustainable development. In M. Freise & T. Hallmann (Eds.), *Modernizing democracy: Associations and associating in the 21st century* (pp. 83–96). Springer. https://doi.org/10.1007/978-1-4939-0485-3_7

Girschik, V., Svystunova, L., & Lysova, E. I. (2020). Transforming corporate social responsibilities: Toward an intellectual activist research agenda for micro-CSR research. *Human Relations, 75*(1), 3–32. https://doi.org/10.1177/0018726720970275

Glozer, S., Caruana, R., & Hibbert, S. A. (2019). The never-ending story: Discursive legitimation in social media dialogue. *Organization Studies, 40*(5), 625–650. https://doi.org/10.1177/0170840617751006

Goffman, E. (1963). *Stigma: Notes on the management of spoiled identity*. Prentice-Hall.

Guérard, S., Bode, C., & Gustafsson, R. (2013). Turning point mechanisms in a dualistic process model of institutional emergence: The case of the Diesel particulate filter in Germany. *Organization Studies, 34*(5–6), 781–822. https://doi.org/10.1177/0170840613479237

Harbo, T. F. (2022). Internet memes as knowledge practice in social movements: Rethinking Economics' delegitimization of economists. *Discourse, Context & Media, 50*, online first. https://doi.org/10.1016/j.dcm.2022.100650

Heath, R. L. (2018). Activism. In R. L. Heath & W. Johansen (Eds.), *The international encyclopedia of strategic communication* (pp. 1–13). Wiley. https://doi.org/10.1002/9781119010722.iesc0001
Hein, J. E., & Chaudhri, V. (2019). Delegitimizing the enemy: Framing, tactical innovation, and blunders in the battle for the Arctic. *Social Movement Studies, 18*(2), 171–192. https://doi.org/10.1080/14742837.2018.1555750
Holtzhausen, D. R., & Voto, R. (2002). Resistance from the margins: The postmodern public relations practitioner as organizational activist. *Journal of Public Relations Research, 14*(1), 57–84. https://doi.org/10.1207/S1532754XJPRR1401_3
Illia, L., et al. (2022). Finding the tipping point: When heterogeneous evaluations in social media converge and influence organizational legitimacy. *Business & Society, 62*(1), 117–150. https://doi.org/10.1177/00076503211073516
Joutsenvirta, M., & Vaara, E. (2015). Legitimacy struggles and political corporate social responsibility in international settings: A comparative discursive analysis of a contested investment in Latin America. *Organization Studies, 36*(6), 741–777. https://doi.org/10.1177/0170840615571958
Koch, C. H., & Ulver, S. (2022). Plant versus cow: Conflict framing in the ant/agonistic relegitimization of a market. *Journal of Macromarketing, 42*(2), 247–261. https://doi.org/10.1177/02761467221080442
Kuhlman, T., & Farrington, J. (2010). What is Sustainability? *Sustainability, 2*(11), 3436–3448. https://doi.org/10.3390/su2113436
Lambin, E. F., & Thorlakson, T. (2018). Sustainability standards: Interactions between private actors, civil society, and governments. *Annual Review of Environment and Resources, 43*(1), 369–393. https://doi.org/10.1146/annurev-environ-102017-025931
Lock, I., & Schulz-Knappe, C. (2019). Credible corporate social responsibility (CSR) communication predicts legitimacy: Evidence from an experimental study. *Corporate Communications: An International Journal, 24*(1), 2–20. https://doi.org/10.1108/CCIJ-07-2018-0071
Mast, C., & Stehle, H. (2009). Corporate social responsibility – Modeerscheinung oder mehr? [Corporate social responsibility – a fad or more?] In S. J. Schmidt & J. Tropp (Eds.), *Die Moral der Unternehmenskommunikation. Lohnt es sich, gut zu sein?* (pp. 170–186). Herbert von Halem.
Mast, C., & Stehle, H. (2016). *Energieprojekte im öffentlichen Diskurs: Erwartungen und Themeninteressen der Bevölkerung [Energy projects in public discourse: Expectations and thematic interests of the general public]*. Springer VS. https://doi.org/10.1007/978-3-658-12711-4.
McCown, N. (2007). The role of public relations with internal activists. *Journal of Public Relations Research, 19*(1), 47–68. https://doi.org/10.1080/10627260709336595
Moernaut, R., et al. (2020). Hot weather, hot topic: Polarization and sceptical framing in the climate debate on Twitter. *Information, Communication & Society, 25*(8), 1047–1066. https://doi.org/10.1080/1369118X.2020.1834600
Palazzo, G., & Scherer, A. G. (2006). Corporate legitimacy as deliberation: A communicative framework. *Journal of Business Ethics, 66*(1), 71–88. https://doi.org/10.1007/s10551-006-9044-2
Paré, G., et al. (2015). Synthesizing information systems knowledge: A typology of literature reviews. *Information & Management, 52*(2), 183–199. https://doi.org/10.1016/j.im.2014.08.008
Ross, A. S., & Rivers, D. J. (2019). Internet memes, media frames, and the conflicting logics of climate change discourse. *Environmental Communication, 13*(7), 975–994. https://doi.org/10.1080/17524032.2018.1560347
Ruser, A. (2022). The skeptic who came in from the cold? The formation of a climate-skeptic alliance in Germany. In M. Hoelscher et al. (Eds.), *Civil society: Concepts, challenges, contexts: Essays in honor of Helmut K. Anheier* (pp. 349–358). Springer. https://doi.org/10.1007/978-3-030-98008-5
Sandhu, S. (2012). *Public Relations und Legitimität [Public relations and legitimacy]*. VS Verlag für Sozialwissenschaften. https://doi.org/10.1007/978-3-531-19304-5
Sandhu, S. (2014). Public Relations und gesellschaftliche Kommunikation: Legitimation im Diskurs [Public relations and societal communication: Legitimacy in discourse]. In A. Zerfaß

& M. Piwinger (Eds.), *Handbuch Unternehmenskommunikation* (pp. 1161–1183). Springer. https://doi.org/10.1007/978-3-8349-4543-3_58

Sandhu, S. (2018). Meet the blind spot: Aktivistische Strategiekonzepte als Irritation und Inspiration für die strategische Kommunikation [Meet the blind spot: Activist strategy concepts as irritation and inspiration for strategic communication]. In S. Wehmeier & D. Schoeneborn (Eds.), *Strategische Kommunikation im Spannungsfeld zwischen Intention und Emergenz* (pp. 59–82). Springer Fachmedien. https://doi.org/10.1007/978-3-658-17634-1_4

Sandhu, S. (2022). Gesellschaftsorientierte Unternehmenskommunikation: Stakeholderorientierung und Legitimation als Ziel der Public Relations [Societal corporate communication: Stakeholder orientation and legitimation as public relations objectives]. In A. Zerfaß, M. Piwinger, & U. Röttger (Eds.), *Handbuch Unternehmenskommunikation: Strategie – Management – Wertschöpfung* (pp. 859–880). Springer. https://doi.org/10.1007/978-3-658-22933-7_38

Schifeling, T., & Hoffman, A. J. (2019). Bill McKibben's influence on U.S. climate change discourse: Shifting field-level debates through radical flank effects. *Organization & Environment, 32*(3), 213–233. https://doi.org/10.1177/1086026617744278

Schwarz, A., & Fritsch, A. (2014). Communicating on behalf of global civil society: Management and coordination of public relations in international nongovernmental organizations. *Journal of Public Relations Research, 26*(2), 161–183. https://doi.org/10.1080/1062726X.2013.864242

Smith, M. F. (2005). Activism. In R. L. Heath (Ed.), *Encyclopedia of public relations* (pp. 6–9). SAGE. https://doi.org/10.4135/9781412952545

Smith, M. F., & Ferguson, D. P. (2013). "Fracking democracy": Issue management and locus of policy decision-making in the Marcellus Shale gas drilling debate. *Public Relations Review, 39*(4), 377–386. https://doi.org/10.1016/j.pubrev.2013.08.003

Sovacool, B. K., & Dunlap, A. (2022). Anarchy, war, or revolt? Radical perspectives for climate protection, insurgency and civil disobedience in a low-carbon era. *Energy Research & Social Science, 86*, online first. https://doi.org/10.1016/j.erss.2021.102416

Stehle, H. (2022). *Citizens' perceptions of dialogue: A Q study on the communication of energy projects*. Conference paper, Annual Conference, European Public Relations Education and Research Association (EUPRERA), Vienna.

Stokes, A. Q., & Atkins-Sayre, W. (2018). PETA, rhetorical fracture, and the power of digital activism. *Public Relations Inquiry, 7*(2), 149–170. https://doi.org/10.1177/2046147X18770216

Suchman, M. C. (1995). Managing legitimacy: Strategic and institutional approaches. *The Academy of Management Review, 20*(3), 571–610. https://doi.org/10.2307/258788

Then, V., & Mildenberger, G. (2022). Social innovation: Not without civil society. In M. Hoelscher et al. (Eds.), *Civil society: Concepts, challenges, contexts: Essays in honor of Helmut K. Anheier* (pp. 83–100). Springer. https://doi.org/10.1007/978-3-030-98008-5

Thomsen, C. (2018). Sustainability. In R. L. Heath & W. Johansen (Eds.), *The international encyclopedia of strategic communication* (pp. 1–9). Wiley. https://doi.org/10.1002/9781119010722.iesc0188

United Nations (UN). (1987). *Report of the World Commission on Environment and Development: Our common future*. Annex to Document A/42/427—Development and International Co-operation: Environment. United Nations (UN). http://www.un-documents.net/wced-ocf.htm.

United Nations (UN), Office of the High Commissioner for Human Rights. (2008). *A handbook for civil society: Working with the United Nations human rights programme*. United Nations (UN), Office of the High Commissioner for Human Rights.

Valentini, C., & Kruckeberg, D. (2016). The future role of social media in international crisis communication. In A. Schwarz, M. W. Seeger, & C. Auer (Eds.), *The handbook of international crisis communication research* (pp. 478–488). Wiley. https://doi.org/10.1002/9781118516812.ch43

van Leeuwen, T. (2007). Legitimation in discourse and communication. *Discourse & Communication, 1*(1), 91–112. https://doi.org/10.1177/1750481307071986

van Vuuren-Verkerk, K., Aarts, N., & van der Stoep, J. (2021). Meaning-making on the ground: An empirical study on interactional framing in environmental conflicts. *Journal of Communication Management, 25*(4), 368–384. https://doi.org/10.1108/JCOM-11-2020-0149

Weder, F., et al. (2021). Sustainability communication as critical perspective in media and communication studies—an introduction. In F. Weder, L. Krainer, & M. Karmasin (Eds.), *The sustainability communication reader: A reflective compendium* (pp. 1–12). Springer. https://doi.org/10.1007/978-3-658-31883-3_1

Weder, F. (2023). Communication of CSR. In F. Weder & M. B. Erikson (Eds.), *CSR communication and cultures of sustainability* (pp. 68–77). University of Queensland. https://doi.org/10.14264/428d81b

Weder, F., & Erikson, M. B. (2023). Communication about CSR. In F. Weder & M. B. Erikson (Eds.), *CSR communication and cultures of sustainability* (pp. 78–91). University of Queensland. https://doi.org/10.14264/428d81b

Wolf, K. (2018). Power struggles: A sociological approach to activist communication. *Public Relations Review, 44*(2), 308–316. https://doi.org/10.1016/j.pubrev.2018.03.004

Woods, C. L. (2019). From an "outside group" to a catalyst for corporate change: Examining activists' legitimation strategies. *Public Relations Review, 45*(2), 332–347. https://doi.org/10.1016/j.pubrev.2019.02.001

Further Reading

Feldner, S. B., & Meisenbach, R. J. (2007). SaveDisney.com and activist challenges: A Habermasian perspective on corporate legitimacy. *International Journal of Strategic Communication, 1*(4), 207–226. https://doi.org/10.1080/15531180701623569

Yaziji, M. (2004). Toward a theory of social risk. *International Studies of Management & Organization, 34*(4), 87–107. https://doi.org/10.1080/00208825.2004.11043716

Sonja Fürst, née Walke, B. A., is a master's student at the University of Hamburg, specializing in journalism and communication studies. She focuses her academic pursuits on exploring the intersection of mediaandenvironmentalissues.https://www.linkedin.com/in/sonja-f%C3%BCrst-0975a2245/

Helena Stehle is a professor of communication science with a focus on strategic communication at the University of Münster. Her research interests include strategic and organizational communication, and sustainability and energy communication. https://www.uni-muenster.de/Kowi/en/personen/helena-stehle.shtml

An Ethical Framework for Constructive Sustainability Debate in Political Forums: Aristotle's Principles of "Deliberative Rhetoric" for Analysis of Rhetorical Argumentation

Simon McLaughlin

Abstract This chapter develops an ethical framework for strategic sustainability communication in political speech and applies this framework to an analysis of an Australian parliamentary debate on climate change. In the Australian Parliament, political speakers have been criticised for an impractical and unethical debate surrounding climate legislation. Notwithstanding the multiple political and discursive factors influencing the debate, I contend that in the mechanics of rhetorical argumentation lie potential solutions. Following a brief discussion on sustainability communication and rhetorical argumentation, this chapter draws from Aristotle's instruction on the "deliberative species" of rhetoric to develop a rhetorical framework that promotes practical reasoning for civic decision-making within contemporary pluralistic debate. Key to my Aristotelian approach are three principles to maintain practical arguments: First, the argumentation should be deliberative and the premises of arguments should be derived from relevant civic topics; Second, the argumentation should follow a practical line of reasoning using logical deductions; Finally, emotional appeals are tied to practical arguments. In my qualitative, exploratory, conceptual study, I apply this framework as a novel approach to the analysis of rhetorical argumentation in political speech. I make a case study of two Australian Parliamentary speeches from legislative debate in 2021 in which the speakers take opposing views on the utility of climate change legislation. I determine to what extent their rhetorical argumentation contributes to practical debate based on the three criteria of my Aristotelian framework. I finish by posing this analytical framework as a model for further research in sustainability communication and beyond.

S. McLaughlin (✉)
School of Communication and Arts, University of Queensland, Brisbane, QLD, Australia
e-mail: simon.mclaughlin@uq.edu.au

F. Weder (ed.), *Strategic Sustainability Communication*, CSR, Sustainability, Ethics & Governance, https://doi.org/10.1007/978-3-031-89486-2_27

1 Background and Literature

1.1 The Australian Political Response to Climate Change

In a 2021 report, the UN ranked Australia's response to climate change as the worst of more than 170 nations according to four markers: per capita fossil fuel emissions, per capita CO_2 emissions represented in imports, per capita CO_2 emissions represented in exports, and carbon pricing (Westfall, 2021). Communicative dysfunction in Australia's parliamentary debate around climate change has been identified as a significant barrier to meaningful climate policy (Jones, 2013; Niemeyer, 2013; Gruen, 2019). Former Australian Minister for Science Barry Jones points to inherent ethical problems, citing a "values deficit" in the debate: "There is no better example of the debasement of Australian political discourse and process than that which has surrounded action on climate change" (2013, n.p.). Scholars writing on 'deliberative democracy' (e.g., Neimeyer, 2013; Dryzek, 2010) suggest allowing citizen assemblies the means to formulate practical legislative measures where the parliament has failed. Recognising the status quo of Australian democratic processes, however, I point attention to the communicators who debate, and indeed make decisions on climate change legislation—politicians—to better understand rhetorical features in the Australian parliamentary debate and particularly those related to climate change legislation. In exploring a politician's rhetorical choices, I look at how ethical considerations in strategic sustainability arguments can help bring about a more constructive approach for those aiming for transformative legislative outcomes.

1.2 An Ethical Perspective on Strategic Sustainability Communication

Sustainability communication is a relatively new, inter-disciplinary branch of communication studies with roots in environmental communication, risk communication, and science communication discourses (Adomßent & Godemann, 2011). Over the last decade or so, sustainability communication has been further informed and developed by a range of theoretical and methodological approaches, drawing from, for example, Strategic Communication and Marketing, Media Studies, Psychology, and Pedagogy (Weder et al., 2021). Weder et al. explained the two dominating sustainability communication narratives, one which speaks to the notion of a balance between economic growth and the more altruistic issues of environmental care and social justice and, second, the "rainy side" narrative (4), which speaks to a worsening climate emergency and more complicated communication challenges. The authors, however, point to a third narrative, one characterised by a more transformative viewpoint on sustainability that recognises a need for broad cultural change underlined by "deeper reflection and critical, revolutionary communicators with a profound understanding of ethics" (6). This ethical approach is in keeping with notions of sustainability itself, given the motivations from which it spawned (Godemann & Michelsen, 2011), and it is the narrative that concerns this chapter.

Strategic sustainability communication goes beyond simply explaining sustainable futures; it argues *for* them (Weder, 2022). And here I see an inherently rhetorical nature to strategic sustainability communication; rhetorical argumentation in the civic sphere is the reasoning for and against proposals as they relate to the public good (Kock & Lantz, 2023). The ethical perspective I take, then, is on the *way* arguments are made for sustainable futures, that is, the mechanics of rhetorical arguments. And so, my focus is not on the broad political discourse or the interactive communicative relationships in the Australian climate change debate but on the ways a speaker—in this case, a politician—argues for meaningful climate legislation. According to my framework for ethical and practical political speech, my analysis assesses whether the rhetorical argumentation of a political speaker facilitates or inhibits constructive debate around climate change legislation.

1.3 Ethics of Political Speech and the Benefit of Deliberative Choice

I agree with proponents of virtue ethics in political speech (e.g., MacIntyre, 1981; Borden, 2021; Slattery, 2022) that moral reflection on rhetorical choices provides opportunities to advance the civic good; in virtue ethics, deliberative arguing is the primary characteristic of ethical speech. With rhetoric, wrote Aristotle, we "confer the greatest of benefits by a right use... and inflict the greatest of injuries by using [it] wrongly" (Aristotle, 1984, *Rhetoric*, 1.1.1355b6-7). Given the influence of political speech on the direction of a society, I maintain that politicians have an *obligation* to ruminate on the reasoning in their arguments. I equally contend that the choice to argue civic proposals deliberatively—that is, in a considered, practical way—is ethical in itself; it is important for good legislative outcomes and it encourages deliberative faculty in the general population (Niemeyer, 2013). How, then, can the ethical choice to be deliberative in reasoning promote practical arguments?

In terms of sustainability, Golob et al. (2023) suggested deliberative communication "could be important as it stimulates creative and committed solutions for nature and society" (43). Some look to deliberative processes as a means to reaching consensus-driven outcomes (e.g., Niemeyer, 2013; Dryzek, 2010; Bächtiger et al., 2018). However, an escalating climate crisis and prevailing (if not worsening) political partisanship would suggest that arguing for compromised climate policies is neither effective nor realistic. Hammond (2021) points to the Copenhagen COP15 Climate Change summit as an example of the failures in consensus-driven sustainability discourse. I take Hammond's view that the "post-political" goal of consensus in sustainability discourse has had the effect of "foreclosing debate about alternative futures" (53). Where the two prevailing approaches in political communication—Habermas' deliberative framework (e.g., 2006) and the Mouffean pluralistic, agonistic model (e.g., 1999)—may appear natural contraries, I believe a kind of merging of the two has potential. In an Aristotelian type of deliberative debate, where pluralistic notions of civic good are accepted as inherent, the function of deliberation is not to reach compromise, but to construct rhetorical arguments that demonstrate the most practical course of action amongst competing interests.

1.4 An Aristotelian Approach to Constructive Political Debate

Aristotle frames rhetoric as a *productive* science (Shields, 2008) whose primary activity is to discover persuasive means for arguments (1.1.1355b10). Aristotle's Rhetoric has a civic purpose: "utility... what is most useful to do" (1.5.1362a17-18). His deliberative branch of rhetoric is characterised by a political speaker's ethical choice to weigh the good and bad of legislative proposals and to systematically construct relevant, practical, logical arguments. In Aristotle's *Rhetoric*, I see a methodical approach to constructive rhetorical argumentation in political speech that remains largely unexplored in communication studies. Garver (1994) argued that despite belief to the contrary, Aristotle's *Rhetoric* is a "relatively untouched... untapped resource," and his principles for rhetorical argumentation have had marginal influence on rhetoric's evolution in Western scholarship (4). MacIntyre (1994) notes the profound importance of ethics in our civic debates and points to Aristotle for guidance: "No issue is more central to contemporary concerns than that of the relationship of ethics to rhetoric. No book has more to teach us than Aristotle's *Rhetoric*" (n.p). In this chapter, I examine the rhetorical argumentation from speeches on climate change legislation to demonstrate how an adherence to key Aristotelian deliberative principles can promote constructive debate.

1.5 Rhetorical Argumentation: A Clarification

Before developing my Aristotelian framework, I should reiterate what I mean by the term rhetoric. Much of the communication literature defines rhetoric quite broadly in terms of interactive, meaning-making processes in communication and discursive elements related to symbolism, knowledge, and power (e.g., Foss, 2009; 2017; Peeples, 2015). My interest is in a specialised study of rhetoric theory, one concerned with a political speaker's argumentative attempts to convince an audience, where "*rhetoric provides the reasons for and against various choices, which we exchange when, in some human collective, we are to decide on a course of action*" (Kock & Lantz, 2023, 1). I take the approach of those in "deliberative democracy" (e.g., Dryzek, 2010) and more closely that of the "Copenhagen School" of rhetorical argumentation, whose research aim is "*to define what a meaningful and productive debate is, and what types of debate behaviour help or harm the productivity of a debate*" (Kock & Lantz, 2023, 8). This is the rhetoric of interest to me, rather than the notion of a rhetoric largely equated to communication or discourse (e.g., Foss). Early neo-Aristotelian approaches (e.g., Wichelns, 1924), "did very little that related to Aristotle and his Rhetoric" (Gaines, 2000, 6); this is even more noticeable today (e.g., Foss, 2017), where Aristotle's principles are largely reduced to little more than a handful of misrepresented key terms. Rhetorical argumentation is the *reasoning* for political arguments about courses of action (Strauss, 1964; Kock & Lantz, 2023), and it is primarily concerned with the role of the speaker (Knape, 2012). In this civic decision-making context, "Rhetoric as an activity has its own values, criteria for success and failure, and intriguing internal complexity" (Garver, 1994, 10). So, to be clear, my discussion examines the unique values and criteria for success and failure *within* the activity of rhetorical argumentation.

2 Aristotelian Foundations for an Ethical, Constructive Rhetorical Framework

2.1 *Pistis: Argumentation = Conviction*

The ancient Greek word for persuasion, *pistis,* is at the heart of Aristotle's instruction on rhetorical argumentation. *Pistis* "can be rightly translated as proof, argument, reasoning, persuasion, belief, trust, faith, conviction, obligation, and confidence" (Garver, 2014, 137). These definitions can be split into two distinct themes: the *argument* of the speaker and the *conviction* of the audience; Garver (1994) describes these, respectively, as the "internal" and "external" ends of rhetoric. Working towards the internal end—argumentation—is the chief activity in an Aristotelian rhetorical praxis (1.1.1355b10-11). Garver explains that in developing an approach to rhetoric primarily focused on argumentation, Aristotle makes *pathos* subordinate to *logos* and in so doing ingeniously develops an ethical, practical, civic-minded rhetoric where the cause of persuasion (argument) becomes *equal* to its effect (conviction). That is, through the deliberative process with which the speaker demonstrates argumentative proof for a proposed course of action, the audience is given the opportunity to find conviction through the same deliberative means.

2.2 *The "Deliberative Species" of Rhetoric: Weighing the Benefit and Harm of Legislation*

Aristotle introduces three rhetorical species for political speech, each of them associated with different objectives for the speaker and different choices for the audience. The *judicial* species is for legal arguments about just and unjust actions; *epideictic* rhetoric is used when praising or censuring a person; the *deliberative* species concerns arguments about civic legislation and action (1.3.1358b7-11). The deliberative species of rhetoric is at the core of the activity (Garver, 2014). The primary task in deliberative rhetoric is to weigh consequences related to proposed legislation and civic action and to find argumentative proof that demonstrates the proposed legislation/action will either harm or benefit the civic good (1.3.1358b3-25). Arguing from the epideictic species can support the deliberative argument; the praise or censure of another's character can benefit or diminish a case for an advised future action (1.9.1367b36). Crucially, however, Aristotle says in the same passage that when a speaker uses epideictic argument in deliberative speech, they should do so in terms of their legislative "advice"; that is, the character praise or attack should be associated with argument about the best course of action. In other words, deliberative speech should be relevant to the proposition under debate; thus, the deliberative species provides the speaker an ethical footing for practical argumentation.

2.3 *Deliberative "Topics" Make Political Debate Ethical and Practical*

Within Aristotle's three species of rhetoric are topics, the relevant sources from which a speaker derives premises for an argument. The deliberative topics provide "the grounds on which we are to base our persuasions in favour of its utility" (1.8.1366a19) (note the word "persuasions" to describe argument). In the epideictic species, the topics relate strictly to types of good and bad character (1.9). Of less interest to our discussion, the judicial topics concern a variety of matters related to criminality and justice (1.10-15). The deliberative species has four categories of topics. From these topics the speaker draws premises concerning generally accepted beliefs (*endoxa*) related to the civic good. The rhetorician deliberates over these truisms in relation to the proposed legislation (1.4.1359a30). The first category Aristotle describes is political topics, that is, matters about which a political assembly can make decisions, for example, resources, economy, trade, national defence, and, above all, "legislation, for it is on a country's laws that its whole welfare depends" (1.4.1360a20). The three other categories of deliberative topics are "*eudaimonia*" (happiness, the "good life" of the citizen) (1.5), civic utility (1.6), and relative good and bad (1.7), the latter for when the utility of a proposition is agreed upon, but to different degrees (for example, the most prudent time to end coal production). A truism related to the political topics could be that healthy trade relations benefit Australian producers. From the topic of *eudaimonia*, a speaker could argue that a citizen has a right to feel safe in their own home. These are *endoxa* we might see in an Australian political debate.

How can deliberative topics promote constructive climate change debate in the Australian Parliament? Garver (1994) maintains, "rhetorical topics are the means of making arguments ethical" (100-101). Deliberative topics constrain legislative arguments to that which concerns the civic good. By observing whether or not premises are drawn from deliberative topics, I can assess if a speaker is relevant to the legislation under debate. I note here a differentiation between my approach to rhetorical analysis and that which considers wider discursive elements, such as prevailing beliefs and narratives, cultural, and social factors. I would never suggest that these things are not important in political discourse. However, the rhetoric I examine concerns the technical aspects of argumentation from the speaker's point of view to discover the way rhetorical choices, in and of themselves, can enable or disable constructive debate.

2.4 Argumentative Pathos Is Ethical Because It Promotes Civic Deliberation

Aristotle observes that in political decision-making, emotions affect judgement (1.2.1356a14-15). Thus, the rhetorician should understand how emotions function (2.1.1378a21-30) and put the audience in the "right frame of mind" to decide between one speaker and another (2.1.1377b21-25). Depew (1996) explained the rhetorical role of emotions: "Their arousal throws substantive light on the situation and persons being discussed, and they give evidence of the speaker's good will (*eunoia*) to the audience of his loving and hating or more generally valuing the same things they do" (455). Yack (2006) explained further: "deliberation about what serves the common advantage requires a living reason, reasoning informed by the emotions that interest us in the consequences of our decisions" (433). Important, however, is that Aristotle's *pathos* is subordinate to the internal end of argument (Garver, 1994); thus, like epideixis, *pathos* provides a support for deliberative argument, not an alternative. Indeed, *pathos* can help an argument make *more* sense: "Rhetoric shows us how reason can be contingent, emotional, and interested without ceasing to be rational... an ethical or an emotional argument can sometimes be more rational than an argument that tries to rely on reason alone" (Garver, 2004, 3-4). And so, we have an ethical place for *pathos* as a constructive tool for engaging citizens in civic deliberation.

2.5 Logical Deliberative Enthymemes Demonstrate Practical Reason

The goal of deliberative rhetoric is argumentative proof: "the substance" of which is found in the enthymeme (1.1.1354a15). With the enthymeme, a speaker demonstrates practical reason. Strauss makes clear, "[the enthymeme is] the core of Aristotle's *Rhetoric*, whether we like it or not. Someone might find the discussion of the passions much more interesting and rewarding, but from Aristotle's point of view there can be no doubt" (1964, 246). Jörke (2014) explains the enthymeme, the rhetorical deduction:

> a kind of logical argument based on commonly held beliefs [*endoxa*]. An enthymeme is 'a sort of syllogism' (1.1.1355a8), a deductive argument that is not strictly logical like in the method of dialectic but still convincing in a commonsensical way (72).

Note the "commonsensical" here (i.e., practical reason). The enthymeme consists of two parts: the premise and the logical reasoning that supports the premise. Rhetorical premises are understood to be true by most if not all in the audience, or by those considered experts (2.22.1396a1). "*A child needs a nutritious diet to reach their potential as an adult*" is an example of deliberative *endoxa* that might premise an enthymeme arguing for the provision of school lunches for underprivileged children. The reasoning in the enthymeme would provide specific support for this general premise. Aristotle describes rhetorical premises as "roughly" equivalent to maxims (2.21.1394a27-8). The *endoxa* in maxims assists persuasion by connecting a specific line of reasoning to a general truism. Maxims carry *ethos* because they indicate a speaker's deliberative choice (2.21.1395b12-18). Strauss's (1964) writes, "[The maxim] lacks theoretical certainty but it has moral certainty... you can be sure that you act rightly, justly in taking this course" (263). For the premise to show practical reason, it must represent genuine *endoxa*, actual truisms (2.21.1395b12-18). And here is the technical function of *endoxa* in practical argumentation. How can a speaker expect to make a commonsensical case if the very *grounds* of the argument are widely disputed? So, the premise in a deliberative enthymeme is a truism about a topic relevant to the civic good.

2.6 Deliberative Character: Logos = Ethos

Aristotle maintains it is ultimately the speaker's character that creates audience conviction (1.2.1356a4-5). Each of the three modes carry their own character: the deliberative argument of *logos* demonstrates the virtue of practical reason, *pathos* shows the audience a speaker's goodwill toward them, and *ethos* displays moral character (2.1.1378a7-19). But how does one speak to moral character from within a deliberative activity, where argument's premises concern civic concerns, not character? Garver (1994) clarifies *ethos's* role in the argumentative art of rhetoric: "Good practical reasoning is the discursive embodiment of good character. The character of the speaker… is revealed in the reasoning of the speech… to deal with one is to deal with the other (151). Aristotle tells us, "There is character in every speech in which the [deliberative] choice is conspicuous" (2.21.1395b12-18). To show *ethos* from within the deliberative species, then, is to demonstrate the moral decision to argue logically and practically about the most prudent course of action in relation to the civic good.

2.7 *The Apparent Enthymeme: Argument Without Logos or Ethos*

In the genuine enthymeme, the speaker shows practical reason through logical deduction. An 'apparent' enthymeme cannot be genuinely practical, however, because it relies on selective evidence to prove a supposedly categorical premise (Rapp, 2022). This fallacious enthymeme reaches only an illusion of proof. Rapp explains the root of the apparent enthymeme is in the false suggestion that the premise is absolutely a truism. To be an effective rhetorician, according to Aristotle, the speaker must learn the mechanics of both genuine and apparent enthymemes in order to best recognise and refute fraudulent arguments (1.1.1135b16-17). With the identification of false *endoxa* (i.e., questionable truisms) and the absence of logical reasoning, I can determine where, by design, argumentation fails to be constructive.

My Aristotelian framework for constructive rhetorical argumentation, then, can be characterised by three principles: deliberative species and topics, argumentative *pathos*, and genuine enthymemes. This framework represents a practical, logical, civic-minded type of argumentation that demonstrates ethical character through deliberative choice. The next section will contextualise this framework for analysis of rhetorical argumentation in two Australian Parliamentary climate change speeches.

3 From Theory to Method: A Novel Analytical Framework for Political Speech

3.1 *Developing an Aristotelian Methodological Approach*

My framework allows for mixed method analysis, which Davies (2020) writes is a common approach in exploratory research designs, those which "involve discovering and describing" (6). While my case study leans towards a qualitative type of explorative analysis, I will discuss later the potential for further insights through the identification of quantitative patterns. My method is not a typical rhetorical analysis, because I adhere to an atypical understanding of rhetoric. In addition, methodologies in rhetorical analysis tend to be characterised by a subjective, "do-it-yourself" approach, as in Croucher and Cronn-Mills (2021): "Don't worry about selecting the best theory. No particular theory is best (285)... we do not have objective standards of analysis in rhetorical criticism" (289). In my case, however, it seems there is indeed an ideal, "best" theoretical approach; Aristotle's principles speak directly to communicative problems in Australian political sustainability debate I wish to address.

Content analysis is recognised as a useful methodical approach both for quantitative and qualitative research (Drisko & Maschi, 2016; Krippendorf, 2019). In both cases, the researcher isolates patterns, themes, and frequencies within a text and makes inferences about them through the application of a coding process (Drisko & Maschi, 2016). Importantly, though, the categories for my coding framework are not, as is typical, defined by the data set. My method, then, is not empirically grounded, which Krippendorff (2019) considers one of three distinguishing characteristic of content analysis (my approach does meet Krippendorff's other two criteria; it is exploratory and inferential). In addition, where qualitative approaches to content analysis tend to be interpretative (Neuendorf, 2017), my theory-based framework looks for objective findings.

Given my analytical approach is a novel one, with characteristics and goals unique to Aristotelian rhetorical concepts and without reliance on contemporary approaches, I believe differentiating it from conventional rhetorical analysis and content analysis is appropriate. My analytical framework is uniquely concerned with the technical constituents of rhetorical argumentation (I note that "rhetorical argumentation" here is distinctly political in nature), my research method could be defined as *Aristotelian analysis of rhetorical argumentation*.

3.2 Data and Analytical Categories

The data for my case study is retrieved from the *Hansard*—the publicly available record of parliamentary debate since Australian federation in 1901. I have chosen two speeches from a 2021 debate in the week before the Glasgow Climate Change Conference (COP26). The conservative Coalition Government had just announced Australia's first-ever commitment to net-zero emissions. The two speakers under analysis are the Government's Energy and Emissions spokesperson, Tim Wilson, and Leader of the Australian Greens, Adam Bandt. The speakers argue (respectively) *against* and *for* meaningful climate change legislation. My analysis examines and compares rhetorical features of their argumentation.

The three analytical categories (Fig. 1) come from my Aristotelian framework for constructive rhetorical argumentation. Each demonstrate a speaker's deliberative choice.

The first category represents the speaker's opportunity to argue the civic utility of legislation and civic action within the bounds of the deliberative species of rhetoric. Failure to do so, to my mind, renders practical argumentation unachievable. How can there be a constructive debate about a proposition if there is little attempt to address that which the proposition concerns? The second category is focused on the substance of rhetorical argumentation: enthymemes. I can determine an

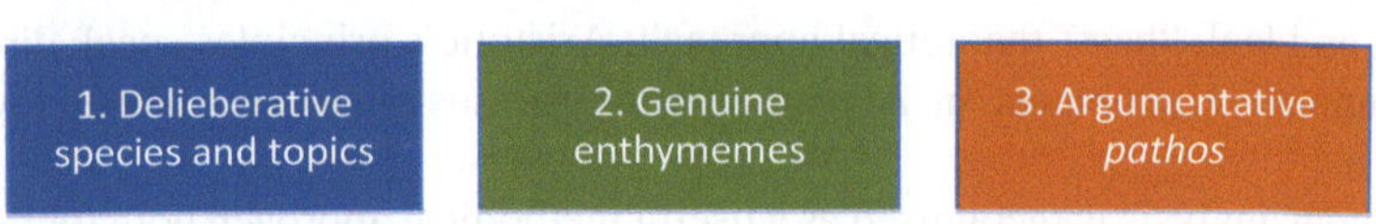

Fig. 1 Categories for Aristotelian analysis of rhetorical argumentation

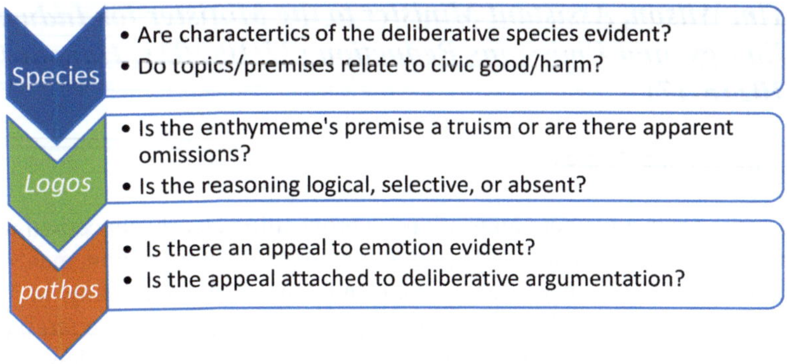

Fig. 2 Guiding questions for analysis

enthymeme as fallacious (i.e., not constructive) by the limitations to its premise (as an absolute truism) and by illogical reasoning. The final category examines *pathos*. I contend *pathos* can be dealt with ethically only by tying it to deliberative argumentation, otherwise there is no line of practical advice for the audience to deliberate on. Given emotional engagement can promote practical deliberation, it is also important to note whether or not an appeal to *pathos* is made at all.

My analytical framework is guided by the following questions (Fig. 2):

4 Case Study: Aristotelian Analysis of Rhetorical Argumentation in Climate Change Debate in the Australian Parliament

On the 26th of October 2021, the Australian Government announced their plan to achieve net-zero greenhouse emissions by 2050, though the plan would not be legislated and coal production would continue. The following day in the House of Representatives, in a 10-min speech, Government spokesperson Tim Wilson makes his case *against* the need for legislation. After my analysis of Wilson's argumentation, I turn to Greens Leader Adam Bandt, whose 15-min speech argues *for* a proposed renewables bill in the context of the Government's unlegislated plan. The complete speeches can be found at the Parliament of Australia's *Hansard* webpage: https://www.aph.gov.au/Parliamentary_Business/Hansard.

4.1 Tim Wilson, Assistant Minister to the Minister for Industry, Energy, and Emissions Reduction (27/10/2021, Hansard, 10216-17)

4.1.1 Species and Topics

Wilson's case against net-zero legislation—a proposition clearly concerned with the deliberative species of rhetoric—largely employs epideictic enthymemes. Targets for Wilson's character attacks are varied and numerous. According to Wilson, Albanese "*takes more of a Donald Trump approach to engagement in international fora,*" the Greens want to "*burn down the village to save it,*" the Shadow Treasurer "*before the last election... came after Australian retirees and tried to slash their incomes,*" and Wilson takes aim at "*a particularly shrill and hostile*" female BBC journalist who, in an interview, had sought further information from Wilson about the government's net-zero plan. Where Wilson does refer to the legislation and to considerations of the civic good, my analysis of enthymemes will assess whether his arguments demonstrate practical reason.

4.1.2 Enthymemes

Wilson argued that his government is dedicated to having the world "*follow our leadership.*" This is a bold claim given the UN's last-place ranking for Australia's response to climate change. Former UN Secretary-General Ban Ki-moon maintains that the Australian Government is "out of step with its states, its trading partners, and other comparable nations" (Clarke, 2021). Wilson's premise, then, is not a genuine truism and so, inevitably, fallacious reasoning must follow. Wilson's "proof" that Australia is a world leader on climate action is in the mere *existence* of the Government's new emissions plan. However, this is hardly world-leading; many countries had a policy for net-zero emissions at the time of the Government's announcement; and a number of these had legislated their commitment.

Wilson provides two arguments that speak directly to net-zero emissions legislation. One is hyperbolic to say the least and is best left for the later examination of *pathos*. In the other, Wilson asserts that emissions legislation is unnecessary because good emissions outcomes happen without the intent legislation provides: "*We want to make sure we deliver a solution for the Australian people and what matters in this debate isn't intent, it's outcomes.*" On the matter of *endoxa*, could it be considered universally established that legislative intent is not important for good climate outcomes? Wilson's reasoning follows:

> Without a legislated target, this government hasn't just reached a 20.8 per cent emissions reduction on 2005 levels... the OECD average for emissions reduction over the same time frame is only seven per cent. So we're beating it by a factor of three.

These figures would suggest the Government achieved relatively impressive emissions outcomes. How does this fit with Australia's lowly world ranking for climate change action? Wilson's figures are, of course, incorrect. Australia's emissions reductions are understood to be considerably, *intentionally* inflated. On the Australian Government's deceptive calculation of emissions data, Christiana Figueres (who led the 2015 Paris Agreement negotiations) said, "It is not legal, it is not correct, it is not moral. It is cheating" (O'Malley, 2020, para. 19). Aristotle overtly states that the deliberative rhetorician must be informed by the facts as they relate to the subject in question (2.22.1396a4-7). How can the rhetorical product be genuinely practical if the matter that constitutes it is false?

The fallacious enthymeme is prevalent in Wilson's speech. In another example, Wilson argues that the USA, China, and New Zealand need to be part of the "*global solution,*" asserting, "*They all ran away from the Kyoto protocol.*" In fact, while China did not sign the treaty, the US had an 8% reduction target for the first Kyoto commitment period and New Zealand argued for 0%. The Australian Government negotiated for an 18% *increase* in emissions and left the treaty unsigned. The EU's spokesman for environment policy at the time, Peter Jorgensen, described the Australian negotiations as "wrong and immoral … a disgrace" (Slezak, 2021). As logical and relevant deliberation demonstrates a speaker's practical reason and moral virtue, I suggest it follows that a disregard for deliberative choice exhibits a *lack* of these character and trust-building qualities and leaves the speaker only with *pathos* to find conviction.

4.1.3 *Pathos* Appeal

Wilson claims that by introducing net zero legislation, the Labour Party's chief objective is to destroy jobs, industry, the economy, and people's livelihoods. What rhetorical role does this rather outlandish suggestion play? In his exposition of emotions key to rhetorical argumentation, Aristotle describes *fear*, writing that a speaker attempts to evoke this emotion by putting into the audience's imagination "some destructive or painful disturbance in the future" (2.5.1382a22-3). Wilson not only contends that legislating for net-zero emissions will destroy Australia. He also claims that in reducing fossil fuel use, "*Northern Europeans will literally die in the tens of thousands*" and Labour will "*increase retirees' bills, particularly their electricity bills.*" Clearly, Wilson's statements meet the criteria described by Aristotle as an appeal to fear.

Aristotle also writes we feel the emotion of *friendship* towards those with shared goals and ideals and those who treat us kindly (2.4.1381a5-15). And we feel the opposite emotion—*hostility*—towards those with contrary traits. Aristotle explains that we can feel hostility not simply because of someone's words or actions but "merely because of what we take to be their (deficient) character" (2.4.1328a4). We can view Wilson's statements as appealing to hostility towards an Opposition party who would cause terrible harm to the Australian people. As contrast, Wilson introduces Aristotle's criteria for the emotion of friendship with assurances that his government intends to take an "*Australian approach*" to the emissions problem, that it will "*back Australians and their jobs.*"

According to the principles of my Aristotelian framework, Wilson demonstrably fails to make rhetorical choices that allow for the demonstration of practical reason. Thus, this speech is unable to contribute to constructive deliberation about the best course of action in regard to legislating net-zero emission. Wilson's case against the proposal is overwhelmingly reliant on apparent enthymemes and emotional appeals that sit in isolation from practical, logical argument.

4.2 *Adam Bandt, Leader of the Australian Greens (27/10/21, Hansard, 10248-50)*

4.2.1 Species and Topics

In his opening line, Bandt concedes the Government's proposed "Offshore Electricity Infrastructure Bill" (legislation to expand wind farming) could be beneficial for renewable energy. He then embarks on a sustained epideictic attack on the Government, Prime Minister Morrison, and Deputy Prime Minister Joyce that continues, without reference to the Bill, until well into the second half of this 15-min speech. Bandt claims his opponents are climate denialists, that their emissions plan "is a fraud." When Bandt comes to the matter of the proposed Bill, his argumentation relies on deliberative topics related to the civic good. On the topic of "relative good," for example, Bandt argues the utility of the Bill will be improved with amendments, including: "*criteria for feasibility licences, commercial licences and transmission licences... local procurement of labour and goods right through the logistics and supply chains. This could unleash a massive domestic employment potential.*"

Bandt's is a speech of two parts with the relatively smaller section addressing the legislative matter under debate. Thus, I cannot determine this speech as *absolutely* falling within the constraints of deliberative debate. However, with the deliberative species evident in a significant portion of the speech, and with the Aristotelian understanding that epideictic rhetoric can assist a deliberative argument (though, admittedly, in a subordinate role), it is worth examining enthymemes for evidence of practical reason and to gather if emotional appeals appear as deliberative support.

4.2.2 Enthymemes

Most of the deliberative argumentation here tends not to address the legislative Bill at hand, though it might be argued that it offers indirect support for the legislation by speaking to broader issues in the Australian climate policy debate: "*$50 million of public money is now going to help open up a new climate time-bomb... Inside those Northern Territory gas basins there's the equivalent of 70 years of all of Australia's pollution.*" However, evidence such as this tends to support arguments related to the upcoming election, as made explicit in the following:

> If you want more legislation that's going to fast track renewables and phase out coal and gas, don't waste your vote on a Liberal or Labor Party.... Vote to put the Greens in the balance of power so that we can kick out the Liberals and push the next government.

This kind of argumentation does not provide for the audience practical advice in relation to the benefits of the proposed Offshore Electricity Bill.

The following, on the other hand, shows relevant, logical argumentation that acts directly to support the Bill: "*There's 40 times Australia's current energy use in our offshore waters. That's why the Greens have been saying for so long that we need legislation that will help unlock that potential.*" In another clear demonstration of practical reason, Brandt deliberates on the civic benefits of expanding offshore renewables: "*These projects... help workers transition as we phase out coal and gas... we could replace those coal-fired power stations and turn them off in a way that allows workers to move directly into secure industries with well-paid jobs.*"

4.2.3 *Pathos* Appeal

Like Wilson, Bandt appeals to hostility by means of epideictic rhetoric. Bandt paints his opposition as uncaring about the impact of climate change on Australians, and as people who actively sabotage climate change action through the funding of the fossil fuel industry and the approval of future coal mines:

> Our Prime Minister came in here and held up a lump of coal and said, 'This is coal. It's amazing. It won't hurt you,' when the whole world was screaming at him and saying that coal and gas are the major causes of the climate crisis, and we've got to stop burning them.

Bandt suggests the Government's change of heart (i.e., the net-zero announcement) is a pretence motivated by the upcoming election, given polls show safe conservative seats are in danger due to the Government's climate change response. Bandt's arguments work to arouse hostility not simply by portraying individuals with flawed values and fundamentally deficient character who act against the good of Australians. Like Wilson, Bandt appeals to friendship with his stated intent to look after working people and ensure their prosperous future.

While Bandt's enthymemes offer something of 'mixed bag' in regard to relevancy, there is certainly far more here than in Wilson's speech to demonstrate genuine deliberative choice through logical, practical argumentation. Interestingly, in arguments related to the Offshore Electricity Bill in question, no *pathos* is evident, thus emotional appeal is not used to support enthymemes specific to the legislation. One could argue, however, some of the emotional appeals here have an indirect capacity to engage the audience with the Bill by pointing them towards broader climate concerns and the Government's position in regard to them.

5 Further Research and Conclusion

Applying my novel method for analysis of larger data sets allows for research questions that could help uncover more precise empirical findings. To what extent does a disregard for practical reasoning characterise arguments that stymie meaningful legislation? And, perhaps more importantly, how often do those arguing *for* climate action lose sight of practical reason? What emotions are most relied upon both by those arguing for and against meaningful legislation? To what degree? I believe that with my Aristotelian framework, there is the potential to discover prevailing rhetorical 'templates' in our sustainability debates beyond merely commonly stated propositions of the different sides.

Middleton (2023) notes a recent Opposition leader's "blistering rhetoric" (para. 2). Descriptive adjectives such as these tell us nothing about a speaker's rhetorical choices. I go deeper into the technical realm of rhetorical argumentation, beneath, for example, "value statements" (para. 13) to observe how practical reason (or a lack of it) become manifest. We might often have a "sense" that a political debate lacks ethical integrity without being able to justify this feeling with any great specificity. Aristotle's principles allow us to delve into the speaker's choices and, to a fair extent, *prove* whether or not their argumentation is practical and concerned with the civic good.

I see my analytical framework as offering benefit not only to parliamentary debates around sustainability, but to *any* process of civic decision-making that seeks to find the most useful course of action. How do we argue at the community level? In town halls, in schools, online? In an expansive study of public writing over the last 170 years, Scheffer et al. (2021) found in recent decades "a marked shift in public interest from the collective to the individual, and from rationality toward emotion" (1). From a more strategic, practical perspective, could an Aristotelian framework be legitimately applied to contemporary debate such that attention is redirected to the civic good through practical reason supported by *pathos*, not replaced by it? When *pathos* is argumentative, are there particular emotions that best achieve sustainable outcomes? Broader discourse studies on rhetorical effect would enable insight into the effectiveness of these new approaches.

From Aristotle's principles in *Rhetoric*, I have developed an argumentative framework for constructive speech in strategic sustainability debate. An argumentative framework allows those with meaningful sustainability proposals to reason consciously and methodically—not randomly—according to guidelines directing the speaker towards practical, reasoned argumentation. In short, the speaker can reliably aim for utility, the civic good. My case study demonstrates that rhetorical argumentation has unique properties and values by which we can judge political arguments "good" or "bad" in terms of their capacity to enable practical deliberation. This chapter aimed to show that a more considered approach to rhetorical choices has transformative potential for the way we argue about climate change and other critical issues surrounding sustainability.

References

Aristotle. (1984). Rhetoric (W. R. Roberts, Trans.). In J. Barnes (Ed.), *The complete works of Aristotle* (Vol. 2, pp. 2152–2269, The Revised Oxford edition). Princeton University Press.

Adomßent, M., & Godemann, J. (2011). Sustainability communication: An integrative approach. In J. Godemann & G. Michelsen (Eds.), *Sustainability communication: Interdisciplinary perspectives and theoretical foundation* (pp. 27–37). Springer Netherlands.

Bächtiger, A., Dryzek, J. S., Mansbridge, J., & Warren, M. E. (2018). *The Oxford handbook of deliberative democracy*. Oxford University Press.

Borden, S. (2021). Five questions about ethics in political communication – Sandra Borden. *Project on ethics in political communication*. https://ethicsinpoliticalcommunication.org/blog/2021/10/21/five-questions-about-ethics-in-political-communication-sandra-borden

Clarke, M. (2021, August 17). Australia faces international punishment for lagging on climate change action, former UN chief warns. *ABC*. https://www.abc.net.au/news/2021-08-17/un-ban-ki-moon-australia-behind-climate-action/100383384

Croucher, S. M., & Cronn-Mills, D. R. (2021). *Understanding communication research methods: A theoretical and practical approach* (3rd ed.). Routledge.

Davies, C. (2020). *A quick guide to quantitative research in the social sciences*. The University Of Wales Trinity Saint David Research Repository. https://open.umn.edu/opentextbooks/textbooks/822

Depew, D. J. (1996). [Review] Aristotle's "Rhetoric": An art of character. *Journal of the History of Philosophy, 34*(3), 454–456.

Drisko, J. W., & Maschi, T. (2016). *Content analysis*. Oxford University Press.

Dryzek, J. S. (2010). Rhetoric in democracy: A systemic appreciation. *Political Theory, 38*(3), 319–339. https://doi.org/10.1177/0090591709359596

Foss, S. K. (2009). Rhetorical theory. In *Encyclopedia of communication theory* (pp. 854–858). Thousand Oaks, CA.

Foss, S. K. (2017). *Rhetorical criticism: Exploration and practice* (5th ed.). Waveland Press.

Gaines, R. N. (2000). Aristotle's Rhetoric and the contemporary arts of practical discourse. In A. G. Gross & A. E. Walzer (Eds.), *Rereading Aristotle's Rhetoric*. Southern Illinois University Press.

Garver, E. (1994). *Aristotle's Rhetoric: An art of character*. The University of Chicago Press.

Garver, E. (2004). *For the sake of argument: Practical reasoning, character, and the ethics of belief*. University of Chicago Press.

Garver, E. (2014). Aristotle's Rhetoric in theory and practice. In M. J. MacDonald (Ed.), *The Oxford handbook of rhetorical studies* (pp. 133–142). Oxford University Press.

Godemann, J., & Michelsen, G. (2011). Sustainability communication – An introduction. In J. Godemann & G. Michelsen (Eds.), *Sustainability communication: Interdisciplinary perspectives and theoretical foundation* (pp. 3–11). Springer Netherlands.

Golob, U., Podnar, K., & Zabkar, V. (2023). Sustainability communication. *International Journal of Advertising, 42*(1), 42–51. https://doi.org/10.1080/02650487.2022.2144035

Gruen, N. (2019, February 16). Polarisation and the case for citizens' juries. *Quillette*. https://quillette.com/2019/02/16/polarisation-and-the-case-for-citizens-juries/

Habermas, J. (2006). Political communication in media society: Does democracy still enjoy an epistemic dimension? The impact of normative theory on empirical research. *Communication Theory, 16*(4), 411–426.

Hammond, P. (2021). Post-political communication and sustainability. In F. Weder, L. Krainer, & M. Karmasin (Eds.), *The sustainability communication reader: A reflective compendium* (pp. 51–69). Springer Fachmedien Wiesbaden.

Jones, B. (2013, November 11). A values deficit, toxic politics, and the climate change debacle. *The Conversation*. https://theconversation.com/a-values-deficit-toxic-politics-and-the-climate-change-debacle-20024

Jörke, D. (2014). Rhetoric as deliberation or manipulation? About Aristotle's Rhetoric and its misuse in recent literature. *Redescriptions: Political Thought, Conceptual History and Feminist Theory, 17*(1), 68–85. https://doi.org/10.7227/R.17.1.4

Knape, J. (2012). *Modern Rhetoric in Culture, Arts, and Media: 13 Essays* (trans: Alan, I.F.). Berlin/Boston: De Gruyter.

Kock, C., & Lantz, M. (Eds.). (2023). *Rhetorical argumentation: The Copenhagen school.* The University of Windsor's Digital Press. Windsor Studies in Argumentation. https://doi.org/10.22329/wsia.13.2023

Krippendorff, K. (2019). *Content analysis: An introduction to its methodology* (4th ed.). https://doi.org/10.4135/9781071878781

MacIntyre, A. (1981). *After virtue: A study in moral theory.* University of Notre Dame Press.

MacIntyre, A. (1994). *[Review] In Eugene Garver Aristotle's Rhetoric: An art of Character.* The University of Chicago Press.

Mouffe, C. (1999). Deliberative democracy or agonistic pluralism? *Social Research, 66*(3), 745–758. http://www.jstor.org/stable/40971349

Neuendorf, K. A. (2017). Defining content analysis. In *The content analysis guidebook* (2nd ed.). https://doi.org/10.4135/9781071802878

Niemeyer, S. (2013). Democracy and climate change: What can deliberative democracy contribute? *Australian Journal of Politics & History, 59*(3), 429–448. https://doi.org/10.1111/ajph.12025

O'Malley, N. (2020, Dec 8). What are Kytoto carryover credits and why do Australia's critics say they don't exist? *The Sydney Morning herald.* Retrieved from https://www.smh.com.au/environment/climate change

Peeples, J. (2015). Discourse/rhetorical analysis approaches to environment, media, and communication. In A. Hansen & R. Cox (Eds.), *The Routledge handbook of environment and communication* (pp. 59–68). Routledge.

Rapp, C. (2022). Aristotle's Rhetoric. In E. N. Z. U. Nodelman (Ed.), *The Stanford encyclopedia of philosophy* (Winter 2023 Edition). Retrieved December 12, 2023, from https://plato.stanford.edu/archives/win2023/entries/aristotle-rhetoric/

Scheffer, M., van de Leemput, I., Weinans, E., & Bollen, J. (2021). The rise and fall of rationality in language. *Proceedings of the National Academy of Sciences, 118*(51). https://doi.org/10.1073/pnas.2107848118

Shields, C. (2008). Aristotle. In E. N. Zalta (Ed.), *The Stanford encyclopedia of philosophy* (Spring 2022 Edition). Retrieved August 13, 2022, from https://plato.stanford.edu/entries/aristotle/

Slattery, L. (2022, February 10). How to build a more virtuous politics, the Aristotelian way. *The Weekend Australian.* https://www.theaustralian.com.au/commentary/how-to-build-a-more-virtuous-politics-thearistotelian-way/news-story/d6e22706aeaf20c4de94288bc69b02d5

Slezak, M. (2021, November 1). How Australia earned its climate change reputation. ABC News. http://www.abc.net.au/news/2021-11-01/

Strauss, L. (1964). *Aristotle's Rhetoric* [Lecture transcript, Spring 1964, University of Chicago]. https://leostrausscenter.uchicago.edu/aristotle-rhetoric-spring-1964/

Wichelns, H. A. (1924). The literary criticism of oratory. In A. M. Drummond (Ed.), *Studies in rhetoric and public speaking in honor of James A. Winans* (pp. 181–183). Century.

Weder, F., Karmasin, M., Krainer, L., & Voci, D. (2021). Sustainability communication as critical perspective in media and communication studies—An introduction. In F. Weder, L. Krainer, & M. Karmasin (Eds.), *The sustainability communication reader: A reflective compendium* (pp. 1–12). Springer Fachmedien Wiesbaden.

Weder, F. (2022). Strategic problematization of sustainability reframing dissent in strategic communication for transformation. *Public Relations Inquiry, 11*(3), 337–360. https://doi.org/10.1177/2046147X211026857

Westfall, S. (2021, July 2). Australia ranks last on climate action in U.N. report. *The Washington Post*. https://www.washingtonpost.com/world/2021/07/02/australia-climate-action-un-sustainable-development/
Yack, B. (2006). Rhetoric and public reasoning: An Aristotelian understanding of political deliberation. *Political Theory, 34*(4), 417–438. https://doi.org/10.1177/0090591706288232

Simon McLaughlin is from the University of Queensland. With a background in professional writing, Simon moved to researching political debate in Australia, and particularly the rhetorical argumentation around climate change. His current research looks at how deliberative principles of argumentation from Aristotle's Rhetoric can be applied to contemporary political speechmaking. Simon is interested in how ethical approaches to argumentation can facilitate constructive civic debate.

Communicating Sustainable Futures for Collective Action

Cassandra L. C. Troy

Abstract In order to unite for collective action and shift the status quo, scholars have suggested that people require shared visions of a positive future. However, many media depictions of environmental futures are bleak, focused on negative outcomes to be avoided rather than desirable possibilities to strive for. This chapter begins by examining media genres such as speculative, cli-fi, utopian, and solarpunk fiction that depict positive, sustainable futures. Next, psychological theories of future thinking and social change are reviewed, including the social identity theory of intergroup behavior, model of utopian thinking, and collective futures framework. These theories offer insight into a number of cognitive mechanisms that could motivate constructive responses to depictions of desirable futures. Furthermore, this scholarship can provide guidance for the operationalization of sustainable futures messaging in strategic communication contexts. Altogether, the review of these research areas presents numerous opportunities for strategic communication practitioners and scholars to examine the use of positive messaging about sustainable futures to motivate collective action in the present through a variety of media channels and among a range of stakeholder groups.

Scholars have argued that positive visions or imaginings of better possible futures are necessary in order to motivate transformative action on grand sustainability challenges posed by threats such as climate change (David Tábara et al., 2018; Wright et al., 2013). However, some suggest humanity suffers from a dangerous "failure of imagination" (p.1) not only in understanding the effects of environmental problems, but also when it comes to conceiving of desirable, sustainable outcomes that could be achieved through collective effort (Moore & Milkoreit, 2020). Media is one force that could inform people's conceptions of sustainable futures. However, media may also be afflicted by a failure of imagination. Both news and entertainment media focus more prominently on negative potential environmental outcomes than solutions or positive futures (Feldman et al., 2017; Svoboda, 2016; Tavares

C. L. C. Troy (✉)
College of Media, Journalism, University of Illinois Urbana-Champaign, Champaign, IL, USA
e-mail: clt5484@psu.edu

F. Weder (ed.), *Strategic Sustainability Communication*, CSR, Sustainability, Ethics & Governance, https://doi.org/10.1007/978-3-031-89486-2_28

et al., 2020). Overall, this chapter considers how strategic sustainability communication might support positive outcomes and motivate collective action among audiences through depictions of sustainable futures.

In this chapter, sustainability means society operating in such a way that present and future generations are able to lead healthy, meaningful lives while protecting the environment and responsibly using natural resources (Brundtland, 1987; The Natural Step, 2024). Meanwhile, strategic communication is communication undertaken among some segment of the public in order to achieve specific goals. Building on that foundational definition, strategic communication has developed into an interdisciplinary practice situated at the intersection of organizational communication, political communication, marketing, and strategic management that can facilitate a number of outcomes beneficial to an organization and broader society (Goldberg & Gustafson, 2023; Seiffert-Brockman, 2018; Werder et al., 2018). Strategic communication can help organizations' weather crises, strengthen relationships with internal and external stakeholders, cultivate a desirable brand image and more within a complex societal context (Goldberg & Gustafson, 2023; Seiffert-Brockman, 2018). Drawing on both of these definitions, strategic sustainability communication in this chapter refers to the intentional process of an organization communicating with stakeholders to raise awareness of sustainability efforts or to achieve sustainable outcomes.

This chapter will begin by reviewing current research on media representations of environmental futures. Next, drawing on the idea that positive visions for the future are an important component of achieving sustainable communities, I review a number of psychological theories relevant to future thinking and social change. Finally, I highlight practical implications based on current scholarship alongside pressing theoretical questions that can inform a future research agenda.

1 Media Depictions of Sustainable Futures

Negative representations of environmental futures or a focus on future consequences of environmental problems are common in news and entertainment media (Feldman et al., 2017; Svoboda, 2016; Tavares et al., 2020). Within news media, this can take the form of a heavy focus on present and future environmental, health, economic, and national security threats to many regions around the world (Feldman et al., 2017). Often, reporting on negative environmental futures involves portraying doomsday scenarios at the middle and end of the twenty-first century resulting from issues such as biodiversity loss and increasingly frequent extreme weather events based on prediction put forth in scientific reports (Guenther et al., 2022). In fictional media, a wide variety of stories present vivid and terrible environmental futures plagued by dire ecological as well as societal impacts (Schneider-Mayerson, 2018). Fictional environmental dystopias may act as cautionary tales, detailing dangerous postapocalyptic societies resulting from a failure to prioritize sustainability (Schneider-Mayerson, 2017; Svoboda, 2016). Too often, bleak depictions fail to offer a "clear route to a better destination" (Schneider-Mayerson, 2017, p. 315) and therefore may rob such media of its capacity to motivate audiences (Svoboda, 2016).

Meanwhile, there is scant research on news media or strategic messaging that does describe a way forward by highlighting positive environmental futures. An analysis of news coverage around climate talks suggests that even when future expectations are hopeful, they are often extremely vague. One examination of magazine stories on climate change highlighted a shift from pessimistic to solutions-oriented content that emphasized avenues for current action in order to avoid negative future climate impacts (Guenther et al., 2022). Considering news coverage of sustainability topics more generally, scholars have noted an increase in solutions-focused coverage (Ford & King, 2015; Stecula & Merkley, 2019), although this does not necessarily mean these articles described specific sustainability futures. One specific example of a recent, positive approach to news is constructive or solutions journalism, which typically focuses on responses to problems (McIntyre & Lough, 2021) and has the capacity to impact efficacy beliefs and collective action intentions within the context of climate reporting (Thier & Lin, 2022).

Fictional and entertainment media, in contrast to news media, offer much more detailed and vivid examples of desirable environmental futures. There are a number of fictional genres in which representations of sustainable futures are more common, including speculative, climate, utopian, and solarpunk fiction. Notably, these genres do overlap, such that some pieces of climate fiction, for example, may also be considered speculative or utopian fiction (Adamson, 2018; Wuebben et al., 2023). Speculative fiction stories present possible futures extrapolated from current circumstances (Dasilva, 2019). Over the past few decades, popular speculative fiction films have often contained environmental themes, with some offering visions of sustainable societies, like cities with "green infrastructure" (p. 4) or societies that are deeply connected with nature (Dasilva, 2019). Meanwhile, climate fiction, also called cli-fi, can describe negative futures warning of the need for change or can present alternative futures in which humanity has shifted the status quo to create societies less dependent on fossil fuels (Adamson, 2018; Schneider-Mayerson, 2017; Wuebben et al., 2023).

Utopian fiction is another broad umbrella that often, but not exclusively, focuses on environmental themes (Garforth, 2021). Distinct from speculative fiction, outcomes in utopian fiction are necessarily positive and these stories intentionally "reach beyond what is towards a better alternative" (p. 292) for societies (Garforth, 2021). For example, utopian stories of climate adaptation could include sustainable energy systems in near future Earth societies (Wuebben et al., 2023). Meanwhile, one specific genre of utopian fiction focused specifically on optimistic environmental and social futures is solarpunk (Gillam, 2023). Solarpunk narratives often involve extensive use of renewable energy, sustainable use of natural resources, more equitable and diverse societies, and rebuilding thriving communities from difficult or even apocalyptic circumstances (Klata, 2022). Altogether, scholars suggest that these future representations offer a tool to help people contemplate current environmental challenges and possible paths forward (Adamson, 2018; Dasilva, 2019; Garforth, 2021).

While valuable work has been done analyzing and interrogating the content of speculative, utopian, cli-fi, and solarpunk fictional futures, substantially less research has been done on the effects of specific sustainable future narratives or depictions among the public. Researchers have examined methods for promoting utopian future thinking about energy systems among the general public through collective storytelling activities at public events (Smith et al., 2017). Findings suggest that utopian storytelling exercises could be an effective way to engage a range of stakeholders while supporting participants' efficacy to contribute to sustainable change (Smith et al., 2017). Meanwhile, extant research on the effects of reading or watching cli-fi suggests that short stories can affect concern for climate change impacts, support imagination of possible futures, and encourage positive emotions like hope (Schneider-Mayerson, 2018; Schneider-Mayerson et al., 2023; Shreedhar et al., 2024). However more research is required to understand effects of stories focused primarily on positive environmental futures, rather than cl-fi narratives more broadly.

Other written or film works may fall outside the genres discussed above, but still present desirable, sustainable visions of the future. For example, the film *2040*[1] is conceptualized as a documentary examining existing environmental solutions and considering what the world could look like if these solutions are implemented (Gameau, 2020).

Likewise, the book *The Future Earth*,[2] imagines how societies around the globe could implement changes in coming decades to address climate change and establish more sustainable communities without using narratives, characters, or plot in the way that the fictional genres described above would (Holthaus, 2020).

[1] https://www.imdb.com/title/tt7150512/

[2] https://www.google.com/books/edition/The_Future_Earth/UA6qDwAAQBAJ?hl=en&gbpv=0

Finally, games can offer a unique way to interact with possible sustainable futures. Specifically, serious games seek to go beyond entertainment and promote outcomes such as learning (Dörner et al., 2016) and behavior change (Douglas & Brauer, 2021). Serious games can take many forms, including board games and video games (Stanitsas et al., 2019). Sustainability is a popular topic within serious gaming research (Dallaqua et al., 2024) and such games have been used to educate players on a range of sustainability topics, ranging from natural resource use to social sustainability (Hallinger et al., 2020). In an environmental context, serious games help players understand environmental risks and imagine what the future could look like, for better or for worse, based on specific economic and policy decisions (Imagine Earth, n.d.; Van Beek et al., 2022). Furthermore, serious games can be effective in encouraging the adoption of pro-environmental behaviors (Douglas & Brauer, 2021). Altogether, a range of media may discuss sustainable futures, but detailed descriptions are most common in fictional entertainment media. More research is needed to understand what kinds of portrayals exist in non-fiction media and what persuasive effects these depictions can have on audiences. The following section introduces key scholarly theories that can help further explain why and how media focused on sustainable futures could promote beneficial outcomes among stakeholders.

2 Theories of Imagined Futures and Social Change

A number of psychological theories can shed light on how desirable conceptions of possible futures could impact both groups and individuals. The following section is not a systematic review of all applicable theories, rather a selection of theories and concepts that could be particularly helpful for studying the effects of sustainable futures messaging in a strategic communication context. Many of these theories consider how shared visions of the future contribute to collective action. However, they vary widely in the scope and timeframe of imagined goals.

2.1 *Social Identity Theory and Cognitive Alternatives*

Social identity theory (SIT), or the social identity theory of intergroup behavior, examines how beliefs about social mobility, social change, and group identity may lead to various kinds of action in the event that one group feels subordinate to or in competition with another (Tajfel & Turner, 1986). Most relevant to the present chapter, a prerequisite for constructive collective action to address tension between groups is the availability of cognitive alternatives (Tajfel & Turner, 1986). A cognitive alternative is a conception of another feasible future and Tajfel and Turner (1986) provide the hypothetical example of a sports team that loses one season but is still motivated to play because they believe they could win in the future. On a more serious note, cognitive alternatives could also involve imagined futures in which marginalized racial groups gain more power in society (Tajfel & Turner, 1986).

SIT and the concept of cognitive alternatives have been foundational not only in broad theories of social change via collective action (e.g., the Social Identity Model of Collective Action), but also specifically in research on motivators for pro-environmental outcomes. Wright et al. (2020) argued that cognitive alternatives may motivate action for change not only in instances where relations between groups of people are unsatisfactory, but also when relations between people and the natural world are undesirable. In order to examine this idea, they created a scale to measure how easily people are able to imagine environmental cognitive alternatives, such as "a world in which we supply all of our energy needs without harming the natural world" and "a harmonious relationship between humans and the natural world" (Wright et al., 2020, p. 5). Research utilizing this scale has found that access to cognitive alternatives positively predicts willingness to engage in environmental activism and actual activism behavior (Troy et al., 2024; Wright et al., 2020, 2022). Furthermore, depictions of innovative sustainability technologies can make desirable environmental cognitive alternatives easier to imagine, which in turn can reduce feelings of hopelessness (Neale et al., 2023). Future strategic communication research could expand on existing work by examining how persuasive messaging impacts cognitive alternatives and subsequent engagement in behavior to support sustainable outcomes.

2.2 *Model of Utopian Thinking*

The model of utopian thinking also emphasizes the importance of positive future possibilities for social change, but specifically considers effects of contemplating utopias (Badaan et al., 2020). Utopia is a much more specific concept than cognitive alternatives and can be defined as "visions for better societies that emphasize equality, progress, and social justice" (Badaan et al., 2022, p. 78). This model theorizes that utopian thinking should motivate collective action by positively affecting hope and abstract thinking about the future, while reducing comfort with the status quo.

Initial experimental research into this model has found support for the motivating effects of increased hope and decreased justification of current circumstances (Badaan et al., 2022). Notably, this study manipulated rather than measured utopian thinking, through a brief textual description of a future sustainable, diverse, equitable, and technologically advanced society (Badaan et al., 2022).

Other work on utopian imaginings offers insight into possible approaches for conceptualizing and measuring how people think about utopias. Some scholars have recently created a scale to measure the transformative utopian impulse for planetary health (Basso & Krpan, 2022). This measure includes dimensions related to ability to think about utopias, beliefs about utopias, affective responses to utopian thinking, and impacts of utopian thinking on current actions (Basso & Krpan, 2022). Other researchers have also offered more abbreviated measures of utopian thinking (Fernando et al., 2018). Some work has found that offering descriptions of a sustainable utopia and asking people to write about what their life would look like in that scenario can motivate desire to help the environment and likelihood of engaging in pro-environmental behavior. Based on this scholarship, strategic communication campaigns that provide vivid descriptions of utopian, sustainable societies could be a promising avenue for mobilizing activism or support for a particular sustainability cause.

2.3 *Collective Futures Framework*

A third relevant theoretical model is the collective futures framework, which proposes that imagining improved futures, including considering how specific aspects of society and individuals could differ from the present, can motivate collectively oriented behaviors that could contribute to achieving such a future (e.g., policy support) (Bain et al., 2013). Notably the authors divide future imaginings into a number of key categories—ideas about forces that could broadly impact humanity (e.g., climate change), ideas about power dynamics in society (e.g., which political party is in power), and ideas about structural changes (e.g., different policies being implemented by the government) (Bain et al., 2013). These categorizations could provide helpful detail for communicators considering important areas of emphasis in strategic futures messaging or for researchers investigating how to operationalize and test depictions of sustainable futures.

An initial test of this theory prompted people to read about a topic and imagine a different future across eight experiments, one of which focused on projected negative effects of climate change in Australia and a possible future in which the country engaged in climate action (Bain et al., 2013). Altogether, when participants anticipated that a future scenario would result in reduced societal dysfunction (e.g., less crime) or more benevolent (e.g., more considerate) societies, people were more willing to act to bring about that future through actions like signing petitions or making donations in support of causes (Bain et al., 2013). Other work rooted in this framework qualitatively and quantitatively examined effects of contemplating a future in which people consume more plant-based foods (Judge & Wilson, 2015). Findings revealed a range of appealing and undesirable imagined outcomes, although some were utopian (Judge & Wilson, 2015). Moreover, participants anticipated that vegan or vegetarian societies would be more caring and concerned for conservation, and some of these beliefs positively predicted support for policies promoting plant-based diets (Judge & Wilson, 2015).

Thus far, research related to these three theories identifies various types of future conceptions (e.g., environmental cognitive alternatives, utopias), provides scales for measuring future thinking, and examines cognitive mechanisms that could be relevant for sustainability strategic communication (e.g., hope, perceptions of status quo, expectations about future societies). Notably, these theories do not specify how to effectively encourage people to engage in future thinking, presenting an opportunity for strategic communication scholars and practitioners to further examine media's role in stakeholders' contemplation of the future in order to motivate sustainable behaviors.

2.4 *Other Helpful Theories*

A number of other theories have been utilized to examine effects of future expectations or imagined futures on action, however, with more of a focus on individuals or immediate possible outcomes. Thus, I discuss them only briefly here, based on the idea that they could support fruitful opportunities in the realm of sustainability strategic communication if applied to collective and farther off future contexts. First, fantasy realization theory posits that desirable imagined futures are only motivating under specific circumstances (Oettingen, 2012). Namely, fantasies translate into meaningful action when people consciously contrast the future with the present to identify changes necessary to achieve a different outcome and when people believe that action to achieve a desired goal (e.g., losing weight or completing an assignment) is possible (Oettingen, 2012). Some advertising research has found that purchase intentions were highest when participants considered both the benefits of owning a product as well as the downsides of currently not owning it, and when they were provided with strong arguments about the ease of owning an item, lending support to this theory (Handley & Goss, 2012).

Next, the theory of persuasive hope suggests that messages presenting important, feasible, and desirable possible futures in line with individuals' goals can motivate message engagement and behavior via feelings of hope (Chadwick, 2015). These propositions have been tested in the context of environmental messaging, indicating that messages about likely climate protection that could benefit both the environment and individuals' financial well-being can motivate hope, which in turn can prompt interest in climate action and more favorable message evaluations (Chadwick, 2015).

Although the theory of persuasive hope specifically has not been extensively tested, a robust body of scholarship has pointed to the importance of hope as a motivator for sustainable outcomes. One recent meta-analysis found that hope was positively associated with climate engagement (e.g., learning about climate change, engaging in pro-environmental behaviors, policy support), but this relationship was particularly strong when people considered hope related to specific climate actions that they might take part in (Geiger et al., 2023). Notably, hope was negatively associated with climate engagement when referring to feelings of hope in climate denial; in other words, hope prompted by beliefs that climate change is not a serious threat (Geiger et al., 2023). Finally, this meta-analysis found that hope messaging focused on personal efficacy and the importance of climate action, as well as opportunities to engage in in-depth experiences were most strongly positively related to feelings of hope. Nationally representative survey research in the U.S. has shed light on sources of hope within the context of climate change, indicating that action, information, and science and technology were some of the most common sources of hope (Marlon et al., 2019).

A final helpful theoretical perspective is gain versus loss framing, which has been used in the past to examine effects of messaging about desirable or undesirable environmental outcomes (e.g., Bilandzic et al., 2017; Nabi et al., 2018). Framing means emphasizing particular information or aspects of a topic (Entman, 1993) and gain/loss framing specifically refers to emphasizing either benefits or consequences (Davis, 1995). In an environmental context, this could mean highlighting the possibility of improved environmental and human health if change occurs (gain framing) or suffering from pollution and other environmental problems if the status quo is upheld (loss framing) (Davis, 1995). Gain framing may be a helpful umbrella under which to categorize a range of positive future manipulations in experimental, from detailed utopian depictions (e.g., a solarpunk society in a short piece of fiction) to writing prompts (e.g., encouraging people to imagine benefits in a society run completely on renewable energy). Loss framing would provide a natural comparison group to such stimuli. However, it is important to note that not all gain-framed content is necessarily relevant to the kinds of future thinking discussed earlier, as it can focus on shorter-term outcomes or personal rather than collective benefits. Overall, there are a multitude of opportunities to test strategic messaging about sustainable futures based on these theories.

3 Implications for Strategic Sustainability Communication

Much of the media research mentioned above was primarily descriptive (i.e., analyzing futures presented in media) while the theories reviewed rarely mentioned how external stimuli might influence conceptions of the future. Notably, research on persuasive hope (e.g., Chadwick, 2015) and framing (e.g., Nabi et al., 2018) illustrate that messages can jumpstart thinking about future outcomes. Thus, the question remains: what role can strategic communication play in catalyzing contemplation of desirable futures in order to mobilize widespread support for sustainability efforts?

Based on the theories reviewed, sustainable futures messaging could be relevant to activist organizations seeking support for a cause, nonprofits fundraising for sustainability efforts, governments hoping to engage citizens in sustainability planning, and companies striving to achieve impact through corporate social responsibility or corporate social advocacy. Myriad questions arise about strategic sustainability communication among these groups. For example, could advocacy organizations and campaigns benefit from vivid depictions of the worlds they seek to bring about? What effect would it have if cities put up billboards or commissioned murals showing what sustainable neighborhoods of the future could look like? How could online tools like social media videos or interactive maps be used to offer compelling visualizations of the possibilities for environmentally and socially flourishing communities?

There are some examples of sustainable futures content within the realm of strategic sustainability communication, though they have not been thoroughly studied by scholars. Municipal planning documents, like New York City's *OneNYC 2050* textually describe how life could improve for residents in the coming decades if particular actions are taken by the city government (De Blasio et al., 2019). Meanwhile, promotional posters for the Green New Deal show thriving natural and urban environments, presumably presenting the sustainable future that could be achieved if the plan is enacted (Creative Action Network, 2024). Finally, the annual short story competition *Imagine 2200* run by Grist has an explicit goal of motivating readers to achieve "a more sustainable and just world" (Grist, 2024). Communicators may look to examples such as these to inspire strategic messaging campaigns. However, because research on sustainable future messaging for strategic outcomes is so sparse, it is essential for practitioners to assess impacts of their efforts and solicit feedback on messaging from stakeholders.

Many studies on future thinking require some type of reflection from their participants, whether that be an individual writing exercise (e.g., Judge & Wilson, 2015) or group discussion (e.g., Smith et al., 2017). Meanwhile, fantasy realization theory suggests that people should consciously contrast a desired future with their present reality in order to motivate behavior change (Oettingen, 2012). How might strategic messages in the media, which could easily be passively consumed, prompt the kind of active engagement needed to spur future thinking among stakeholders? Communicators might consider strategies employed in campaigns focused on promoting interpersonal conversation around health topics (e.g., ITV, 2024). Alternatively, more engaging strategies like serious games could be promising.

Meanwhile, a number of critical theoretical questions remain for researchers. First, there is a need to examine what message characteristics are essential to trigger audience thinking about cognitive alternatives or utopian futures. Scholars may benefit from defining utopian appeals or positive future appeals based on inherent message features (Chadwick, 2015; O'Keefe, 2003) rather than audience reactions. Coulter et al. (2019) have proposed a Future Climate Narrative typology, in which adaptive future narratives exist at the intersection of climate knowledge, future thinking, and narrative communication. Specifically, discussions of adaptive futures often involved a focus on personal resilience, connecting concerned individuals, avoiding adaptation choices with high costs and uncertain benefits, and feelings of cautious optimism (Coulter et al., 2019). Although this typology was developed in the context of narratives among climate professionals, this framework could be a helpful starting point for operationalizing components of a positive climate futures in strategic messaging. For example, organizations might consider whether any planned strategic messaging concerning sustainable futures touches on ideas of imagined, dangerous, and unfamiliar futures, or whether one or more of those elements is missing. A message might offer imagined and unfamiliar futures, but without acknowledging the prospect of dangerous futures that could occur absent sufficient and effective actions, such messaging may not be very motivational for audiences. Alternatively, a message focused only on dangerous futures could discourage audiences. Similar to how Coulter et al. (2019) used this framework to analyze conversations with climate experts, organizations could also use this typology to take stock of past strategic messaging to better understand the kind of future (or futures) they have been portraying to stakeholders or to guide analysis of stakeholder feedback that could be solicited through focus groups or open-ended survey questions.

Moreover, future research might examine whether some media formats have advantages over others when communicating sustainable futures. How vivid or detailed should sustainable futures messaging be? For example, some work has found that people can more easily imagine cognitive alternatives when exposed to video rather than audio or textual messages (Shreedhar et al., 2024). From a practitioner perspective, strategic communication campaigns may garner different results based on the amount of information a particular medium is able to provide. Thus, it is imperative to compare stakeholder reactions to campaign materials that make use of varying combinations of text, images, video, and sound. Social media analytics and comments could be particularly helpful for such assessments. Researchers could also consider investigating the timeframe specified in future messaging. Sustainability goals and environmental reports often consider 2050 or 2100 as benchmark dates (e.g., IPCC, 2018; The White House, 2021) but nearer or more distant timeframes could have varying effects based on perceptions of psychological distance (Spence et al., 2012). Practitioners creating sustainable futures messaging should consider choosing milestone dates that align with organizational goals to call out in messages. For example, if an organization has sustainability goals set for 2040 or if an organization will have a major anniversary in 2055, futures messaging could feature those dates as well as benchmark dates along the way that an organization will use to track progress.

While the theories reviewed in this research highlighted a number of important cognitive mechanisms, such as hope (Badaan et al., 2020) and access to cognitive alternatives (Wright et al., 2020) as important motivators of action, these theories hypothesized few roadblocks to action. Specifically, reactance could be triggered by sustainable futures messaging, which might inhibit constructive outcomes. Reactance involves pushing back against what is presented in a persuasive message by becoming angry and arguing against a message's assertions (Quick et al., 2013). Systemic solutions to sustainability issues can be polarizing, with people recognizing the existence of a problem but feeling averse to policy solutions that are perceived to be too extreme (Kantack & Paschall, 2023). Thus, sustainable future messaging showing positive outcomes achieved through broad-reaching policy change or federal programs could alienate stakeholders who do not favor government intervention. Future research would benefit from exploring when sustainable future messaging tends to trigger unintended reactions among different stakeholder groups and how proposed futures could appeal to a broad audience.

4 Conclusion

One avenue for motivating collective action to achieve positive social and environmental change could be depicting sustainable futures in media. While such an approach is not currently common in news media or strategic communication campaigns, there are rich examples of sustainable future depictions in various fictional genres, such as cli-fi (Schneider-Mayerson, 2017) and solarpunk (Klata, 2022), as well as in serious games (Van Beek et al., 2022) and some nonfiction media (e.g., Gameau, 2020; Holthaus, 2020). A number of psychological theories offer insight into how sustainable future messaging could motivate audience action for societal change by presenting cognitive alternatives (Tajfel & Turner, 1986), motivating utopian thinking (Badaan et al., 2020), fostering contemplation of how specific aspects of society might change (Bain et al., 2013), promoting feelings of hope (Badaan et al., 2020; Chadwick, 2015), and contrasting present circumstances with desired future changes (Oettingen, 2012).

Based on this scholarship, strategic communication practitioners could benefit from incorporating vivid depictions of desirable futures into sustainability campaigns. However, these may be most effective if they foster audience reflection beyond brief media exposure, for example by encouraging stakeholders to talk about possible futures with others. Ultimately, further research is needed to more rigorously define message categories such as utopian appeals, consider impacts of medium on message effects, investigate the role of psychological distance in message processing, and examine possible unintended consequences such as reactance. Altogether, it is as of yet uncertain how sustainable societies will be in 2050, 2100, or beyond. Scholars and practitioners should consider the benefits of showing positive futures through media that the public could desire to work toward.

References

Adamson, J. (2018). Environmental fiction. In N. Castree, M. Hulme, & J. D. Proctor (Eds.), *Companion to environmental studies* (1st ed., pp. 611–615). Routledge. https://doi.org/10.4324/9781315640051-119

Badaan, V., Akil, C., Zebian, Y., & Jost, J. T. (2022). Envisioning change: An empirical test of the social psychological model of Utopian thinking and collective action. *Journal of Social Psychology Research*, 77–96. https://doi.org/10.37256/jspr.1120221140

Badaan, V., Jost, J. T., Fernando, J., & Kashima, Y. (2020). Imagining better societies: A social psychological framework for the study of utopian thinking and collective action. *Social and Personality Psychology Compass, 14*(4), e12525. https://doi.org/10.1111/spc3.12525

Bain, P. G., Hornsey, M. J., Bongiorno, R., Kashima, Y., & Crimston, C. R. (2013). Collective futures: How projections about the future of society are related to actions and attitudes supporting social change. *Personality and Social Psychology Bulletin, 39*(4), 523–539. https://doi.org/10.1177/0146167213478200

Basso, F., & Krpan, D. (2022). Measuring the transformative utopian impulse for planetary health in the age of the Anthropocene: A multi-study scale development and validation. *The Lancet Planetary Health, 6*(3), e230–e242. https://doi.org/10.1016/S2542-5196(22)00004-3

Bilandzic, H., Kalch, A., & Soentgen, J. (2017). Effects of goal framing and emotions on perceived threat and willingness to sacrifice for climate change. *Science Communication, 39*(4), 466–491. https://doi.org/10.1177/1075547017718553

Brundtland, G. (1987). *Report of the World Commission on Environment and Development: Our common future*. United Nations General Assembly document A/42/427. http://www.un-documents.net/our-common-future.pdf

Chadwick, A. E. (2015). Toward a theory of persuasive hope: Effects of cognitive appraisals, hope appeals, and hope in the context of climate change. *Health Communication, 30*(6), 598–611. https://doi.org/10.1080/10410236.2014.916777

Coulter, L., Serrao-Neumann, S., & Coiacetto, E. (2019). Climate change adaptation narratives: Linking climate knowledge and future thinking. *Futures, 111*, 57–70. https://doi.org/10.1016/j.futures.2019.05.004

Creative Action Network. (2024). *Green new deal.* https://creativeaction.network/collections/green-new-deal

Dallaqua, M. F., Nunes, B., & Carvalho, M. M. (2024). Serious games research streams for social change: Critical review and framing. *British Journal of Educational Technology, 55*(2), 460–483. https://doi.org/10.1111/bjet.13404

Dasilva, C. (2019). Imagining decline or sustainability: Hope, fear, and ideological discourse in Hollywood speculative fiction. *Elementa: Science of the Anthropocene, 7*, 7. https://doi.org/10.1525/elementa.344

David Tàbara, J., Frantzeskaki, N., Hölscher, K., Pedde, S., Kok, K., Lamperti, F., Christensen, J. H., Jäger, J., & Berry, P. (2018). Positive tipping points in a rapidly warming world. *Current Opinion in Environmental Sustainability, 31*, 120–129. https://doi.org/10.1016/j.cosust.2018.01.012

Davis, J. J. (1995). The effects of message framing on response to environmental communications. *Journalism & Mass Communication Quarterly, 72*(2), 285–299. https://doi.org/10.1177/107769909507200203

De Blasio, B., Fuleihan, D., Williams, D., & Zarrilli, D. A. (2019). *OneNYC 2050: Building a strong and fair city*. https://climate.cityofnewyork.us/wp-content/uploads/2022/10/OneNYC-2050-Summary.pdf

Dörner, R., Göbel, S., Effelsberg, W., & Wiemeyer, J. (2016). Introduction. In R. Dörner, S. Göbel, W. Effelsberg, & J. Wiemeyer (Eds.), *Serious games: Foundations, concepts and practice* (pp. 1–34). Springer International. https://doi.org/10.1007/978-3-319-40612-1_1

Douglas, B. D., & Brauer, M. (2021). Gamification to prevent climate change: A review of games and apps for sustainability. *Current Opinion in Psychology, 42*, 89–94. https://doi.org/10.1016/j.copsyc.2021.04.008

Entman, R. M. (1993). Framing: Toward clarification of a fractured paradigm. *Journal of Communication, 43*(4), 51–58. https://doi.org/10.1111/j.1460-2466.1993.tb01304.x

Feldman, L., Hart, P. S., & Milosevic, T. (2017). Polarizing news? Representations of threat and efficacy in leading US newspapers' coverage of climate change. *Public Understanding of Science, 26*(4), 481–497. https://doi.org/10.1177/0963662515595348

Fernando, J. W., Burden, N., Ferguson, A., O'Brien, L. V., Judge, M., & Kashima, Y. (2018). Functions of Utopia: How Utopian thinking motivates societal engagement. *Personality and Social Psychology Bulletin, 44*(5), 779–792. https://doi.org/10.1177/0146167217748604

Ford, J. D., & King, D. (2015). Coverage and framing of climate change adaptation in the media: A review of influential North American newspapers during 1993–2013. *Environmental Science & Policy, 48*, 137–146. https://doi.org/10.1016/j.envsci.2014.12.003

Gameau, D. (2020). 2040: A solutions narrative. *FORUM, 62*(2), 173–176. https://doi.org/10.15730/forum.2020.62.2.173

Garforth, L. (2021). Utopia: Futurity, realism and the social. In C. López Galviz & E. Spiers (Eds.), *Routledge handbook of social futures* (pp. 292–303). Routledge.

Geiger, N., Dwyer, T., & Swim, J. K. (2023). Hopium or empowering hope? A meta-analysis of hope and climate engagement. *Frontiers in Psychology, 14*. https://www.frontiersin.org/articles/10.3389/fpsyg.2023.1139427

Gillam, W. J. (2023). A solarpunk manifesto: Turning imaginary into reality. *Philosophies, 8*(4), Article 4. https://doi.org/10.3390/philosophies8040073

Goldberg, M. H., & Gustafson, A. (2023). A framework for understanding the effects of strategic communication campaigns. *International Journal of Strategic Communication, 17*(1), 1–20. https://doi.org/10.1080/1553118X.2022.2137674

Grist. (2024). *Imagine 2200*. https://grist.org/imagine2200-climate-fiction-2024/

Guenther, L., Brüggemann, M., & Elkobros, S. (2022). From global doom to sustainable solutions: International news magazines' multimodal framing of our future with climate change. *Journalism Studies, 23*(1), 131–148. https://doi.org/10.1080/1461670X.2021.2007162

Hallinger, P., Wang, R., Chatpinyakoop, C., Nguyen, V.-T., & Nguyen, U.-P. (2020). A bibliometric review of research on simulations and serious games used in educating for sustainability, 1997–2019. *Journal of Cleaner Production, 256*, 120358. https://doi.org/10.1016/j.jclepro.2020.120358

Handley, I. M., & Goss, R. J. (2012). How mental simulations of the future and message-induced expectations influence purchasing goals. *Psychology & Marketing, 29*(6), 401–410. https://doi.org/10.1002/mar.20529

Holthaus, E. (2020). *The future earth: A radical vision for what's possible in the age of warming*. HarperOne.

Imagine Earth. (n.d.). *Imagine Earth: Planetary colonization*. https://www.imagineearth.info/

IPCC. (2018). Summary for policymakers. In V. Masson-Delmotte, P. Zhai, H.-O. Pörtner, D. Roberts, J. Skea, P. R. Shukla, A. Pirani, W. Moufouma-Okia, C. Péan, R. Pidcock, S. Connors, J. B. R. Matthews, Y. Chen, X. Zhou, M. I. Gomis, E. Lonnoy, T. Maycock, M. Tignor, & T. Waterfield (Eds.), *Global Warming of 1.5°C. An IPCC Special Report on the impacts of global warming of 1.5°C above pre-industrial levels and related global greenhouse gas emission pathways, in the context of strengthening the global response to the threat of climate change, sustainable development, and efforts to eradicate poverty* (pp. 3–24). Cambridge University Press. https://doi.org/10.1017/9781009157940.001.

ITV. (2024). *Britain get talking*. https://www.itv.com/britaingettalking/

Judge, M., & Wilson, M. S. (2015). Vegetarian Utopias: Visions of dietary patterns in future societies and support for social change. *Futures, 71*, 57–69. https://doi.org/10.1016/j.futures.2015.07.005

Kantack, B. R., & Paschall, C. E. (2023). Perceptions of policy problems and solutions: Climate change and structural racism. *Public Understanding of Science, 32*(2), 247–256. https://doi.org/10.1177/09636625221133520

Klata, M. (2022). New maps of hope: Common motifs and narrative structures in solarpunk stories. *Zagadnienia Rodzajów Literackich / The Problems of Literary Genres, 65*(3), Article 3. https://doi.org/10.26485/ZRL/2022/65.3/4

Marlon, J. R., Bloodhart, B., Ballew, M. T., Rolfe-Redding, J., Roser-Renouf, C., Leiserowitz, A., & Maibach, E. (2019). How hope and doubt affect climate change mobilization. *Frontiers in Communication, 4*. https://www.frontiersin.org/articles/10.3389/fcomm.2019.00020

McIntyre, K. E., & Lough, K. (2021). Toward a clearer conceptualization and operationalization of solutions journalism. *Journalism, 22*(6), 1558–1573. https://doi.org/10.1177/1464884918820756

Moore, M.-L., & Milkoreit, M. (2020). Imagination and transformations to sustainable and just futures. *Elementa: Science of the Anthropocene, 8*(1), 081. https://doi.org/10.1525/elementa.2020.081

Nabi, R. L., Gustafson, A., & Jensen, R. (2018). Framing climate change: Exploring the role of emotion in generating advocacy behavior. *Science Communication, 40*(4), 442–468. https://doi.org/10.1177/1075547018776019

Neale, C., Austin, M. M. K., Roe, J., & Converse, B. A. (2023). Making people aware of eco-innovations can decrease climate despair. *Climatic Change, 176*(12), 162. https://doi.org/10.1007/s10584-023-03635-9

Oettingen, G. (2012). Future thought and behaviour change. *European Review of Social Psychology, 23*(1), 1–63. https://doi.org/10.1080/10463283.2011.643698

O'Keefe, D. J. (2003). Message properties, mediating states, and manipulation checks: Claims, evidence, and data analysis in experimental persuasive message effects research. *Communication Theory, 13*(3), 251–274. https://doi.org/10.1111/j.1468-2885.2003.tb00292.x

Quick, B. L., Shen, L., & Dillard, J. P. (2013). Reactance theory and persuasion. In J. P. Dillard & L. Shen (Eds.), *The SAGE handbook of persuasion: Developments in theory and practice* (2nd ed., pp. 167–183). SAGE. https://doi.org/10.4135/9781452218410

Schneider-Mayerson, M. (2017). Climate change fiction. In R. G. Smith (Ed.), *American literature in transition, 2000-2010* (pp. 309–321). Cambridge University Press.

Schneider-Mayerson, M. (2018). The influence of climate fiction. *Environmental Humanities, 10*(2), 473–500. https://doi.org/10.1215/22011919-7156848

Schneider-Mayerson, M., Gustafson, A., Leiserowitz, A., Goldberg, M. H., Rosenthal, S. A., & Ballew, M. (2023). Environmental literature as persuasion: An experimental test of the effects of reading climate fiction. *Environmental Communication, 17*(1), 35–50. https://doi.org/10.1080/17524032.2020.1814377

Seiffert-Brockmann, J. (2018). Evolutionary psychology: A framework for strategic communication research. *International Journal of Strategic Communication, 12*(4), 417–432. https://doi.org/10.1080/1553118X.2018.1490291

Shreedhar, G., Sabherwal, A., & Maldonado, R. (2024). Cli-fi videos can increase charitable donations: Experimental evidence from the United Kingdom. *Frontiers in Psychology, 14*. https://doi.org/10.3389/fpsyg.2023.1176077

Smith, J., Butler, R., Day, R. J., Goodbody, A. H., Llewellyn, D. H., Rohse, M., Smith, B. T., Tyszczuk, R. A., Udall, J., & Whyte, N. M. (2017). Gathering around stories: Interdisciplinary experiments in support of energy system transitions. *Energy Research & Social Science, 31*, 284–294. https://doi.org/10.1016/j.erss.2017.06.026

Spence, A., Poortinga, W., & Pidgeon, N. (2012). The psychological distance of climate change. *Risk Analysis, 32*(6), 957–972. https://doi.org/10.1111/j.1539-6924.2011.01695.x

Stanitsas, M., Kirytopoulos, K., & Vareilles, E. (2019). Facilitating sustainability transition through serious games: A systematic literature review. *Journal of Cleaner Production, 208*, 924–936. https://doi.org/10.1016/j.jclepro.2018.10.157

Stecula, D. A., & Merkley, E. (2019). Framing climate change: Economics, ideology, and uncertainty in American news media content from 1988 to 2014. *Frontiers in Communication, 4*. https://www.frontiersin.org/article/10.3389/fcomm.2019.00006

Svoboda, M. (2016). Cli-fi on the screen(s): Patterns in the representations of climate change in fictional films. *WIREs Climate Change, 7*(1), 43–64. https://doi.org/10.1002/wcc.381

Tajfel, H., & Turner, J. C. (1986). The social identity theory of intergroup behavior. In S. Worchel & W. G. Austin (Eds.), *Psychology of intergroup relations* (2nd ed., pp. 7–24). Nelson-Hall Publishers.

Tavares, A. O., Areia, N. P., Mellett, S., James, J., Intrigliolo, D. S., Couldrick, L. B., & Berthoumieu, J.-F. (2020). The European media portrayal of climate change: Implications for the social mobilization towards climate action. *Sustainability, 12*(20), Article 20. https://doi.org/10.3390/su12208300

The Natural Step. (2024). *Our approach*. https://thenaturalstep.org/approach/

Thier, K., & Lin, T. (2022). How solutions journalism shapes support for collective climate change adaptation. *Environmental Communication*, 1–19. https://doi.org/10.1080/17524032.2022.2143842

Troy, C. L. C., Norman, M., Kim, N., Skurka, C., & Myrick, J. G. (2024). Can you picture it? Effects of positive and negative depictions of climate futures on climate action intentions. *Journal of Environmental Psychology, 96*, 102312. https://doi.org/10.1016/j.jenvp.2024.102312

Van Beek, L., Milkoreit, M., Prokopy, L., Reed, J. B., Vervoort, J., Wardekker, A., & Weiner, R. (2022). The effects of serious gaming on risk perceptions of climate tipping points. *Climatic Change, 170*(3–4), 31. https://doi.org/10.1007/s10584-022-03318-x

The White House. (2021). *ICYMI: President Biden signs executive order catalyzing America's clean energy economy through federal sustainability*. https://www.whitehouse.gov/ceq/news-updates/2021/12/13/icymi-president-biden-signs-executive-order-catalyzing-americas-clean-energy-economy-through-federal-sustainability/#:~:text=Net%2DZero%20Emissions%20by%202050&text=President%20Joe%20Biden%20on%20Wed

Werder, K. P., Nothhaft, H., Verčič, D., & Zerfass, A. (2018). Strategic communication as an emerging interdisciplinary paradigm. *International Journal of Strategic Communication, 12*(4), 333–351. https://doi.org/10.1080/1553118X.2018.1494181

Wright, J. D., Schmitt, M. T., Mackay, C. M. L., & Neufeld, S. D. (2020). Imagining a sustainable world: Measuring cognitive alternatives to the environmental status quo. *Journal of Environmental Psychology, 72*, 101523. https://doi.org/10.1016/j.jenvp.2020.101523

Wright, J. D., Schmitt, M. T., & Mackay, C. M. L. (2022). Access to environmental cognitive alternatives predicts pro-environmental activist behavior. *Environment and Behavior, 54*(3), 712–742. https://doi.org/10.1177/00139165211065008

Wright, C., Nyberg, D., De Cock, C., & Whiteman, G. (2013). Future imaginings: Organizing in response to climate change. *Organization, 20*(5), 647–658. https://doi.org/10.1177/1350508413489821

Wuebben, D., Meinhold, R., & Frigo, G. (2023). SDGs and fictional energy utopias: Gauging sustainable energy transitions against Ecotopia (1975) and The Ministry for the Future (2020). *Energy & Environment*. https://doi.org/10.1177/0958305X231186849

Cassandra L. C. Troy is a PhD candidate at Penn State University's Bellisario College of Communications and will be an incoming Assistant Professor at the University of Illinois Urbana-Champaign's Department of Journalism in Fall 2024. She conducts mixed methods research on environmental communication, including impacts of solutions-oriented climate change messaging.

Last Thoughts

Franzisca Weder

Sustainable development as a normative framework is increasingly manifested not only as a narrative in economic and political discourses but also as a guiding principle for individual behavior. It also stimulates academic disciplines to reassess their paradigms, theories, and concepts and think about their contribution to global transformation. On the one hand, current conflicts and a multiple crisis scenario make it obvious that there is the need for new stories, new narratives, new ways to communicate, and new values that need to be cultivated (Weder, 2023). On the other hand, transformation is a cultural project and sustainability is not a topic, theme, or issue. The sustainable development goals are accordingly often misinterpreted as issue fields with a specific goal for all action directed toward reaching this goal.

After a first decade of communication scholars studying sustainability related issues and who communicates them to whom in which channels and with what potential effect(s), especially Public Relations researchers and teachers still feel itchy; as much as we see the contribution of strategic communication thinking to the formation of sustainability communication as new field of research and professional communication practice, we also feel that our concepts are not "enough" to grasp the complexity, path dependency and ambiguity of sustainability as a normative principle and related collective and individual behavior. Especially the reduction in communication to being interpreted and "applied" as a tool or means to an end understood as supporting business in achieving their goals, stirs up the feeling that there is more to it!

First attempts to take a broader perspective on communication in the context of socio-economic transformation processes and sustainable development led to a holistic view complementing the functionalist perspective on communication of

F. Weder
Department of Business Communication, WU Vienna University of Business and Economics, Vienna, Austria
e-mail: franzisca.weder@wu.ac.at

F. Weder (ed.), *Strategic Sustainability Communication*, CSR, Sustainability, Ethics & Governance, https://doi.org/10.1007/978-3-031-89486-2

sustainability with a constructivist and constitutive understanding of communication, including processes of value co-creation within organizations as much as public discourses about sustainability. But somchow, we still feel achy. Doesn't sustainability force us to go beyond our concepts, to build new theories, rethink the way we research and teach? Doesn't just the idea of a socio-economic or even socio-cultural transformation stimulate us to dare something new, leave our comfort zone, and deconstruct our own concepts and have a thorough look at their parts?

Therefore, this volume started with a—potentially provocative—question: what happens with strategic communication if we take away the orientation toward a specific goal but ask instead for a—not specifically predefined—purpose? Purposeful communication is different compared to "help the organization to fulfill its mission." The introduction and several chapters within this volume follow the idea that the purpose of strategic communication in this "roaring 20s" of a new century is to be sustainable, to be regenerative, and thus keep the sustainability related conversations going. In this scenario, strategic sustainability communication necessarily goes beyond information and coordination into a new space of deliberations and sense- and meaning-making processes, even a space of agonistic debate. In this scenario, sustainability is an ethical foil, a principle of action itself. It is a *postulate*! Therefore, planetary boundaries are the normative reference point that therefore should also direct and drive strategic communication. Then, *sustainability has the potential to develop into a normative horizon for all forms of strategic communication*, a new basic value—comparable to care, kindship, or democracy. Not every regulatory idea or concept has the chance to complement our catalogue of basic values; however, looking at the chapters in this volume, sustainability apparently has already turned into something that has a certain validity and offers stability and security for everyday communication practices.

This volume on strategic sustainability communication therefore does not only point to the need for making sustainability negotiable again—within the discipline and beyond. It also advocates for treating communication as a resource that we take care of following the principle of sustainability. Strategic communication needs to be dealt with in a sustainable way.

This opens more questions for future research, theory development, and mainly pedagogical thinking: who holds the responsibility for sustainable strategic communication and who feels the agency to spark sustainability related conversations and re-negotiations? Who can and should open spaces for the co-construction of sustainability experiences, interventions, and contestation? Here new emerging roles like sustainability managers or sustainability commissioners or consultants might play a significant role as communicative agents which needs to be further studied in the future. This also requires to re-think the competencies and skills we teach in our strategic communication programs and the conversational spaces we open for imaginatory and imagination-enabling narrative thinking, places where negotiations and the in(de)finition of sustainability can take place. Therefore, the volume opens doors and invites researchers and communication professionals to walk through the doors and keep on daring the new—as we all need to do!

References

Weder, F. (2023). Cultivation of sustainability in a discourse of change: Perspectives on communication for sustainability as new "norm" and principle of action in socio-ecological transformation processes. *Journal of Language and Politics, 22*(5), 577–600.

MIX
Papier aus verantwortungsvollen Quellen
Paper from responsible sources
FSC® C105338

If you have any concerns about our products,
you can contact us on
ProductSafety@springernature.com

In case Publisher is established outside the EU,
the EU authorized representative is:
Springer Nature Customer Service Center GmbH
Europaplatz 3, 69115 Heidelberg, Germany

Printed by Libri Plureos GmbH
in Hamburg, Germany